STATA USER'S GUIDE
RELEASE 8

A Stata Press Publication
STATA CORPORATION
College Station, Texas

Stata Press, 4905 Lakeway Drive, College Station, Texas 77845

ISBN 1-881228-74-6

The suggested citation for this software is

StataCorp. 2003. *Stata Statistical Software: Release 8.0*. College Station, TX: Stata Corporation.

Contents of User's Guide

BASICS

1 Read this — it will help 3
    1.1   Getting Started with Stata 4
    1.2   The User's Guide and the Reference manuals 4
        1.2.1   Cross-referencing 5
        1.2.2   The index 5
        1.2.3   The subject table of contents 5
        1.2.4   Typography 5
    1.3   What's new 6
        1.3.1   What's big 7
        1.3.2   What's useful 9
        1.3.3   What's convenient 10
        1.3.4   What was needed 10
        1.3.5   What's faster 10
        1.3.6   What's new in time series analysis 11
        1.3.7   What's new in cross-sectional time-series analysis 11
        1.3.8   What's new in survival analysis 12
        1.3.9   What's new in survey analysis 13
        1.3.10   What's new in cluster analysis 14
        1.3.11   What's new in statistics useful in all fields 14
        1.3.12   What's new in data management 19
        1.3.13   What's new in expressions and functions 22
        1.3.14   What's new in display formats 25
        1.3.15   What's new in programming 25
        1.3.16   What's new in the user interface 27
        1.3.17   What's more 28

2 Resources for learning and using Stata 29
    2.1   Overview 29
    2.2   The http://www.stata.com web site 30
    2.3   The http://www.stata-press.com web site 30
    2.4   The Stata listserver 31
    2.5   The Stata Journal and the Stata Technical Bulletin 31
    2.6   Updating and adding features from the web 32
        2.6.1   Official updates 32
        2.6.2   Unofficial updates 32
    2.7   NetCourses 33
    2.8   Books and other support materials 33
    2.9   Technical support 34
        2.9.1   Register your software 34
        2.9.2   Before contacting technical support 34
        2.9.3   Technical support by email 34
        2.9.4   Technical support by phone or fax 35
        2.9.5   Comments and suggestions for our technical staff 35

3 A brief description of Stata 37

4   Flavors of Stata 41
    4.1   Platforms 41
    4.2   Stata/SE, Intercooled Stata, and Small Stata 41
           4.2.1   Determining which version you own 42
           4.2.2   Determining which version is installed 42
    4.3   Size limits comparison of Stata/SE, Intercooled Stata, and Small Stata 42
    4.4   Speed comparison of Stata/SE, Intercooled Stata, and Small Stata 43
    4.5   Features of Stata/SE 43

5   Starting and stopping Stata 45
    5.1   Starting Stata 45
    5.2   Verifying that Stata is correctly installed 46
    5.3   Exiting Stata 47
    5.4   Features worth learning about 47
           5.4.1   Windows 47
           5.4.2   Macintosh 48
           5.4.3   Unix 48

6   Troubleshooting starting and stopping Stata 51
    6.1   If Stata does not start 51
    6.2   verinst problems 52
    6.3   Troubleshooting Windows 52
    6.4   Troubleshooting Macintosh 52
    6.5   Troubleshooting Unix 52

7   Setting the size of memory 53
    7.1   Memory size considerations 53
    7.2   Setting memory size on the fly: Stata/SE 53
           7.2.1   Advice on setting maxvar 54
           7.2.2   Advice on setting matsize 55
           7.2.3   Advice on setting memory 56
    7.3   Setting memory size on the fly: Intercooled Stata 57
    7.4   The memory command 57
    7.5   Virtual memory and speed considerations 59
    7.6   An issue when returning memory to Unix 59

8   Stata's online help and search facilities 61
    8.1   Introduction 61
    8.2   help: Stata's online manual pages 61
    8.3   search: Stata's online index 63
    8.4   Accessing help and search from the Help menu 65
    8.5   More on search 65
    8.6   More on help 66
    8.7   help contents: Stata's online table of contents 67
    8.8   search: All the details 67
           8.8.1   How search works 67
           8.8.2   Author searches 68
           8.8.3   Entry id searches 68
           8.8.4   FAQ searches 69
           8.8.5   Return codes 69
    8.9   net search: Searching net resources 70

9   Stata's sample datasets 71

10 –more– conditions 73
    10.1 Description 73
    10.2 set more off 73
    10.3 The more programming command 73

11 Error messages and return codes 75
    11.1 Making mistakes 75
          11.1.1 Mistakes are forgiven 75
          11.1.2 Mistakes stop user-written programs and do-files 75
          11.1.3 Advanced programming to tolerate errors 76
    11.2 The return message for obtaining command timings 76

12 The Break key 79
    12.1 Making Stata stop what it is doing 79
    12.2 Side effects of pressing Break 80
    12.3 Programming considerations 80

13 Keyboard use 81
    13.1 Description 81
    13.2 F-keys 81
    13.3 Editing keys in Stata for Windows, Macintosh, and Unix(GUI) 83
    13.4 Editing keys in Stata for Unix(console) 84
    13.5 Editing previous lines in Stata for all operating systems 85
    13.6 Tab expansion of variable names 86

ELEMENTS

14 Language syntax 89
    14.1 Overview 89
          14.1.1 varlist 89
          14.1.2 by varlist: 90
          14.1.3 if exp 92
          14.1.4 in range 93
          14.1.5 =exp 94
          14.1.6 weight 95
          14.1.7 options 96
          14.1.8 numlist 98
    14.2 Abbreviation rules 99
          14.2.1 Command abbreviation 99
          14.2.2 Option abbreviation 100
          14.2.3 Variable-name abbreviation 100
    14.3 Naming conventions 101
    14.4 varlists 101
          14.4.1 Lists of existing variables 101
          14.4.2 Lists of new variables 102
          14.4.3 Time-series varlists 104
    14.5 by varlist: construct 106

14.6   File-naming conventions 109
        14.6.1   A special note for Macintosh users 110
        14.6.2   A special note for Unix users 111

15 Data 113
    15.1   Data and datasets 113
    15.2   Numbers 114
            15.2.1   Missing values 114
            15.2.2   Numeric storage types 118
    15.3   Dates 118
    15.4   Strings 119
            15.4.1   Strings containing identifying data 119
            15.4.2   Strings containing categorical data 119
            15.4.3   Strings containing numeric data 120
            15.4.4   String storage types 120
    15.5   Formats: controlling how data are displayed 120
            15.5.1   Numeric formats 120
            15.5.2   European numeric formats 123
            15.5.3   Date formats 124
            15.5.4   Time-series formats 125
            15.5.5   String formats 126
    15.6   Dataset, variable, and value labels 127
            15.6.1   Dataset labels 127
            15.6.2   Variable labels 128
            15.6.3   Value labels 129
    15.7   Notes attached to data 134
    15.8   Characteristics 135

16 Functions and expressions 137
    16.1   Overview 137
    16.2   Operators 138
            16.2.1   Arithmetic operators 138
            16.2.2   String operators 138
            16.2.3   Relational operators 139
            16.2.4   Logical operators 140
            16.2.5   Order of evaluation, all operators 141
    16.3   Functions 141
    16.4   System variables (_variables) 142
    16.5   Accessing coefficients and standard errors 142
            16.5.1   Simple models 143
            16.5.2   ANOVA and MANOVA models 143
            16.5.3   Multiple-equation models 143
    16.6   Accessing results from Stata commands 144
    16.7   Explicit subscripting 145
            16.7.1   Generating lags and leads 145
            16.7.2   Subscripting within groups 146
    16.8   Time-series operators 149
            16.8.1   Generating lags and leads 149
            16.8.2   Operators within groups 149
    16.9   Label values 150
    16.10   Precision and problems therein 150

17 Matrix expressions 153
    17.1   Overview 153
            17.1.1   Definition of a matrix 153
            17.1.2   matsize 154
    17.2   Row and column names 154
            17.2.1   The purpose of row and column names 155
            17.2.2   Three-part names 157
            17.2.3   Setting row and column names 158
            17.2.4   Obtaining row and column names 159
    17.3   Vectors and scalars 159
    17.4   Inputting matrices by hand 159
    17.5   Accessing matrices created by Stata commands 160
    17.6   Creating matrices by accumulating data 161
    17.7   Matrix operators 161
    17.8   Matrix functions 162
    17.9   Subscripting 163
    17.10   Using matrices in scalar expressions 164

18 Printing and preserving output 165
    18.1   Overview 165
            18.1.1   Starting and closing logs 166
            18.1.2   Appending to an existing log 168
            18.1.3   Temporarily suspending and resuming logging 168
    18.2   Placing comments in logs 169
    18.3   Logging only what you type 169
    18.4   The log-button alternative 170
    18.5   Printing logs 170

19 Do-files 171
    19.1   Description 171
            19.1.1   Version 172
            19.1.2   Comments and blank lines in do-files 173
            19.1.3   Long lines in do-files 174
            19.1.4   Error handling in do-files 175
            19.1.5   Logging the output of do-files 177
            19.1.6   Preventing –more– conditions 177
    19.2   Calling other do-files 178
    19.3   Ways to run a do-file (Stata for Windows) 178
    19.4   Ways to run a do-file (Stata for Macintosh) 179
    19.5   Ways to run a do-file (Stata for Unix) 179
    19.6   Programming with do-files 180
            19.6.1   Argument passing 180
            19.6.2   Suppressing output 181

20 Ado-files 183
    20.1   Description 183
    20.2   What is an ado-file? 183
    20.3   How can I tell if a command is built-in or an ado-file? 184
    20.4   Can I look at an ado-file? 184

20.5   Where does Stata look for ado-files? 185
       20.5.1   Where are the official ado directories? 185
       20.5.2   Where is my personal ado directory? 185
20.6   How do I install an addition? 186
20.7   How do I add my own ado-files? 187
20.8   How do I install official updates? 187

21   Programming Stata 189
21.1   Description 190
21.2   Relationship between a program and a do-file 190
21.3   Macros 193
       21.3.1   Local macros 194
       21.3.2   Global macros 194
       21.3.3   The difference between local and global macros 195
       21.3.4   Macros and expressions 196
       21.3.5   Double quotes 198
       21.3.6   Extended macro functions 199
       21.3.7   Macro expansion operators 203
       21.3.8   Advanced local macro manipulation 205
       21.3.9   Advanced global macro manipulation 207
       21.3.10 Constructing Windows filenames using macros 208
       21.3.11 Accessing system values 208
       21.3.12 Referencing characteristics 209
21.4   Program arguments 209
       21.4.1   Renaming positional arguments 212
       21.4.2   Incrementing through positional arguments 213
       21.4.3   Using macro shift 214
       21.4.4   Parsing standard Stata syntax 215
       21.4.5   Parsing immediate commands 217
       21.4.6   Parsing nonstandard syntax 217
21.5   Scalars and matrices 218
21.6   Temporarily destroying the data in memory 218
21.7   Temporary objects 219
       21.7.1   Temporary variables 219
       21.7.2   Temporary scalars and matrices 219
       21.7.3   Temporary files 220
21.8   Accessing results calculated by other programs 220
21.9   Accessing results calculated by estimation commands 223
21.10 Saving results 224
       21.10.1 Saving results in r() 225
       21.10.2 Saving results in e() 226
       21.10.3 Saving results in s() 228
21.11 Ado-files 229
       21.11.1 Version 231
       21.11.2 Comments and long lines in ado-files 231
       21.11.3 Debugging ado-files 231
       21.11.4 Local subroutines 232
       21.11.5 Development of a sample ado-command 233
       21.11.6 Writing online help 239
21.12 A compendium of useful commands for programmers 243
21.13 References 243

22   Immediate commands 245
     22.1   Overview 245
            22.1.1   Examples 246
            22.1.2   A list of the immediate commands 247
     22.2   The display command 248

23   Estimation and post-estimation commands 249
     23.1   All estimation commands work the same way 249
     23.2   Standard syntax 251
     23.3   Replaying prior results 252
     23.4   Cataloging estimation results 253
     23.5   Specifying the estimation subsample 255
     23.6   Specifying the width of confidence intervals 255
     23.7   Obtaining the variance–covariance matrix 256
     23.8   Obtaining predicted values 257
            23.8.1   predict can be used after any estimation command ... 258
            23.8.2   Making in-sample predictions 259
            23.8.3   Making out-of-sample predictions 259
            23.8.4   Obtaining standard errors, tests, and confidence intervals for predictions 260
     23.9   Accessing estimated coefficients 261
     23.10 Performing hypothesis tests on the coefficients 263
            23.10.1 Linear tests 263
            23.10.2 test can be used after any estimation command . 264
            23.10.3 Likelihood-ratio tests 266
            23.10.4 Nonlinear Wald tests 267
     23.11 Obtaining linear combinations of coefficients 267
     23.12 Obtaining nonlinear combinations of coefficients 269
     23.13 Obtaining marginal effects 269
     23.14 Obtaining robust variance estimates 270
     23.15 Obtaining scores 275
     23.16 Weighted estimation 278
            23.16.1 Frequency weights 278
            23.16.2 Analytic weights 279
            23.16.3 Sampling weights 280
            23.16.4 Importance weights 282
     23.17 A list of post-estimation commands 282
     23.18 References 283

ADVICE

24   Commands to input data 287
     24.1   Six ways to input data 287
     24.2   Eight rules for determining which input method to use . 287
            24.2.1   If you wish to enter data interactively: Rule 1 ... 288
            24.2.2   If the dataset is in binary format: Rule 2 288
            24.2.3   If the data are simple: Rule 3 289
            24.2.4   If the dataset is formatted and the formatting is significant: Rule 4 290
            24.2.5   If there are no string variables: Rule 5 291
            24.2.6   If all the string variables are enclosed in quotes: Rule 6 291
            24.2.7   If the undelimited strings have no blanks: Rule 7 292
            24.2.8   If you make it to here: Rule 8 293

24.3 If you run out of memory 293
24.4 Transfer programs 293
24.5 ODBC sources 294
24.6 References 294

25 Commands for combining data 295

26 Commands for dealing with strings 297
26.1 Description 297
26.2 Categorical string variables 297
26.3 Mistaken string variables 298
26.4 Complex strings 299

27 Commands for dealing with dates 301
27.1 Overview 301
27.2 Dates 302
      27.2.1 Inputting dates 302
      27.2.2 Conversion into elapsed dates 303
              27.2.2.1 The mdy() function 303
              27.2.2.2 The date() function 304
      27.2.3 Displaying dates 306
      27.2.4 Other date functions 309
      27.2.5 Specifying particular dates (date literals) 309
27.3 Time-series dates 310
      27.3.1 Inputting time variables 311
      27.3.2 Specifying particular dates (date literals) 312
      27.3.3 Time-series formats 313
      27.3.4 Translating between time units 313
      27.3.5 Extracting components of time 314
      27.3.6 Creating time variables 314
      27.3.7 Setting the time variable 314
      27.3.8 Selecting periods of time 314
      27.3.9 The %tg format 315

28 Commands for dealing with categorical variables 317
28.1 Continuous, categorical, and indicator variables 317
      28.1.1 Converting continuous to indicator variables 318
      28.1.2 Converting continuous to categorical variables 318
      28.1.3 Converting categorical to indicator variables 320
28.2 Using indicator variables in estimation 322
      28.2.1 Testing the significance of indicator variables 323
      28.2.2 Importance of omitting one of the indicators 324

29 Overview of Stata estimation commands 327
29.1 Introduction 327
29.2 Linear regression with simple error structures 328
29.3 ANOVA, ANCOVA, MANOVA, and MANCOVA 320
29.4 Generalized linear models 330
29.5 Binary outcome qualitative dependent variable models 330

29.6   Conditional logistic regression 332
29.7   Multiple outcome qualitative dependent variable models 332
29.8   Simple count dependent variable models 333
29.9   Linear regression with heteroskedastic errors 333
29.10   Stochastic frontier models 333
29.11   Linear regression with systems of equations (correlated errors) 334
29.12   Models with endogenous sample selection 334
29.13   Models with time-series data 335
29.14   Panel-data models 335
        29.14.1 Linear regression with panel data 335
        29.14.2 Censored linear regression with panel data 337
        29.14.3 Generalized linear models with panel data 337
        29.14.4 Qualitative dependent variable models with panel data 337
        29.14.5 Count dependent variable models with panel data 338
        29.14.6 Random-coefficient models with panel data 338
29.15   Survival-time (failure-time) models 338
29.16   Commands for estimation with survey data 338
29.17   Multivariate analysis 339
29.18   Pharmacokinetic data 340
29.19   Cluster analysis 340
29.20   Not elsewhere classified 341
29.21   Have we forgotten anything? 341
29.22   References 341

30 Overview of survey estimation 343
    30.1   Introduction 343
    30.2   Accounting for the sample design in survey analyses 345
            30.2.1 Multistage sample designs 346
            30.2.2 Finite population corrections 347
            30.2.3 Design effects: deff and deft 348
    30.3   Example of the effects of weights, clustering, and stratification 348
            30.3.1 Halfway isn't enough: the importance of stratification and clustering .. 349
    30.4   Linear regression and other models 351
    30.5   Pseudo-likelihoods 352
    30.6   Building your own survey estimator using ml 353
    30.7   Hypothesis testing using test and testnl 353
    30.8   Estimation of linear and nonlinear combinations of parameters 354
    30.9   Two-way contingency tables 355
    30.10   Differences between the svy commands and other commands 356
    30.11   References 358

31 Commands everyone should know 360
    31.1   Forty-one commands 360
    31.2   The by construct 361

32 Using the Internet to keep up to date 362
    32.1   Overview 363
    32.2   Sharing datasets (and other files) 363

32.3   Official updates 363
        32.3.1   Example 364
        32.3.2   Updating ado-files 365
        32.3.3   Frequently asked questions about updating the ado-files 366
        32.3.4   Updating the executable 366
        32.3.5   Frequently asked questions about updating the executable 367
        32.3.6   Updating both ado-files and the executable 367
32.4   Downloading and managing additions by users 368
        32.4.1   Downloading files 368
        32.4.2   Managing files 370
        32.4.3   Finding files to download 370
32.5   Making your own download site 371

Author index 373

Subject index 387

Subject Table of Contents

This is the complete contents for all of the Reference manuals.

Getting Started

[GS]   Getting Started manual .... Getting Started with Stata for Macintosh
[GS]   Getting Started manual .... Getting Started with Stata for Unix
[GS]   Getting Started manual .... Getting Started with Stata for Windows
[U]   User's Guide, Chapter 2 .... Resources for learning and using Stata
[R]   help .... Obtain online help

Data manipulation and management

Basic data commands

[R]   describe .... Describe contents of data in memory or on disk
[R]   display ,,,,,,,,,,,,,,,,,,,,,,,,, ... Substitute for a hand calculator
[R]   drop .... Eliminate variables or observations
[R]   edit .... Edit and list data using Data Editor
[R]   egen .... Extensions to generate
[R]   generate .... Create or change contents of variable
[R]   list .... List values of variables
[R]   memory .... Memory size considerations
[R]   obs .... Increase the number of observations in a dataset
[R]   sort .... Sort data

Functions and expressions

[U]   User's Guide, Chapter 16 .... Functions and expressions
[R]   egen .... Extensions to generate
[R]   functions .... Functions

Dates

[U]   User's Guide, Section 15.5.3 .... Date formats
[U]   User's Guide, Chapter 27 .... Commands for dealing with dates
[R]   functions .... Functions

Inputting and saving data

[U]   User's Guide, Chapter 24 .... Commands to input data
[R]   edit .... Edit and list data using Data Editor
[R]   infile .... Quick reference for reading data into Stata
[R]   insheet .... Read ASCII (text) data created by a spreadsheet
[R]   infile (free format) .... Read unformatted ASCII (text) data
[R]   infix (fixed format) .... Read ASCII (text) data in fixed format
[R]   infile (fixed format) .... Read ASCII (text) data in fixed format with a dictionary
[R]   input .... Enter data from keyboard
[R]   odbc .... Load data from ODBC sources
[R]   outfile .... Write ASCII-format dataset
[R]   outsheet .... Write spreadsheet-style dataset
[R]   save .... Save and use datasets

[R]   sysuse . . . . Use shipped dataset
[R]   webuse . . . . Use dataset from web

Combining data

[U]   User's Guide, Chapter 25 . . . . Commands for combining data
[R]   append . . . . Append datasets
[R]   merge . . . . Merge datasets
[R]   joinby . . . . Form all pairwise combinations within groups

Reshaping datasets

[R]   collapse . . . . Make dataset of means, medians, etc.
[R]   contract . . . . Make dataset of frequencies
[R]   compress . . . . Compress data in memory
[R]   cross . . . . Form every pairwise combination of two datasets
[R]   expand . . . . Duplicate observations
[R]   fillin . . . . Rectangularize dataset
[R]   obs . . . . Increase the number of observations in dataset
[R]   reshape . . . . Convert data from wide to long and vice versa
[R]   separate . . . . Create separate variables
[R]   stack . . . . Stack data
[R]   statsby . . . . Collect statistics for a command across a by list
[R]   xpose . . . . Interchange observations and variables

Labeling, display formats, and notes

[U]   User's Guide, Section 15.5 . . . . Formats: controlling how data are displayed
[U]   User's Guide, Section 15.6 . . . . Dataset, variable, and value labels
[R]   format . . . . Specify variable display format
[R]   label . . . . Label manipulation
[R]   labelbook . . . . Label utilities
[R]   notes . . . . Place notes in data

Changing and renaming variables

[U]   User's Guide, Chapter 28 . . . . Commands for dealing with categorical variables
[R]   destring . . . . Change string variables to numeric
[R]   encode . . . . Encode string into numeric and vice versa
[R]   generate . . . . Create or change contents of variable
[R]   mvencode . . . . Change missing to coded missing value and vice versa
[R]   order . . . . Reorder variables in dataset
[R]   recode . . . . Recode categorical variable
[R]   rename . . . . Rename variable
[R]   split . . . . Split string variables into parts

Examining data

[R]   cf . . . . Compare two datasets
[R]   codebook . . . . Produce a codebook describing the contents of data
[R]   compare . . . . Compare two variables
[R]   count . . . . Count observations satisfying specified condition
[R]   duplicates . . . . Detect and delete duplicate observations
[R]   gsort . . . . Ascending and descending sort
[R]   inspect . . . . Display simple summary of data's characteristics

[R]   isid .... Check for unique identifiers
[R]   pctile .... Create variable containing percentiles
[ST]   stdes .... Describe survival-time data
[R]   summarize .... Summary statistics
[SVY]   svytab .... Tables for survey data
[R]   table .... Tables of summary statistics
[P]   tabdisp .... Display tables
[R]   tabstat .... Display table of summary statistics
[R]   tabsum .... One- and two-way tables of summary statistics
[R]   tabulate .... One- and two-way tables of frequencies
[XT]   xtdes .... Describe pattern of xt data

Miscellaneous data commands

[R]   corr2data .... Create a dataset with a specified correlation structure
[R]   drawnorm .... Draw a sample from a normal distribution
[R]   icd9 .... ICD-9-CM diagnostic and procedures codes
[R]   ipolate .... Linearly interpolate (extrapolate) values
[R]   range .... Numerical ranges, derivatives, and integrals
[R]   sample .... Draw random sample

Utilities

Basic utilities

[U]   User's Guide, Chapter 8 .... Stata's online help and search facilities
[U]   User's Guide, Chapter 18 .... Printing and preserving output
[U]   User's Guide, Chapter 19 .... Do-files
[R]   about .... Display information about my version of Stata
[R]   by .... Repeat Stata command on subsets of the data
[R]   copyright .... Display copyright information
[R]   do .... Execute commands from a file
[R]   doedit .... Edit do-files and other text files
[R]   exit .... Exit Stata
[R]   help .... Obtain online help
[R]   level .... Set default confidence level
[R]   log .... Echo copy of session to file or device
[R]   obs .... Increase the number of observations in dataset
[R]   #review .... Review previous commands
[R]   search .... Search Stata documentation
[R]   translate .... Print and translate logs
[R]   view .... View files and logs

Error messages

[U]   User's Guide, Chapter 11 .... Error messages and return codes
[R]   error messages .... Error messages and return codes
[P]   error .... Display generic error message and exit
[P]   rmsg .... Return messages

Saved results

[U]   User's Guide, Section 16.6 .... Accessing results from Stata commands
[U]   User's Guide, Section 21.8 .... Accessing results calculated by other programs

[U]   User's Guide, Section 21.9 . . . .   Accessing results calculated by estimation commands
[U]   User's Guide, Section 21.10 . . . .   Saving results
[P]   creturn . . . .   Return c-class values
[R]   estimates . . . .   Manage estimation results
[P]   return . . . .   Return saved results
[R]   saved results . . . .   Saved results

Internet

[U]   User's Guide, Chapter 32 . . . .   Using the Internet to keep up to date
[R]   checksum . . . .   Calculate checksum of file
[R]   net . . . .   Install and manage user-written additions from the net
[R]   net search . . . .   Search Internet for installable packages
[R]   netio . . . .   Commands to control Internet connections
[R]   news . . . .   Report Stata news
[R]   sj . . . .   Stata Journal and STB installation instructions
[R]   ssc . . . .   Install and uninstall packages from SSC
[R]   update . . . .   Update Stata

Data types and memory

[U]   User's Guide, Chapter 7 . . . .   Setting the size of memory
[U]   User's Guide, Section 15.2.2 . . . .   Numeric storage types
[U]   User's Guide, Section 15.4.4 . . . .   String storage types
[U]   User's Guide, Section 16.10 . . . .   Precision and problems therein
[U]   User's Guide, Chapter 26 . . . .   Commands for dealing with strings
[R]   compress . . . .   Compress data in memory
[R]   data types . . . .   Quick reference for data types
[R]   limits . . . .   Quick reference for limits
[R]   matsize . . . .   Set the maximum number of variables in a model
[R]   memory . . . .   Memory size considerations
[R]   missing values . . . .   Quick reference for missing values
[R]   recast . . . .   Change storage type of variable

Advanced utilities

[R]   assert . . . .   Verify truth of claim
[R]   cd . . . .   Change directory
[R]   checksum . . . .   Calculate checksum of file
[R]   copy . . . .   Copy file from disk or URL
[R]   db . . . .   Launch dialog
[P]   dialogs . . . .   Dialog programming
[R]   dir . . . .   Display filenames
[P]   discard . . . .   Drop automatically loaded programs
[R]   erase . . . .   Erase a disk file
[P]   hexdump . . . .   Display hexadecimal report on file
[R]   mkdir . . . .   Create directory
[R]   more . . . .   The —more— message
[R]   query . . . .   Display system parameters
[P]   quietly . . . .   Quietly and noisily perform Stata command
[R]   set . . . .   Quick reference for system parameters
[R]   shell . . . .   Temporarily invoke operating system
[P]   smcl . . . .   Stata markup and control language
[P]   sysdir . . . .   Set system directories

[R] type . . . . Display contents of files
[R] which . . . . Display location and version for an ado-file

Graphics

[G]   Graphics manual . . . . Stata Graphics Reference Manual
[R]   boxcox . . . . Box–Cox regression models
[TS] corrgram . . . . Correlogram
[TS] cumsp . . . . Cumulative spectral distribution
[R]   cumul . . . . Cumulative distribution
[R]   cusum . . . . Cusum plots and tests for binary variables
[R]   diagnostic plots . . . . Distributional diagnostic plots
[R]   dotplot . . . . Comparative scatterplots
[R]   factor . . . . Factor analysis
[R]   grmeanby . . . . Graph means and medians by categorical variables
[R]   histogram . . . . Histograms for continuous and categorical variables
[R]   kdensity . . . . Univariate kernel density estimation
[R]   lowess . . . . Lowess smoothing
[ST] ltable . . . . Life tables for survival data
[R]   lv . . . . Letter-value displays
[R]   mkspline . . . . Linear spline construction
[R]   pca . . . . Principal component analysis
[TS] pergram . . . . Periodogram
[R]   qc . . . . Quality control charts
[R]   regression diagnostics . . . . Regression diagnostics
[R]   roc . . . . Receiver-Operating-Characteristic (ROC) analysis
[R]   serrbar . . . . Graph standard error bar chart
[R]   smooth . . . . Robust nonlinear smoother
[R]   spikeplot . . . . Spike plots and rootograms
[ST] stphplot . . . . Graphical assessment of the Cox proportional hazards assumption
[ST] streg . . . . Graph estimated survivor, hazard, and cumulative hazard functions
[ST] sts graph . . . . Graph the survivor, hazard, and cumulative hazard functions
[R]   stem . . . . Stem-and-leaf displays
[TS] wntestb . . . . Bartlett's periodogram-based test for white noise
[TS] xcorr . . . . Cross-correlogram for bivariate time series

Statistics

Basic statistics

[R]   egen . . . . Extensions to generate
[R]   anova . . . . Analysis of variance and covariance
[R]   bitest . . . . Binomial probability test
[R]   ci . . . . Confidence intervals for means, proportions, and counts
[R]   correlate . . . . Correlations (covariances) of variables or estimators
[R]   logistic . . . . Logistic regression
[R]   oneway . . . . One-way analysis of variance
[R]   prtest . . . . One- and two-sample tests of proportions

[R] regress .... Linear regression

    [R] predict .... Obtain predictions, residuals, etc. after estimation
    [R] predictnl ... Obtain nonlinear predictions, standard errors, etc. after estimation
    [R] regression diagnostics .... Regression diagnostics
    [R] test .... Test linear hypotheses after estimation
    [R] testnl .... Test nonlinear hypotheses after estimation

[R] sampsi .... Sample size and power determination
[R] sdtest .... Variance comparison tests
[R] signrank .... Sign, rank, and median tests
[R] statsby .... Collect statistics for a command across a by list
[R] summarize .... Summary statistics
[R] table .... Tables of summary statistics
[R] tabstat .... Display table of summary statistics
[R] tabsum .... One- and two-way tables of summary statistics
[R] tabulate .... One- and two-way tables of frequencies
[R] ttest .... Mean comparison tests

ANOVA and related

[U] User's Guide, Chapter 29 .... Overview of Stata estimation commands
[R] anova .... Analysis of variance and covariance
[R] loneway .... Large one-way ANOVA, random effects, and reliability
[R] manova .... Multivariate analysis of variance and covariance
[R] oneway .... One-way analysis of variance
[R] pkcross .... Analyze crossover experiments
[R] pkshape .... Reshape (pharmacokinetic) Latin square data

Linear regression and related maximum-likelihood regressions

[U] User's Guide, Chapter 29 .... Overview of Stata estimation commands
[U] User's Guide, Chapter 23 .... Estimation and post-estimation commands
[U] User's Guide, Section 23.14 .... Obtaining robust variance estimates
[R] estimation commands .... Quick reference for estimation commands
[R] areg .... Linear regression with a large dummy-variable set
[R] cnsreg .... Constrained linear regression
[R] eivreg .... Errors-in-variables regression
[R] fracpoly .... Fractional polynomial regression
[R] frontier .... Stochastic frontier models
[R] glm .... Generalized linear models
[R] heckman .... Heckman selection model
[R] impute .... Impute data for missing values
[R] ivreg .... Instrumental variables and two-stage least squares regression
[R] mfp .... Multivariable fractional polynomial models
[R] mvreg .... Multivariate regression
[R] nbreg .... Negative binomial regression
[TS] newey .... Regression with Newey–West standard errors
[R] nl .... Nonlinear least squares
[R] orthog .... Orthogonal variables and orthogonal polynomials
[R] poisson .... Poisson regression
[TS] prais .... Prais–Winsten regression and Cochrane–Orcutt regression
[R] qreg .... Quantile (including median) regression
[R] reg3 .... Three-stage estimation for systems of simultaneous equations

[R]   regress .... Linear regression
[R]   regression diagnostics .... Regression diagnostics
[R]   roc .... Receiver-Operating-Characteristic (ROC) analysis
[R]   rreg .... Robust regression
[ST]   stcox .... Fit Cox proportional hazards model
[ST]   streg .... Fit parametric survival models
[R]   sureg .... Zellner's seemingly unrelated regression
[SVY] svy estimators .... Estimation commands for complex survey data
[R]   sw .... Stepwise maximum-likelihood estimation
[R]   tobit .... Tobit, censored-normal, and interval regression
[R]   treatreg .... Treatment effects model
[R]   truncreg .... Truncated regression
[R]   vwls .... Variance-weighted least squares
[XT]   xtabond .... Arellano–Bond linear, dynamic panel-data estimator
[XT]   xtfrontier .... Stochastic frontier models for panel data
[XT]   xtgee .... Fit population-averaged panel-data models using GEE
[XT]   xtgls .... Fit panel-data models using GLS
[XT]   xtintreg .... Random-effects interval data regression models
[XT]   xthtaylor .... Hausman–Taylor estimator for error components models
[XT]   xtivreg ..... Instrumental variables and two-stage least squares for panel-data models
[XT]   xtnbreg Fixed-effects, random-effects, and population-averaged negative binomial models
[XT]   xtpcse .... OLS or Prais–Winsten models with panel-corrected standard errors
[XT]   xtpoisson .... Fixed-effects, random-effects, and population-averaged Poisson models
[XT]   xtrchh .... Hildreth–Houck random coefficients models
[XT]   xtreg ... Fixed-, between-, and random-effects and population averaged linear models
[XT]   xtregar .... Fixed- and random-effects linear models with an AR(1) disturbance
[R]   zip .... Zero-inflated Poisson and negative binomial models

Logistic and probit regression

[U]   User's Guide, Chapter 29 .... Overview of Stata estimation commands
[U]   User's Guide, Chapter 23 .... Estimation and post-estimation commands
[U]   User's Guide, Section 23.14 .... Obtaining robust variance estimates
[R]   biprobit .... Bivariate probit models
[R]   clogit .... Conditional (fixed-effects) logistic regression
[R]   cloglog .... Maximum-likelihood complementary log-log estimation
[R]   constraint .... Define and list constraints
[R]   glogit .... Logit and probit on grouped data
[R]   heckprob .... Maximum-likelihood probit estimation with selection
[R]   hetprob .... Maximum-likelihood heteroskedastic probit estimation
[R]   logistic .... Logistic regression
[R]   logit .... Maximum-likelihood logit estimation
[R]   mlogit .... Maximum-likelihood multinomial (polytomous) logistic regression
[R]   nlogit .... Maximum-likelihood nested logit estimation
[R]   ologit .... Maximum-likelihood ordered logit estimation
[R]   oprobit .... Maximum-likelihood ordered probit estimation
[R]   probit .... Maximum-likelihood probit estimation
[R]   rologit .... Rank-ordered logistic regression
[R]   scobit .... Maximum-likelihood skewed logit estimation
[SVY] svy estimators .... Estimation commands for complex survey data
[R]   sw .... Stepwise maximum-likelihood estimation
[XT]   xtcloglog .... Random-effects and population-averaged cloglog models

[XT]   xtgee .... Fit population-averaged panel-data models using GEE
[XT]   xtlogit .... Fixed-effects, random-effects, and population-averaged logit models
[XT]   xtprobit .... Random-effects and population-averaged probit models

Pharmacokinetic statistics

[U]   User's Guide, Section 29.18 .... Pharmacokinetic data
[R]   pk .... Pharmacokinetic (biopharmaceutical) data
[R]   pkcollapse .... Generate pharmacokinetic measurement dataset
[R]   pkcross .... Analyze crossover experiments
[R]   pkexamine .... Calculate pharmacokinetic measures
[R]   pkequiv .... Perform bioequivalence tests
[R]   pkshape .... Reshape (pharmacokinetic) Latin square data
[R]   pksumm .... Summarize pharmacokinetic data

Survival analysis

[U]   User's Guide, Chapter 29 .... Overview of Stata estimation commands
[U]   User's Guide, Chapter 23 .... Estimation and post-estimation commands
[U]   User's Guide, Section 23.14 .... Obtaining robust variance estimates
[ST]   ct .... Count-time data
[ST]   ctset .... Declare data to be count-time data
[ST]   cttost .... Convert count-time data to survival-time data
[ST]   ltable .... Life tables for survival data
[ST]   snapspan .... Convert snapshot data to time-span data
[ST]   st .... Survival-time data
[ST]   st_is .... Survival analysis subroutines for programmers
[ST]   stbase .... Form baseline dataset
[ST]   stci .... Confidence intervals for means and percentiles of survival time
[ST]   stcox .... Fit Cox proportional hazards model
[ST]   stdes .... Describe survival-time data
[ST]   stfill .... Fill in by carrying forward values of covariates
[ST]   stgen .... Generate variables reflecting entire histories
[ST]   stir .... Report incidence-rate comparison
[ST]   stphplot .... Graphical assessment of the Cox proportional hazards assumption
[ST]   stptime .... Calculate person-time, incidence rates, and SMR
[ST]   strate .... Tabulate failure rates and rate ratios
[ST]   streg .... Fit parametric survival models
[ST]   sts ...... Generate, graph, list, and test the survivor and cumulative hazard functions
[ST]   sts generate .... Create survivor, hazard, and other variables
[ST]   sts graph .... Graph the survivor and the cumulative hazard functions
[ST]   sts list .... List the survivor and the cumulative hazard functions
[ST]   sts test .... Test equality of survivor functions
[ST]   stset .... Declare data to be survival-time data
[ST]   stsplit .... Split and join time-span records
[ST]   stsum .... Summarize survival-time data
[ST]   sttocc .... Convert survival-time data to case–control data
[ST]   sttoct .... Convert survival-time data to count-time data
[ST]   stvary .... Report which variables vary over time
[R]   sw .... Stepwise maximum-likelihood estimation

Time series

[U]	User's Guide, Section 14.4.3	Time-series varlists
[U]	User's Guide, Section 15.5.4	Time-series formats
[U]	User's Guide, Section 16.8	Time-series operators
[U]	User's Guide, Section 27.3	Time-series dates
[U]	User's Guide, Section 29.12	Models with time-series data
[TS]	time series	Introduction to time-series commands
[TS]	arch	Autoregressive conditional heteroskedasticity (ARCH) family of estimators
[TS]	arima	Autoregressive integrated moving average models
[TS]	corrgram	Correlogram
[TS]	cumsp	Cumulative spectral distribution
[TS]	dfgls	Perform DF-GLS unit-root test
[TS]	dfuller	Augmented Dickey Fuller test for a unit root
[TS]	newey	Regression with Newey–West standard errors
[TS]	pergram	Periodogram
[TS]	pperron	Phillips–Perron test for unit roots
[TS]	prais	Prais–Winsten regression and Cochrane–Orcutt regression
[TS]	regression diagnostics	Regression diagnostics for time series
[TS]	tsappend	Add observations to time-series dataset
[TS]	tsreport	Report time series aspects of dataset or estimation sample
[TS]	tsrevar	Time-series operator programming command
[TS]	tsset	Declare dataset to be time-series data
[TS]	tssmooth	Smooth and forecast univariate time-series data
[TS]	tssmooth dexponential	Double exponential smoothing
[TS]	tssmooth exponential	Exponential smoothing
[TS]	tssmooth hwinters	Holt–Winters nonseasonal smoothing
[TS]	tssmooth ma	Moving-average filter
[TS]	tssmooth nl	Nonlinear filter
[TS]	tssmooth shwinters	Holt–Winters seasonal smoothing
[TS]	var intro	An introduction to vector autoregression models
[TS]	var	Vector autoregression models
[TS]	var svar	Structural vector autoregression models
[TS]	varbasic	Fit a simple VAR and graph impulse response functions
[TS]	varfcast compute	Compute dynamic forecasts of dependent variables after var or svar
[TS]	varfcast graph	Graph forecasts of dependent variables after var or svar
[TS]	vargranger	Perform pairwise Granger causality tests after var or svar
[TS]	varirf	An introduction to the varirf commands
[TS]	varirf add	Add VARIRF results from one VARIRF file to another
[TS]	varirf cgraph	Make combined graphs of impulse response functions and FEVDs
[TS]	varirf create	Obtain impulse response functions and forecast error decompositions
[TS]	varirf ctable	Make combined tables of impulse response functions and FEVDs
[TS]	varirf describe	Describe a VARIRF file
[TS]	varirf dir	List the VARIRF files in a directory
[TS]	varirf drop	Drop VARIRF results from the active VARIRF file
[TS]	varirf erase	Erase a VARIRF file
[TS]	varirf graph	Graph impulse response functions and FEVDs
[TS]	varirf ograph	Graph overlaid impulse response functions and FEVDs
[TS]	varirf rename	Rename a VARIRF result in a VARIRF file
[TS]	varirf set	Set active VARIRF file
[TS]	varirf table	Create tables of impulse response functions and FEVDs
[TS]	varlmar	Obtain LM statistics for residual autocorrelation after var or svar

[TS]   varnorm .... Test for normally distributed disturbances after var or svar
[TS]   varsoc .... Obtain lag-order selection statistics for a set of VARs
[TS]   varstable .... Check stability condition of var or svar estimates
[TS]   varwle .... Obtain Wald lag exclusion statistics after var or svar
[TS]   wntestb .... Bartlett's periodogram-based test for white noise
[TS]   wntestq .... Portmanteau (Q) test for white noise
[TS]   xcorr .... Cross-correlogram for bivariate time series

Cross-sectional time series (panel data)

[U]   User's Guide, Chapter 29 .... Overview of Stata estimation commands
[U]   User's Guide, Chapter 23 .... Estimation and post-estimation commands
[XT]   xt .... Introduction to xt commands
[XT]   quadchk .... Check sensitivity of quadrature approximation
[XT]   xtabond .... Arellano–Bond linear, dynamic panel-data estimation
[XT]   xtcloglog .... Random-effects and population-averaged cloglog models
[XT]   xtdata .... Faster specification searches with xt data
[XT]   xtdes .... Describe pattern of xt data
[XT]   xtfrontier .... Stochastic frontier models for panel data
[XT]   xtgee .... Fit population-averaged panel-data models using GEE
[XT]   xtgls .... Fit panel-data models using GLS
[XT]   xthtaylor .... Hausman–Taylor estimator for error component models
[XT]   xtintreg .... Random-effects interval data regression models
[XT]   xtivreg ..... Instrumental variables and two-stage least squares for panel-data models
[XT]   xtlogit .... Fixed-effects, random-effects, and population-averaged logit models
[XT]   xtnbreg Fixed-effects, random-effects, and population-averaged negative binomial models
[XT]   xtpcse .... OLS or Prais–Winsten models with panel-corrected standard errors
[XT]   xtpoisson .... Fixed-effects, random-effects, and population-averaged Poisson models
[XT]   xtprobit .... Random-effects and population-averaged probit models
[XT]   xtrchh .... Hildreth–Houck random-coefficients models
[XT]   xtreg ... Fixed-, between-, and random-effects and population-averaged linear models
[XT]   xtregar .... Fixed- and random-effects linear models with an AR(1) disturbance
[XT]   xtsum .... Summarize xt data
[XT]   xttab .... Tabulate xt data
[XT]   xttobit .... Random-effects tobit models

Auxiliary regression and related commands

[U]   User's Guide, Section 16.5 .... Accessing coefficients and standard errors
[U]   User's Guide, Chapter 23 .... Estimation and post-estimation commands
[R]   estimation commands .... Quick reference for estimation commands
[R]   adjust .... Tables of adjusted means and proportions
[R]   constraint .... Define and list constraints
[R]   correlate .... Correlations (covariances) of variables or estimators
[R]   hausman .... Hausman specification test
[R]   lincom .... Linear combinations of estimators
[R]   linktest .... Specification link test for single-equation models
[R]   lrtest .... Likelihood-ratio test after estimation
[R]   maximize .... Details of iterative maximization
[R]   mfx .... Obtain marginal effects or elasticities after estimation
[R]   mkspline .... Linear spline construction
[R]   nlcom .... Nonlinear combinations of estimators
[R]   predict .... Obtain predictions, residuals, etc. after estimation

[R]   predictnl .... Obtain nonlinear predictions, standard errors, etc. after estimation
[R]   regression diagnostics .... Regression diagnostics
[TS]   regression diagnostics .... Regression diagnostics for time series
[P]   _robust .... Robust variance estimates
[R]   suest .... Seemingly unrelated estimation
[R]   test .... Test linear hypotheses after estimation
[R]   testnl .... Test nonlinear hypotheses after estimation
[R]   vce .... Display covariance matrix of the estimators
[R]   xi .... Interaction expansion

Commands for epidemiologists

[R]   binreg .... Generalized linear models: extensions to the binomial family
[R]   dstdize .... Direct and indirect standardization
[ST]   epitab .... Tables for epidemiologists
[R]   icd9 .... ICD-9-CM diagnostic and procedures codes
[R]   roc .... Receiver-Operating-Characteristic (ROC) analysis
[ST]   st .... Survival-time data
[R]   symmetry .... Symmetry and marginal homogeneity tests
[R]   tabulate .... One- and two-way tables of frequencies

Analysis of survey data

[U]   User's Guide, Chapter 30 .... Overview of survey estimation
[U]   User's Guide, Chapter 29 .... Overview of Stata estimation commands
[U]   User's Guide, Chapter 23 .... Estimation and post-estimation commands
[SVY] svy .... Introduction to survey commands
[SVY] lincom for svy .... Estimate linear combinations after survey estimation
[SVY] ml for svy .... Pseudo-maximum-likelihood estimation for survey data
[SVY] nonlinear for svy ..... Nonlinear combinations, predictions, and tests for survey data
[SVY] svy estimators .... Estimation commands for complex survey data
[SVY] svydes .... Describe survey data
[SVY] svymean .... Estimate means, totals, ratios, and proportions for survey data
[SVY] svyset .... Set variables for survey data
[SVY] svytab .... Tables for survey data
[SVY] test for svy .... Test linear hypotheses after survey estimation
[P]   _robust .... Robust variance estimates

Transforms and normality tests

[R]   boxcox .... Box–Cox regression models
[R]   fracpoly .... Fractional polynomial regression
[R]   ladder .... Ladder of powers
[R]   lnskew0 .... Find zero-skewness log or Box–Cox transform
[R]   mfp .... Multivariable fractional polynomial models
[R]   sktest .... Skewness and kurtosis test for normality
[R]   swilk .... Shapiro–Wilk and Shapiro–Francia tests for normality

Nonparametric statistics

[R]   kdensity .... Univariate kernel density estimation
[R]   ksmirnov .... Kolmogorov–Smirnov equality-of-distributions test
[R]   kwallis .... Kruskal–Wallis equality-of-populations rank test
[R]   lowess .... Lowess smoothing
[R]   nptrend .... Test for trend across ordered groups

[R]   qreg .... Quantile (including median) regression
[R]   roc .... Receiver-Operating-Characteristic (ROC) analysis
[R]   runtest .... Test for random order
[R]   signrank .... Sign, rank, and median tests
[R]   smooth .... Robust nonlinear smoother
[R]   spearman .... Spearman's and Kendall's correlations
[R]   symmetry .... Symmetry and marginal homogeneity tests

Simulation/resampling

[R]   bootstrap .... Bootstrap sampling and estimation
[R]   jknife .... Jackknife estimation
[R]   permute .... Monte Carlo permutation tests
[R]   simulate .... Monte Carlo simulations

Cluster analysis

[U]   User's Guide, Section 29.19 .... Cluster analysis
[CL]   cluster .... Introduction to cluster analysis commands
[CL]   cluster averagelinkage .... Average linkage cluster analysis
[CL]   cluster centroidlinkage .... Centroid linkage cluster analysis
[CL]   cluster completelinkage .... Complete linkage cluster analysis
[CL]   cluster dendrogram .... Dendrograms for hierarchical cluster analysis
[CL]   cluster generate .... Generate summary or grouping variables from a cluster analysis
[CL]   cluster kmeans .... Kmeans cluster analysis
[CL]   cluster kmedians .... Kmedians cluster analysis
[CL]   cluster medianlinkage .... Median linkage cluster analysis
[CL]   cluster notes .... Place notes in cluster analysis
[CL]   cluster programming subroutines .... Add cluster analysis routines
[CL]   cluster programming utilities .... Cluster analysis programming utilities
[CL]   cluster singlelinkage .... Single linkage cluster analysis
[CL]   cluster stop .... Cluster analysis stopping rules
[CL]   cluster utility .... List, rename, use, and drop cluster analyses
[CL]   cluster wardslinkage .... Ward's linkage cluster analysis
[CL]   cluster waveragelinkage .... Weighted-average linkage cluster analysis

Factor analysis and principal components

[R]   alpha .... Cronbach's alpha
[R]   canon .... Canonical correlations
[R]   factor .... Factor analysis
[R]   impute .... Impute data for missing values
[R]   pca .... Principal component analysis

Do-it-yourself maximum likelihood estimation

[P]   matrix .... Introduction to matrix commands
[R]   ml .... Maximum likelihood estimation

Quality control

[R]   qc .... Quality control charts
[R]   serrbar .... Graph standard error bar chart

Other statistics

[R]	alpha	....	Cronbach's alpha
[R]	brier	....	Brier score decomposition
[R]	canon	....	Canonical correlations
[R]	centile	....	Report centile and confidence interval
[R]	hotelling	....	Hotelling's T-squared generalized means test
[R]	impute	....	Impute data for missing values
[R]	kappa	....	Interrater agreement
[R]	means	....	Arithmetic, geometric, and harmonic means
[R]	pcorr	....	Partial correlation coefficients
[R]	pctile	....	Create variable containing percentiles
[R]	range	....	Numerical ranges, derivatives, and integrals

Matrix commands

Basics

[U]	User's Guide, Chapter 17	....	Matrix expressions
[P]	matrix	....	Introduction to matrix commands
[P]	matrix define	....	Matrix definition, operators, and functions
[P]	matrix utility	....	List, rename, and drop matrices

Programming

[P]	matrix accum	....	Form cross-product matrices
[R]	ml	....	Maximum likelihood estimation
[P]	ereturn	....	Post-estimation results
[P]	matrix rowname	....	Name rows and columns
[P]	matrix score	....	Score data from coefficient vectors

Other

[P]	matrix constraint	....	Constrained estimation
[P]	matrix eigenvalues	....	Eigenvalues of nonsymmetric matrices
[P]	matrix get	....	Access system matrices
[P]	matrix mkmat	....	Convert variables to matrix and vice versa
[P]	matrix svd	....	Singular value decomposition
[P]	matrix symeigen	....	Eigenvalues and eigenvectors of symmetric matrices

Programming

Basics

[U]	User's Guide, Chapter 21	....	Programming Stata
[U]	User's Guide, Section 21.3	....	Macros
[U]	User's Guide, Section 21.11	....	Ado-files
[P]	comments	....	Add comments to programs
[P]	program	....	Define and manipulate programs
[P]	macro	....	Macro definition and manipulation
[P]	return	....	Return saved results

Program control

[U]	User's Guide, Section 21.11.1	....	Version
[P]	version	....	Version control
[P]	continue	....	Break out of loops
[P]	foreach	....	Loop over items
[P]	forvalues	....	Loop over consecutive values
[P]	if	....	if programming command
[P]	while	....	Looping
[P]	error	....	Display generic error message and exit
[P]	capture	....	Capture return code

Parsing and program arguments

[U]	User's Guide, Section 21.4	....	Program arguments
[P]	syntax	....	Parse Stata syntax
[P]	confirm	....	Argument verification
[P]	gettoken	....	Low-level parsing
[P]	numlist	....	Parse numeric lists
[P]	tokenize	....	Divide strings into tokens

Console output

| [P] | display | .... | Display strings and values of scalar expressions |
| [P] | tabdisp | .... | Display tables |

Commonly used programming commands

[P]	byable	....	Make programs byable
[P]	#delimit	....	Change delimiter
[P]	exit	....	Exit from a program or do-file
[P]	quietly	....	Quietly and noisily perform Stata command
[P]	mark	....	Mark observations for inclusion
[P]	more	....	Pause until key is depressed
[P]	preserve	....	Preserve and restore data
[P]	matrix	....	Introduction to matrix commands
[P]	scalar	....	Scalar variables
[P]	smcl	....	Stata markup and control language
[P]	sortpreserve	....	Sort with programs
[TS]	tsrevar	....	Time-series operator programming command

Debugging

| [P] | pause | .... | Program debugging command |
| [P] | trace | .... | Debug Stata programs |

Advanced programming commands

[P]	break	....	Suppress Break key
[P]	char	....	Characteristics
[P]	class	....	Class programming
[P]	class exit	....	Exit class member program and return result
[P]	classutil	....	Class programming utility
[P]	_estimates	....	Manage estimation results
[P]	file	....	Read and write ASCII test and binary files
[P]	findfile	....	Find file in path

[P]   macro .... Macro definition and manipulation
[P]   macro lists .... Manipulate lists
[R]   ml .... Maximum likelihood estimation
[P]   postfile .... Save results in Stata dataset
[P]   _predict .. Obtain predictions, residuals, etc. after estimation programming command
[P]   _return .... Preserve saved results
[P]   _rmcoll .... Remove collinear variables
[P]   _robust .... Robust variance estimates
[P]   serset .... Create and manipulate sersets
[P]   unab .... Unabbreviate variable list
[P]   unabcmd .... Unabbreviate command name
[P]   window fopen .... Display open/save dialog box
[P]   window menu .... Create menus
[P]   window stopbox .... Display message box

Special interest programming commands

[CL]   cluster programming subroutines .... Add cluster analysis routines
[CL]   cluster programming utilities .... Cluster analysis programming utilities
[ST]   st_is .... Survival analysis subroutines for programmers
[TS]   tsrevar .... Time-series operator programming command

File formats

[P]   file formats .dta .... Description of .dta file format

Interface features

[P]   dialogs .... Dialog programming
[R]   doedit .... Edit do-files and other text files
[R]   edit .... Edit and list data using Data Editor
[P]   sleep .... Pause for a specified time
[P]   smcl .... Stata markup and control language
[P]   window fopen .... Display open/save dialog box
[P]   window manage .... Manage window characteristics
[P]   window menu .... Create menus
[P]   window push .... Copy command into Review window
[P]   window stopbox .... Display message box

Stata Basics

Chapters

1   Read this—it will help 3

2   Resources for learning and using Stata 29

3   A brief description of Stata 37

4   Flavors of Stata . . . . , , , , 41

5   Starting and stopping Stata 45

6   Troubleshooting starting and stopping Stata 51

7   Setting the size of memory 53

8   Stata's online help and search facilities 61

9   Stata's sample datasets 71

10   –more– conditions 73

11   Error messages and return codes 75

12   The Break key 79

13   Keyboard use 81

1 Read this—it will help

Contents

1.1   Getting Started with Stata
1.2   The User's Guide and the Reference manuals
    1.2.1   Cross-referencing
    1.2.2   The index
    1.2.3   The subject table of contents
    1.2.4   Typography
1.3   What's new
    1.3.1   What's big
    1.3.2   What's useful
    1.3.3   What's convenient
    1.3.4   What was needed
    1.3.5   What's faster
    1.3.6   What's new in time-series analysis
    1.3.7   What's new in cross-sectional time-series analysis
    1.3.8   What's new in survival analysis
    1.3.9   What's new in survey analysis
    1.3.10   What's new in cluster analysis
    1.3.11   What's new in statistics useful in all fields
    1.3.12   What's new in data management
    1.3.13   What's new in expressions and functions
    1.3.14   What's new in display formats
    1.3.15   What's new in programming
    1.3.16   What's new in the user interface
    1.3.17   What's more

The Stata Documentation Set contains over 4,000 pages of information. There are ten parts to the full set:

[GS]	*Getting Started with Stata*
[U]	*Stata User's Guide*
[R]	*Stata Base Reference Manual*
	Volume 1, A–F
	Volume 2, G–M
	Volume 3, N–R
	Volume 4, S–Z
[G]	*Stata Graphics Reference Manual*
[P]	*Stata Programming Reference Manual*
[CL]	*Stata Cluster Analysis Reference Manual*
[XT]	*Stata Cross-Sectional Time-Series Reference Manual*
[SVY]	*Stata Survey Data Reference Manual*
[ST]	*Stata Survival Analysis & Epidemiological Tables Reference Manual*
[TS]	*Stata Time-Series Reference Manual*

Detailed information about each of these manuals, and the other Stata manuals, may be found online at

http://www.stata-press.com/manuals/

1.1   Getting Started with Stata

There are three *Getting Started* manuals:

Getting Started with Stata for Windows
Getting Started with Stata for Macintosh
Getting Started with Stata for Unix

1. Locate your *Getting Started* manual.

2. Install Stata. The instructions are found in the *Getting Started* manual.

3. Learn how to invoke Stata and use it—read the *Getting Started* manual.

4. Now turn to the other manuals; see [U] **1.2 The User's Guide and the Reference manuals**.

1.2   The User's Guide and the Reference manuals

The *User's Guide* is divided into three sections: Basics, Elements, and Advice. At the beginning of each section is a list of the chapters found in that section. In addition to exploring the fundamentals of Stata—information that all users should know—this manual will guide you to other sources for Stata information.

The other manuals are the *Reference* manuals. The Stata *Reference* manuals are each arranged like an encyclopedia—alphabetically. Look at the *Base Reference Manual*. Look under the name of the command. If you do not find the command, look in the index. There are a few commands that are so closely related that they are documented together, such as signrank, signtest, ranksum, and median, which are all documented in [R] **signrank**.

Not all the entries in the *Base Reference Manual* are Stata commands; some contain technical information, like [R] **maximize**, which details Stata's iterative maximization process, or [R] **error messages**, which provides information on error messages and return codes. Others are a "quick reference", like [R] **estimation commands**. The quick reference entries summarize information that is discussed in more detail in either the *User's Guide* or one of the other *Reference* manuals.

Like an encyclopedia, the *Reference* manuals are not designed to be read from cover(s) to cover(s). When you want to know what a command does, complete with all the details, qualifications, and pitfalls, or when a command produces an unexpected result, read its description. Each entry is written at the level of the command. The descriptions assume little knowledge of Stata's features when explaining simple commands, such as those for using and saving data. For more complicated commands, they assume that you have a firm grasp of Stata's other features.

If a Stata command is not in the *Base Reference Manual*, it can be found in one of the other *Reference* manuals. The titles of the manuals indicate the commands that they contain. The *Programming Reference Manual*, however, contains not only commands for programming Stata, but also the matrix-manipulation commands and commands that are of such a technical nature that you would nearly have to be a programmer to use them.

1.2.1   Cross-referencing

The *Getting Started* manual, the *User's Guide*, and the *Reference* manuals cross-reference each other.

[R]   **regress**
[P]   **matrix define**
[XT]   **xtabond**

The first is a reference to the `regress` entry in the *Base Reference Manual*; the second is a reference to the `matrix define` entry in the *Stata Programming Reference Manual*; and the third is a reference to the `xtabond` entry in the *Cross-Sectional Time-Series Reference Manual*.

[GSW] **A. More on starting and stopping Stata**
[GSM] **A. Starting and stopping Stata for Macintosh**
[GSU] **A. Starting and stopping Stata for Unix**

are instructions to see the appropriate section of the *Getting Started with Stata for Windows*; *Getting Started with Stata for Macintosh*; or *Getting Started with Stata for Unix* manuals.

1.2.2   The index

The *User's Guide* and the *Base Reference Manual* contain a combined index for the *Guide* and all the *Reference* manuals, except the *Graphics Reference Manual*. This index is located at the end of this manual and at the end of Volume 4 of the *Base Reference Manual*.

The other *Reference* manuals each have their own index.

1.2.3   The subject table of contents

At the beginning of this manual and at the beginning of the *Base Reference Manual* Volume 1 is a subject table of contents for the *Guide* and for the *Reference* manuals, except the *Graphics Reference Manual*.

If you look under "Functions and expressions", you will see

[U]   User's Guide, Chapter 16 . . . . Functions and expressions
[R]   egen . . . . Extensions to generate
[R]   functions . . . . Functions

1.2.4   Typography

You will note that we mix the ordinary printing that you are reading now with a typewriter-style typeface `that looks like this`. When something is printed in the typewriter-style typeface, it means that something is a command or an option—it is something that Stata understands and something that you might actually type into your computer. Differences in typeface are important. If a sentence reads, "You could list the result . . . ", it is just an English sentence—you *could* list the result, but the sentence provides no clue as to how you might actually do that. On the other hand, if the sentence reads, "You could `list` the result . . . ", it is telling you much more—you could list the result and you could do that using the `list` command.

We will occasionally lapse into periods of inordinate cuteness and write, "We described the data and then listed the data." You get the idea. describe and list are Stata commands. We purposely began the previous sentence with a lowercase letter. Because describe is a Stata command, it must be typed in lowercase letters. The ordinary rules of capitalization are temporarily suspended in favor of preciseness.

You will also notice that we mix in words printed in italic type, like "To perform the rank-sum test, type ranksum *varname* , by(*groupvar*)". Italicized words are not supposed to be typed; instead, you are to substitute another word for them.

We would also like users to note our rule for punctuation of quotes. We follow a rule that is often used in mathematics books and British literature. The punctuation mark at the end of the quote is only included in the quote if it is a part of the quote. For instance, the pleased Stata user said she thought that Stata was a "very powerful program". Another user simply said, "I love Stata."

In this manual, however, there is little dialog, and we follow this rule to make precise what you are to type, as in, type "cd c:". The period is outside the quotation mark since you should not type the period. If we had wanted you to type the period, we would have included two periods at the end of the sentence; one inside the quotation and one outside, as in, type "use myfile.".

We have tried not to violate the other rules of English. If you find such violations, they were unintentional and due to our own ignorance or carelessness. We would appreciate hearing about them.

We have heard from Nicholas J. Cox of the Department of Geography at the University of Durham in the U.K. and wish to express our appreciation. Nicholas' efforts have gone far beyond dropping us a note and—without Nicholas' assistance—there is no way with words that we can express our gratitude.

1.3   What's new

This section is intended for previous Stata users. If you are new to Stata, skip to [U] **1.3.17 What's more**.

As always, the new Stata is 100% compatible with the previous release of Stata, but we remind programmers that it is vitally important that you put version 7 at the top of your old do-files and ado-files if they are to work. You were supposed to do that when you wrote them but, if you did not, go back and do it now. We have made a lot of changes (improvements) to Stata.

In addition, Stata's dataset format has changed because of the new longer data storage types and the fact that Stata now has multiple representations for missing values. You will not care because Stata automatically reads old-format datasets, but if you need to send a dataset to someone still using Stata 7, remember to use the saveold command; see [R] **save**.

(Continued on next page)

1.3.1   What's big

The big news is the new GUI and the new Graphics. There is no putting them in an order.

1. Graphics. You can create graphs that look like this

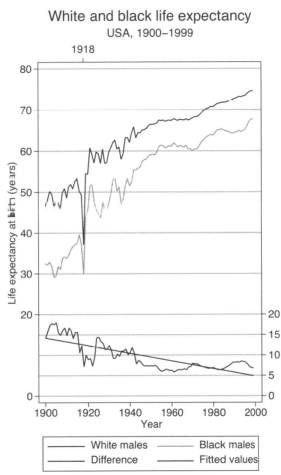

White and black life expectancy
USA, 1900–1999

Source: National Vital Statistics, Vol 50, No. 6
(1918 dip caused by 1918 Influenza Pandemic)

or this

(Continued on next page)

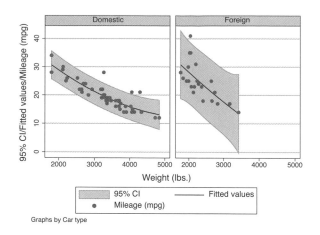

Graphs by Car type

There is a whole manual dedicated to the subject—the *Stata Graphics Reference Manual*. To learn about the graphics, see *What's new* in [G] **intro** or, even easier, type `whelp graph_intro` in the Command window. Everything you need to know is online.

All existing statistical commands that produce graphs have been updated to take advance of the new graphics.

2. GUI stands for Graphical User Interface, and to try it, you do not need to read a thing. Pull down **Data**, **Graphics**, or **Statistics**, find what you are looking for, and click. You will see

(*Continued on next page*)

Fill in the dialog box and click to submit. Do not ignore tabs at the top—there are very useful things hidden under them.

If you know the command you want, you can skip the menus and type db followed by the command name. You can jump to the above dialog box by typing db stcox.

1.3.2   What's useful

Stata 8 has so many features that finding what you are looking for can be a challenge. We have addressed that:

1. Pull down **Help** and select **Contents**. You will be presented with the categories Basics, Data management, Statistics, Graphics, and Programming. Click on one of them—say, Statistics— and you will be presented with another set of categories: Summary statistics and tests, Tables, Estimation, Multivariate analysis, Resampling and simulation, Statistical hand calculations, and Special topics. Click on one of those and, well, you get the idea. With the new help contents, it never takes long to find what you need.

2. Help files now have hyperlinks in the header for launching the dialog associated with the command. So, there are three ways to launch a dialog box: (1) use the menus (pull down **Data**, **Graphics**, or **Statistics**); (2) use the new db command (see [R] **db**); or (3) pick the command from the online help.

3. When you do need to search, `findit` is the key. `findit` searches everywhere: Stata itself, the Stata web site, the FAQs, the *Stata Journal*, and even user-written programs available on the web. An earlier version of `findit` was made available as an update to Stata 7, but the new version is better. You can also access `findit` by pulling down **Help** and selecting **Search**. If you do that, be sure to click *Search all* in the dialog box. See [R] **search**.

4. The new `ssc` command lists and installs user-written packages from the Statistical Software Components (SSC) archive, also known as the Boston College Archive, located at *http://www.repec.org*. See [R] **ssc**.

5. The new `net sj` command makes loading files from the new *Stata Journal* easier; see [R] **net**.

1.3.3 What's convenient

The existing `set` command has a new `permanently` option that allows you to make the setting permanent. This does away with the necessity of having a `profile.do` file for most users.

1.3.4 What was needed

Stata now has multiple missing values! In addition to the previously existing ., there is now .a, .b, ..., .z, and you can attach value labels to the new missing codes!

One thing to watch out for: Do not type

.   *stata_command* ... if x != .

Instead, type

.   *stata_command* ... if x < .

You need remember this only if you use the new missing values, but better to have good habits. The way things now work,

$$\text{all numbers} < . < .a < .b < ... < .z$$

So, if you wanted to list all observations for which x is missing, you would type

.   list if x >= .

See [U] **15.2.1 Missing values**.

1.3.5 What's faster

Stata 8 executes programming commands in half the time of Stata 7, on average. This results in commands implemented as ado-files running about 17 to 43% faster.

1. This speed-up is due to a new, faster memory manager that reduces the time needed to find, access, and store results. Thus, the improvement does not change much the time to run built-in, heavily computational commands. `regress`, for instance, runs only 1.43% faster. Nevertheless, the effect can be marked on other commands. `poisson` runs up to 31% faster, and `heckman` runs up to 43% faster. The larger the dataset, the less will be the improvement: `heckman` runs 17% faster on a 4,000 observations.

2. That statistical commands run faster is a happy side effect. The big advantage of the speed-up is that it allows some problems to be approached using ado-files that previously would have required internal code, such as Stata's new graphics, which is an ado-file implementation! Some programming commands run up to 400% faster. Implementing features as ado-files is part of the effort to keep Stata open and extendable by users.

1.3.6   What's new in time-series analysis

1. Stata now can fit vector autoregression (VAR) and structural vector autoregression (SVAR) models. New commands var, varbasic, and svar perform the estimation; see [TS] **var intro**.

   a. A suite of varirf commands estimate, tabulate, and graph impulse–response functions, cumulative impulse–response functions, orthogonalized impulse–response functions, structural impulse–response functions, and their confidence intervals, along with forecast-error variance decompositions and structural forecast-error variance decompositions; see [TS] **varirf**. This suite allows graphical comparisons of IRFS and variance decompositions across models and orderings.

   b. varfcast produces dynamic forecasts from a previously fitted var or svar model.

   c. There is also a full suite of diagnostic and testing tools including

      i. vargranger, that performs Granger causality tests.

      ii. varlmar, that performs a Lagrangian multiplier (LM) test for residual autocorrelation.

      iii. varnorm, that performs a series of tests for normality of the disturbances.

      iv. varsoc, that reports a series of lag order selection statistics.

      v. varstable, that checks the eigenvalue stability condition.

      vi. varwle, that performs a Wald test that all the endogenous variables of a given lag are zero, both for each equation separately and for all equations jointly.

2. The new tssmooth command smooths and predicts univariate time series using weighted or unweighted moving average, single exponential smoothing, double exponential smoothing, Holt–Winters nonseasonal smoothing, Holt–Winters seasonal smoothing, or nonlinear smoothing. See [TS] **tssmooth**.

3. The new tsappend command appends observations to a time-series dataset, automatically filling in the time variable and the panel variable, if set, by using the information contained in tsset. See [TS] **tsappend**.

4. The new archlm command computes a Lagrange multiplier test for autoregressive conditional heteroskedasticity (ARCH) effects in the residuals after regress; see [TS] **regression diagnostics**.

5. The new bgodfrey command computes the Breusch–Godfrey Lagrange multiplier (LM) test for serial correlation in the disturbances after regress; see [TS] **regression diagnostics**.

6. The new durbina command computes the Durbin (1970) alternative statistic to test for serial correlation in the disturbances after regress when some of the regressors are not strictly exogenous; see [TS] **regression diagnostics**.

7. The new dfgls command performs the modified Dickey–Fuller t test for a unit root (proposed by Elliot, Rothenberg, and Stock (1996)) using models with 1 to *maxlags* lags of the first differenced variable in an augmented Dickey–Fuller regression; see [TS] **dfgls**.

8. The existing arima command may now be used with the by prefix command, and it now allows prediction in loops over panels; see [TS] **arima**.

9. The existing newey command now allows (and requires) that you tsset your data; see [TS] **newey**.

1.3.7   What's new in cross-sectional time-series analysis

1. The new xthtaylor command fits panel-data random-effects models using the Hausman–Taylor and the Amemiya–MaCurdy instrumental-variables estimators; see [XT] **xthtaylor**.

2. The new `xtfrontier` command fits stochastic production or cost frontier models for panel data allowing two different parameterizations for the inefficiency term: a time-invariant model and the Battese–Coelli (1992) parameterization of time effects; see [XT] **xtfrontier**.

3. The existing `xtabond` command now allows endogenous regressors; see [XT] **xtabond**.

4. The existing `xtivreg` command will now optionally report first stage results of Baltagi's EC2SLS random-effects estimator; see [XT] **xtivreg**.

5. The existing `xttobit` and `xtintreg` commands have new `predict` options:

   a. `pr0(#_a,#_b)` produces the probability of the dependent variable being uncensored $P(\#_a < y < \#_b)$.

   b. `e0(#_a,#_b)` produces the corresponding expected value $E(y|\#_a < y < \#_b)$.

   c. `ystar(#_a,#_b)` produces the expected value of the dependent variable truncated at the censoring point(s), $E(y^*)$, where $y^* = \max(\#_a, \min(y, \#_b))$.

   See [XT] **xttobit** and [XT] **xtintreg**.

6. Existing commands `xtgee` and `xtlogit` have a new `nodisplay` option that suppresses the header and table of coefficients; `xtregar, fe` now allows `aweights` and `fweights`; and `xtpcse` now has no restrictions on how `aweights` are applied. See [XT] **xtgee**, [XT] **xtlogit**, and [XT] **xtpcse**.

7. Two commands have been renamed: `xtpois` is now called `xtpoisson` and `xtclog` is now `xtcloglog`. The old names continue to work. See [XT] **xtpoisson** and [XT] **xtcloglog**.

1.3.8   What's new in survival analysis

1. Existing command `stcox` has an important new feature and some minor improvements:

   a. `stcox` will now fit models with gamma-distributed frailty. In this model, frailty is assumed to be shared across groups of observations. Previously, if one wanted to analyze multivariate survival data using the Cox model, one would fit a standard model and account for the correlation within groups by adjusting the standard errors for clustering. Now, one may directly model the correlation by assuming a latent gamma-distributed random effect or frailty; observations within group are correlated because they share the same frailty. Estimation is via penalized likelihood. An estimate of the frailty variance is available and group-level frailty estimates can be retrieved.

   b. `fracpoly`, `sw`, and `linktest` now work after `stcox`.

   See [ST] **stcox**.

2. Existing command `streg` has an important new feature and some minor improvements:

   a. `streg` has new option `shared(varname)` for fitting parametric shared frailty models, analogous to random effects models for panel data. `streg` could, and still can, fit frailty models where the frailties are assumed to be randomly distributed at the observation level.

   b. `fracpoly`, `sw`, and `linktest` now work after `streg`.

   c. `streg` has four other new options: `noconstant`, `offset()`, `noheader`, and `nolrtest`.

   See [ST] **streg**.

3. `predict` after `streg, frailty()` has two new options:

   a. `alpha1` generates predictions conditional on a frailty equal to 1.

   b. `unconditional` generates predictions that are "averaged" over the frailty distribution.

   These new options may also be used with `stcurve`. See [ST] **streg**.

4. `sts graph` and `stcurve` (after `stcox`) can now plot estimated hazard functions, which are calculated as weighted kernel smooths of the estimated hazard contributions; see [ST] **sts graph**.

5. `streg, dist(gamma)` is now faster and more accurate. In addition, you can now predict mean time after gamma; see [ST] **streg**.

6. Old commands `ereg`, `ereghet`, `llogistic`, `llogistichet`, `gamma`, `gammahet`, `weibull`, `weibullhet`, `lnormal`, `lnormalhet`, `gompertz`, `gompertzhet` are deprecated (they continue to work) in favor of `streg`. Old command `cox` is now removed (it continues to work) in favor of `stcox`.

1.3.9 What's new in survey analysis

1. Stata's `ml` user-programmable likelihood-estimation routine has new options that automatically handle the production of survey estimators, including stratification and estimation on a subpopulation; see [R] **ml**.

2. Four new survey estimation commands are available:

   a. `svynbreg` for negative-binomial regression.

   b. `svygnbreg` for generalized negative binomial regression.

   c. `svyheckman` for the Heckman selection model.

   d. `svyheckprob` for probit regression with selection.

   See [SVY] **svy estimators**.

3. Use of the survey commands has been made more consistent.

   a. `svyset` has new syntax. Before it was

      `svyset` *thing_to_set* [, `clear`]

      and now it is

      `svyset` [*weight*] [, `strata`(*varname*) `psu`(*varname*) `fpc`(*varname*)]

      See [SVY] **svyset** for details. In addition, you must now `svyset` your data prior to using the survey commands; no longer can you set the data via options to the other survey commands.

   b. Two survey estimation commands have been renamed: `svyreg` to `svyregress` and `svypois` to `svypoisson`; see [SVY] **svy estimators**.

   c. `svyintreg` now applies constraints in the same manner as all other estimation commands; see [SVY] **svy estimators**.

   d. `lincom` now works after all `svy` estimators; see [R] **lincom**. (`svylc` is now deprecated.)

   e. `testnl` now works after all `svy` estimators; see [R] **testnl**.

   f. `testparm` now works after all `svy` estimators; see [R] **test**.

   g. The new `nlcom` and `predictnl` commands, which form nonlinear combinations of estimators and generalized predictions, work after all `svy` estimators; see [R] **nlcom** and [R] **predictnl**.

4. Existing command `svytab` has three new options: `cellwidth()`, `csepwidth()`, and `stub-width()`; they specify the widths of table elements in the output. See [SVY] **svytab**.

1.3.10   What's new in cluster analysis

1. The new `cluster wardslinkage` command provides Ward's linkage hierarchical clustering and can produce Ward's method, also known as minimum-variance clustering. See [CL] **cluster wardslinkage**.

2. The new `cluster waveragelinkage` command provides weighted-average linkage hierarchical clustering to accompany the previously available average linkage clustering. See [CL] **cluster waveragelinkage**.

3. The new `cluster centroidlinkage` command provides centroid linkage hierarchical clustering. This differs from the previously available `cluster averagelinkage` in that it combines groups based on the average of the distances between observations of the two groups to be combined. See [CL] **cluster centroidlinkage**.

4. The new `cluster medianlinkage` command provides median linkage hierarchical clustering, also known as Gower's method. See [CL] **cluster medianlinkage**.

5. The new `cluster stop` command provides stopping rules. Two popular stopping rules are provided, the Caliński & Harabasz pseudo-F index (Caliński and Harabasz (1974)) and the Duda & Hart Je(2)/Je(1) index with associated pseudo T-squared (Duda and Hart (1973)). See [CL] **cluster stop**.

    Additional stopping rules can be added; see [CL] **cluster programming subroutines**.

6. Two new dissimilarity measures have been added: L2squared and Lpower(#). L2squared provides squared Euclidean distance. Lpower(#) provides the Minkowski distance metric with argument # raised to the # power. See *Similarity and dissimilarity measures* in [CL] **cluster**.

7. A list of the variables used in the cluster analysis is now saved with the cluster analysis structure, which is useful for programmers; see [CL] **cluster programming subroutines**.

1.3.11   What's new in statistics useful in all fields

1. The following new estimators are available:

    a. `manova` fits multivariate analysis-of-variance (MANOVA) and multivariate analysis-of-covariance (MANCOVA) models for balanced and unbalanced designs, including designs with missing cells; and for factorial, nested, or mixed designs. See [R] **manova**. (`manovatest` provides multivariate tests involving terms from the most recently fitted `manova`.)

    b. `rologit` fits the rank-order logit model, also known as the exploded logit model. This model is a generalized McFadden's choice model as fitted by `clogit`. In the choice model, only the alternative that maximizes utility is observed. `rologit` fits the corresponding model in which the preference ranking of the alternatives is observed, not just the alternative that is ranked first. `rologit` supports incomplete rankings and ties ("indifference"). See [R] **rologit**.

    c. `frontier` fits stochastic frontier models with technical or cost inefficiency effects. `frontier` can fit models in which the inefficiency error component is assumed to be from one of the three distributions: half-normal, exponential, or truncated-normal. In addition, when the inefficiency term is assumed to be either half-normal or exponential, `frontier` can fit models in which the error components are heteroskedastic, conditional on a set of covariates. `frontier` can also fit models in which the mean of the inefficiency term is modeled as a linear function of a set of covariates. See [R] **frontier**.

    These new estimators are in addition to the new estimators listed in previous sections.

2. New command `mfp` selects the fractional polynomial model that best predicts the dependent variable from the independent variables; see [R] **mfp**.

3. The new `nlcom` command computes point estimates, standard errors, t and Z statistics, p-values, and confidence intervals for nonlinear combinations of coefficients after any estimation command. Results are displayed in the table format that is commonly used for displaying estimation results. The standard errors are based on the delta method, an approximation appropriate in large samples. See [R] **nlcom**.

4. The new `predictnl` command produces nonlinear predictions after any Stata estimation command, and optionally, can calculate the variance, standard errors, Wald test-statistics, significance levels, and point-wise confidence intervals for these predictions. Unlike `testnl` and `nlcom`, the quantities generated by `predictnl` are allowed to vary over the observations in the data. The standard errors and other inference-related quantities are based on the "delta method", an approximation appropriate in large samples. See [R] **predictnl**.

5. The new `bootstrap` command replaces the old `bstrap` and `bs` commands. `bootstrap` has an improved syntax and allows for stratified sampling. See [R] **bootstrap**.

   Existing command `bsample` also now accepts the `strata()` option, and it has a new `weight()` option that allows the user to save the sample frequency instead of changing the data in memory. See [R] **bootstrap**.

6. The existing `bstat` command can now construct bias-corrected and accelerated (BCa) confidence intervals. In addition, `bstat` is now an e-class command, meaning all the post-estimation commands can be used on bootstrap results. See [R] **bootstrap**.

7. Existing command `jknife` now accepts the `cluster()` option; see [R] **jknife**.

8. New command `permute` estimates p-values for permutation tests based on Monte Carlo simulations. These estimates can be one sided or two sided. See [R] **permute**.

9. Existing command `sample` has new option `count` that allows samples of the specified number of observations (rather than a percentage) to be drawn. In addition, `sample` now allows the by *varlist*: prefix as an alternative to the already existing by(*varlist*) option; both do the same thing. See [R] **sample**.

10. New command `simulate` replaces `simul` and provides improved syntax for specifying simulations; see [R] **simulate**.

11. Existing command `statsby` has a new syntax, new options, and now allows time-series operators; see [R] **statsby**.

12. The new `estimates` command provides a new, consistent way to store and refer to estimation results. Post-estimation commands that make comparisons across models, such as `lrtest` and `hausman`, previously had their own idiosyncratic ways to store and refer to estimation results. These commands now support a unified way of retrieving estimation results utilizing the new `estimates` suite.

    Under the new scheme, after fitting a model, you can type

        . estimates store *name*

    to save the results. At some point later in the session, you can type

        . estimates restore *name*

    to get back the estimates. You can redisplay estimates (without restoring them) by typing

        . estimates replay *name*

    Other estimation manipulation commands are provided; see [R] **estimates**.

a. Existing command `lrtest` have been modified to have syntax

     `lrtest` *name name*

b. Existing command `hausman` has been modified to have syntax

     `hausman` *name name*

c. The new `estimates for` command can be used in front of any post-estimation command, such as `test` or `predict`, to perform the action on the specified set of estimation results, without disturbing the current estimation results. With `estimates for`, you can type such things as

     . `estimates for` *earlierresults*: `test mpg`

See [R] **estimates**.

d. The new `estimates stats` command displays the Akaike Information Criterion (AIC) and Schwarz Information Criterion (BIC) model selection indexes. See [R] **estimates**.

13. Existing command `lrtest` now supports composite models specified by a parenthesized list of model names. In a composite model, it is assumed that the log likelihood and dimension of the full model are obtained as the sum of the log likelihoods and the sum of the dimensions of the constituent models.

`lrtest` has a new `stats` option to display statistical information about the unrestricted and restricted models, including the AIC and BIC model selection statistics. See [R] **lrtest**.

14. `test` has improved syntax:

a. You may now type

     . `test` $a = b$

for expressions a and b, or you may type

     . `test` $a == b$

The use of `==` is more consistent with Stata's syntax that treats `==` as indicating comparison and `=` as meaning assignment.

b. You may now specify multiple tests on one line:

     . `test` $(a == b == c)$

     . `test` $(a == b) (c == d)$

c. `test` has new option `coef`, which specifies that the constrained coefficients are to be displayed.

d. `test` has two new options for use with the `test [eq1==eq2]` syntax: `constant` and `common`. `constant` specifies that `_cons` should be included in the list of coefficients to be tested. `common` specifies that `test` restrict itself to the coefficient in common between *eq1* and *eq2*.

e. `test` may now be used after survey estimation.

f. `test` has a new programmer's option `matvlc(`*matname*`)`, which saves the variance–covariance matrix of the linear combination(s).

See [R] **test**.

15. `testnl` now allows typing `testnl` *exp*`==` *exp* `==` ... `==` *exp* to test whether two or more expressions are equal. Single equal signs may be used: `testnl` *exp*`=` *exp* `=` ... `=` *exp*.

In addition, `testnl` has new option `iterate(#)` for specifying the maximum number of iterations used to find the optimal step size in the calculation of the numerical derivatives of the expressions to be tested. See [R] **testnl**.

16. `testparm` has new option `equation()` for use after fitting multiple-equation models such as `mvreg`, `mlogit`, `heckman`, etc. It specifies the equation for which the all-zero or all-equal hypothesis is to be tested. See [R] **test**.

17. `lincom` now works after `anova` and after all survey estimators; see [R] **lincom**.

18. `bitest`, `prtest`, `ttest`, and `sdtest` now allow == to be used wherever = is allowed in their syntax; See [R] **bitest**, [R] **prtest**, [R] **prtest**, [R] **ttest**, and [R] **sdtest**.

19. New command `suest` is a post-estimation command that combines multiple estimation results (parameter vectors and their variance–covariance matrices) into simultaneous results with a single stacked parameter vector and a robust (sandwich) variance–covariance matrix. The estimation results to be combined may be based on different, overlapping, or even the same data. After creating the simultaneous estimation results, one can use `test` or `testnl` to obtain Hausman-type tests for cross-model hypotheses. `suest` supports survey data. See [R] **suest**.

20. New command `imtest` performs the information matrix test for an a regression model. In addition, it provides the Cameron–Trevedi decomposition of the IM-test in tests for heteroskedasticity, skewness, and kurtosis, and White's original heteroskedasticity test. See [R] **regression diagnostics**.

21. New command `szroeter` performs Szroeter's test for heteroskedasticity in a regression model; see [R] **regression diagnostics**.

22. Existing command `hettest` now provides option `rhs` to test for heteroskedasticity in the independent variables. It now also supports multiple comparison testing. See [R] **regression diagnostics**.

23. Existing command `tabulate` has output changes, new features, and expanded limits.

    a. Three new statistics are available for twoway tabulations: `expected`, `cchi2`, and `clrchi2`. `expected` reports the expected number in each cell. `cchi2` reports the contribution to Pearson's χ^2. `clrchi2` reports the contribution to the likelihood-ratio $\chi2$.

    b. New options `key` and `nokey` force or suppress a key explaining the entries in the table.

    c. Twoway tabulations now respect `set linesize`, meaning you can produce wide tables.

    d. Both oneway and twoway tabulations now put commas in the reported frequency counts.

    e. `tabulate` for oneway tabulations has new option `sort`, which puts the table in descending order of frequency.

    f. `tabulate` has expanded limits:

	1-way	2-way
Stata/SE:	12,000	12,000 x 80
Intercooled Stata:	3,000	300 x 20
Small Stata:	500	160 x 20

    See [R] **tabulate**.

24. Existing command `tabstat` has new options `statistics(variance)` and `statistics(semean)` which display the variance and the standard error of the mean. (Also provided is new option `varwidth(#)`, specifying the number of characters used to display variable names.) See [R] **tabstat**.

25. Existing command `roctab` has new option `specificity` to graph sensitivity versus specificity, instead of the default sensitivity versus $(1 - \text{specificity})$; see [R] **roc**.

26. Existing command `ologit` now has option `or` to display results as odds ratios (display exponentiated coefficients); see [R] **ologit**.

27. New command `lowess` replaces old command `ksm`. `lowess` allows graph twoway's `by()` option and is much faster than `ksm`; see [R] **lowess**.

28. Existing command kdensity has been rewritten so that it executes faster; see [R] **kdensity**.

29. Existing command intreg now applies constraints in the same manner as all other estimation commands, and existing command mlogit now allows constraints with constants; see [R] **tobit** and [R] **mlogit**.

30. New command pca performs principal components analysis, replacing factor, pc; see [R] **pca**.

31. Existing command ml maximize and all estimators using ml have a new tolerance option nrtolerance(#) for determining convergence. Convergence is declared when $\mathbf{g}\mathbf{H}^{-1}\mathbf{g}' <$ nrtolerance(#), where g represents the gradient vector and $\mathbf{H}$ the Hessian matrix; see [R] **maximize**.

32. Existing command mfx will now use pweights or iweights when calculating the means or medians for the *atlist* following an estimation command that used pweights or iweights. Previously, only fweights and aweights were supported. See [R] **mfx**.

33. Existing command adjust now allows the pr option to display predicted probabilities when used after svylogit, svyprobit, xtlogit, and xtprobit. See [R] **adjust**.

34. The existing regression diagnostics commands acprplot, cprplot, hettest, lvr2plot, ovtest, rvfplot, and rvpplot have been extended to work after anova. In addition, cprplot and acprplot have new options lowess and mspline that allow putting a lowess curve or median spline through the data. See [R] **regression diagnostics**.

35. Existing command ranksum has new option porder that estimates $P(x_1 > x_2)$; see [R] **signrank**.

36. Existing command poisgof has new option pearson to request the Pearson χ^2 goodness-of-fit statistic; see [R] **poisson**.

37. Existing command binreg now respects the init() option; see [R] **binreg**.

38. Existing command boxcox now accepts iweights; see [R] **boxcox**.

39. Existing commands zip and zinb now accept the *maximize_option* from() to provide starting values; see [R] **zip**.

40. Existing command cnsreg now accepts the noconstant option; see [R] **cnsreg**.

41. Existing command hotel has been renamed hotelling; hotel is now an abbreviation for hotelling; see [R] **hotelling**.

42. The score() option is now unified across all estimation commands. You must specify the correct number of score variables, and, in multiple-equation estimators, you may specify *stub** to mean create new variables named *stub*1, *stub*2, ...

    Estimation commands now save in e(scorevars) the names of the score variables if score() was specified.

43. Existing command summarize without the detail option now allows iweights; see [R] **summarize**.

44. Existing commands ci and summarize have new option separator(#) that specifies how frequently separation lines should be inserted into the output; see [R] **ci** and [R] **summarize**.

45. Existing command impute has three new options, regsample, all, and copyrest that control the sample used for forming the imputation and how out-of-sample values are treated; see [R] **impute**.

46. Existing command collapse now takes time-series operators; see [R] **collapse**.

1.3.12   What's new in data management

1. New command odbc allows Stata for Windows to act as an ODBC client, meaning you can fetch data directly from ODBC sources; see [R] **odbc**.

2. Existing command generate has new, more convenient syntax. Now you can type

   . generate a = 2 + 3

   or

   . generate b = "this" + "that"

   without specifying whether new variable b is numeric or string of a particular length. If you wish, you can also type

   . generate str b = "this" + "that"

   which asserts that b is a string but leaves it to generate to determine the length of the string. This is useful in programming situations because it helps to prevent bugs. Of course, you can continue to type

   . generate double a = pi/2

   and

   . generate str10 b = "this" + "that"

   See [R] **generate**.

3. Existing command list has been completely redone. Not only is output far more readable—and even pretty—but programmers will want to use list to format tables. See [R] **list**.

4. Existing command merge has been improved:

   a. New options unique, uniqmaster, and uniqusing ensure that the merge goes as you intend. These options amount to assertions that, if false, cause merge to stop. unique specifies that there should not be repeated observations within match variables, and that if you say "merge *id* using *myfile*", there should be one observation per *id* value in the master data (the data in memory) and one observation per *id* in the using data. If observations are not unique; merge will complain.

      Options uniqmaster and uniqusing make the same claim for one or the other half of the merge; uniq is equivalent to specifying uniqmaster and uniqusing.

   b. merge no longer has a limit on the number of match (key) variables.

   c. merge has new option keep(*varlist*) that specifies the variables to be kept from the using data.

   See [R] **merge**.

5. Existing command append has new option keep(*varlist*) that specifies the variables to be kept from the using data; see [R] **append**.

6. New command tsappend appends observations in a time-series context. tsappend uses the information set by tsset, automatically fills in the time variable, and fills in the panel variable if the panel variable was set. See [TS] **tsappend**.

7. Existing command describe using will now allow you to specify a *varlist*, so you can check whether a variable exists in a dataset before merging or appending. Programmers will be interested in the new varlist option, which will leave in r() the names of the variables in the dataset. See [R] **describe**.

8. New command isid verifies that a variable or set of variables uniquely identify the observations and so are suitable for use with merge; see [R] **isid**.

9. Existing command `codebook` has new option `problems` to report potential problems in the data; see [R] **codebook**.

10. New command `labelbook` is like `codebook`, but for value labels. In addition to providing documentation, the output includes a list of potential problems.

    New command `numlabel` prefixes numerical values onto value labels and removes them. For example, the mapping 2 → "Catholic" becomes "2. Catholic" and vice versa.

    See [R] **labelbook**.

11. New command `duplicates` reports on, gives examples of, lists, browses, tags, and/or drops duplicate observations; see [R] **duplicates**.

12. Existing command `recode` has three new features:

    a. `recode` now allows a *varlist* rather than a *varname*, so several variables can be recoded at once.

    b. `recode` has new option `generate()` to specify that the transformed variables be stored under different names than the originals.

    c. `recode` has new option `prefix()`, an alternative to `generate`, to specify that the transformed variables are to be given their original names, but with a prefix.

13. Existing command `sort` has new option `stable` that says, within equal values of the sort keys, the observations are to appear in the same order as they did originally. See [R] **sort**.

14. New command `webuse` loads the specified dataset, obtaining it over the web. By default, datasets are obtained from *http://www.stata-press.com/data/r8/*, but you can reset that. See [R] **webuse**.

    New command `sysuse` loads the specified dataset that was shipped with Stata, plus any other datasets stored along the ado-path; see [R] **sysuse**.

15. Existing command `insheet` has a new `delimiter(`*char*`)` option that allows you to specify an arbitrary character as the value separator; see [R] **insheet**.

16. Existing commands `infile` and `infix` no longer treat `^Z` as the end of a file; see [R] **infile (free format)**, [R] **infile (fixed format)**, and [R] **infix (fixed format)**.

17. Existing command `save` has features:

    a. New option `orphans` specifies that all value labels, including those not attached to any variables, are to be saved in the file.

    b. New option `emptyok` specifies that the dataset is to be saved even if it contains no variables and no observations.

    c. Existing option `old` is removed. To save datasets in Stata 7 format, use the new `saveold` command.

    See [R] **save**. By the way, Stata 8 now has a single `.dta` dataset format used by both Stata/SE and Intercooled Stata, meaning that sharing data with colleagues is easy.

18. Existing command `outfile` has new features:

    a. New options `rjs` and `fjs` specify how strings are to be aligned in the output file. The default is left alignment. Option `rjs` specifies right alignment. Option `fjs` specifies alignment as specified by the variables' formats.

    b. New option `runtogether` is for use by programmers; it specifies that all string variables be run together without extra spaces in between or quotes.

    See [R] **outfile**.

19. You may attach value labels to the new extended missing values (`.a`, `.b`, . . . , `.z`); see [R] **label**.

20. As a consequence of the 26 new missing value codes, the maximum value that can be stored in a byte, int, and long is reduced to 100, 32,740, and 2,147,483,620; see [R] **data types**.

21. New command split splits the contents of a string variable into one or more parts and is useful for separating words into multiple variables; see [R] **split**.

22. In the way of minor improvements are

    a. Existing command egen now allows longer *numlists* in the values() option for the eqany() and neqany() functions; see [R] **egen**.

    b. Existing command destring now allows an abbreviated *newvarlist* in the generate() option; see [R] **destring**.

    c. Existing commands icd9 and icd9p have been updated to use the V18 and V19 codes; V16, V18, and V19 codes have been merged so that icd9 and icd9p work equally well with old and new datasets; see [R] **icd9**.

    d. Existing command egen mtr() has been updated to include the marginal tax rates for the years 2000 and 2001; see [R] **egen**.

    e. Existing command mvdecode's mv() option now allows a *numlist*; see [R] **mvencode**.

    f. Existing command mvencode has a new, more versatile syntax to accommodate extended missing values; see [R] **mvencode**.

    g. Existing command xpose has three new options: format, format(*%fmt*), and promote. The format option finds the largest numeric display format in the pretransposed data and applies it to the transposed data. The format(*%fmt*) option sets the transposed data to the specified format. The promote option causes the transposed data to have the most compact numeric data type that preserves the original data accuracy. See [R] **xpose**.

    h. Existing command notes now allows the individual notes to include SMCL directives; see [R] **notes**.

    i. Existing command mkmat has new nomissing option that causes observations with missing values to be excluded (because matrices can now contain missing values). mkmat has also been made faster. See [P] **matrix mkmat**.

    j. Existing command ds has three new options: alpha, varwidth(*#*), and skip(*#*). alpha sorts the variables in alphabetic order. varwidth(*#*) specifies the display width of the variable names. skip(*#*) specifies the number of spaces between variables. See [R] **describe**.

    k. Existing commands label dir now returns the names of the defined value labels in r(names) and label list now returns the minimum and maximum of the mapped values in r(min) and r(max); see [R] **label**.

(*Continued on next page*)

1.3.13   What's new in expressions and functions

1. First, a warning: Do not type

    . generate *newvar* = . . . if *oldvar* != .

    . replace *oldvar* = . . . if *oldvar* != .

    . list . . . if *var* !=.

    Type

    . generate *newvar* = . . . if *oldvar* < .

    . replace *oldvar* = . . . if *oldvar* < .

    . list . . . if =it var < .

    or type

    . generate *newvar* = . . . if !mi(*oldvar*)

    . replace *oldvar* = . . . if !mi(*oldvar*)

    . list . . . if !mi(*var*)

    Stata has new missing values and the ordering is *all numbers* $<$. $<$ .a $<$ .b . . . $<$ .z. If you do not use the new missing values, then your old habits will work, but better to be safe.

    It is a hot topic of debate at StataCorp whether *varname*<. or !mi(*varname*) is the preferred way of excluding missing values, and therefore both constructs are deemed to be equally stylish; use whichever appeals to you.

    New function mi() is a synonym for existing function missing(); it returns 1 (true) if missing and false otherwise. See the *Programming functions* section of [R] **functions**.

2. By the same token, do not type

    . list . . . if *var* == .

    To list observations with missing values of *var*, type

    . list . . . if *var* >=.

    or type

    . list . . . if mi(*var*)

3. Matrices can now contain missing values, both the standard one (.) and the extended ones (.a, .b, . . . , .z). See [U] **17 Matrix expressions**.

4. The following new density functions are provided:

    a. tden(n,t), the density of Student's t distribution.

    b. Fden(n_1,n_2,F), the density F.

    c. nFden(n_1,n_2,λ,F), the noncentral F density.

    d. betaden(a,b,x), the 2-parameter Beta density.

    e. nbetaden(a,b,g,x), the noncentral Beta density.

    f. gammaden(a,b,g,x), the 3-parameter Gamma density.

    See the *Probability distributions and density functions* section of [R] **functions**.

5. The following new cumulative density functions are provided:

    a. $\texttt{nFtail}(n_1, n_2, \lambda, f)$, the upper-tail of the noncentral F.

    b. $\texttt{nibeta}(a, b, \lambda, x)$, the cumulative noncentral ibeta probability.

    See the *Probability distributions and density functions* section of [R] **functions**.

6. The following new inverse cumulative density functions are provided:

    a. $\texttt{invnFtail}(n_1, n_2, \lambda, p)$, the noncentral F corresponding to upper-tail p.

    b. $\texttt{invibeta}(a, b, p)$, the incomplete beta value corresponding to p.

    c. $\texttt{invnibeta}(a, b, \lambda, p)$, the noncentral beta value corresponding to p.

    In addition, existing function $\texttt{invbinomial}(n, k, p)$ has improved accuracy. See the *Probability distributions and density functions* section of [R] **functions**.

7. A suite of new functions provides partial derivatives of the cumulative gamma distribution. In what follows, $P(a, x) = \texttt{gammap}(a, x)$; the following new functions are provided:

    a. $\texttt{dgammapda}(a, x)$, $\partial P(a, x)/\partial a$.

    b. $\texttt{dgammapdx}(a, x)$, $\partial P(a, x)/\partial x$.

    c. $\texttt{dgammapdada}(a, x)$, $\partial^2 P(a, x)/\partial a^2$.

    d. $\texttt{dgammapdxdx}(a, x)$, $\partial^2 P(a, x)/\partial x^2$.

    e. $\texttt{dgammapdadx}(a, x)$, $\partial^2 P(a, x)/\partial a \partial x$.

    See the *Probability distributions and density functions* section of [R] **functions**.

8. All density and distribution functions have been extended to return nonmissing values over the entire real line; See the *Probability distributions and density functions* section of [R] **functions**.

9. The following new string functions are provided:

    a. $\texttt{word}(s, n)$ returns the nth word in s.

    b. $\texttt{wordcount}(s)$ returns the number of words in s.

    c. $\texttt{char}(n)$ returns the character corresponding to ASCII code n.

    d. $\texttt{plural}(n, s_1)$ returns the plural of s_1 if $n \neq \pm 1$, and otherwise returns s_1.

    e. $\texttt{plural}(n, s_1, s_2)$ returns the plural of s_1 if $n \neq \pm 1$, forming the plural by adding or removing suffix s_2, and returns the singular.

    f. $\texttt{proper}(s)$ capitalizes the first letter of a string and any other letters immediately following characters that are not letters; remaining letters are converted to lowercase.

    See the *String functions* section of [R] **functions**.

10. The following new mathematical functions are provided:

    a. $\texttt{logit}(x)$, the log of the odds ratio.

    b. $\texttt{invlogit}(x)$, the inverse logit.

    c. $\texttt{cloglog}(x)$, the complementary log-log.

    d. $\texttt{invcloglog}(x)$, the inverse of the complementary log-log.

    e. $\texttt{tanh}(x)$, the hyperbolic tangent.

    f. $\texttt{atanh}(x)$, the arc-hyperbolic tangent of x.

g. floor(x), the integer n such that $n \leq x < n + 1$.

h. ceil(x), the integer n such that $n < x \leq n + 1$.

In addition, the following existing mathematical functions have been modified:

i. round(x,y) now allows the second argument be optional and defaults it to 1, so round(x) returns x rounded to the closest integer.

j. lngamma(x) and gammap(a,x) now have improved accuracy.

See the *Mathematical functions* section of [R] **functions**.

11. Existing function uniform() will now allow you to capture and reset its seed. The seed value, in encrypted form, is now shown by query. You can store its value by typing

        local seed = c(seed)

    Later, you can reset it by typing

        set seed 'seed'

    See the *Probability distributions and density functions* section of [R] **functions**.

12. The following new matrix functions are provided:

    a. issym(M) returns 1 if matrix M is symmetric and returns 0 otherwise; issym() may be used in any context.

    b. matmissing(M) returns 1 if any elements of M are missing and returns 0 otherwise; matmissing() may be used in any context.

    c. vec(M) returns the column vector formed by listing the elements of M, starting with the first column and proceeding column by column.

    d. hadamard(M,N) returns a matrix whose i, j element is $M[i,j] \cdot N[i,j]$.

    e. matuniform(r,c) returns the $r \times c$ matrix containing uniformly distributed pseudo-random numbers on the interval $[0,1)$.

    See the *Matrix functions returning matrices* and the *Matrix functions returning scalars* sections of [R] **functions**.

    In addition, the new command matrix eigenvalues returns the complex eigenvalues of an $n \times n$ nonsymmetric matrix; see [P] **matrix eigenvalues**.

13. The following new programming functions have been added:

    a. clip(x,a,b) returns x if $a \leq x \leq b$, a if $x \leq a$, b if $x \geq b$, and *missing* if x is missing.

    b. chop(x,ϵ) returns round(x) if abs$\{x - \text{round}(x)\} < \epsilon$, otherwise returns x.

    c. irecode($z,x_1,x_2,\ldots,x_n$) returns the index of the range in which z falls.

    d. maxbyte(), maxint(), maxint(), maxlong(), maxfloat(), and maxdouble() return the maximum value allowed by the storage type.

    e. minbyte(), minint(), minint(), minlong(), minfloat(), and mindouble() return the minimum value allowed by the storage type.

    f. epsfloat() and epsdouble() return the precision associated with the storage type.

    g. byteorder() returns 1 if the computer stores numbers in most-significant-byte-first format and 0 if in least-significant-byte-first format.

The following programming functions have been modified or extended:

h. missing(x) now optionally allows multiple arguments so that it becomes missing($x_1, x_2, \ldots, x_n$). The extended function returns 1 (true) if any of the x_i are missing and returns 0 (false) otherwise.

i. cond(x, a, b) now optionally allows a fourth argument so that it becomes cond(x, a, b, c). c is returned if x evaluates to missing.

See the *Programming functions* section of [R] **functions**.

1.3.14 What's new in display formats

1. The %g format has been modified: %#.0g still means the same as previously, but %#.#g has a new meaning. For instance, %9.5g means to show approximately 5 significant digits. We say approximately because, given the number 123,456, %9.5g will show 123456 rather than 1.2346e+05, as would strictly be required if only five digits are to be shown. Other than that, it does what you would expect, and we think, in all cases, does what you want.

2. %[-]0#.#f formats, note the leading 0, now specify that leading zeros are to be included in the result. 1.2 in %09.2f format is 000001.20.

3. Stata has a new %21x hexadecimal format that will mainly be of interest to numerical analysts. In %21x, 123,456 looks like +1.e240000000000X+010, which you read as the hexadecimal number 1.e24 multiplied by 2^{10}. The period in 1.e24 is the base-16 point. The beauty of this format is that it reveals numbers exactly as the binary computer thinks of it. For instance, the new format shows how difficult numbers like 0.1 are for binary computers: +1.999999999999aX-004.

You can use this hexadecimal way of writing numbers in expressions; Stata will understand, for instance,

```
. generate xover4 = x / 1.0x+2
```

but it is unlikely you would want to do that. The notation will even by understood by input, infix, and infile. There is no %21x input format, but wherever a number appears, Stata will understand #.##...#x±###.

See [U] **15.5 Formats: controlling how data are displayed**.

1.3.15 What's new in programming

Lots of programming improvements have been made; see *What's new* in [P] **intro**. Here we will just touch on a few highlights.

1. The two big features are the ability to program dialog boxes and the addition of class programming; see [P] **dialogs** and [P] **class**. Stata's new GUI and new graphics have been programmed using these new features.

2. The new c-class collects where settings are found. Type creturn list and all will become clear. Recorded in c(*settingname*) are all the system settings, so no longer do you have to wonder whether the setting is in $S_*something*, obtained as a result of an extended macro function, or found somewhere else. See [P] **creturn**.

3. Program debugging is now easier thanks to the new trace facilities.

   a. Trace output now shows the line with macros expanded as well as unexpanded. This makes spotting errors easier.

   b. Separators are drawn and output indented when one program calls another, making it easier to
      see where you are.

   c. `set trace` is now pushed-and-popped, so the original value will be restored when a program
      ends.

   d. The new command `set tracedepth` allows you to specify how deeply calls to subroutines
      should be traced, so you can eliminate unwanted output.

   See [P] **trace**.

4. One change will bite you: With if *exp*, `while` *exp*, `forvalues`, and all the other commands that
   take a brace, no longer can the open brace and close brace be on the same line as the command.
   You may not code

   > if (*exp*) { ...}

   You must instead code

   > if (*exp*) {
   >
   > ...
   >
   > }

   In the case of `if`, you may omit the braces altogether:

   > if (*exp*) ...

   Under version control, Stata continues to tolerate the old, all on one line syntax, but the new
   syntax makes Stata considerably faster. See [P] **if**.

5. Existing commands `postfile`, `post`, and `postclose` will now save string variables; see [P] **postfile**.

6. Do-files and ado-files now allow `//` comments and `///` continuation lines. `//` on a line says that
   from here to the end of the line is a comment. `///` does the same, but also says that the next line
   is to be joined with the current line (and not treated as a comment). See [P] **comments**.

7. Existing command `which` will now not only locate `.ado` files, but other system files as well. You
   can type, for instance, `which anova.hlp` to discover the location of the help file for `anova`. See
   [R] **which**.

   New command `findfile` will find look for any file along the adopath; see [P] **findfile**.

8. The `sysdir` directory STBPLUS is now called PLUS; see [P] **sysdir**.

9. `net` `.pkg` files have new features:

   a. F *filename* is a variation on `f` *filename* that specifies the file is to be installed into the system
      directories, even if it ordinarily would not. This is useful for installing `.dta` datasets that
      accompany ado-files.

   b. g *platformname* *filename* is another variation on `f` *filename*. It specifies that the file is to be
      installed only if the user's computer is of type *platformname*.

   c. G *platformname* *filename* is variation on F *filename*. The file is installed only if the user's
      computer is of type *platformname*, and, if it is installed, it is installed in the system directories.

   d. h *filename* asserts that *filename* must be loaded or else this package cannot be installed.

   e. The maximum number of description lines in a `.pkg` file has been increased from 20 to 100.

   See [R] **net**.

There are lots of new programming features, and the ones we have chosen to mention may not be
of the most interest to you. Do see *What's new* in [P] **intro**.

1.3.16   What's new in the user interface

1. The GUI, of course, but we have already mentioned that; see *Stata's interface* in Chapter 3 of the *Getting Started with Stata* manual.

2. Stata now has tab-name completion. When typing a command, type the first few letters of a variable name and press tab. See [U] **13.6 Tab expansion of variable names**.

3. Existing commands set and query have been redone. set now has a permanently option that makes the setting permanent across sessions, alleviating the need for creating profile.do files. query has a new output format. See [R] **set** and [R] **query**.

4. There are lots of new set parameters. Do not even try to dig them out of the manual. Instead, type query. The new query output shows you where you can find out about each and what values you can set.

5. Almost all windows now have contextual menus; right-click when you are in the window to try them.

6. Under Windows and Macintosh, the following improvements have been made:

   a. If an http proxy is needed, Stata will attempt to get the proper settings from the operating system; see [R] **netio**.

   b. You are no longer limited to a maximum of 10 nested do-files. The limit is now 64, the same as Stata for Unix.

7. Under Windows, the following improvements have been made:

   a. Shortcuts for .smcl files have been added. By default, double-clicking on the shortcut will open the file in the Viewer, and right-clicking on the shortcut and choosing **Edit** will open the file in the Do-file Editor.

   b. Multiple instances of Stata for Windows running at the same time are now clearly marked in their title bar with an instance number.

   c. You can now set the maximum number of lines recorded in the Review window using set reventries; see [R] **set**.

8. Under Macintosh, the following improvements have been made:

   a. Stata is now a native Mach-O application. It may be launched from a terminal with command line options in addition to the usual double-clicking on Stata from the Finder.

   b. Stata can now change the amount of memory allocated on the fly just as Stata can on other operating systems; see [R] **memory**.

   c. Stata can now pass commands to the operating system for execution; see [R] **shell**.

   d. The filename separator is now forward slash (/) rather than colon (:) in keeping with changes made by Apple. For backwards compatibility, Stata still recognizes a colon (:) as a filename separator.

   e. You can now open more than one file simultaneously in the Do-file Editor.

   f. Stata honors and sets file permissions when creating files.

   g. Stata now uses /tmp for its temporary files.

   h. You can now select all of the contents of the Results or Viewer windows by selecting **Select All** from the **Edit** menu.

   i. There is a new menu item, **Bring All to Front**, in the Window menu that brings all Stata windows to the front.

9. Stata for Unix now looks for the environment variable STATATMP in addition to the environment variable TMPDIR for the location of the directory where temporary files are stored. STATATMP takes precedence over TMPDIR.

1.3.17 What's more

We have not listed all the changes, but we have listed the important ones. The remaining changes—a list of about equal length as the one above—are all implications of what has been listed.

What is important to know is that Stata is continually being updated and those updates are available for free over the Internet. All you have to do is type

. update query

and follow the instructions.

To learn what has been added since this manual was printed, pull down **Help** and select **What's new?** or type

. help whatsnew

We hope you enjoy Stata 8.

2 Resources for learning and using Stata

Contents

2.1   Overview
2.2   The http://www.stata.com web site
2.3   The http://www.stata-press.com web site
2.4   The Stata listserver
2.5   The Stata Journal and the Stata Technical Bulletin
2.6   Updating and adding features from the web
        2.6.1   Official updates
        2.6.2   Unofficial updates
2.7   NetCourses
2.8   Books and other support materials
2.9   Technical support
        2.9.1   Register your software
        2.9.2   Before contacting technical support
        2.9.3   Technical support by email
        2.9.4   Technical support by phone or fax
        2.9.5   Comments and suggestions for our technical staff

2.1   Overview

The *Getting Started* manual, *User's Guide*, and *Reference* manuals are the primary tools for learning about Stata; however, there are many other sources of information. A few are

1. Stata itself. Stata has a subject table-of-contents online with links to the help system and dialog boxes that make it easy to find and to execute a Stata command. See [U] **8 Stata's online help and search facilities**.

2. The Stata web site. Visit *http://www.stata.com*. Much of the site is dedicated to user support; see [U] **2.2 The http://www.stata.com web site**.

3. The Stata Press web site. Visit *http://www.stata-press.com*. This site contains the datasets used throughout the Stata manuals; see [U] **2.3 The http://www.stata-press.com web site**.

4. The Stata listserver. An active group of Stata users communicate over an Internet listserver, which you can join for free; see [U] **2.4 The Stata listserver**.

5. The Stata software distribution site and other user-provided software distribution sites. Stata itself can download and install updates and additions. We provide official updates to Stata—type `update` or pull down **Help** and select **Official Updates**. We also provide user-written additions to Stata and links to other user-provided sites—type `net` or pull down **Help** and select **SJ and User-written Programs**; see [U] **2.6 Updating and adding features from the web**.

6. The *Stata Journal* and the *Stata Technical Bulletin*. The *Stata Journal* contains reviewed papers, regular columns, book reviews, and other material of interest to researchers applying statistics in a variety of disciplines. The *Stata Technical Bulletin*, the predecessor to the *Stata Journal*, contains articles and user-written commands. See [U] **2.5 The Stata Journal and the Stata Technical Bulletin**.

7. NetCourses. We offer training via the Internet. Details are in [U] **2.7 NetCourses** below.

8. Books and support materials. Supplementary Stata materials are available; see [U] **2.8 Books and other support materials**.

9. Technical support. We provide technical support by email, telephone, and fax; see [U] **2.9 Technical support**.

2.2 The http://www.stata.com web site

Point your browser to *http://www.stata.com* and click on **User Support**. Over half of our web site is dedicated to providing support to users.

1. The web site provides FAQs (Frequently Asked Questions) on Windows, Macintosh, Unix, statistics, programming, Internet capabilities, graphics, and data management. These FAQs run the gamut from "I cannot save/open files" to "What does 'completely determined' mean in my logistic-regression output?" Everyone will find something of interest.

2. Visiting the web site is one way that you can subscribe to the Stata listserver; see [U] **2.4 The Stata listserver**.

3. The web site provides detailed information about NetCourses, along with the current schedule; see [U] **2.7 NetCourses**.

4. The web site provides information about Stata courses and meetings, both in the United States and elsewhere.

5. The web site provides an online bookstore for Stata-related books and other supplementary materials; see [U] **2.8 Books and other support materials**.

6. The web site provides links to information about statistics: other statistical software providers, book publishers, statistical journals, statistical organizations, and statistical listservers.

7. The web site provides links to Stata resources for learning Stata at *http://www.stata.com/links/resources.html*. Be sure to look at these materials, as many valuable resources about Stata are listed here, including the UCLA Stata portal, which includes a set of links about Stata, and the SSC archive, which has become the premier Stata download site for user-written software on the web.

In short, the web site provides up-to-date information on all support materials and, where possible, provides the materials themselves. Visit *http://www.stata.com* if you can.

2.3 The http://www.stata-press.com web site

Point your browser to *http://www.stata-press.com*. This site is devoted to the publications and activities of Stata Press.

1. Datasets that are used in the Stata *Reference* manuals and other books published by Stata Press may be downloaded. Visit *http://www.stata-press.com/data*. These datasets can be used in Stata by simply typing use `http://www.stata-press.com/data/r8/auto`. Alternatively, you could type `webuse auto`, see [R] **webuse**.

2. An online catalog of all of our books and multimedia products is at *http://www.stata-press.com/catalog.html*. We have tried to include enough information, such as table of contents, preface material, etc., so that you may tell if the book is appropriate for you.

3. Information about forthcoming publications is posted at *http://www.stata-press.com/forthcoming.html*.

2.4 The Stata listserver

The Stata listserver (Statalist) is an independently operated, real-time list of Stata users on the Internet. Anyone may join. Instructions for doing so can be found at *http://www.stata.com* by clicking on User Support and then Statalist or by emailing *stata@stata.com*.

Many knowledgeable users are active on the list, as are the StataCorp technical staff. We recommend that new users subscribe, observe the exchanges, and, if it turns out not to be useful, unsubscribe.

2.5 The Stata Journal and the Stata Technical Bulletin

The *Stata Journal* (SJ) is a printed and electronic journal, published quarterly, containing articles about statistics, data analysis, teaching methods, and effective use of Stata's language. The *Journal* publishes reviewed papers together with shorter notes and comments, regular columns, book reviews, and other material of interest to researchers applying statistics in a variety of disciplines. The *Journal* is a publication for all Stata users, both novice and experienced, with different levels of expertise in statistics, research design, data management, graphics, reporting of results, and of Stata, in particular.

Tables of contents for past issues and abstracts of the articles are available at *http://www.stata-journal.com/archives.html*.

We recommend that all users subscribe to the SJ. Visit *http://www.stata-journal.com* to learn more about the *Stata Journal* and to order your subscription.

To obtain any programs associated with articles in the SJ, type

```
. net from http://www.stata.com
. net link sj
. net cd software
```

or

1. Pull down **Help**
2. Select **SJ and User-written Programs**
3. Click on *http://www.stata-journal.com/software*

The Stata Technical Bulletin

For ten years, the *Stata Technical Bulletin* (STB) served as the means of distributing new commands and Stata upgrades, both user-written and "official". After ten years of continuous publication, the STB evolved into the *Stata Journal*. The Internet provided an alternative delivery mechanism for user-written programs, so the emphasis shifted from user-written programs to more expository articles. Although the STB is no longer published, many of the programs and articles that appeared in it are still valuable today. Reprints of past issues are available from *http://www.stata.com/bookstore/stbr.html*. To obtain the programs that were published in the STB, type

```
. net from http://www.stata.com
. net cd stb
```

(Continued on next page)

2.6   Updating and adding features from the web

Stata itself is web-aware.

First, try this:

```
. use http://www.stata.com/manual/oddeven.dta, clear
```

That will load an uninteresting dataset into your computer from our web site. If you have a homepage, you can use this feature to share datasets with coworkers. Save a dataset on your homepage, and researchers worldwide can use it. See [R] **net**.

2.6.1   Official updates

Although we follow no formal schedule for the release of updates, we typically provide updates to Stata approximately every two weeks. Installing the updates is easy. Type

```
. update
```

or pull down **Help** and select **Official Updates**. Do not be concerned; nothing will be installed unless and until you say so. Once you have installed the update, you can type

```
. help whatsnew
```

or pull down **Help** and select **What's New** to find out what has changed. We distribute official updates to fix bugs and to add new features.

2.6.2   Unofficial updates

There are also "unofficial" updates—additions to Stata written by Stata users, which includes members of the StataCorp technical staff. Stata is programmable, and even if you never write a Stata program, you may find these additions useful, and some of them spectacularly so. You start by typing

```
. net from http://www.stata.com
```

or pull down **Help** and select **SJ and User-written Programs**.

Be sure to try visit the SSC-Archive. The `ssc` command makes it easy for you to install and uninstall packages from SSC. Type

```
. ssc whatsnew
```

to find out what's new at the site. If you find something that interests you, type

```
. ssc describe pkgname
```

for more information.

Periodically, you can type

```
. news
```

or pull down **Help** and select **News** to display a short message from our web site telling you what is newly available.

See [U] **32 Using the Internet to keep up to date**.

2.7 NetCourses

We offer courses on Stata at the introductory and advanced levels. Courses on software are typically expensive and time-consuming. They are expensive because, in addition to the direct costs of the course, participants must travel to the course site. We have found it is better to organize courses over the Internet—saving everyone time and money.

We offer courses over the Internet and call them Stata NetCourses^TM.

1. **What is a NetCourse?**
   A NetCourse is a course offered through the Stata web site that varies in length from seven to eight weeks. You must have an email address and web browser to participate.

2. **How does it work?**
   Every Friday a "lecture" is posted on a password-protected web site. After reading the lecture over the weekend or perhaps on Monday, participants then post questions and comments on a message board. Course leaders typically respond to the questions and comments on the same day they are posted. The other participants are encouraged to amplify or otherwise respond to the questions or comments as well. The next lecture is then posted on Friday, and the process repeats.

3. **How much of my time does it take?**
   It depends on the course, but the introductory courses are designed to take roughly 3 hours per week.

4. **There are three of us here—can just one of us enroll and we redistribute the NetCourse materials ourselves?**
   We ask that you not. NetCourses are priced to cover the substantial time input of the Course Leaders. Moreover, enrollment is typically limited to prevent the discussion from becoming unmanageable. The value of a NetCourse, just like a real course, is the interaction of the participants, both with each other and with the Course Leaders.

5. **I've never taken a course by Internet before. I can see that it might work, but then again, it might not. How do I know I'll benefit?**
   All Stata NetCourses come with a 30-day satisfaction guarantee. The 30 days begin after the conclusion of the final lecture.

You can learn more about the current NetCourse offerings by visiting *http://www.stata.com*. Our offerings include

NC-101 An introduction to Stata
NC-151 An introduction to Stata programming
NC-152 Advanced Stata programming
NC-631 Introduction to survival analysis with Stata

2.8 Books and other support materials

There are books published on Stata, both by us and by others. Visit the bookstore at *http://www.stata.com* for an up-to-date list and for the table of contents of each. For books that we carry in the Stata bookstore, we post a comment written by a member of out technical staff, explaining why we think this book might interest you.

2.9   Technical support

We are committed to providing superior technical support for Stata software. In order to assist you as efficiently as possible, please follow the procedures listed below.

2.9.1   Register your software

You must register your software in order to be eligible for technical support, updates, special offers, and other benefits. By registering, you will receive the *Stata News*, and you may access our support staff for free with any question that you encounter. You may register your software either electronically or by mail.

Electronic registration: After installing Stata and successfully entering your License and Authorization Key, your default web browser will open to the online registration form at the Stata web site. You may also manually point your web browser to *http://www.stata.com/register/* if you wish to register your copy of Stata at a later time.

Mail-in registration: Fill in the registration card that came with Stata and mail it to Stata Corporation.

2.9.2   Before contacting technical support

Before you spend the time gathering the information our technical support department needs, make sure that the answer does not already exist in the help files. You can use the `help` and `search` commands to find all the entries in Stata that address a given subject. Be sure to try pulling down **Help** and selecting **Contents**. Check the manual for a particular command. There are often examples that address questions and concerns. Another good source of information is our web site. You should keep a bookmark of our frequently asked questions page (*http://www.stata.com/support/faqs/*) and check it occasionally for new information.

Our technical department will need some information from you in order to provide detailed assistance. Most important is your serial number, but they will also need the following information:

1. The system information on the computer that you are using is especially important if you are having hardware problems. This includes the make and model of various hardware components such as the computer manufacturer, the video driver, the operating system and its version number, relevant peripherals, and the version number of any other software with which you experience a conflict. See [U] **6 Troubleshooting starting and stopping Stata**.

2. The version of Stata that you are running. Type `about` at the Stata prompt, and Stata will display this information.

3. The types of variables in your dataset and the number of observations.

4. The command that is causing the error along with the exact error message and return code (error number).

2.9.3   Technical support by email

This is the preferred method of asking a technical support question. It has the following advantages:

- You will receive a prompt response from us saying that we have received your question and that it has been forwarded to *Technical Services* to answer.

- We are able to route your question to a specialist for your particular question.

- Questions submitted via email may be answered after normal business hours, or even on weekends or holidays. Although we cannot promise that this will happen, it may, and your email inquiry is bound to receive a faster response than leaving a message on Stata's voice mail.

- If you are receiving an error message or an unexpected result, it is easy to include a log file that demonstrates the problem.

Please see the FAQ at *http://www.stata.com/support/faqs/techsup/* for some suggestions to follow that will aid *Technical Services* in promptly answering your question.

2.9.4   Technical support by phone or fax

Our technical support telephone number is 979-696-4600. Please have your serial number handy. It is also best if you are at your computer when you call. If your question involves an error message from a command, please note the error message and number, as this will greatly help us in assisting you. Telephone support is reserved for nonstatistical questions. If your question requires the attention of a statistician, the question should be submitted via email or fax.

Send fax requests to 979 696 4601. If possible, collect the relevant information in a log file and include the file in your fax.

Please see the FAQ at *http://www.stata.com/support/faqs/techsup/* for some suggestions to follow that will aid *Technical Services* in promptly answering your question.

2.9.5   Comments and suggestions for our technical staff

By all means, send in your comments and suggestions. Your input is what determines the changes that occur in Stata between releases, so if we don't hear from you, we may not include your most desired new estimation command! Email is preferred, as this provides us with a permanent copy of your request. When requesting new commands, please include any references that you would like us to review should we develop those new commands. Email your suggestions to *servicemail@stata.com*.

3 A brief description of Stata

Stata is a statistical package for managing, analyzing, and graphing data.

Stata is available for a variety of platforms. Stata may be used either as a point-and-click application or as a command-driven package.

Stata's GUI provides an easy interface for those new to Stata, and for experienced Stata users who wish to execute a command that they seldom use.

The command-drive language provides a fast way to communicate with Stata, and a way to communicate more complex ideas.

Here is an extract of a Stata session using the GUI:

(Throughout the Stata manuals, we will refer to various datasets. These datasets are all available from *http://www.stata-press.com/data/r8/*. For easy access to them from within Stata, you may type webuse *dataset_name*.)

```
. webuse lbw2
(Hosmer & Lemeshow data)
```

We select **Data—Describe data—Summary statistics**, and choose to summarize variables low, age, and smoke, which names we obtained from the Variables window. We press **OK**.

```
. summarize low age smoke
    Variable |       Obs        Mean    Std. Dev.       Min        Max
-------------+--------------------------------------------------------
         low |       189    .3121693    .4646093          0          1
         age |       189     23.2381    5.298678         14         45
       smoke |       189    .3915344    .4893898          0          1
```

Stata shows us the command that we could have typed in command mode—summarize low age smoke—before displaying the results of our request.

Next, we fit a logistic regression model of low on age and smoke. We select **Statistics—Binary outcomes—Logistic regression**, fill in the fields, and press **OK**.

```
. logistic low age smoke
Logistic regression                             Number of obs   =        189
                                                LR chi2(2)      =       7.40
                                                Prob > chi2     =     0.0248
Log likelihood = -113.63815                     Pseudo R2       =     0.0315

-------------------------------------------------------------------------------
         low | Odds Ratio   Std. Err.      z    P>|z|     [95% Conf. Interval]
-------------+-----------------------------------------------------------------
         age |   .9514394   .0304194    -1.56   0.119     .8936481    1.012968
       smoke |   1.997405    .642777     2.15   0.032     1.063027    3.753081
-------------------------------------------------------------------------------
```

Here's an extract of a Stata session using the command language:

```
. use http://www.stata-press.com/data/r8/auto
(1978 Automobile Data)

. summarize mpg weight

    Variable |       Obs        Mean   Std. Dev.       Min        Max
-------------+--------------------------------------------------------
         mpg |        74     21.2973    5.785503        12         41
      weight |        74    3019.459    777.1936      1760       4840
```

The user typed summarize mpg weight and Stata responded with a table of summary statistics. Other commands would produce different results:

```
. correlate mpg weight
(obs=74)

             |      mpg    weight
-------------+------------------
         mpg |   1.0000
      weight |  -0.8072    1.0000

. gen w_sq = weight^2

. regress mpg weight w_sq

      Source |       SS       df       MS              Number of obs =      74
-------------+------------------------------           F(  2,    71) =   72.80
       Model | 1642.52197        2  821.260986           Prob > F      =  0.0000
    Residual | 800.937487       71  11.2808097           R-squared     =  0.6722
-------------+------------------------------           Adj R-squared =  0.6630
       Total | 2443.45946       73  33.4720474           Root MSE      =  3.3587

-------------+----------------------------------------------------------------
         mpg |      Coef.   Std. Err.      t    P>|t|     [95% Conf. Interval]
-------------+----------------------------------------------------------------
      weight |  -.0141581   .0038835    -3.65   0.001    -.0219016   -.0064145
        w_sq |   1.32e-06   6.26e-07     2.12   0.038     7.67e-08    2.57e-06
       _cons |   51.18308   5.767884     8.87   0.000     39.68225    62.68392
------------------------------------------------------------------------------

. scatter mpg weight, by(foreign, total row(1))
```

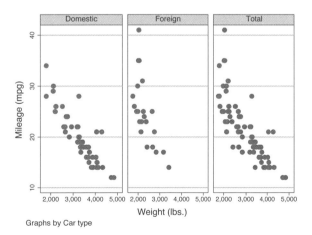

The user-interface model is type a little, get a little, etc., so that the user is always in control.

Stata's model for a dataset is that of a table—the rows are the observations and the columns are the variables:

```
. list mpg weight in 1/10
```

	mpg	weight
1.	22	2,930
2.	17	3,350
3.	22	2,640
4.	20	3,250
5.	15	4,080
6.	18	3,670
7.	26	2,230
8.	20	3,280
9.	16	3,880
10.	19	3,400

Observations are numbered; variables are named.

Stata is very fast. Partly, that speed is due to clever programming, and, partly, it is because Stata keeps the data in memory. Stata's data model is that of a word processor: a dataset may exist on disk, but that is just a copy. The dataset is loaded into memory, where it is worked on, analyzed, changed, and then perhaps stored back on disk.

Working on a copy of the data in memory makes Stata safe for interactive use. The only way to harm the permanent copy of your data on disk is if you explicitly save over it.

Having the data in memory means that the dataset size is limited by the amount of memory. Stata stores the data in memory in a very compressed format—you will be surprised how much data can fit into a given region of memory. Nevertheless, if you work with large datasets, you may run into memory constraints. There are two solutions to this problem:

1. By default, Stata/SE allocates 10 megabytes to Stata's data areas, and you can change it; see [U] **7 Setting the size of memory**.

   By default, Intercooled Stata allocates 1 megabyte to Stata's data areas, and you can change it; see [U] **7 Setting the size of memory**.

   By default, Small Stata allocates about 300K to Stata's data areas, and you cannot change it.

2. You will want to learn how to compress your data as much as possible; see [R] **compress**.

4 Flavors of Stata

Contents

4.1   Platforms
4.2   Stata/SE, Intercooled Stata, and Small Stata
          4.2.1   Determining which version you own
          4.2.2   Determining which version is installed
4.3   Size limits comparison of Stata/SE, Intercooled Stata, and Small Stata
4.4   Speed comparison of Stata/SE, Intercooled Stata, and Small Stata
4.5   Features of Stata/SE

4.1   Platforms

Stata is available for a variety of different computers, including

Stata for Windows

Stata for Macintosh

Stata for AIX
Stata for Digital Unix
Stata for HP-UX
Stata for Linux (Intel)
Stata for SGI Irix
Stata for Solaris (SPARC)

At this time, three 64-bit systems are supported by Stata: Solaris (SPARC), SGI Irix, and AIX.

Which version of Stata you run does not matter—Stata is Stata. You instruct Stata in the same way and Stata produces the same results, right down to the random-number generator.

Even files can be shared. For instance, a dataset created with Stata for Macintosh can be used on any other computer, and the same goes for graphs, programs, and any other file Stata uses or produces.

Moving a dataset or any other file across platforms requires no translation. If you do share datasets or graphs with other users across platforms, be sure that you make exact binary copies.

4.2   Stata/SE, Intercooled Stata, and Small Stata

Stata for Windows and Stata for Macintosh are available in three "flavors": Stata/SE (Special Edition), Intercooled Stata, and Small Stata. Stata for Unix is available only in the Stata/SE and Intercooled Stata flavors. All three flavors of Stata have the same features, but Stata/SE and Intercooled Stata are able to work with larger datasets and are faster. How much faster depends on the platform, but the advantage ranges from 50 to 600 percent. Stata/SE has much larger limits for matrix sizes and string lengths, and will accommodate up to 16 times the number of variables as Intercooled Stata.

Stata/SE is the version that we recommend to users who frequently analyze large datasets. Intercooled Stata is the version that we recommend for a typical user doing serious data analysis and statistics with small to moderate size datasets.

Small Stata would perhaps be better named Stata for Small Computers.

4.2.1   Determining which version you own

Included with every copy of Stata is a paper license that contains important codes that you will input during installation. This license also determines which flavor of Stata you have — SE, Intercooled, or Small. Look at the license to see if it says Stata/SE, Intercooled Stata, or Small Stata.

If you purchased Intercooled Stata and you want Stata/SE, your Intercooled Stata version can be upgraded to Stata/SE. In fact, we put all three flavors on the same installation media, so you won't have to wait to receive another CD. All you need is an upgraded paper license with the appropriate codes.

By the way, even if you purchased Stata/SE or Intercooled Stata, you may use Small Stata with your Stata/SE or Intercooled Stata license. This might be useful if you had a large computer at work and a smaller computer at home. Please remember, however, that you have only one license (or however many licenses you purchased). You may, both legally and morally, use one, the other, or both, but you should not subject the pair to simultaneous use.

4.2.2   Determining which version is installed

If Stata is already installed, you can find out which Stata you are using by entering Stata as you normally do and typing about:

```
. about

Intercooled Stata 8.0 for Windows
Born 01 Dec 2002
Copyright (C) 1985-2003

10-user Windows (network) perpetual license:
        Serial number:  198040000
        Licensed to:    Marsha Martinez
                        StataCorp
```

You are running Intercooled Stata 8.0 for Windows.

4.3   Size limits comparison of Stata/SE, Intercooled Stata, and Small Stata

Here are some of the different size limits for Stata/SE, Intercooled Stata, and Small Stata. See [R] **limits** for a longer list.

Maximum size limits for Stata/SE, Intercooled Stata, and Small Stata

	Stata/SE	Intercooled	Small
Number of observations	limited only by memory	limited only by memory	fixed at approx. 1,000
Number of variables	32,767	2,047	fixed at 99
Width of a dataset	393,192	24,564	200
Maximum matrix size (matsize)	11,000	800	fixed at 40
Number of characters in a macro	1,081,511	67,784	8,681
Number of characters in a command	1,081,527	67,800	8,697

That is, Stata/SE allows more variables, larger matrices, longer macros (and longer strings), and a longer command line than Intercooled Stata. Intercooled Stata allows larger datasets, fits models with more independent variables, has longer macros, and allows a longer command line (required because of the increased number of variables allowed) than Small Stata.

4.4   Speed comparison of Stata/SE, Intercooled Stata, and Small Stata

Why the difference in speed for Stata/SE, Intercooled Stata, and Small Stata? In part, it is due to different options we specified when we compiled Stata. However, it is also due to differences in the code.

For instance, in Stata's test command, there comes a place where it must compute the matrix calculation $\mathbf{RZR}'$ (where $\mathbf{Z} = (\mathbf{X}'\mathbf{X})^{-1}$). Stata/SE and Intercooled Stata make the calculation in a straightforward way, which is to form $\mathbf{T} = \mathbf{RZ}$ and then calculate $\mathbf{TR}'$. This requires temporarily storing the matrix $\mathbf{T}$. Small Stata, on the other hand, goes into more complicated code to form the result directly—code that requires temporary storage of only one scalar! This code, in effect, recalculates intermediate results over and over again, and so it is slower.

Another difference is that Small Stata, since it is designed to work with smaller datasets, uses different memory-management routines. These memory-management routines use 2-byte rather than 4-byte offsets, and therefore require only half the memory to track locations.

In any case, the differences are all technical and internal. From the user's point of view, Stata/SE, Intercooled Stata, and Small Stata work the same way.

4.5   Features of Stata/SE

For those familiar with Stata, a table will say it all:

Parameter	Intercooled Stata			Stata/SE		
	Default	min	max	Default	min	max
maxvar	2,047	2,047	2,047	**5,000**	2,047	**32,767**
matsize	40	10	800	**400**	10	**11,000**
memory	1M	500K	...	**10M**	500K	...
str#	.	1	80	.	1	**244**

In other words, this means that Stata/SE

1. Allows datasets with more variables—up to 32,767 variables.

2. Allows datasets to contain longer string variables—variables up to 244 characters long.

3. Allows larger matrices—matrices up to 11,000 x 11,000. (An implication of this is that Stata/SE can fit models with more independent variables and can fit certain panel-data models with longer time series within panel.)

To learn how to exploit Stata/SE to its fullest, type whelp SpecialEdition after invoking Stata.

5 Starting and stopping Stata

Contents

5.1 Starting Stata
5.2 Verifying that Stata is correctly installed
5.3 Exiting Stata
5.4 Features worth learning about
    5.4.1   Windows
    5.4.2   Macintosh
    5.4.3   Unix

Below we assume that Stata is already installed. If you have not yet installed Stata, follow the installation instructions in the *Getting Started* manual.

5.1 Starting Stata

To start Stata:

Stata for Windows:	Click on **Start**
	Select **Programs**
	Select **Stata**
	Click on **Stata/SE**, **Intercooled Stata**, or **Small Stata**
	as appropriate; see [U] **4 Flavors of Stata**
Stata for Macintosh:	Double-click on the **Stata.do** file icon
Stata for Unix:	Type `xstata` at the Unix command prompt for Intercooled Stata(GUI)
	Type `stata` at the Unix command prompt for Intercooled Stata(console)
	Type `xstata-se` at the Unix command prompt for Stata/SE(GUI)
	Type `stata-se` at the Unix command prompt for Stata/SE(console)

Unix(console) users will see something like this:

```
% stata

  ___  ____  ____  ____  ____ tm
 /__    /   ____/   /   ____/
___/   /   /___/   /   /___/    8.0    Copyright 1984-2003
   Statistics/Data Analysis            Stata Corporation
                                       4905 Lakeway Drive
                                       College Station, Texas 77845 USA
                                       800-STATA-PC        http://www.stata.com
                                       979-696-4600        stata@stata.com
                                       979-696-4601 (fax)

 Registration information appears

 Other information appears, such as Notes

 . _
```

Stata for Windows users will see something like

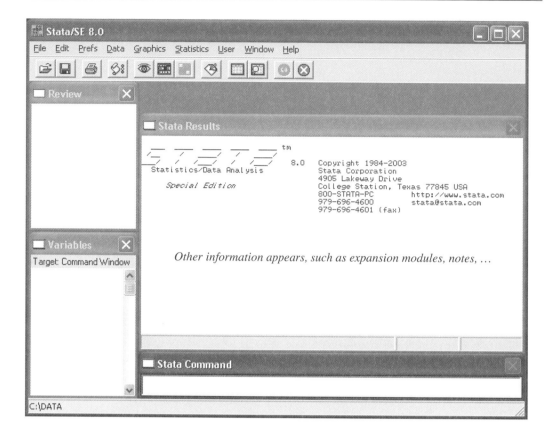

Stata for Unix(GUI) and Stata for Macintosh users will see something similar.

If Stata does not come up, see [U] **6.1 If Stata does not start**.

Stata is now waiting for you to type something in the Command window. From now on, when we say, for instance, type verinst, which we might even show as

```
. verinst
```

we mean type verinst and press *Enter* in the Command window.

5.2   Verifying that Stata is correctly installed

The first time you start Stata, you should verify that it is installed correctly. Type verinst and you should see something like

```
. verinst
You are running Stata/SE 8.0 for ...
Stata is correctly installed.
You can type exit to exit Stata.
```

If you do not see the message "Stata is correctly installed", see [U] **6.2 verinst problems**.

Remember the verinst command. If you ever change your computer setup and are worried that you somehow damaged Stata in the process, you can type verinst and obtain the reassuring "Stata is correctly installed" message.

5.3   Exiting Stata

To exit Stata,

Stata for Windows:	Click on the close box or type `exit`
Stata for Macintosh:	Click on the close box or type `exit`
Stata for Unix(GUI):	Click on the close box or type `exit`
Stata for Unix(console):	Type `exit`

Stata works with a copy of the data in memory. If (1) there are data in memory, (2) the data have changed, and (3) you click to close, a box will appear asking if it is okay to exit without saving the changes.

If you instead type `exit` and there are changed data in memory, Stata will refuse and say "no; data in memory would be lost". In this case, you must either save the data on disk (see [R] **save**) or type `exit, clear` if you do not want to save the changes.

All of this is designed to prevent you from accidentally losing your data.

As you will discover, the `clear` option is allowed with all potentially destructive commands, of which `exit` is just one example. One command to bring data into memory, `use`, is another example. `use` is destructive since it loads data into memory and, in the process, eliminates any data already there.

If you type a destructive command and the data in memory have been safely stored on disk, Stata performs your request. If your data have changed in some way since the data were last saved, Stata responds with the message "no; data in memory would be lost". If you want to go ahead anyway, you can retype the command and add the `clear` option. Once you become familiar with Stata's editing keys, you will discover that it is not necessary to physically retype the line. You can press the *PrevLine* key (*PgUp*) to retrieve the last command you typed and append `, clear`. (On some Unix computers, you may have to press *Ctrl-R* instead of *PgUp*; see [U] **13 Keyboard use** for more details.)

Of course, you need not wait for Stata to complain—you can add the `clear` option the first time you issue the command—if you do not mind living dangerously.

5.4   Features worth learning about

5.4.1   Windows

1. The Windows Properties Sheet for Stata controls how Stata comes up. The properties sheet determines which will be the working (start-in) directory when Stata comes up and how much memory will be allocated to Stata, among other things. You can change this; see

> [GSW] **A.4 The Windows Properties Sheet**
> [GSW] **A.5 Starting Stata from other folders**
> [GSW] **A.6 Specifying the amount of memory allocated**

You can also change the amount of memory Stata has during a session by using the `set memory` command; see [U] **7 Setting the size of memory**.

2. Stata can execute commands every time it is invoked if you record the commands in the ASCII file `profile.do`, which you could put in Stata's working (start-in) directory; see [GSW] **A.7 Executing commands every time Stata is started**.

3. You can arrange to start Stata without going through the **Start** button. On your standard Windows screen, do you see how **My Computer** just sits on the desktop? You can put a Stata icon on the desktop, too. See [GSW] **A.8 Making shortcuts**.

4. You can run large jobs in Stata in batch mode; see [GSW] **A.9 Executing Stata in background (batch) mode**.

5. You can launch Stata by double-clicking on a Stata `.dta` dataset; see [GSW] **A.10 Launching by double-clicking on a .dta dataset**.

6. You can launch Stata and run a do-file by double-clicking on the do-file; see [GSW] **A.11 Launching by double-clicking on a do-file**.

7. Each time you launch Stata, you invoke a new instance of it, so if you want, you may run multiple Stata sessions simultaneously.

5.4.2 Macintosh

1. You can change the amount of memory allocated to Stata; see [GSM] **A.5 Specifying the amount of memory allocated**.

2. You can start Stata from other folders; simply copy `Stata.do`—a small file—to whatever folders you wish; see [GSM] **A.4 Starting Stata from other folders**.

3. You can start Stata from the Dock if you add `Stata.do` (not the Stata application) as an alias; see [GSM] **A.7 Creating aliases**.

4. Stata can execute commands every time it is invoked if you record the commands in the ASCII file `profile.do`, which you could put in your home directory; see [GSM] **A.6 Executing commands every time Stata is started**.

5. You can launch Stata by double-clicking on a Stata `.dta` dataset.

6. You can launch Stata by double-clicking on a do-file.

7. Pressing the up/down cursor keys and the page up/down while the Results or Viewer windows are in front scrolls the respective window. Pressing shift and the up/down cursor keys or page up/down while the Command window is in front also scrolls the Results window.

8. You may have multiple Do-file Editor windows open.

9. Graphs may be dragged and dropped to the Desktop or to another application.

10. Double-clicking on a variable from the Variables window when the keyboard focus is on a `VARLIST` or `VARNAME` input control will advance the keyboard focus to the next edit control.

11. Do not have more than one Stata application installed on the same computer.

    Do not install the current version in one folder and an older version in another.

    Do not install more than one of Intercooled Stata or Small Stata (`Stata` or `SmStata`) on the same computer.

    If you do, when you double-click on `Stata.do`, your Macintosh will randomly choose one of the Statas to run and you will never know which.

    If you installed more than one Stata on your computer, drag the extra Statas to the Trash, empty the Trash, and restart your Macintosh. All will be well.

5.4.3 Unix

1. You must add `/usr/local/stata` or `/usr/local/stata8` to your shell's PATH before you can use Stata.

2. There are two possible user interfaces available for Unix users. The first, a Graphical User Interface (GUI), looks and functions much like the interfaces of the Windows and Macintosh versions of Stata. The second, the console interface, is strictly command driven and runs in a Unix console. See the *Selecting a Stata user interface* section of [GSU] **1 Installation**.

3. You can specify the amount of memory to be allocated to Stata when you invoke it, or, once Stata is running, at any time using the `set memory` command. To learn about the start-up options, see [GSU] **A.4 Advanced starting of Stata for Unix**. To learn about `set memory`, see [U] **7 Setting the size of memory**.

4. Stata can execute commands every time it is invoked if you record the commands in the ASCII file `profile.do` and place the file someplace along your PATH; see [GSU] **A.7 Executing commands every time Stata is started**.

5. Stata has a background (batch) mode; see [GSU] **A.8 Executing Stata in background (batch) mode**.

6. For notes on using X Windows remotely (or from another computer), see [GSU] **A.9 Using X Windows remotely**.

7. For a summary of environment variables that affect Stata, see [GSU] **A.10 Summary of environment variables**.

6 Troubleshooting starting and stopping Stata

Contents

6.1   If Stata does not start
6.2   verinst problems
6.3   Troubleshooting Windows
6.4   Troubleshooting Macintosh
6.5   Troubleshooting Unix

6.1 If Stata does not start

You tried to start Stata and it refused; Stata or your operating system presented a message explaining that something is wrong. Here are the possibilities:

Bad command or filename or **Command not found**
You are using Stata for Unix and forgot to add Stata to your shell's PATH.

Cannot find license file
This message means just what it says; nothing is too seriously wrong, Stata simply could not find what it is looking for, probably because you did not complete the installation process or Stata is not installed where it should be.

Did you insert the codes printed on your paper license to unlock Stata? If not, go back and complete the installation; see the *Getting Started* manual.

Assuming you did unlock Stata, Stata is merely mislocated or the location has not been filled in.

Stata for Unix must be installed in /usr/local/stata8 or /usr/local/stata; verify that you have done this.

Error opening or reading the file
Something is distinctly wrong and for purely technical reasons. Stata found the file it was looking for, but either the operating system refused to let Stata open it or there was an I/O error. On Windows and Macintosh computers, about the only way this could happen would be a hard-disk error. Under Unix, the stata.lic file could have incorrect permissions. Verify that stata.lic is in /usr/local/stata or /usr/local/stata8 and that everybody has been granted read permission. To change the permissions, as superuser, type chmod a+r /usr/local/stata/stata.lic.

This is a *n*-user license and *n* users are currently using Stata
You are trying to use Stata for Unix, but the number of people using it at your site is already at the maximum allowed by the license at your site; see *How Stata counts license positions* in [GSU] **1 Installation** for more information.

License not applicable
Stata has determined that you have a valid Stata license, but it is not applicable to the version of Stata that you are trying to run. You would get this message if, for example, you tried to run Stata for Macintosh using a Stata for Windows license.

The most common reason for this message is that you have a license for Small Stata, but you are trying to run Intercooled Stata, or you have a license for Intercooled Stata, but you are trying to run Stata/SE. If this is the case, reinstall Stata and choose the appropriate version.

Other messages

The other messages indicate that Stata thinks you are attempting to do something you are not licensed to do. Most commonly, you are attempting to run Stata over a network when you do not have a network license, but there are a host of other alternatives. There are two possibilities: either you really are attempting to do something you are not licensed to do or Stata is wrong. In either case, you are going to have to call us. Your license can be upgraded, or, if Stata is wrong, we can provide codes over the telephone to make Stata stop thinking you are violating the license.

6.2   verinst problems

Once Stata is running, you can type `verinst` to check if it is correctly installed. If the installation is correct, you will see something like

```
. verinst

You are running Intercooled Stata 8.0 for Windows.

Stata is correctly installed.
You can type exit to exit Stata.
```

If, however, there is a problem, `verinst` will report it. In most cases, `verinst` itself tells you what is wrong and how to fix it. There is one exception:

```
. verinst
unrecognized command
r(199);
```

This indicates that Stata could not even find the `verinst` command. The most likely cause is that, somehow, you have more than one `stata.lic` file on your computer. There can only be one, and it must be in the directory in which Stata is installed. Exit Stata (type `exit`), erase the copies, and try again.

6.3   Troubleshooting Windows

If you experience a problem not mentioned above, first look at the Frequently Asked Questions (FAQ) for Windows in the User Support section of the Stata web site (*http://www.stata.com*). You may find the answer to the problem there. If not, we can help, but you must give us as much information as possible. See [GSW] **B. Troubleshooting starting and stopping Stata**.

6.4   Troubleshooting Macintosh

If you experience a problem not mentioned above, first look at the Frequently Asked Questions (FAQ) for Macintosh in the User Support section of the Stata web site (*http://www.stata.com*). You may find the answer to the problem there. If not, we can help, but you must give us as much information as possible. See [GSM] **B. Troubleshooting starting and stopping Stata**.

6.5   Troubleshooting Unix

If you experience a problem not mentioned above, first look at the Frequently Asked Questions (FAQ) for Unix in the User Support section of the Stata web site (*http://www.stata.com*). You may find the answer to the problem there. If not, we can help, but you must give us as much information as possible. See [GSU] **B. Troubleshooting starting and stopping Stata**.

7 Setting the size of memory

Contents

7.1   Memory size considerations
7.2   Setting memory size on the fly: Stata/SE
        7.2.1   Advice on setting maxvar
        7.2.2   Advice on setting matsize
        7.2.3   Advice on setting memory
7.3   Setting memory size on the fly: Intercooled Stata
7.4   The memory command
7.5   Virtual memory and speed considerations
7.6   An issue when returning memory to Unix

7.1   Memory size considerations

Stata works with a copy of the data that it loads into memory.

By default, Stata/SE allocates 10 megabytes and Intercooled Stata allocates 1 megabyte to Stata's data areas, and you can change it.

By default, Small Stata allocates about 300K to Stata's data areas, and you cannot change it.

You can even change the allocation to be larger than the physical amount of memory on your computer because Windows, Macintosh, and Unix systems provide virtual memory.

Virtual memory is slow but adequate in rare cases when you have a dataset that is too large to load into real memory. If you use large datasets frequently, we recommend that you add more memory to your computer.

One way to change the allocation is when you start Stata. Instructions for doing this are provided in

Windows	[GSW]	**A.6 Specifying the amount of memory allocated**
Macintosh	[GSM]	**A.5 Specifying the amount of memory allocated**
Unix	[GSU]	**A.6 Specifying the amount of memory allocated**

In addition, if you use Stata/SE or Intercooled Stata for Windows, Unix, or Macintosh, you can change the total amount of memory allocated while Stata is running. That is the topic of this chapter.

Understand that it does not matter which method you use. Being able to change the total on the fly is convenient, but even if you cannot do this, it just means that you specify it ahead of time. If later you need more, simply exit Stata and reinvoke it with the larger total.

7.2   Setting memory on the fly: Stata/SE

There are three limits in Stata/SE that affect memory allocation and usage. The three limits are

1. `maxvar`, the maximum number of variables allowed in a dataset. This limit is initially set to 5,000; you can increase it up to 32,767.

2. `matsize`, the largest dimension of a matrix: this limit is initially set to 400, and you can increase it up to 11,000. In most cases, this relates to the maximum number of independent variables allowed in the models that you fit, and thus, the dimension of the estimated variance–covariance matrix. However, in some panel-data models, covariance or correlation matrices must be fitted, and their dimensions depend on either the number of panels (groups) in your data or on the number of observations in your dataset.

3. `memory`, the amount of memory Stata requests from the operating system to store your data. This limit is initially set to 10 megabytes in Stata/SE. You may set it to as large a number as your operating system will allow.

You set the limits using the

$$\texttt{set maxvar } \# \qquad \left[\texttt{, } \underline{\texttt{permanently}} \right]$$

$$\texttt{set matsize } \# \qquad \left[\texttt{, } \underline{\texttt{permanently}} \right]$$

$$\texttt{set memory } \# \left[\texttt{b}|\texttt{k}|\texttt{m}|\texttt{g} \right] \left[\texttt{, } \underline{\texttt{permanently}} \right]$$

commands. For instance, you might type

```
. set maxvar   5000
. set matsize 900
. set memory   50m
```

The order in which you set the limits does not matter. If you specify the `permanently` option when you set a limit, in addition to making the change right now, Stata will remember the new limit and use it in the future when you invoke Stata:

```
. set maxvar   5000, permanently
. set matsize 900, permanently
. set memory   50m, permanently
```

You can reset the current or permanent limits whenever and as often as you wish.

7.2.1   Advice on setting maxvar

$$\texttt{set maxvar } \# \left[\texttt{, } \underline{\texttt{permanently}} \right]$$

where $2{,}048 \le \# \le 32{,}767$

Why is there a limit on `maxvar`? Why not just set `maxvar` to 32,767 and be done with it? Because simply allowing room for variables, even if they do not exist, causes Stata to consume memory and, if you will only be using datasets with a lot fewer variables, you will be wasting memory.

The formula for the amount of memory consumed by `set maxvar` is approximately

$$\text{megs} = .3147 * (\text{maxvar}/1000) + .002$$

For instance, if you set `maxvar` to 20,000, the memory would be approximately

$$\text{megs} = .3147 * 20 + .002 = 6.296 \text{ megs}$$

and if you left it at the default, the memory use would be roughly

$$\text{megs} = .3147 * 5 + .002 = 1.575 \text{ megs}$$

Thus, how big you set `maxvar` does not dramatically affect memory usage. Still, at `maxvar=32,000`, memory use is 10.072M.

Recommendation: Think about datasets with the most variables that you typically use. `set maxvar` to a few hundred or even a 1,000 above that. (Note that the memory cost of an extra 1,000 variables is only .315 megs.)

Remember, you can always reset `maxvar` temporarily by typing `set maxvar`.

❏ Technical Note

The formula above is only approximate, and the formula given is the formula appropriate for 32-bit computers. When you `set maxvar`, Stata/SE will give you a memory report showing the exact amount of memory used:

```
. set maxvar 10000
```

Current memory allocation

settable	current value	description	memory usage (1M = 1024k)
set maxvar	10000	max. variables allowed	3.149M
set memory	10M	max. data space	10.000M
set matsize	400	max. RHS vars in models	1.254M
			14.403M

❏

7.2.2 Advice on setting matsize

set matsize # [, permanently]

where $10 \le \# \le 11{,}000$

Although `matsize` can theoretically be set up to 11,000, on all but 64-bit computers, you will be unable to do that, and, even if you succeeded, Stata/SE would probably run out of memory subsequently. The value of `matsize` has a dramatic effect on memory usage, the formula being

$$\text{megs} = (8 * \text{matsize}^2 + 88 * \text{matsize})/(1024^2)$$

This formula is valid across all computers, 32-bit and 64-bit. For instance, the above formula states

matsize	memory use
400	1.254M
800	4.950M
1,600	19.666M
3,200	78.394M
6,400	313.037M
11,000	924.080M

The formula, in fact, is an understatement of the amount of memory certain Stata commands use, and is an understatement of what you will certainly use yourself if you use matrices directly. The formula gives the amount of memory required for one matrix and 11 vectors. If two matrices are required, the numbers above are nearly doubled. When you `set matsize`, if you specify too large a value, Stata will refuse, but remember that just because Stata does not complain, you still may run into problems later. What might happen is that Stata could be running a command and then complain, "op. sys. refuses to provide memory"; r(909).

For `matsize` = 11,000, nearly 1 gigabyte of memory is required, and doubling that would require nearly 2 gigabytes of memory. On most 32-bit computers, 2 gigabytes is the maximum amount of memory the operating system will allocate to a single task, so nearly nothing would be left for all the rest of Stata.

Why, then, is `matsize` allowed to be set so large? Because on 64-bit computers, such large amounts cause no difficulty.

For "reasonable" values of matsize (say up to 3,200), memory consumption is not too great. Choose a reasonable value given the kinds of models you fit, and remember that you can always reset the value.

7.2.3   Advice on setting memory

set memory #[b|k|m|g] [, permanently]

where # ≥ 500k and (k, m, and g may be typed in uppercase.)

The advice for setting memory is the same as for Intercooled Stata: set enough so that your datasets fit easily, and do not set so much that you exceed physical memory present on your computer, except in emergencies.

You may set memory in bytes (b) kilobytes (k), megabytes (m), or gigabytes (g), but the number specified must be an integer, so if you want to set 1.5g, you set 1500m. Actually, 1.5g is 1536m because the formulas for a kilobyte, megabyte, and gigabyte are

```
1024 bytes     = 1 kilobyte
1024 kilobytes = 1 megabyte
1024 megabytes = 1 gigabyte
```

This detail does not matter, but this is the rule that Stata uses when presenting numbers, so do not be surprised when 2000k is not displayed as 2M, or 2M is displayed as 2048k.

If you have a very large 32-bit computer, the maximum amount of memory you can set may surprise you. Many people think that 32-bit computers can allow up to 4 gigabytes of memory and that, in a sense, is true. Some 32-bit computers will even allow you to install 4 gigabytes of physical memory. Nevertheless, most modern operating systems will allocate a maximum of one-half the theoretical maximum to individual tasks, which is to say, most operating systems will allow Stata only 2 gigabytes of memory (even if they have 4 gigabytes of memory)!

The same one-half rule applies to 64-bit computers, but half the theoretical limit is still 536,870,912 gigabytes, so no one much cares.

This one-half limit is imposed by the operating system, not by Stata, and the operating system developers have good technical reasons for imposing the rule.

7.3 Setting memory size on the fly: Intercooled Stata

You can reallocate memory on-the-fly. Assume you have changed nothing about how Stata starts, so you get the default 1 megabyte of memory allocated to Stata's data areas. You are working with a large dataset and now wish to increase it to 32 megabytes. You can type

```
. set memory 32m
(32768k)
```

and, if your operating system can provide the memory to Stata, Stata will work with the new total. Later in the session, if you want to release that memory and work with only 2 megabytes, you could type

```
. set memory 2m
(2048k)
```

There is only one restriction on the set memory command: whenever you change the total, there cannot be any data already in memory. If you have a dataset in memory, save it, clear memory, reset the total, and then use it again. We are getting ahead of ourselves, but you might type

```
. save mydata, replace
file mydata.dta saved

. drop _all

. set memory 32m
(32768k)

. use mydata
```

When you request the new allocation, your operating system might refuse to provide it:

```
. set memory 128m
op. sys. refuses to provide memory
r(909);
```

If that happens, you are going to have to take the matter up with your operating system. In the above example, Stata asked for 128 megabytes and the operating system said no.

For most 32-bit computers, the absolute maximum amount of memory that can theoretically be allocated will be approximately $2^{(32-1)}$ bytes (2 gigabytes), regardless of the operating system. In practice, the amount of memory available to any one application on a system is affected by many factors, and may be somewhat less than the theoretical maximum.

64-bit computers can theoretically have up to $2^{(64-1)}$ bytes (over 2 billion gigabytes). In practice, 64-bit computers will have as much memory as is affordable. We are aware of sites using Stata with datasets consuming over 10 gigabytes of memory.

7.4 The memory command

memory helps you figure out whether you have sufficient memory to do something.

(Continued on next page)

```
. use http://www.stata-press.com/data/r8/regsmpl
(NLS Women 14-26 in 1968)

. memory
```

	bytes	
Details of set memory usage		
overhead (pointers)	114,136	10.88%
data	913,088	87.08%
data + overhead	1,027,224	97.96%
free	21,344	2.04%
Total allocated	1,048,568	100.00%
Other memory usage		
system overhead	677,289	
set matsize usage	16,320	
programs, saved results, etc.	505	
Total	694,114	
Grand total	1,742,682	

21,344 bytes free is not much. You might increase the amount of memory allocated to Stata's data areas by specifying set memory 2m.

```
. save regsmpl
file regsmpl.dta saved

. clear

. set memory 2m
(2048k)

. use regsmpl
(NLS Women 14-26 in 1968)

. memory
```

	bytes	
Details of set memory usage		
overhead (pointers)	114,136	5.44%
data	913,088	43.54%
data + overhead	1,027,224	48.98%
free	1,069,920	51.02%
Total allocated	2,097,144	100.00%
Other memory usage		
system overhead	677,289	
set matsize usage	16,320	
programs, saved results, etc.	667	
Total	694,276	
Grand total	2,791,420	

Over 1 megabyte free; that's better. See [R] **memory** for more information.

7.5   Virtual memory and speed considerations

When you use more memory than is physically available on your computer, Stata slows down. If you are only using a little more memory than on your computer, performance is probably not too bad. On the other hand, when you are using a lot more memory than is on your computer, performance will be noticeably affected. In these cases, we recommend that you

```
. set virtual on
```

Virtual memory systems exploit locality of reference, which means that keeping objects closer together allows virtual memory systems to run faster. set virtual controls whether Stata should perform extra work to arrange its memory to keep objects close together. By default, virtual is set off. set virtual can be used with Stata/SE and Intercooled Stata on all supported operating systems.

In general, you want to leave set virtual set to the default of off so that Stata will run faster.

When you set virtual on, you are asking Stata to arrange its memory so that objects are kept closer together. This requires Stata to do a substantial amount of work. We recommend setting virtual on only when the amount of memory in use drastically exceeds what is physically available. In these cases, setting virtual on will help, but keep in mind that performance will still be slow. If you are using virtual memory frequently, you should consider adding memory to your computer.

7.6   An issue when returning memory to Unix

There is a surprising issue of returning memory that Unix users need to understand. Let's say that you set memory to 128 megabytes, went along for a while, and then, being a good citizen, returned most of it:

```
. set memory 2m
(2048k)
```

Theoretically, 126 megabytes just got returned to the operating system for use by other processes. If you use Windows, that is exactly what happens and, with some Unixes, that is what happens, too.

Other Unixes, however, are strange about returned memory in a misguided effort to be efficient: they do not really take the memory back. Instead, they leave it allocated to you in case you ask for it back later. Still other Unixes sort of take the memory back: they put it in a queue for your use, but, if you do not ask for it back in 5 or 10 minutes, then they return it to the real system pool!

The unfortunate situation is that we at Stata cannot force the operating system to take the memory back. Stata returns the memory to Unix, and then Unix does whatever it wants with it.

So, let's review: You make your Stata smaller in an effort to be a good citizen. You return the memory so that other users can use it, or perhaps so you can use it with some other software.

If you use Windows, the memory really is returned and all works exactly as you anticipated.

If you use Unix, it might go back immediately, it might go back in 5 or 10 minutes, or it might never go back. In the last case, the only way to really return the memory is to exit Stata. All Unixes agree on that: when a process ends, the memory really does go back into the pool.

To find out how your Unix works, you need to experiment. We would publish a table and just tell you, but we have found that within manufacturer the way their Unix works will vary by subrelease! The experiment is tedious but not difficult:

1. Bring up a Stata and make it really big; use a lot of memory, so much that you are virtually hogging the computer.

2. Go to another window or session and bring up another Stata. Verify that you cannot make it big—that you get the "system limit exceeded" message.

3. Go back to the first Stata, leaving the second running, and make it smaller.

4. Go to the second Stata and try again to make it big. If you succeed, then your Unix returns memory instantly.

5. If you still get the "system limit exceeded" message, wait 5 minutes and try again. If it now works, your system delays accepting returned memory for about 5 minutes.

6. If you still get the "system limit exceeded" message, wait another 5 minutes and try again. If it now works, your system delays accepting returned memory for about 10 minutes.

7. Go to the first Stata and exit from it.

8. Go to the second Stata and try to make it big again. If it now works, your system never really accepts returned memory. If it still does not work, start all over again. Some other process took memory and corrupted your experiment.

If you are one of the unfortunates who have a Unix that never accepts returned memory, you will just have to remember that you must exit and reenter Stata to really give memory back.

8 Stata's online help and search facilities

Contents

8.1   Introduction
8.2   help: Stata's online manual pages
8.3   search: Stata's online index
8.4   Accessing help and search from the Help menu
8.5   More on search
8.6   More on help
8.7   help contents: Stata's online table of contents
8.8   search: All the details
        8.8.1   How search works
        8.8.2   Author searches
        8.8.3   Entry id searches
        8.8.4   FAQ searches
        8.8.5   Return codes
8.9   net search: Searching net resources

8.1 Introduction

Stata has help online and a lot of it. Users have two ways to access this:

1. They can pull down **Help**.

2. They can type the `help` and `search` commands.

Stata for Unix(console) users only have the second approach available to them.

Understand that both methods access the same underlying information, but the first method of accessing is better because it invokes the Viewer, displaying the help file in a separate window. In either case, blue text indicates a hypertext link, so you can click on it to go directly to the help. Stata for Unix(console) users must type out `help` followed by the command name or topic.

8.2 help: Stata's online manual pages

The `help` command provides access to Stata's interactive help files. These files are a shortened version of what is in the printed manuals.

It is easier to describe the line-by-line access on paper, so begin by entering Stata and typing `help help`. That will show you the help file for the `help` command itself:

(*Continued on next page*)

```
. help help
```

---

help for **help**, **whelp**   manual: **[R] help**
                                                                dialog: **help**

What to do when you see —more—

The characters —more— now appear at the bottom of the screen. Stata pauses
and displays this message whenever the output from a command is about to scroll
off the screen.

Action	Result
Press **Enter** or **Return**	One more line of text is displayed
Press **b**	The previous screen of text is displayed
Press any other key (such as space bar)	The next screen of text is displayed
PCs:	
Press **Ctrl-Break**	Stata stops processing the command ASAP
Macintosh:	
Press **Command-.**	" " " " " "
Unix:	
Press **Ctrl-C**	" " " " " "

—more—

If you now press the space bar or click on the blue —more—, you will see the next page. Pressing the *Enter* or the *l* (lowercase *L*) key will allow you to advance a single line at a time. If you instead press the *Break* key, Stata will stop showing you help on help and issue another dot prompt; see [U] **12 The Break key**. We are going to press *Break*:

```
—Break—
r(1);
. _
```

Rule: If you know the name of the command and want to learn more about it, type help followed by the command name.

Try it. Two of Stata's commands are use and regress. Type help use. Type help regress. Also, see [U] **10 —more— conditions** for additional information of the more condition.

8.3   search: Stata's online index

Our Rule works fine when you know the name of the command, but what if you do not?

In that case, you use search. search's syntax is

search *anything you want*

or you may pull down **Help**, select **Search...**, check **Search documentation and FAQs**, and then type *anything you want* in the Keywords input field. Either way, search is very understanding. Here is what happens when you search logistic regression:

(Continued on next page)

```
. search logistic regression
```

Keyword search

      Keywords: **logistic regression**
  Search: **(1) Official help files, FAQs, SJs, and STBs**

Search of official help files, FAQs, SJs, and STBs

[U]   Chapter 29 . . . . Overview of Stata estimation commands
          (help **est**)

[U]   Chapter 30 . . . . Overview of survey estimation
          (help **svy**)

[R]   clogit . . . . Conditional (fixed-effects) logistic regression
          (help **clogit**)
   (*output omitted*)

[R]   logistic . . . . Logistic regression
          (help **logistic**, **lfit**, **lstat**, **lroc**, **lsens**)

[R]   logit . . . . Maximum-likelihood logit estimation
          (help **logit**)
   (*output omitted*)

FAQ   . Standard error of the predicted probability with logistic regression
          . . . . R. Gutierrez
   3/01   How do I obtain the standard error of the predicted
          probability with logistic regression analysis?
          http://www.stata.com/support/faqs/stat/delta.html

FAQ   . . . . Logistic regression with grouped data
          . . . . W. Sribney
   1/00   How can I do logistic regression or multinomial
          logistic regression with grouped data?
          http://www.stata.com/support/faqs/stat/grouped.html

   (*output omitted*)

Example . . . . Capabilities: logistic regression
          http://www.stata.com/info/capabilities/binary/logistic.html

   (*output omitted*)

SJ-2-3   st0021 . . . . Measuring effect size
          . . . . R. M. Conroy
      Q3/03   SJ 2(3):290--295   (no commands)
          case study showing superiority of regression over t tests,
          exploratory scatterplot smoothing as a key method of
          checking form of relationship, and the value of logistic
          regression followed by adjust

SJ-2-3   st0022 . . Least likely observations in reg. models for cat. outcomes
          (help leastlikely if installed) . . . . J. Freese
      Q3/03   SJ 2(3):296--300
          command for identifying poorly fitting observations for
          maximum-likelihood regression models for categorical
          dependent variables

STB-61   sg163 . . . . Stereotype ordinal regression
          (help soreg if installed) . . . . M. Lunt
      5/01   pp.12--18; STB Reprints Vol 10, pp.298--307
          implementation of Stereotype Ordinal Regression (SOR) which
          can be thought of as imposing constraints on a multinomial
          model

   (*output omitted*)

search responds by providing a list of references—references to the online help, references to the printed documentation, references to FAQs at the *http://www.stata.com* web site, references to articles that have appeared in the *Stata Journal* (SJ), and references to articles that have appeared in the *Stata Technical Bulletin* (STB). Moreover, if you install the official updates—see [U] **32 Using the Internet to keep up to date**—the references to the FAQs, SJs, and STBs will even be up to date.

Anyway, you are supposed to look over the list and find what looks relevant to you. If you are really interested in logistic regression, `logistic` and `logit` seem particularly appropriate, so you might next type `help logistic` and `help logit`, or click on the blue hypertext links for the `logistic` and `logit` help files.

8.4   Accessing help and search from the Help menu

Users will get exactly the same output if they pull down **Help**, select **Search...**, click on **Search documentation and FAQs**, and type `logistic regression` in the Keyword(s) Search box.

There is an advantage because a separate window, the Viewer, is invoked instead of presenting the resulting output in the Results window. The Viewer can remain open as a reference while you continue to use Stata.

Some of the resulting output will be highlighted in blue to indicate hyperlinks. *help logit* and *help logistic*, for instance, will be displayed in blue. Click on either one and you will go to the help file for the command.

There will be other places you can click, too. When `search` mentions a *Stata Journal* or STB article, the insert number (such as *st0001* or *pr0006* for the *Stata Journal* or *sbe14* or *sg63* for the STB) will be in blue. Click on one and Stata will go to *http://www.stata-journal.com* (or *http://www.stata.com*) and show you a detailed description of the addition, leaving you just one click away from installing the addition if it interests you.

When `search` mentions a FAQ, the URL will be in blue. Click on that and Stata will launch your browser, and you will be looking right at the answer to the frequently asked question.

You can pull down **Help** at any time, not just when Stata is idle. You can leave the Viewer up while you use Stata, which is especially convenient when viewing syntax diagrams and options.

8.5   More on search

However you access `search`—command or menu—it does the same thing. You tell `search` what you want information on and it searches for relevant entries. If you want `search` to look for the topic across all sources, including the online help, the FAQs at the Stata web site, the *Stata Journal*, and all Stata-related Internet sources including user-written additions, then use `findit`, which is a synonym for `search, all`.

`search` can be used broadly or narrowly. For instance, if you want to perform the Kolmogorov–Smirnov test for equality of distributions, you could type

```
. search Kolmogorov-Smirnov test of equality of distributions
[R]     ksmirnov . . . . . . Kolmogorov-Smirnov equality of distributions test
        (help ksmirnov)
```

In fact, we did not have to be nearly so complete—typing `search Kolmogorov-Smirnov` would have been adequate. Had we specified our request more broadly—looking up `equality of distributions`—we would have obtained a longer list that included ksmirnov.

Here are guidelines on using `search`.

1. Capitalization does not matter. Look up `Kolmogorov-Smirnov` or `kolmogorov-smirnov`.

2. Punctuation does not matter. Look up `kolmogorov smirnov`.

3. Order of words does not matter. Look up `smirnov kolmogorov`.

4. You may abbreviate, but how much depends. Break at syllables. Look up `kol smir`. `search` tends to tolerate a lot of abbreviation; it is better to abbreviate than to misspell.

5. The prepositions for, into, of, on, to, and with are ignored. Use them—look up `equality of distributions`—or omit them—look up `equality distributions`—it makes no difference.

6. `search` is tolerant of plurals, especially when they can be formed by adding an *s*. Even so, it is better to look up the singular. Look up `normal distribution`, not `normal distributions`.

7. Specify the search criterion in English, not computer jargon.

8. Use American spellings. Look up `color`, not `colour`.

9. Use nouns. Do not use -ing words. Look up `median tests`, not `testing medians`.

10. Use few words. Every word specified further restricts the search. Look up `distribution` and you get one list; look up `normal distribution` and the list is a sublist of that.

11. Sometimes words have more than one context. The following words can be used to restrict the context:

   a. `data`, meaning in the context of data management. Order could refer to the order of data or to order statistics. Look up `order data` to restrict order to the data management sense.

   b. `statistics` (abbreviation `stat`), meaning in the context of statistics. Look up `order statistics` to restrict order to the statistical sense.

   c. `graph` or `graphs`, meaning in the context of statistical graphics. Look up `median graphs` to restrict the list to commands for graphing medians.

   d. `utility` (abbreviation `util`), meaning in the context of utility commands. The `search` command itself is not data management, not statistics, and not graphics; it is a utility.

   e. `programs` or `programming` (abbreviation `prog`), to mean in the context of programming. Look up `programming scalar` to obtain a sublist of scalars in programming.

`search` has other features as well; see [U] **8.8 search: All the details**.

8.6   More on help

Both `help` and `search` are understanding of some mistakes. For instance, you may abbreviate a command name. If you type either `help regres` or `help regress`, you will bring up the help file for `regress`.

When `help` cannot find the command you are looking for, try the `search` feature. In this case, typing `search regres` will also find the command (because 'regres' is an abbreviation of the word regression), but, in general, that will not be the case.

Stata can run into some problems with abbreviations. For instance, Stata has a command with the inelegant name `ksmirnov`. You forget and think the command is called `ksmir`:

```
. help ksmir
help for ksmir not found
try help contents or search ksmir
```

This is a case where `help` gives bad advice because typing `search ksmir` will do you no good. You should type `search` followed by what you are really looking for: search kolmogorov smirnov.

8.7   help contents: Stata's online table of contents

Typing help contents, or pulling down **Help** and selecting **Contents**, provides another way of locating entries in the documentation and online help. Either way, you will be presented with a long table of contents, organized topically.

```
. help contents
```

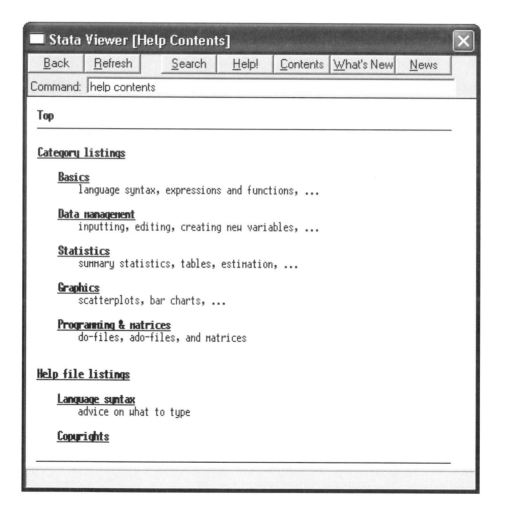

8.8 search: All the details

The `search` command actually provides a few features not available from the **Help** menu. The full syntax of the `search` command is

search *word* [*word* ...] [, [local | net | all] author entry exact faq

historical or manual sj]

where underlining indicates minimum allowable abbreviation and [brackets] indicate optional.

local, the default (unless changed by set searchdefault), specifies that the search is to be performed using only Stata's keyword database.

net specifies that the search is to be performed across the materials available via Stata's net command. Using search word [*word* ...], net is equivalent to typing net search word [*word* ...] (without options); see [R] **net**.

all specifies that the search is to be performed across both the local keyword database and the net materials.

author specifies that the search is to be performed on the basis of author's name rather than keywords.

entry specifies that the search is to be performed on the basis of entry ids rather than keywords.

exact prevents matching on abbreviations.

faq limits the search to entries found in the FAQs at *http://www.stata.com*.

historical adds to the search entries that are of historical interest only. By default, such entries are not listed. Past entries are classified as historical if they discuss a feature that later became an official part of Stata. Updates to historical entries will always be found, even if `historical` is not specified.

or specifies that an entry should be listed if any of the words typed after `search` are associated with the entry. The default is to list the entry only if all the words specified are associated with the entry.

manual limits the search to entries in the *User's Guide* and all the *Reference* manuals.

sj limits the search to entries in the *Stata Journal* and the *Stata Technical Bulletin*.

8.8.1 How search works

`search` has a database—a file—containing the titles, etc. of every entry in the *User's Guide*, *Reference* manuals, articles in the *Stata Journal* and in the *Stata Technical Bulletin*, and FAQs at *http://www.stata.com*. In this file is a list of words associated with each entry, called keywords.

When you type search *xyz*, `search` reads this file and compares the list of keywords with *xyz*. If it finds *xyz* in the list or a keyword that allows an abbreviation of *xyz*, it displays the entry.

When you type search *xyz abc*, `search` does the same thing, but displays an entry only if it contains both keywords. The order does not matter, so you can search linear regression or search regression linear.

Obviously, how many entries `search` finds depends on how the search database was constructed. We have included a plethora of keywords under the theory that, for a given request, it is better to list too much rather than risk listing nothing at all. Still, you are in the position of guessing the keywords. Do you look up normality test, normality tests, or tests of normality? Answer: normality test would be best, but all would work. In general, use the singular and strike the unnecessary words. We provide guidelines for specifying keywords in [U] **8.5 More on search** above.

8.8.2 Author searches

search ordinarily compares the words following search with the keywords for the entry. If you specify the author option, however, it compares the words with the author's name. In the search database, we have filled in author names for *Stata Journal* and STB articles and for FAQs.

For instance, in [R] **kdensity**, you will discover that Isaías H. Salgado-Ugarte wrote the first version of Stata's kdensity command and published it in the STB. Assume you read his original and find the discussion useful. You might now wonder what else he has written in the STB. To find out, you type

```
. search Salgado-Ugarte, author
(output omitted )
```

Names like Salgado-Ugarte are confusing to some people. search does not require you specify the entire name; what you type is compared with each "word" of the name, and, if any part matches, the entry is listed. The dash is a special character, and you can omit it. Thus, you can obtain the same list by looking up Salgado, Ugarte, or Salgado Ugarte without the dash.

Actually, to find all entries written by Salgado-Ugarte, you need to type

```
. search Salgado-Ugarte, author historical
(output omitted )
```

Prior inserts in the STB that provide a feature that later was superseded by a built-in feature of Stata are marked as historical in the search database and, by default, are not listed. The historical option ensures that all entries are listed.

8.8.3 Entry id searches

If you specify the entry option, search compares what you have typed with the entry id. The entry id is not the title—it is the reference listed to the left of the title that tells you where to look. For instance, in

```
[R]      regress  . . . . . . . . . . . . . . . . . . . . . . . . Linear regression
         (help regress)
```

"[R] regress" is the entry id. In

```
GS       . . . . . . . . . . . . . . . . . . . . . . . . . . Getting Started manual
```

"GS" is the entry id. In

```
STB-28   dm36 . . . . . . . . . . . . . . . . . Comparing two Stata data sets
         (help compdta if installed)  . . . . . . . . . . . . John R. Gleason
         11/95   pp.10--13; STB Reprints Vol 5, pages 39--43
         compares the varlist from the dataset in memory with like-named
         variables in the Stata-format dataset on disk; alternative to cf
         command
```

"STB-28 dm36" is the entry id.

search with the entry option searches these entry ids.

Thus, one could generate a table of contents for the *Reference* manuals by typing

```
. search [R], entry
(output omitted )
```

One could generate a table of contents for the 16th issue of the STB by typing

```
. search STB-16, entry historical
(output omitted )
```

The historical option in this case is possibly important. STB-16 was published in November 1993, and perhaps some of its inserts have been marked as historical.

One could obtain a complete list of all inserts associated with *dm36* by typing

```
. search dm36, entry historical
  (output omitted)
```

Again, we include the historical option in case any of the relevant inserts have been marked historical.

8.8.4   FAQ searches

To search across the FAQs, specify the faq option:

```
. search logistic regression, faq
  (output omitted)
```

8.8.5   Return codes

In addition to indexing the entries in the *User's Guide* and all of the *Stata Reference manuals*, search also can be used to look up return codes.

To see information on return code 131, type

```
. search rc 131

[R]     error messages  . . . . . . . . . . . . . . . . . . . . Return code 131
        not possible with test;
        You requested a test of a hypothesis that is nonlinear in the
        variables.  test tests only linear hypotheses.  Use
        testnl.
```

To get a list of all Stata return codes, type

```
. search rc
  (output omitted)
```

8.9   net search: Searching net resources

When you select **Search...** from the **Help** menu, you will notice there are two types of searches to choose. The first, which has been discussed in the previous sections, is to **Search documentation and FAQs**. The second is to **Search net resources**. This feature of Stata searches resources over the Internet.

When you choose **Search net resources** in the search dialog box and enter *keywords* in the field, Stata searches all user-written programs on the Internet, including user-written additions published in the *Stata Journal* and the STB. The results are displayed in the Viewer, and you can click to go to any of the matches found.

Equivalently, you can type net search *keywords* on the Stata command line to display the results in the Results window. For the full syntax on how to use the net search command, see [R] **net search**.

9 Stata's sample datasets

Various examples in this manual use what is referred to as the automobile dataset `auto.dta`. We previously created a dataset on the prices, mileages, weights, and other characteristics of 74 automobiles and saved it in a file called `auto.dta`. (These data originally came from the April 1979 issue of *Consumer Reports* and from the United States Government EPA statistics on fuel consumption; they were compiled and published by Chambers et al., 1983.)

In our examples, you will often see us type

```
. use http://www.stata-press.com/data/r8/auto
```

We include the `auto.dta` file with Stata. If you want to use it from your own computer rather than via the Internet, you can type

```
. sysuse auto
```

See [R] **sysuse**.

You can also load `auto.dta` by pulling down **File** and choosing **Open**, but you will need to know where it is stored. This all depends on where you installed Stata. The easy way to find out is to type

```
. sysdir
    STATA:  D:\STATA\
  UPDATES:  D:\STATA\ado\updates\
     BASE:  D:\STATA\ado\base\
     SITE:  D:\STATA\ado\site\
     PLUS:  D:\ado\plus\
 PERSONAL:  D:\ado\personal\
 OLDPLACE:  D:\ado\
```

Stata is installed in the STATA directory, so, on the Windows computer that we are using, we could type 'use d:\stata\auto'.

You could copy `auto.dta` to the current directory so that you do not have to specify its path:

```
. copy d:\stata\auto.dta auto.dta
```

Here is a list of the example datasets included with Stata:

auto.dta	1978 Automobile data
autornd.dta	Subset of 1978 Automobile data
bplong.dta	Blood pressure data, fictional
bpwide.dta	Blood pressure data, fictional
cancer.dta	Patient survival in drug trial
census.dta	1980 Census data by state
citytemp.dta	US city temperature data
educ99gdp.dta	Education and GDP
gnp96.dta	US GNP, 1967–2002
lifeexp.dta	Life expectancy
nlsw88.dta	US National Longitudinal Study of Young Women
pop2000.dta	US Census, 2000, extract
sp500.dta	S&P 500 historical data
uslifeexp.dta	US life expectancy, 1900-1999
uslifeexp2.dta	US life expectancy, 1900-1940
voter.dta	1992 presidential voter data

Most of the datasets that are used in the *Reference* manuals can be found at the Stata Press web site (*http://www.stata-press.com/data/*). You can download the datasets using your browser, or you can use them directly from the Stata command line; that is,

```
. use http://www.stata-press.com/data/r8/nlswork
```

An alternative to the use command is webuse.

```
. webuse auto
```

For additional information, see [R] **webuse**.

10 –more– conditions

Contents

10.1   Description
10.2   set more off
10.3   The more programming command

10.1 Description

When you see —more— at the bottom of the screen,

Press ...	and Stata...
letter *l* or *Enter*	displays the next line
letter *q*	acts as if you pressed *Break*
space bar or any other key	displays the next screen

In addition, you can press the *clear –more– condition* button, the button labeled **Go** with a circle around it.

—more— is Stata's way of telling you that it has something more to show you, but showing you that something more will cause the information on the screen to scroll off.

10.2 set more off

If you type `set more off`, —more— conditions will never arise and Stata's output will scroll by at full speed.

If you type `set more on`, —more— conditions will be restored at the appropriate places.

Programmers: Do-file writers sometimes include `set more off` in their do-files because they do not care to interactively watch the output. They want Stata to proceed at full speed because they plan on making a log of the output that they will review later. Do-filers need not bother to `set more on` at the conclusion of their do-file. Stata automatically restores the previous `set more` when the do-file (or program) concludes.

10.3 The more programming command

Ado-file programmers need take no special action to have —more— conditions arise when the screen is full. Stata handles that automatically.

If, however, you wish to force a —more— condition early, you can include the `more` command in your program. The syntax of `more` is

```
more
```

`more` takes no arguments.

For additional information, see [P] **more**.

11 Error messages and return codes

Contents

11.1   Making mistakes
    11.1.1   Mistakes are forgiven
    11.1.2   Mistakes stop user-written programs and do-files
    11.1.3   Advanced programming to tolerate errors
11.2   The return message for obtaining command timings

11.1   Making mistakes

When an error occurs, Stata produces an error message and a *return code*. For instance,

```
. list myvar
no variables defined
r(111);
```

We ask Stata to list the variable named `myvar`. Since we have no data in memory, Stata responds with the message "no variables defined" and a line that reads "r(111)"

The "no variables defined" is called the error message.

The 111 is called the return code. You can click on blue return codes to get a detailed explanation of the error.

11.1.1   Mistakes are forgiven

Having said "no variables defined" and `r(111)`, all is forgiven; it is as if the error never occurred.

Typically, the message will be enough to guide you to a solution, but, if it is not, the numeric return codes are documented in [P] **error**.

11.1.2   Mistakes stop user-written programs and do-files

Whenever an error occurs in a user-written program or do-file, the program or do-file immediately stops execution and the error message and return code are displayed.

For instance, consider the following do-file:

```
———————————————————————————————— top of myfile.do ————————
use http://www.stata-press.com/data/r8/auto
decribe
list
———————————————————————————————— end of myfile.do ————————
```

Note the second line—you meant to type `describe` but typed `decribe`. Here is what happens when you execute this do-file by typing `do myfile`:

```
. do myfile

. use http://www.stata-press.com/data/r8/auto
(1978 Automobile Data)
```

```
. decribe
unrecognized command:  decribe
r(199);

end of do-file
r(199);

. _
```

The first error message and return code were caused by the illegal `decribe`. This then caused the do-file itself to be aborted; the valid `list` command was never executed.

11.1.3   Advanced programming to tolerate errors

Errors are not only of the typographical kind; some are substantive. A command that is valid in one dataset might not be valid in another. Moreover, in advanced programming, errors are sometimes anticipated: use one dataset if it is there, but use another if you must.

Programmers can access the return code to determine whether an error occurred, which they can then ignore, or, by examining the return code, code their programs to take the appropriate action. This is discussed in [P] **capture**.

You can also prevent do-files from stopping when errors occur by using the do command's `nostop` option.

```
. do myfile, nostop
```

11.2   The return message for obtaining command timings

In addition to error messages and return codes, there is something called a return message, which you normally do not see. Normally, if you typed `summarize tempjan`, you would see

```
. use http://www.stata-press.com/data/r8/citytemp
(City Temperature Data)
. summarize tempjan
```

Variable	Obs	Mean	Std. Dev.	Min	Max
tempjan	954	35.74895	14.18813	2.2	72.6

If you were to type

```
. set rmsg on
r; t=0.00 10:21:22
```

sometime during your session, Stata would display return messages:

```
. summarize tempjan
```

Variable	Obs	Mean	Std. Dev.	Min	Max
tempjan	954	35.74895	14.18813	2.2	72.6

```
r; t=0.01 10:21:26
```

The line that reads `r; t=0.01 10:21:26` is called the return message.

The `r;` indicates that Stata successfully completed the command.

The `t=0.01` shows the amount of time, in seconds, it took Stata to perform the command (timed from the point you pressed *Return* to the time Stata typed the message). This command took a hundredth of a second. In addition, Stata shows the time of day using a 24-hour clock. This command completed at 10:21 a.m.

You will learn that Stata has the ability to run commands stored in files (called do-files) and the ability to log output. Some users find the detailed return message helpful with do-files. They construct a lengthy program and let it run overnight, logging the output. They come back the next morning, look at the output, and discover a mistake in some portion of the job. They can look at the return messages to determine how long it will take to rerun that portion of the program.

You may `set rmsg on` whenever you wish.

When you want Stata to stop displaying the detailed return message, type `set rmsg off`.

12 The Break key

Contents

12.1   Making Stata stop what it is doing
12.2   Side-effects of pressing Break
12.3   Programming considerations

12.1 Making Stata stop what it is doing

When you want to make Stata stop what it is doing and return to the Stata dot prompt, you press *Break*:

Stata for Windows:	click the **Break** button (it is the button with the big red X), hold down *Ctrl* and press the *Pause/Break* key
Stata for Macintosh:	click the **Break** button or hold down *Command* and press period
Stata for Unix(GUI):	click the **Break** button or hold down *Ctrl* and press k
Stata for Unix(console):	hold down *Ctrl* and press c or press q

Elsewhere in this manual, we describe this action as simply pressing *Break*. Break tells Stata to cancel what it is doing and return control to you as soon as possible.

If you press *Break* in response to the input prompt or while you are typing a line, Stata ignores it, since you are already in control.

If you press *Break* while Stata is doing something—creating a new variable, sorting a dataset, making a graph, etc.—Stata stops what it is doing, undoes it, and issues an input prompt. The state of the system is the same as if you had never issued the command.

▷ Example

You are estimating a logit model, type the command, and, as Stata is working on the problem, realize that you omitted an important variable:

```
. logit foreign mpg weight
Iteration 0:  log likelihood =-1801.3284
Iteration 1:  log likelihood =-1197.7089
—Break—
r(1);

. _
```

When you pressed *Break*, Stata responded by typing —Break— and then typing r(1);. Pressing *Break* always results in a return code of 1—that is why return codes are called return codes and not error codes. The 1 does not indicate an error, but it does indicate that the command did not complete its task.

◁

12.2 Side-effects of pressing Break

In general, there are none. We said above that Stata undoes what it is doing so that the state of the system is the same as if you had never issued the command. There are two exceptions to this.

If you are reading data from disk using insheet, infile, or infix, whatever data have already been read will be left behind in memory, the theory being that perhaps you stopped the process so you could verify that you were reading the right data correctly before sitting through the whole process. If not, you can always drop _all.

```
. infile v1-v9 using workdata
(eof not at end of obs)
(4 observations read)
—Break—
r(1);
```

The other exception is sort. You have a large dataset in memory, decide to sort it, and then change your mind.

```
. sort price
—Break—
r(1);
```

If the dataset was previously sorted by, say, the variable prodid, it is no longer. When you press *Break* in the middle of a sort, Stata marks the data as unsorted.

12.3 Programming considerations

There are basically no programming considerations for handling Break because Stata handles it all automatically. If you write a program or do-file, execute it, and then press *Break*, Stata stops execution just as it would with an internal command.

Advanced programmers may be concerned about cleaning up after themselves; perhaps, they have generated a temporary variable they intended to drop later or a temporary file they intended to erase later. If a Stata user presses *Break*, how can you ensure that these temporary variables and files will be erased?

If you obtain names for such temporary items from Stata's tempname, tempvar, and tempfile commands, Stata will automatically erase the temporary items; see [U] **21.7 Temporary objects**.

There are instances, however, when a program must commit to executing a group of commands without interruption, or the user's data would be left in an intermediate or undefined state. In these instances, Stata provides a

```
nobreak {
        ...
}
```

construct; see [P] **break**.

13 Keyboard use

Contents

13.1   Description
13.2   F-keys
13.3   Editing keys in Stata for Windows, Macintosh, and Unix(GUI)
13.4   Editing keys in Stata for Unix(console)
13.5   Editing previous lines in Stata for all operating systems
13.6   Tab expansion of variable names

13.1   Description

The keyboard should operate very much the way you would expect, with a few additions:

1. There are some unexpected keys you can press to obtain previous commands you have typed. In addition, you can click once on a command in the Review window to reload it, or click it twice to reload and execute; this feature is discussed in the *Getting Started* manuals.

2. There are a host of command-editing features for Stata for Unix(console) users since their user interface does not offer such features.

3. Regardless of operating system or user interface, if there are *F*-keys on your keyboard, they have special meaning and you can change the definitions of the keys.

13.2   F-keys

Note to Macintosh users: Not all Macintosh keyboards have *F*-keys.

Note to Windows users: *F10* is reserved internally by Windows; you cannot program this key.

By default, Stata defines the *F*-keys to mean

F-key	Definition
F1	help
F2	#review;
F3	describe;
F7	save
F8	use

The semicolons at the end of some of the entries indicate the presence of an implied *Return*.

help shows a Stata help file—you use it by typing help followed by the name of a Stata command; see [U] **8 Stata's online help and search facilities**. You can type out help or you can press *F1*, type the Stata command, and press *Return*.

#review is the command to show the last few commands you issued. It is described in [U] **13.6 Editing previous lines**. Rather than typing out #review and pressing *Return*, you can simply press *F2*. You do not press *Return* following *F2* because the definition of F2 ends in a semicolon—Stata presses the *Return* key for you.

describe is the Stata command to report the contents of data loaded into memory. It is explained in [R] **describe**. Normally, you type describe and press *Return*. Alternatively, you can press *F3*.

save is the command to save the data in memory into a file, and use is the command to load data. Both are described in [R] **save** and the syntax of each is the same: save or use followed by a filename. You can type out the commands or you can press *F7* or *F8* followed by the filename.

You can change the definitions of the *F*-keys. For instance, the command to list data is list; you can read about it in [R] **list**. The syntax is list to list all the data, or list followed by the names of some variables to list just those variables (there are other possibilities).

If you wanted *F3* to mean list, you could type

. global F3 "list "

In the above, F3 refers to the letters *F* followed by *3*, not the *F3* key. Note the capitalization and spacing of the command.

You type global in lowercase, type out the letters F3, and then type "list ". The blank at the end of list is important. In the future, rather than typing out list mpg weight, you want to be able to press the *F3* key and then type only mpg weight. You put a blank in the definition of F3 so that you would not have to type a blank in front of the first variable name following pressing *F3*.

Now say you wanted *F5* to mean list all the data—list followed by *Return*. You could define

. global F5 "list;"

Now you would have two ways of listing all the data: (1) press *F3*, then press *Return*, or (2) press *F5*. The semicolon at the end of the definition of *F5* will press *Return* for you.

If you really want to change the definitions of *F3* and *F5*, you will probably want to change the definition every time you invoke Stata. One way would be to type out the two global commands every time you invoked Stata. Another way would be to type the two commands into an ASCII text file called profile.do. Stata executes the commands in profile.do every time it is launched if profile.do is placed in the appropriate directory:

Windows:	put profile.do in the "start-in" directory;
	see [GSW] **A.7 Executing commands every time Stata is started**
Macintosh:	put profile.do in your home directory;
	see [GSM] **A.6 Executing commands every time Stata is started**
Unix:	put profile.do someplace along your shell's PATH;
	see [GSU] **A.7 Executing commands every time Stata is started**

You can use the *F*-keys any way you desire: They contain a string of characters, and pressing the *F*-key is equivalent to typing out those characters.

❑ Technical Note

(*Stata for Unix users.*) Sometimes Unix assigns a special meaning to the *F*-keys, and, if it does, those meanings will take precedence over our meanings. Stata provides a second way to get to the *F*-keys. Hold down *Ctrl*, press F, release the keys, and then press a number from 0 through 9. Stata interprets *Ctrl-F* plus 1 as equivalent to the *F1* key, *Ctrl-F* plus 2 as *F2*, and so on. *Ctrl-F* plus 0 means *F10*.

❑

❑ Technical Note

On some international keyboards, the left single quote is used as an accent character. In this case, we recommend mapping this character to one of your function keys. In fact, you might find it convenient to map both the left single quote (`) and right single quote (') characters so that they are next to each other.

Within Stata, open the Do-file Editor. Type the following two lines in the Do-file Editor:

```
macro define F4 `
macro define F5 '
```

Save the file as `profile.do` into your Stata directory. If you already have a `profile.do` file, append the two lines to your existing `profile.do` file.

Exit Stata, and restart it. You should see the startup message

running /Users/*myhome*/profile.do ...

or some variant of it depending on where your Stata is installed. Press the function keys *F4* and *F5* to verify that they work.

If you did not see the startup message, you did not save the `profile.do` in your home folder.

You can of course map to any other function keys, but *F1*, *F2*, *F3*, *F7*, and *F8* are already used.

❑

13.3   Editing keys in Stata for Windows, Macintosh, and Unix(GUI)

Users have available to them the standard editing keys for their operating system. So, Stata should just edit what you type in the natural way—the Stata Command window is a standard edit window.

In addition, you can fetch commands from the Review window into the Command window. Click on a command in the Review window, and it is loaded into the Command window, where you can edit it. Alternatively, if you double-click on a line in the Review window, it is loaded and executed.

Another way to get lines from the Review window into the Command window is with the *PgUp* and *PgDn* keys. Tap *PgUp* and Stata loads the last command you typed into the Command window. Tap it again and Stata loads the line before that, and so on. *PgDn* goes the opposite direction.

Another editing key that may interest users is *Esc*. This key clears the Command window.

In summary,

Press	Result
PgUp	Steps back through commands and moves command from Review window to Command window
PgDn	Steps forward through commands and moves command from Command window to Review window
Esc	Clears Command window

13.4   Editing keys in Stata for Unix(console)

Certain keys allow you to edit the line you are typing. Since Stata supports a variety of computers and keyboards, the location and the names of the editing keys are not the same for all Stata users.

Every keyboard has the standard alphabet keys (*QWERT* and so on), and every keyboard has a *Ctrl* key. Some keyboards go further and have extra keys located to the right, above, or left, with names like *PgUp* and *PgDn*.

Throughout this manual we will refer to Stata's editing keys using names that appear on nobody's keyboard. For instance, PrevLine is one of the Stata editing keys—it retrieves a previous line. Hunt all you want, but you will not find it on your keyboard. So, where is PrevLine? We have tried to put it where you would naturally expect it. On keyboards with a key labeled *PgUp*, *PgUp* is the PrevLine key, but on everybody's keyboard, no matter which version of Unix, brand of keyboard, or anything else, *Ctrl-R* also means PrevLine.

When we say press PrevLine, now you know what we mean: press *PgUp* or *Ctrl-R*. With that introduction, the editing keys are

Name for Editing Key	Editing Key	Function
Kill	*Esc* on PCs and *Ctrl-U*	Deletes the line and lets you start over.
Dbs	*Backspace* on PCs and *Backspace*, *Rubout*, or *Delete* on other computers	Backs up and deletes one character.
Lft	←, *4* on the numeric keypad for PCs, and *Ctrl-H*	Moves the cursor left one character without deleting any characters.
Rgt	→, *6* on the numeric keypad for PCs, and *Ctrl-L*	Moves the cursor forward one character.
Up	↑, *8* on the numeric keypad for PCs, and *Ctrl-O*	Moves the cursor up one physical line on a line that takes more than one physical line. Also see PrevLine
Dn	↓, *2* on the numeric keypad for PCs, and *Ctrl-N*	Moves the cursor down one physical line on a line that takes more than one physical line. Also see NextLine.
PrevLine	*PgUp* and *Ctrl-R*	Retrieves a previously typed line. You may press PrevLine multiple times to step back through previous commands.
NextLine	*PgDn* and *Ctrl-B*	The inverse of PrevLine.
Seek	*Ctrl-Home* on PCs and *Ctrl-W*	Goes to the line number specified. Before pressing Seek, type the line number. For instance, typing *3* and then pressing Seek is the same as pressing PrevLine three times.
Ins	*Ins* and *Ctrl-E*	Toggles insert mode. In insert mode, characters typed are inserted into the line at the position of the cursor.
Del	*Del* and *Ctrl-D*	Deletes the character at the position of the cursor.
Home	*Home* and *Ctrl-K*	Moves the cursor to the start of the line.
End	*End* and *Ctrl-P*	Moves the cursor to the end of the line.
Hack	*Ctrl-End* on PCs, and *Ctrl-X*	Hacks off the line at the cursor.
Tab	⊣ on PCs, *Tab*, and *Ctrl-I*	Moves the cursor forward eight spaces.
Btab	⊢ on PCs, and *Ctrl-G*	The inverse of Tab.

▷ Example

It is difficult to demonstrate the use of editing keys on paper. You should try each of them. Nevertheless, here is an example:

. summarize price wa̲ht

You typed summarize price waht and then tapped the *Lft* key (← key or *Ctrl-H*) three times to maneuver the cursor back to the a of waht. If you were to press *Return* right now, Stata would see the command summarize price waht, so where the cursor is does not matter when you press *Return*. If you wanted to execute the command summarize price, you could back up one more character and then press the Hack key. We will assume, however, that you meant to type weight.

If you were now to press the letter e on the keyboard, an e would appear on the screen to replace the a, and the cursor would move under the character h. We now have we̲ht. You press *Ins*, putting Stata into insert mode, and press *i* and *g*. The line now says summarize price weig̲ht, which is correct, so you press *Return*. Notice that we did not have to press *Ins* before every character we wanted to insert. The *Ins* key is a toggle: If we press it again, Stata turns off insert mode, and what we type replaces what was there. When we press *Return*, Stata forgets all about insert mode, so we do not have to remember from one command to the next whether we are in insert mode.

◁

❑ Technical Note

Stata performs its editing magic based on the information about your terminal recorded in /etc/termcap(5) or, under System V, /usr/lib/terminfo(4). If some feature does not appear to work, it is probable that the entry for your terminal in the termcap file or terminfo directory is incorrect. Contact your system administrator.

❑

13.5   Editing previous lines in Stata for all operating systems

In addition to what is said below, remember that the Review window also shows the contents of the review buffer.

You may edit previously typed lines, or at least any of the last 25 or so lines. Stata records every line you type in a wraparound buffer. A wraparound buffer is a buffer of finite length in which the most recent thing you type replaces the oldest thing stored in the buffer. Stata's buffer is 8,697 characters long for Small Stata, 67,800 characters long for Intercooled Stata, and can range from 67,800 to 1,081,527 for Stata/SE.

One way to retrieve lines is with the PrevLine and NextLine keys. Remember, PrevLine and NextLine are the names we attach to these keys—there are no such keys on your keyboard. You have to look back at the previous section to find out which keys correspond to PrevLine and NextLine on your computer. To save you the effort this time, PrevLine probably corresponds to *PgUp* and NextLine probably corresponds to *PgDn*.

Suppose you wanted to reissue the third line back. You could press PrevLine three times and then press *Return*. If you made a mistake and pressed PrevLine four times, you could press NextLine to go forward in the buffer. You do not have to count lines because, each time you press PrevLine or NextLine, the current line is displayed on your monitor. Simply tap the key until you find the line you want.

Another method for reviewing previous lines, `#review`, is convenient when you want to see the lines in context.

▷ **Example**

Typing `#review` by itself causes Stata to list the last five commands you typed. (You need not type out `#review`—pressing *F2* has the same effect.) For instance,

```
. #review
5 list make mpg weight if abs(res)>6
4 list make mpg weight if abs(res)>5
3 tabulate foreign if abs(res)>5
2 regress mpg weight weight2
1 test weight2=0
. _
```

We can see from the listing that the last command typed by the user was `test weight2=0`.

◁

▷ **Example**

Perhaps the command you are looking for is not among the last five commands you typed. You can tell Stata to go back any number of lines. For instance, typing `#review 15` tells Stata to show you the last 15 lines you typed:

```
. #review 15
15 replace resmpg=mpg-pred
14 summarize resmpg, detail
13 drop predmpg
12 describe
11 sort foreign
10 by foreign: summarize mpg weight
9 * lines that start with a * are comments.
8 * they go into the review buffer too.
7 summarize resmpg, detail
6 list make mpg weight
5 list make mpg weight if abs(res)>6
4 list make mpg weight if abs(res)>5
3 tabulate foreign if abs(res)>5
2 regress mpg weight weight2
1 test weight2=0
. _
```

If you wanted to resubmit the tenth previous line, you could type 10 and press Seek, or you could press PrevLine ten times. No matter which of the above methods you prefer for retrieving lines, you may edit previous lines using the editing keys.

◁

13.6   Tab expansion of variable names

Another way to quickly enter a variable name is to take advantage of Stata's variable name completion feature. Simply type the first few letters of the variable name in the Command window and press the *Tab* key. Stata will automatically type the rest of the variable name for you. If more than one variable name matches the letters you have typed, Stata will complete as much as it can and beep at you to let you know that you have typed a nonunique variable abbreviation.

Elements of Stata

Chapters

14 Language syntax 89

15 Data 113

16 Functions and expressions 137

17 Matrix expressions 153

18 Printing and preserving output 165

19 Do-files 171

20 Ado-files 183

21 Programming Stata 189

22 Immediate commands 245

23 Estimation and post-estimation commands 249

14 Language syntax

Contents

14.1   Overview
    14.1.1   varlist
    14.1.2   by varlist:
    14.1.3   if exp
    14.1.4   in range
    14.1.5   =exp
    14.1.6   weight
    14.1.7   options
    14.1.8   numlist
14.2   Abbreviation rules
    14.2.1   Command abbreviation
    14.2.2   Option abbreviation
    14.2.3   Variable-name abbreviation
14.3   Naming conventions
14.4   varlists
    14.4.1   Lists of existing variables
    14.4.2   Lists of new variables
    14.4.3   Time-series varlists
14.5   by varlist: construct
14.6   File-naming conventions
    14.6.1   A special note for Macintosh users
    14.6.2   A special note for Unix users

14.1   Overview

With few exceptions, the basic Stata language syntax is

$$\big[\texttt{by }\textit{varlist}:\big] \; \textit{command} \; \big[\textit{varlist}\big] \; \big[\texttt{=}\textit{exp}\big] \; \big[\texttt{if }\textit{exp}\big] \; \big[\texttt{in }\textit{range}\big] \; \big[\textit{weight}\big] \; \big[\texttt{, }\textit{options}\big]$$

where square brackets denote optional qualifiers. In this diagram, *varlist* denotes a list of variable names, *command* denotes a Stata command, *exp* denotes an algebraic expression, *range* denotes an observation range, *weight* denotes a weighting expression, and *options* denotes a list of options.

14.1.1   varlist

Most commands that take a subsequent *varlist* do not require that one be explicitly typed. If no *varlist* appears, these commands assume a *varlist* of _all, the Stata shorthand for indicating all the variables in the dataset. In commands that alter or destroy data, Stata requires that the *varlist* be specified explicitly. See [U] **14.4 varlists** for a complete description.

▷ Example

The `summarize` command lists the mean, standard deviation, and range of the specified variables. In [R] **summarize**, we see that the syntax diagram for summarize is

<u>su</u>mmarize [*varlist*] [*weight*] [if *exp*] [in *range*] [, [<u>d</u>etail | <u>mean</u>only]

<u>f</u>ormat <u>sep</u>arator(*#*)]

by ... : may be used with `summarize`; see [R] **by**.

Since everything but the word `summarize` is enclosed in square brackets, the simplest form of the command is 'summarize'. Typing `summarize` without arguments is equivalent to typing `summarize _all`; all the variables in the dataset are summarized. Underlining denotes the shortest allowed abbreviation, so we could have typed just `su`; see [U] **14.2 Abbreviation rules**.

The `drop` command eliminates variables or observations from a dataset. The syntax diagram for the version that drops variables is

drop *varlist*

Typing `drop` by itself results in the error message "varlist or in range required". To drop all the variables in the dataset, you must type `drop _all`.

Even before looking at the syntax diagram, we could have predicted that the *varlist* would be required—`drop` is destructive, and hence we are required to spell out our intent. The syntax diagram informs us that the *varlist* is required since *varlist* is not enclosed in square brackets. Since `drop` is not underlined, it cannot be abbreviated.

◁

14.1.2   **by varlist:**

The by *varlist*: prefix causes Stata to repeat a command for each subset of the data for which the values of the variables in the *varlist* are equal. When prefixed with by *varlist*:, the result of the command will be the same as if you had formed separate datasets for each group of observations, saved them, and then given the command on each dataset separately. The data must already be sorted by *varlist*, although by has a `sort` option; see [U] **14.5 by varlist: construct** for more information.

▷ Example

Typing `summarize marriage_rate divorce_rate` produces a table of the mean, standard deviation, and range of `marriage_rate` and `divorce_rate`, using all the observations in the data:

```
. use http://www.stata-press.com/data/r8/census12
(1980 Census data by state)

. summarize marriage_rate divorce_rate
```

Variable	Obs	Mean	Std. Dev.	Min	Max
marriage_r~e	50	.0133221	.0188122	.0074654	.1428282
divorce_rate	50	.0056641	.0022473	.0029436	.0172918

Typing by `region: summarize marriage_rate divorce_rate` produces one table for each region of the country:

```
. sort region
. by region: summarize marriage_rate divorce_rate
```

```
-> region = N Cntrl
    Variable |       Obs        Mean    Std. Dev.        Min         Max
-------------+-----------------------------------------------------------
 marriage_r~e |        12    .0099121    .0011326    .0087363    .0127394
 divorce_rate |        12    .0046974    .0011315    .0032817    .0072868
```

```
-> region = NE
    Variable |       Obs        Mean    Std. Dev.        Min         Max
-------------+-----------------------------------------------------------
 marriage_r~e |         9    .0087811     .001191    .0075757    .0107055
 divorce_rate |         9     .004207    .0010264    .0029436    .0057071
```

```
-> region = South
    Variable |       Obs        Mean    Std. Dev.        Min         Max
-------------+-----------------------------------------------------------
 marriage_r~e |        16    .0114654    .0025721    .0074654    .0172704
 divorce_rate |        16     .005633    .0013355    .0038917    .0080078
```

```
> region = West
    Variable |       Obs        Mean    Std. Dev.        Min         Max
-------------+-----------------------------------------------------------
 marriage_r~e |        13    .0218987    .0363775    .0087365    .1428282
 divorce_rate |        13    .0076037    .0031486    .0046004    .0172918
```

As mentioned, the dataset must be sorted on the by variable(s):

```
. use http://www.stata-press.com/data/r8/census12
(1980 Census data by state)
. by region: summarize marriage_rate divorce_rate
not sorted
r(5);
. sort region
. by region: summarize marriage_rate divorce_rate
( output appears)
```

Alternatively, we could have asked that by sort the data:

```
. by region, sort: summarize marriage_rate divorce_rate
( output appears)
```

by *varlist*: can be used with most Stata commands; you can tell which ones by looking at their syntax diagrams. For instance, we could obtain the correlations, by region, between marriage_rate and divorce_rate by typing by region: correlate marriage_rate divorce_rate.   ◁

❑ Technical Note

The *varlist* in by *varlist*: may contain up to 32,766 variables with Stata/SE or 2,047 variables with Intercooled Stata; the maximum allowed in the dataset. For instance, if you had data on automobiles and wished to obtain means according to market category (market) broken down by manufacturer (origin), you could type by market origin: summarize. That *varlist* contains two variables: market and origin. If the data were not already sorted on market and origin, you would first type sort market origin.   ❑

❑ Technical Note

The *varlist* in by *varlist*: may contain string variables, numeric variables, or both. In the example above, `region` is a string variable, in particular, a `str7`. The example would have worked, however, if `region` were a numeric variable with values 1, 2, 3, and 4, or even 12.2, 16.78, 32.417, and 152.13.

❑

14.1.3   if exp

The `if` *exp* qualifier restricts the scope of a command to those observations for which the value of the expression is *true* (which is equivalent to the expression being nonzero; see [U] **16 Functions and expressions**).

▷ Example

Typing `summarize marriage_rate divorce_rate if region=="West"` produces a table for the western region of the country:

```
. summarize marriage_rate divorce_rate if region == "West"
```

Variable	Obs	Mean	Std. Dev.	Min	Max
marriage_r~e	13	.0218987	.0363775	.0087365	.1428282
divorce_rate	13	.0076037	.0031486	.0046004	.0172918

The double equal sign in `region=="West"` is not an error. Stata uses *double* equal signs to denote equality testing and a *single* equal sign to denote assignment; see [U] **16 Functions and expressions**.

A command may have at most one `if` qualifier. If you want the summary for the West restricted to observations with values of `marriage_rate` in excess of 0.015, do *not* type `summarize marriage_rate divorce_rate if region=="West" if marriage_rate>.015`. Type

```
. summarize marriage_rate divorce_rate if region == "West" & marriage_rate >.015
```

Variable	Obs	Mean	Std. Dev.	Min	Max
marriage_r~e	1	.1428282	.	.1428282	.1428282
divorce_rate	1	.0172918	.	.0172918	.0172918

You may not use the word 'and' in place of the symbol '&' to join conditions. To select observations that meet one condition *or* another, use the '|' symbol. For instance, `summarize marriage_rate divorce_rate if region=="West" | marriage_rate>.015` summarizes all observations for which `region` is West *or* `marriage_rate` is greater than 0.015.

◁

▷ Example

`if` may be combined with `by`. Typing `by region: summarize marriage_rate divorce_rate if marriage_rate>.015` produces a set of tables, one for each region, reflecting summary statistics on `marriage_rate` and `divorce_rate` among observations for which `marriage_rate` exceeds 0.015:

```
. by region: summarize marriage_rate divorce_rate if marriage_rate >.015
```

-> region = N Cntrl

Variable	Obs	Mean	Std. Dev.	Min	Max
marriage_r~e	0				
divorce_rate	0				

-> region = NE

Variable	Obs	Mean	Std. Dev.	Min	Max
marriage_r~e	0				
divorce_rate	0				

-> region = South

Variable	Obs	Mean	Std. Dev.	Min	Max
marriage_r~e	2	.0163219	.0013414	.0153734	.0172704
divorce_rate	2	.0061813	.0025831	.0043548	.0080078

-> region - West

Variable	Obs	Mean	Std. Dev.	Min	Max
marriage_r~e	1	.1428282	.	.1428282	.1428282
divorce_rate	1	.0172918	.	.0172918	.0172918

The results indicate that there are no states in the Northeast and North Central regions for which marriage_rate exceeds 0.015, while there are two such states in the South and one state in the West.

◁

14.1.4   in range

The in *range* qualifier restricts the scope of the command to a specific observation range. A range specification takes the form $\#_1 [/\#_2]$, where $\#_1$ and $\#_2$ are positive or negative. Negative integers are understood to mean "from the end of the data", with -1 referring to the last observation. The implied first observation must be less than or equal to the implied last observation.

The first and last observations in the dataset may be denoted by f and l (letter *el*), respectively. A range specifies absolute observation numbers within a dataset. As a result, the in qualifier may not be used when the command is preceded by the by *varlist*: prefix.

▷ Example

Typing summarize marriage_rate divorce_rate in 5/25 produces a table based on the values of marriage_rate and divorce_rate in observations 5 through 25:

```
. summarize marriage_rate divorce_rate in 5/25
```

Variable	Obs	Mean	Std. Dev.	Min	Max
marriage_r~e	21	.0096001	.0013263	.0075757	.0125884
divorce_rate	21	.004726	.0012025	.0029436	.0072868

This is, admittedly, a rather odd thing to want to do. It would not be odd, however, if we substituted list for summarize. If we wanted to see the states with the 10 lowest values of marriage_rate, we could type sort marriage_rate followed by list marriage_rate in 1/10.

Typing summarize marriage_rate divorce_rate in f/l is equivalent to typing summarize marriage_rate divorce_rate—all observations are summarized.

◁

▷ Example

Typing summarize marriage_rate divorce_rate in 5/25 if region=="South" produces a table based on the values of the two variables in observations 5 through 25 for which the value of region is South:

```
. summarize marriage_rate divorce_rate in 5/25 if region=="South"
```

Variable	Obs	Mean	Std. Dev.	Min	Max
marriage_r~e	4	.0108886	.0015061	.0089201	.0125884
divorce_rate	4	.0054448	.0011166	.0041485	.0068685

The ordering of the in and if qualifiers is not significant. The command could also have been specified as summarize marriage_rate divorce_rate if region=="South" in 5/25.

◁

▷ Example

Negative in ranges can be usefully employed with sort. For instance, we have data on automobiles and wish to list the five with the highest mileage rating:

```
. use http://www.stata-press.com/data/r8/auto
(1978 Automobile Data)
. sort mpg
. list make mpg in -5/l
```

	make	mpg
70.	Toyota Corolla	31
71.	Plym. Champ	34
72.	Subaru	35
73.	Datsun 210	35
74.	VW Diesel	41

◁

14.1.5   =exp

The =*exp* specifies the value to be assigned to a variable and is most often used with generate and replace. See [U] **16 Functions and expressions** for details on expressions and [R] **generate** for details on the generate and replace commands.

▷ Example

Expression	Meaning
generate newvar=oldvar+2	creates a new variable named newvar equal to oldvar+2
replace oldvar=oldvar+2	changes the contents of the existing variable oldvar
egen newvar=rank(oldvar)	creates newvar containing the ranks of oldvar (see [R] **egen**)

◁

14.1.6 weight

weight indicates the weight to be attached to each observation. The syntax of *weight* is

$$[weightword=exp]$$

where you actually type the square brackets, and where *weightword* is one of

weightword	Meaning
weight	default treatment of weights
fweight *or* frequency	frequency weights
pweight	sampling weights
aweight *or* cellsize	analytic weights
iweight	importance weights

The underlining indicates the minimum acceptable abbreviation. Thus, weight may be abbreviated w, or we, etc.

▷ Example

Before explaining what the different types of weights mean, let's obtain the population-weighted mean of a variable called median_age from data containing observations on 50 states of the U.S. The dataset also contains a variable named pop, which is the total population of each state.

```
. use http://www.stata-press.com/data/r8/census12
(1980 Census data by state)
. summarize median_age [weight=pop]
(analytic weights assumed)
```

Variable	Obs	Weight	Mean	Std. Dev.	Min	Max
median_age	50	225907472	30.11047	1.66933	24.2	34.7

In addition to telling us that our dataset contains 50 observations, we are informed that the sum of the weight is 225,907,472, which was the number of people living in the U.S. as of the 1980 census. The weighted mean is 30.11. We were also informed that Stata assumed we wanted "analytic" weights.

◁

Stata understands four kinds of weights:

1. fweights, or frequency weights, indicate duplicated observations. fweights are always integers. If the fweight associated with an observation is 5, that means there are really 5 such observations, each identical.

2. pweights, or sampling weights, denote the inverse of the probability that this observation is included in the sample due to the sampling design. A pweight of 100, for instance, indicates that this observation is representative of 100 subjects in the underlying population. The scale of these weights does not matter in terms of estimated parameters and standard errors, except when estimating totals and computing finite-population corrections with the svy commands; see [U] **30 Overview of survey estimation**.

3. aweights, or analytic weights, are inversely proportional to the variance of an observation; i.e., the variance of the jth observation is assumed to be σ^2/w_j, where w_j are the weights. Typically, the observations represent averages, and the weights are the number of elements that gave rise to the average. For most Stata commands, the recorded scale of aweights is irrelevant; Stata internally rescales them to sum to N, the number of observations in your data, when it uses them.

4. iweights, or importance weights, indicate the relative "importance" of the observation. They have no formal statistical definition; this is a catchall category. Any command that supports iweights will define how they are treated. In most cases, they are intended for use by programmers who want to produce a certain computation.

See [U] **23.16 Weighted estimation** for a thorough discussion of weights and their meaning.

weight is each command's idea of what the "natural" weights are and is one of fweight, pweight, aweight, or iweight. When you specify the vague weight, the command informs you which kind it assumes. Not every command supports every kind of weight. A note below the syntax diagram for a command will tell you which weights the command supports.

❏ Technical Note

When you do not specify a weight, the result is equivalent to specifying [fweight=1]. The emphasis is on equivalent, since Stata may go to more work when you specify a weight.

❏

14.1.7   options

Many commands take command-specific options. These are described along with each command in the *Reference* manuals. Options are indicated by typing a comma at the end of the command, followed by a list of options.

▷ Example

Typing summarize marriage_rate produces a table of the mean, standard deviation, minimum, and maximum of the variable marriage_rate:

```
. summarize marriage_rate
```

Variable	Obs	Mean	Std. Dev.	Min	Max
marriage_r~e	50	.0133221	.0188122	.0074654	.1428282

The syntax diagram for summarize is

> summarize [*varlist*] [*weight*] [if *exp*] [in *range*] [, [detail | meanonly]
>
> format separator(*#*)]

by ... : may be used with summarize; see [R] **by**.

Thus, the options allowed by summarize are detail or meanonly, format, and separator(). The shortest allowed abbreviations for these options are d for detail, mean for meanonly, f for format, and sep for separator(); see [U] **14.2 Abbreviation rules**.

Typing summarize marriage_rate, detail produces a table that also includes selected percentiles, the four largest and four smallest values, the skewness, and the kurtosis.

```
. summarize marriage_rate, detail
                        marriage_rate
```

	Percentiles	Smallest		
1%	.0074654	.0074654		
5%	.0078956	.0075757		
10%	.0080043	.0078956	Obs	50
25%	.0089399	.0079079	Sum of wgt.	50
50%	.0105669		Mean	.0133221
		Largest	Std. Dev.	.0188122
75%	.0122899	.0146266		
90%	.0137832	.0153734	Variance	.0003539
95%	.0153734	.0172704	Skewness	6.718494
99%	.1428282	.1428282	Kurtosis	46.77306

◁

❑ Technical Note

Once you have typed the *varlist* for the command, options may be placed anywhere in the command. You can type summarize marriage_rate divorce_rate if region=="West", detail, or you can type summarize marriage_rate divorce_rate, detail, if region=="West". Note the use of a second comma to indicate return to the command line as opposed to the option list. Leaving out the comma after the word detail would cause an error, since Stata would attempt to interpret the phrase if region=="West" as an option rather than as part of the command.

You may not type an option in the middle of a *varlist*. Typing summarize marriage_rate, detail, divorce_rate will result in an error.

Options need not be contiguously specified. You may type summarize marriage_rate divorce_rate, detail, if region=="South", noformat. Both detail and noformat are options.

❑

❑ Technical Note

Most options are toggles—they indicate that something either is or is not to be done. Sometimes it is difficult to remember which is the default. The following rule applies to all options: If *option* is an option, then no*option* is an option as well, and vice versa. Thus, if we could not remember whether detail or nodetail were the default for summarize but we knew that we did not want the detail, we could type summarize, nodetail. Typing the nodetail option is unnecessary, but Stata will not complain.

Some options take *arguments*. The Stata kdensity command has a n(*#*) option that indicates the number of points at which the density estimate is to be evaluated. When an option takes an argument, it is enclosed in parentheses.

Some options take more than one argument. In such cases, arguments should be separated from one another by commas. For instance, you might see in a syntax diagram

saving(*filename*[, replace])

In this case, replace is the (optional) second argument. *Lists*, such as lists of variables (varlists) and lists of numbers (numlists), are considered to be one argument. If a syntax diagram reported

powers(*numlist*)

the list of numbers would be one argument, so the elements would not be separated by commas. You would type, for instance, powers(1 2 3 4). In fact, Stata will tolerate commas in this case, so you could type powers(1,2,3,4).

Some options take string arguments. regress has an eform() option that works this way—for instance, eform("Exp Beta"). To play it safe, you should type the quotes surrounding the string, although it is not required. If you do not type the quotes, any sequence of two or more consecutive blanks will be interpreted as a single blank. Thus, eform(Exp beta) would be interpreted the same as eform(Exp beta). ❑

14.1.8 numlist

A *numlist* is a list of numbers. Stata allows certain shorthands to indicate ranges:

numlist	meaning
2	just one number
1 2 3	three numbers
3 2 1	three numbers in reversed order
.5 1 1.5	three different numbers
1 3 -2.17 5.12	four numbers in jumbled order
1/3	three numbers, 1, 2, 3
3/1	the same three numbers in reverse order
5/8	four numbers, 5, 6, 7, 8
-8/-5	four numbers -8, -7, -6, -5
-5/-8	four numbers -5, -6, -7, -8
-1/2	four numbers -1, 0, 1, 2
1 2 to 4	four numbers, 1, 2, 3, 4
4 3 to 1	four numbers, 4, 3, 2, 1
10 15 to 30	five numbers 10, 15, 20, 25, 30
1 2:4	same as 1 2 to 4
4 3:1	same as 4 3 to 1
10 15:30	same as 10 15 to 30
1(1)3	three numbers, 1, 2, 3
1(2)9	five numbers, 1, 3, 5, 7, 9
1(2)10	the same five numbers, 1, 3, 5, 7, 9
9(-2)1	five numbers 9, 7, 5, 3, and 1
-1(.5)2.5	the numbers -1, $-.5$, 0, .5, 1, 1.5, 2, 2.5
1[1]3	same as 1(1)3
1[2]9	same as 1(2)9
1[2]10	same as 1(2)10
9[-2]1	same as 9(-2)1
-1[.5]2.5	same as $-1(.5)2.5$
1 2 3/5 8(2)12	eight numbers 1, 2, 3, 4, 5, 8, 10, 12
1,2,3/5,8(2)12	the same eight numbers
1 2 3/5 8 10 to 12	the same eight numbers
1,2,3/5,8,10 to 12	the same eight numbers
1 2 3/5 8 10:12	the same eight numbers

poisson's constraints() option has syntax constraints(*numlist*). Thus, you could type constraints(2 4 to 8), constraints(2(2)8), etc.

14.2 Abbreviation rules

Stata allows abbreviations. In this manual, we usually avoid abbreviating commands, variable names, and options to ensure readability:

    . summarize myvar, detail

Experienced Stata users, on the other hand, tend to abbreviate:

    . sum myv, d

As a general rule, command, option, and variable names may be abbreviated to the shortest string of characters that uniquely identifies them.

This rule is violated if the command or option does something that cannot easily be undone; in that case, the command must be spelled out in its entirety.

In addition, a few common commands and options are allowed to have even shorter abbreviations than the general rule would allow.

The general rule is applied, without exception, to variable names.

14.2.1 Command abbreviation

The shortest allowed abbreviation for a command or option can be determined by looking at the command's syntax diagram. This minimal abbreviation is shown by underlining:

<u>reg</u>ress

<u>ren</u>ame

replace

<u>rot</u>ate

<u>ru</u>n

Lack of underlining means that no abbreviation is allowed. Thus, replace may not be abbreviated; the underlying reason being that replace changes the data.

regress can be abbreviated reg, regr, regre, regres, or can be spelled out in its entirety.

As mentioned above, sometimes very short abbreviations are also allowed. Commands that begin with the letter *d* include decode, describe, destring, dir, discard, display, do, and drop. This suggests that the shortest allowable abbreviation for describe is desc. Since describe is such a commonly used command, you may abbreviate it with the single letter d. You may also abbreviate the list command with the single letter l.

The other exception to the general abbreviation rule concerns commands that alter or destroy data; such commands must be spelled out completely. Two commands that begin with the letter *d*, discard and drop, are destructive in the sense that once you give one of these commands, there is no way you can undo the result. Therefore, both must be spelled out.

The final exceptions to the general rule are commands implemented as ado-files. Such commands may not be abbreviated. Ado-file commands are external, and their names correspond to the names of disk files.

14.2.2   Option abbreviation

Option abbreviation follows the same logic as command abbreviation: you determine the minimum acceptable abbreviation by examining the command's syntax diagram. The syntax diagram for `summarize` reads in part

> <u>su</u>mmarize ... , <u>d</u>etail <u>f</u>ormat

by ... : may be used with `summarize`; see [R] **by**.

Option `detail` may be abbreviated d, de, det, ..., detail. Similarly, option `format` may be abbreviated f, fo, ..., format.

Options `clear` and `replace` occur with many commands. The `clear` option indicates that even though completion of this command will result in the loss of all data in memory, and even though the data in memory have changed since the data were last saved on disk, you are aware of the situation and it is okay to continue. `clear` must be spelled out, as in `use newdata, clear`.

The `replace` option indicates that it's okay to save over an existing dataset. If you type `save mydata` and the file `mydata.dta` already exists, you will receive the message "file mydata.dta already exists" and Stata will refuse to overwrite it. To allow Stata to overwrite the dataset, you type `save mydata, replace`. `replace` may not be abbreviated.

❑ Technical Note

`replace` is a stronger modifier than `clear` and is one you should think about before using. With a mistaken `clear`, you can lose hours of work, but with a mistaken `replace`, you can lose days of work.

❑

14.2.3   Variable-name abbreviation

1. Variable names may be abbreviated to the shortest string of characters that uniquely identifies them given the data currently loaded in memory.

   If your dataset contained four variables, `state`, `mrgrate`, `dvcrate`, and `dthrate`, you could refer to the variable `dvcrate` as `dvcrat`, `dvcra`, `dvcr`, `dvc`, or `dv`. You might type `list dv` to list the data on `dvcrate`. You could not refer to the variable `dvcrate` as `d`, however, since that abbreviation does not distinguish `dvcrate` from `dthrate`. If you were to type `list d`, Stata would respond with the message "ambiguous abbreviation". (If you wanted to refer to *all* variables that started with the letter *d*, you could type `list d*`; see [U] **14.4 varlists**.)

2. The character ~ may be used to mean "zero or more characters go here". For instance, `r~8` might refer to the variable `rep78`, or `rep1978`, or `repair1978`, or just `r8`. (The ~ character is similar to the * character in [U] **14.4 varlists**, except that it adds the restriction "and only one variable matches this specification".)

   In (1), we said you could abbreviate variables. You could type `dvcr` to refer to `dvcrate`, but, if there were more than one variable that started with the letters `dvcr`, you would receive an error. Note that typing `dvcr` is the same as typing `dvcr~`.

14.3   Naming conventions

A name is a sequence of one to thirty-two letters (A–Z and a–z), digits (0–9), and underscores
(_).

Programmers: local macro names can have no more than 31 characters in the name; see [U] **21.3.1 Local macros**.

Stata reserves the following names:

_all	double	long	_rc
_b	float	_n	_se
byte	if	_N	_skip
_coef	in	_pi	using
_cons	int	_pred	with

You may not use these reserved names for your variables.

The first character of a name must be a letter or an underscore. We recommend, however, that
you do not begin your variable names with an underscore. All Stata built-in variables begin with an
underscore, and we reserve the right to incorporate new _variables freely.

Stata respects case; that is, myvar, Myvar, and MYVAR are three distinct names.

All objects in Stata—not just variables—follow this naming convention.

14.4   varlists

A *varlist* is a list of variable names. The variable names in a *varlist* refer either exclusively to new
(not yet created) variables or exclusively to existing variables. A *newvarlist* always refers exclusively
to new (not yet created) variables.

14.4.1   Lists of existing variables

In lists of existing variable names, variable names may be repeated.

▷ Example

If you type `list state mrgrate dvcrate state`, the variable `state` will be listed twice, once
in the leftmost column and again in the rightmost column of the list.

◁

Existing variable names may be abbreviated as described in [U] **14.2 Abbreviation rules**. You may
also use '*' to indicate "zero or more characters go here". For instance, if you suffix * to a partial
variable name, you are referring to all variable names that start with that letter combination. If you
prefix * to a letter combination, you are referring to all variables that end in that letter combination.
If you put * in the middle, you are referring to all variables that begin and end with the specified
letters. You may put more than one * in an abbreviation.

▷ Example

If the variables `poplt5`, `pop5to17`, and `pop18p` are in your dataset, you may type `pop*` as a
shorthand way to refer to all three variables. For instance, `list state pop*` lists the variables `state`,
`poplt5`, `pop5to17`, and `pop18p`.

If you had a dataset with variables inc1990, inc1991, ..., inc1999 along with variables incfarm1990, ..., incfarm1999; pop1990, ..., pop1999; and ms1990, ..., ms1999; then *1995 would be a shorthand way of referring to inc1995, incfarm1995, pop1995, and ms1995. You could type, for instance, list *1995.

In that same dataset, typing list i*95 would be a shorthand way of listing inc1995 and incfarm1995.

Typing list i*f*95 would be a shorthand way of listing to incfarm1995.

◁

~ is an alternative to *, and really, it means the same thing. The difference is that ~ adds the assertion that only one variable will match the specified pattern, and, if you are wrong about that, you want Stata to complain rather than substituting all the variables that match the specification.

▷ Example

In the previous example, we could type list i~f~95 to list incfarm1995. If, however, our dataset also included variable infant1995, then list i*f*95 would list both variables and list i~f~95 would complain that i~f~95 is an ambiguous abbreviation.

◁

You may use ? to specify that one character goes here. Remember, * means zero or more characters go here, so ?* can be used to mean one or more characters goes here, ??* can be used to mean two or more characters go here, and so on.

▷ Example

In a dataset containing variables rep1, rep2, ..., rep78, rep? would refer to rep1, rep2, ..., rep9, and rep?? would refer to rep10, rep11, ..., rep78.

◁

You may place a dash (-) between two variable names to specify all the variables stored between the two listed variables, inclusive. You can determine storage order using describe; it lists variables in the order in which they are stored.

▷ Example

If your dataset contains the variables state, mrgrate, dvcrate, and dthrate, in that order, typing list state-dvcrate is equivalent to typing list state mrgrate dvcrate. In both cases, three variables are listed.

◁

14.4.2   Lists of new variables

In lists of *new variables*, no variable names may be repeated or abbreviated.

You may specify a dash (-) between two variable names that have the same letter prefix and that end in numbers. This form of the dash notation indicates a range of variable names in ascending numerical order.

▷ Example

Typing `input v1-v4` is equivalent to typing `input v1 v2 v3 v4`. Typing `infile state v1-v3 ssn using rawdata` is equivalent to typing `infile state v1 v2 v3 ssn using rawdata`.

◁

You may specify the storage type before the variable name to force a storage type other than the default. The numeric storage types are `byte`, `int`, `long`, `float` (the default), and `double`. The string storage types are `str#`, where # is replaced with an integer between 1 and 80 (244 for Stata/SE), inclusive, representing the maximum length of the string. See [U] **15 Data**.

For instance, the list `var1 str8 var2 var3` specifies that `var1` and `var3` are to be given the default storage type, whereas `var2` is to be stored as a `str8`—a string whose maximum length is eight characters.

The list `var1 int var2 var3` specifies that `var2` is to be stored as an `int`. You may use parentheses to bind a list of variable names. The list `var1 int(var2 var3)` specifies that both `var2` and `var3` are to be stored as `int`s. Similarly, the list `var1 str20(var2 var3)` specifies that both `var2` and `var3` are to be stored as `str20`s. The different storage types are listed in [U] **15.2.2 Numeric storage types** and [U] **15.4.4 String storage types**.

▷ Example

Typing `infile str2 state str10 region v1-v5 using mydata` reads the `state` and `region` strings from the file `mydata.raw` and stores them as `str2` and `str10`, respectively, along with the variables `v1` through `v5`, which are stored with the default storage type `float` (unless you have specified a different default with the `set type` command).

Typing `infile str10(state region) v1-v5 using mydata` would achieve almost the same result, except that the `state` and `region` values recorded in the data would both be assigned to `str10` variables. (You could then use the `compress` command to shorten the strings. See [R] **compress**; it is well worth reading.)

◁

❑ Technical Note

You may append a colon and a *value label name* to numeric variables. (See [U] **15.6 Dataset, variable, and value labels** for a description of value labels.) For instance, `var1 var2:myfmt` specifies that the variable `var2` is to be associated with the value label stored under the name `myfmt`. This has the same effect as typing the list `var1 var2` and then subsequently giving the command `label values var2 myfmt`.

The advantage of specifying the value label association with the colon notation is that value labels can then be assigned by the current command; see [R] **input** and [R] **infile (free format)**.

❑

▷ Example

Typing `infile int(state:stfmt region:regfmt) v1-v5 using mydata, automatic` reads the state and region data from the file `mydata.raw` and stores them as `int`s, along with the variables `v1` through `v5`, which are stored with the default storage type.

In our previous example, both state and region were strings, so how can strings be stored in a numeric variable? See [U] **15.6 Dataset, variable, and value labels** for the complete answer. The colon notation specifies the name of the value label, and the automatic option tells Stata to assign unique numeric codes to all character strings. The numeric code for state, which Stata will make up on the fly, will be stored in the state variable. The mapping from numeric codes to words will be stored in the value label named stfmt. Similarly, region will be assigned numeric codes, which are stored in region, and the mapping will be stored in regfmt.

If you were to list the data, the state and region variables would look like strings. state, for instance, would appear to contain things like AL, CA, and WA, but actually it contains only numbers like 1, 2, 3, and 4.

◁

14.4.3 Time-series varlists

Time-series varlists are a variation on varlists of existing variables. When a command allows a time-series varlist, you may include time-series operators. For instance, L.gnp refers to the lagged value of variable gnp. The time-series operators are

operator	meaning
L.	lag x_{t-1}
L2.	2-period lag x_{t-2}
...	
F.	lead x_{t+1}
F2.	2-period lead x_{t+2}
...	
D.	difference $x_t - x_{t-1}$
D2.	difference of difference $x_t - x_{t-1} - (x_{t-1} - x_{t-2}) = x_t - 2x_{t-1} + x_{t-2}$
...	
S.	"seasonal" difference $x_t - x_{t-1}$
S2.	lag-2 (seasonal) difference $x_t - x_{t-2}$
...	

Time-series operators may be repeated and combined. L3.gnp refers to the third lag of variable gnp. So do LLL.gnp, LL2.gnp, and L2L.gnp. LF.gnp is the same as gnp. DS12.gnp refers to the one-period difference of the 12-period difference. LDS12.gnp refers to the same concept, lagged once.

Note that D1. = S1. but D2. ≠ S2., D3. ≠ S3., and so on. D2. refers to the difference of the difference. S2. refers to the two-period difference. If you wanted the difference of the difference of the 12-period difference of gnp, you would write D2S12.gnp.

Operators may be typed in uppercase or lowercase. Most users would type d2s12.gnp instead of D2S12.gnp.

You may type operators however you wish; Stata internally converts operators to their canonical form. If you typed 1d2l s12d.gnp, Stata would present the operated variable as L2D3S12.gnp.

In addition to *operator#*, Stata understands *operator*(*numlist*) to mean a set of operated variables. For instance, typing L(1/3).gnp in a varlist is the same as typing 'L.gnp L2.gnp L3.gnp'. The operators can also be applied to a list of variables by enclosing the variables in parentheses; e.g.,

```
. list year L(1/3).(gnp cpi)
```

	year	L.gnp	L2.gnp	L3.gnp	L.cpi	L2.cpi	L3.cpi
1.	1989	.	.	.	.	.	.
2.	1990	5452.8	.	.	100	.	.
3.	1991	5764.9	5452.8	.	105	100	.
4.	1992	5932.4	5764.9	5452.8	108	105	100
			(output omitted)				
8.	1996	7330.1	6892.2	6519.1	122	119	112

In *operator#*, making # zero returns the variable itself. L0.gnp is gnp. Thus, the above listing could have been produced by typing list year l(0/3).gnp.

The parentheses notation may be used with any operator. Typing D(1/3).gnp would return the first through third differences.

The parentheses notation may be used in operator lists with multiple operators such as L(0/3)D2S12.gnp

Operator lists may include up to one set of parentheses, and the parentheses may enclose a *numlist*; see [U] **14.1.8 numlist**.

Before you can use time-series operators in varlists, you must set the time variable using the tsset command:

```
. list l.gnp
time variable not set
r(111);
. tsset time
 (output omitted)
. list l.gnp
 (output omitted)
```

See [TS] **tsset**. The time variable must take on integer values. In addition, the data must be sorted on the time variable. tsset handles this, but, later, you might encounter

```
. list l.mpg
not sorted
r(5);
```

In that case, type sort time or type tsset to re-establish the order.

The time-series operators respect the time variable. L2.gnp refers to gnp_{t-2}, regardless of missing observations in the dataset. In the following dataset, the observation for 1992 is missing:

```
. list year gnp l2.gnp, separator(0)
```

	year	gnp	L2.gnp
1.	1989	5,452.8	.
2.	1990	5,764.9	.
3.	1991	5,932.4	5,452.8
4.	1993	6,560.0	5,932.4	← note, filled in correctly
5.	1994	6,922.4	.
6.	1995	7,237.5	6,560.0

Operated variables may be used in expressions:

```
. generate gnplag2 = l2.gnp
(3 missing values generated)
```

Stata also understands panel (cross-sectional time-series) data as well as simple time-series data. If you have cross sections of time series, you indicate this when you `tsset` the data:

```
. tsset country year
```

See [TS] **tsset** and [U] **27.3 Time-series dates**.

14.5   by varlist: construct

by *varlist*: *command*

The `by` prefix causes *command* to be repeated for each unique set of values of the variables in the *varlist*. *varlist* may contain numeric, string, or a mixture of numeric and string variables. (*varlist* may not contain time-series operators.)

`by` is an optional prefix to perform a Stata command separately for each group of observations where the values of the variables in the *varlist* are the same.

During each iteration, the values of the system variables $\_n$ and $\_N$ are set in relation to the first observation in the by-group; see [U] **16.7 Explicit subscripting**. The in *range* qualifier cannot be used in conjunction with by *varlist*: because ranges specify absolute rather than relative observation numbers.

❏ Technical Note

The inability to combine `in` and `by` is not really a constraint, since `if` provides all the functionality of `in` and quite a bit more. If you wanted to perform *command* for the first three observations in each of the by-groups, you could type

```
. by varlist: command if _n<=3
```
❏

The results of *command* will be the same as if you had formed separate datasets for each group of observations, `saved` them, `used` each separately, and issued *command*.

▷ Example

We provide some examples using by in [U] **14.1.2 by varlist:** above. We demonstrate the effect of by on $\_n$, $\_N$, and explicit subscripting in [U] **16.7 Explicit subscripting**.

by requires that the data first be sorted. For instance, if we had data on the average January and July temperatures in degrees Fahrenheit for 420 cities located in the Northeast and West and wanted to obtain the averages, by `region`, across those cities, we might type

```
. use http://www.stata-press.com/data/r8/citytemp3
(City Temperature Data)
. by region: summarize tempjan tempjuly
not sorted
r(5);
```

Stata refused to honor our request since the data are not sorted by `region`. We must either `sort` the data by `region` first (see [R] **sort**) or specify by's `sort` option (which has the same effect):

```
. by region, sort: summarize tempjan tempjuly
```

```
-> region = NE
    Variable |       Obs        Mean   Std. Dev.        Min        Max
-------------+--------------------------------------------------------
     tempjan |       164    27.88537    3.543096       16.6       31.8
    tempjuly |       164       73.35    2.361203       66.5       76.8
```

```
-> region = N Cntrl
    Variable |       Obs        Mean   Std. Dev.        Min        Max
-------------+--------------------------------------------------------
     tempjan |       284    21.69437    5.725392        2.2       32.6
    tempjuly |       284    73.46725    3.103187       64.5       81.4
```

```
-> region = South
    Variable |       Obs        Mean   Std. Dev.        Min        Max
-------------+--------------------------------------------------------
     tempjan |       250     46.1456    10.38646       28.9         68
    tempjuly |       250     80.9896     2.97537         71       87.4
```

```
-> region = West
    Variable |       Obs        Mean   Std. Dev.        Min        Max
-------------+--------------------------------------------------------
     tempjan |       256    46.22539    11.25412         13       72.6
    tempjuly |       256    72.10859    6.483131       58.1       93.6
```

◁

▷ Example

Using the same data as in the example above, we estimate regressions, by `region`, of average January temperature on average July temperature. Both temperatures are specified in degrees Fahrenheit.

```
. by region: regress tempjan tempjuly
```

```
-> region = NE
      Source |       SS       df       MS              Number of obs =     164
-------------+------------------------------           F(  1,   162) =  479.82
       Model | 1529.74026      1  1529.74026           Prob > F      =  0.0000
    Residual | 516.484453    162  3.18817564           R-squared     =  0.7476
-------------+------------------------------           Adj R-squared =  0.7460
       Total | 2046.22471    163  12.5535258           Root MSE      =  1.7855
```

```
     tempjan |      Coef.   Std. Err.      t    P>|t|     [95% Conf. Interval]
-------------+----------------------------------------------------------------
    tempjuly |   1.297424   .0592303    21.90   0.000     1.180461    1.414387
       _cons |  -67.28066   4.346781   -15.48   0.000    -75.86431     -58.697
```

```
-> region = N Cntrl
      Source |       SS       df       MS              Number of obs =     284
-------------+------------------------------           F(  1,   282) =  115.89
       Model | 2701.97917      1  2701.97917           Prob > F      =  0.0000
    Residual | 6574.79175    282  23.3148644           R-squared     =  0.2913
-------------+------------------------------           Adj R-squared =  0.2887
       Total | 9276.77092    283  32.7801093           Root MSE      =  4.8285
```

tempjan	Coef.	Std. Err.	t	P>|t|	[95% Conf. Interval]	
tempjuly	.9957259	.0924944	10.77	0.000	.8136589	1.177793
_cons	-51.45888	6.801344	-7.57	0.000	-64.84673	-38.07103

-> region = South

Source	SS	df	MS		Number of obs =	250
					F(1, 248) =	95.17
Model	7449.51623	1	7449.51623		Prob > F =	0.0000
Residual	19412.2231	248	78.2750933		R-squared =	0.2773
					Adj R-squared =	0.2744
Total	26861.7394	249	107.878471		Root MSE =	8.8473

tempjan	Coef.	Std. Err.	t	P>|t|	[95% Conf. Interval]	
tempjuly	1.83833	.1884392	9.76	0.000	1.467185	2.209475
_cons	-102.74	15.27187	-6.73	0.000	-132.8191	-72.66089

-> region = West

Source	SS	df	MS		Number of obs =	256
					F(1, 254) =	2.84
Model	357.161728	1	357.161728		Prob > F =	0.0932
Residual	31939.9031	254	125.74765		R-squared =	0.0111
					Adj R-squared =	0.0072
Total	32297.0648	255	126.655156		Root MSE =	11.214

tempjan	Coef.	Std. Err.	t	P>|t|	[95% Conf. Interval]	
tempjuly	.1825482	.1083166	1.69	0.093	-.0307648	.3958613
_cons	33.0621	7.84194	4.22	0.000	17.61859	48.5056

The regressions show that a one-degree increase in the average July temperature in the Northeast corresponds to a 1.3 degree increase in the average January temperature. In the West, however, it corresponds to a 0.18 degree increase, which is only marginally significant.

◁

❑ Technical Note

by has a second syntax that is especially useful when you want to play it safe:

by *varlist₁* (*varlist₂*): *command*

What this says is that Stata is to verify that the data are sorted by *varlist₁ varlist₂* and then, assuming that is true, perform *command* by *varlist₁*. For instance,

```
. by subject (time): gen finalval = val[_N]
```

In the above, we want to create new variable finalval that contains, in each observation, the final observed value of val for each subject in the data. The final value will be the last value if, within subject, the data are sorted by time. The above command verifies that the data are sorted by subject and time and then, if they are, performs

```
. by subject: gen finalval = val[_N]
```

If the data are not sorted properly, an error message would instead be issued. Of course, we could have just typed

```
. by subject: gen finalval = val[_N]
```

after verifying for ourselves that the data were sorted properly as long as we were careful to look.

by's second syntax can be used with by's `sort` option, so we can also type

```
. by subject (time), sort: gen finalval = val[_N]
```

which is equivalent to

```
. sort subject time
. by subject: gen finalval = val[_N]
```

❑

14.6   File-naming conventions

Some commands require that you specify a *filename*. Filenames are specified in the way natural for your operating system:

Windows	Unix	Macintosh
mydata	mydata	mydata
mydata.dta	mydata.dta	mydata.dta
b:mydata.dta	~friend/mydata.dta	~friend/mydata.dta
"my data"	"my data"	"my data"
"my data.dta"	"my data.dta"	"my data.dta"
myproj\mydata	myproj/mydata	myproj/mydata
"my project\my data"	"my project/my data"	"my project/my data"
c:\analysis\data\mydata	~/analysis/data/mydata	~/analysis/data/mydata
"c:\my project\my data"	"~/my project/my data"	"~/my project/my data"
..\data\mydata	../data/mydata	../data/mydata
"..\my project\my data"	"../my project/my data"	"../my project/my data"

In most cases (the exceptions being `copy`, `dir`, `ls`, `erase`, `rm`, and `type`), Stata automatically provides a file extension if you do not supply one. For instance, if you type `use mydata`, Stata assumes you mean `use mydata.dta`, since `.dta` is the file extension Stata normally uses for data files.

Stata provides eleven default file extensions that are used by various commands. They are

.ado	automatically loaded do-files
.dct	ASCII data dictionary
.do	do-file
.dta	Stata-format dataset
.gph	graph image
.log	log file in text format
.out	file saved by outsheet
.raw	ASCII-format dataset
.smcl	log file in SMCL format
.sum	checksum files to verify network transfers
.vrf	impulse–response function datasets

You do not have to name your data files with the `.dta` extension—if you type an explicit file extension, it will override the default. For instance, if your dataset was stored as `myfile.dat`, you could type `use myfile.dat`. If your dataset was stored as simply `myfile` with no file extension, type the period at the end of the filename to indicate that you are explicitly specifying the null extension. You type `use myfile.` to use this dataset.

All operating systems allow blanks in filenames and so does Stata. However, if the filename includes a blank, you must enclose the filename in double quotes:

    . save "my data"

would create the file my data.dta. Typing

    . save my data

would be an error.

❏ Technical Note

Stata also makes use of eleven other file extensions. These files are of interest only to advanced programmers or are for Stata's internal use only. They are

.class	class file for object oriented programming; see [P] **class**
.hlp	help files
.dlg	dialog resource file
.idlg	dialog resource include file
.scheme	control file for a graphics scheme
.style	graphics style file
.key	search's keyword database file
.toc	user-site description file
.pkg	user-site package file
.maint	maintenance file (for Stata's internal use only)
.mnu	menu file (for Stata's internal use only)

❏

14.6.1   A special note for Macintosh users

Have you seen the notation myfolder/myfile before? This notation is called a path and describes the location of a file or folder (also called a directory).

You do not have to use this notation if you do not like it. You could instead restrict yourself to using files only in the current folder. If that turns out to be too restricting, Stata for Macintosh provides enough menus and buttons that you can probably get by. You may, however, find the notation convenient. In case you do, here is the rest of the definition.

The character / is called a path delimiter and delimits folder names and file names in a path. If the path starts with no path delimiter, then the path is relative to the current folder.

For example, the path myfolder/myfile refers to the file myfile in the folder myfolder, which is contained in the current folder.

The characters .. refer to the folder containing the current folder. Thus, ../myfile refers to myfile in the folder containing the current folder, and ../nextdoor/myfile refers to myfile in the folder nextdoor in the folder containing the current folder.

If a path starts with a path delimiter, then the path is called an absolute path and describes a fixed location of a file or folder name, regardless of what the current folder is. The leading / in an absolute path refers to the root directory, which is the main hard drive from which the operating system is booted. For example, the path /myfolder/myfile refers to the file myfile in the folder myfolder, which is contained in the main hard drive.

Users familiar with previous versions of Stata for Macintosh will notice that Stata previously used the character : as a path delimiter. In most cases, / and : can be used interchangeably, although we recommend using the former. The case where they differ is in specifying an absolute path. Previous versions of Mac OS had no concept of a root directory, so the hard drive name was required to follow the delimiter in an absolute path. For compatability with older do-files and ado-files, Stata resolves paths that begin with the : path delimiter using the old method of resolving absolute paths, which required the hard drive name.

Thus, the path :Macintosh HD:myfolder:myfile refers to myfile in the folder myfolder, which is contained in the hard drive Macintosh HD. The path to the same file using / as the delimiter would be /myfolder/myfile.

Only the path delimiter used at the beginning of the path is important. Either path delimiter can be used in the rest of the path or relative paths. We do suggest that if you use : at the beginning of the path, you continue using that delimiter in the rest of the path so that it is easier to tell which type of absolute path you are using.

The character ~ refers to the user's home directory. Thus, the path ~/myfolder/myfile refers to myfile in the folder myfolder in the user's home directory.

14.6.2   A special note for Unix users

Stata understands ~ to mean your home directory. Stata understands this even if you do not use csh(1) as your shell.

15 Data

Contents

15.1   Data and datasets
15.2   Numbers
    15.2.1   Missing values
    15.2.2   Numeric storage types
15.3   Dates
15.4   Strings
    15.4.1   Strings containing identifying data
    15.4.2   Strings containing categorical data
    15.4.3   Strings containing numeric data
    15.4.4   String storage types
15.5   Formats: controlling how data are displayed
    15.5.1   Numeric formats
    15.5.2   European numeric formats
    15.5.3   Date formats
    15.5.4   Time-series formats
    15.5.5   String formats
15.6   Dataset, variable, and value labels
    15.6.1   Dataset labels
    15.6.2   Variable labels
    15.6.3   Value labels
15.7   Notes attached to data
15.8   Characteristics

15.1 Data and datasets

Data form a rectangular table of numeric and string values in which each row is an observation on all the variables and each column contains the observations on a single variable. Variables are designated by *variable names*. Observations are numbered sequentially from 1 to _N. The following example of data contains the first five odd and first five even positive integers along with a string variable:

```
       odd  even   name
 1.      1     2   Bill
 2.      3     4   Mary
 3.      5     6    Pat
 4.      7     8  Roger
 5.      9    10   Sean
```

The observations are numbered 1 to 5 and the variables are named odd, even, and name. Observations are referred to by number, variables by name.

A *dataset* is *data* plus labelings, formats, notes, and characteristics.

All aspects of *data* and *datasets* are defined here.

15.2   Numbers

A *number* may contain a sign, an integer part, a decimal point, a fraction part, an e or E, and a signed integer exponent. Numbers may *not* contain commas; for example, the number 1,024 must be typed as 1024 (or 1024. or 1024.0). The following are examples of valid numbers:

```
5
-5
5.2
.5
5.2e+2
5.2e-2
```

❏ Technical Note

As a convenience for Fortran users, you may use d or D, as well as e, to indicate exponential notation. Thus, the number 520 may be written 5.2d+2, 5.2D+2, 5.2d+02, or 5.2D+02.

❏

❏ Technical Note

Stata also allows numbers to be represented in a hexadecimal/binary format, defined as

$$\left[+|-\right]0.0\left[\langle zeros\rangle\right]\{X|x\}-3ff$$

or

$$\left[+|-\right]1.\langle hexdigit\rangle\left[\langle hexdigits\rangle\right]\{X|x\}\{+|-\}\langle hexdigit\rangle\left[\langle hexdigits\rangle\right]$$

The lead digit is always 0 or 1; it is 0 only when the number being expressed is zero. There are a maximum of 13 digits to the right of the hexadecimal point. The power ranges from -3ff to +3ff. The number is expressed in hexadecimal (base 16) digits; the number $aX+b$ means $a \times 2^b$. For instance, 1.0X+3 is 2^3 or 8. 1.8X+3 is 12 because 1.8_{16} is $1 + 8/16 = 1.5$ in decimal and the number is thus $1.5 \times 2^3 = 1.5 \times 8 = 12$.

Stata can also display numbers using this format; see [U] **15.5.1 Numeric formats**. For example,

```
. di 1.81x+2
6.015625

. di %21x 6.015625
+1.8100000000000X+002
```

This hexadecimal format is of special interest to numerical analysts.

❏

15.2.1   Missing values

A number may also take on the special value missing, denoted by a single period (.). You specify a missing value anywhere that you may specify a number. Missing values differ from ordinary numbers in one respect: Any arithmetic operation on a missing value yields a missing value.

In fact, there are 27 missing values in Stata: '.', the one just discussed, and .a, .b, ..., and .z. .a, .b, ..., and .z are known as extended missing values. The missing value '.' is known is as the default or system missing value. In any case, some people use extended missing values to indicate why a certain value is unknown—the question was not asked, the person refused to answer, etc. Other people have no use for extended missing values and just use '.'.

Stata's default or system missing value will be returned when you perform an arithmetic operation on missing values or when the arithmetic operation is not defined, such as division by zero, or the logarithm of a nonpositive number.

```
. display 2/0
.
. list
```

	a
1.	.b
2.	.
3.	.a
4.	3
5.	6

```
. generate x = a + 1
(3 missing values generated)
. list
```

	a	x
1.	.b	.
2.	.	.
3.	.a	.
4.	3	4
5.	6	7

Numeric missing values are represented by "large positive values". The ordering is

$$\text{all numbers} < . < .a < .b < \cdots < .z$$

Thus, the expression

$$\texttt{age} > 60$$

is true if variable age is greater than 60 or is missing. Similarly,

$$\texttt{gender}! = 0$$

is true if gender is not zero or gender is missing.

The way to exclude missing values is to ask whether the value is less than '.', and the way to detect missing values is to ask whether the value is greater than or equal to '.'. For instance,

```
. list if age>60 & age<.
. generate agegt60 = 0 if age<=60
. replace agegt60 = 1 if age>60 & age<.
. generate agegt60 = (age>60) if age<.
```

❏ Technical Note

Old Stata users: Beware! Before Stata 8, Stata only had a single representation for missing values, the period (.). You could test whether an expression or a variable was missing by typing '... if *exp* ==.' or '... if *exp* !=.'.

Now, the statements evaluate to true if *exp* is equal or is not equal to the particular missing value ., excluding the extended missing value cases. Actually, your habits will cause you no harm assuming that your datasets have no extended missing values in them, but it will be safest to start using the proper syntax as soon as possible: an *exp* is not missing if its value is < ., and it is missing if its value is >= ..

In order to ensure that old programs and do-files continue to work properly, when `version` is set less than 8, all missing values are treated as being the same. Thus, . == .a == .b == .z, and so '*exp*==.' and '*exp*!=.' work just as they previously did.

❑

▷ Example

You have data on the income of husbands and wives recorded in the variables `hincome` and `wincome`, respectively. Typing the `list` command, you see that your data contain

```
. list
```

	hincome	wincome
1.	32000	0
2.	35000	34000
3.	47000	.b
4.	.z	50000
5.	.a	.

The values of `wincome` in the third and fifth observations are *missing*, as distinct from the value of `wincome` in the first observation, which is known to be zero.

If you use the `generate` command to create a new variable, income, equal to the sum of `hincome` and `wincome`, three missing values would be produced:

```
.  generate income = hincome + wincome
(3 missing values generated)
. list
```

	hincome	wincome	income
1.	32000	0	32000
2.	35000	34000	69000
3.	47000	.b	.
4.	.z	50000	.
5.	.a	.	.

`generate` produced a warning message that 3 missing values were created, and when we list the data, we see that 47,000 plus *missing* yields *missing*.

◁

❑ Technical Note

Stata stores numeric missing values as the largest 27 numbers allowed by the particular storage type; see [U] **15.2.2 Numeric storage types**. There are two important implications. First, if you `sort` on a variable that has missing values, the missing values will be placed last and the sort order of any missing values will follow the rule regarding the properties of missing values stated above.

```
. sort wincome
. list wincome
```

	wincome
1.	0
2.	34000
3.	50000
4.	.
5.	.b

The second implication concerns relational operators and missing values. Do not forget that a missing value will be larger than any numeric value.

```
. list if wincome > 40000
```

	hincome	wincome	income
3.	.z	50000	.
4.	.a	.	.
5.	47000	.b	.

Observation 4 and observation 5 are listed because '.' and '.b' are both missing, and thus are greater than 40,000. Relational operators are discussed in detail in [U] **16.2.3 Relational operators**.

❑

▷ Example

In producing statistical output, Stata ignores observations with missing values. Continuing with the example above, if we request summary statistics on hincome and wincome using the summarize command, we obtain

```
. summarize hincome wincome
```

Variable	Obs	Mean	Std. Dev.	Min	Max
hincome	3	38000	7937.254	32000	47000
wincome	3	28000	25534.29	0	50000

Some commands will discard the entire observation (known as *casewise deletion*) if one of the variables in the observation is missing. If we use the correlate command to obtain the correlation between hincome and wincome, for instance, we obtain

```
. correlate hincome wincome
(obs=2)
```

	hincome	wincome
hincome	1.0000	
wincome	1.0000	1.0000

Note that the correlation coefficient is calculated over two observations.

◁

15.2.2 Numeric storage types

Numbers can be stored in one of five variable types: byte, int, long, float (the default), or double. bytes are, naturally, stored in 1 byte. ints are stored in 2 bytes, longs and floats in 4 bytes, and doubles in 8 bytes. The table below shows the minimum and maximum values for each storage type.

Storage Type	Minimum	Maximum	Closest to 0 without being 0	bytes
byte	-127	100	± 1	1
int	$-32{,}767$	$32{,}740$	± 1	2
long	$-2{,}147{,}483{,}647$	$2{,}147{,}483{,}620$	± 1	4
float	$-1.70141173319 \times 10^{38}$	$1.70141173319 \times 10^{36}$	$\pm 10^{-36}$	4
double	$-8.9884656743 \times 10^{307}$	$+8.9884656743 \times 10^{308}$	$\pm 10^{-323}$	8

Do not confuse the term *integer*, which is a characteristic of a number, with int, which is a storage type. For instance, the number 5 is an integer no matter how it is stored; thus, if you read that an argument is required to be an integer, that does not mean that it must be stored as an int.

15.3 Dates

You can record dates any way you want, but there is one technique that Stata understands, called an elapsed date. An elapsed date is the number of days from January 1, 1960. In this format,

0	means	January 1, 1960
1		January 2, 1960
31		February 1, 1960
365		December 31, 1960
366		January 1, 1961
12,784		January 1, 1995
-1		December 31, 1959
-2		December 30, 1959
$-12{,}784$		December 31, 1924

Stata understands dates recorded like this from January 1, year 100 (elapsed date $-679{,}350$) to December 31, 9999 (elapsed date 2,936,549), although caution should be exercised in dealing with dates before Friday, October 15, 1582, when the Gregorian calendar went into effect.

Stata provides functions to convert dates into elapsed dates, formats to print elapsed dates in understandable forms, and other functions to manipulate elapsed dates.

In addition to elapsed dates, Stata provides five other date formats:

weekly	monthly	quarterly	half-yearly	yearly
-1 = 1959 week 52	-1 = Dec. 1959	-1 = 1959 quarter 4	-1 = 2nd half 1959	1959 = 1959
0 = 1960 week 1	0 = Jan. 1960	0 = 1960 quarter 1	0 = 1st half 1960	1960 = 1960
1 = 1960 week 2	1 = Feb. 1960	1 = 1960 quarter 2	1 = 2nd half 1960	1961 = 1961
. . .	. . .	. . .	. . .	. . .

For a full discussion of working with date variables, see [U] **27 Commands for dealing with dates**.

15.4 Strings

A *string* is a sequence of printable characters, and is typically enclosed in double quotes. The quotes are not considered a part of the string. They merely delimit the beginning and end of the string. The following are examples of valid strings:

```
"Hello, world"
"String"
"string"
" string"
"string "
""
"x/y+3"
"1.2"
```

All of the strings above are distinct; that is, "String" is different from "string", which is different from " string", which is different from "string ". Also note that "1.2" is a string and not a number because it is enclosed in quotes.

All strings in Stata are of varying length, which means Stata internally records the length of the string and never loses track. There is never a circumstance in which a string cannot be delimited by quotes, but there are rare instances where strings do not have to be delimited by quotes, such as during data input. In those cases, nondelimited strings are stripped of their leading and trailing blanks. Delimited strings are always accepted as is.

The special string "", often called *null string*, is considered by Stata to be a *missing*. No special meaning is given to the string containing a single period, ".".

In addition to double quotes for enclosing strings, Stata also allows compound double quotes: '" and "'. You can type "*string*" or you can type '"*string*"', although users seldom type '"*string*"'. Compound double quotes are of special interest to programmers because they nest and so provide a way for a quoted string to itself contain double quotes (either simple or compound). See [U] **21.3.5 Double quotes**.

15.4.1 Strings containing identifying data

String variables often contain identifying information, such as the patient's name or the name of the city or state. Such strings are typically listed, but are not used directly in statistical analysis, although the data might be sorted on the string or datasets might be merged on the basis of one or more string variables.

15.4.2 Strings containing categorical data

Occasionally, strings contain information that is to be used directly in analysis, such as the patient's sex, which might be coded "male" or "female". Stata shows a decided preference for such information to be numerically encoded and stored in numeric variables. Stata's statistical routines treat string variables as if every observation records a numeric missing value.

All is not lost. Stata provides two commands for converting string variables into numeric codes and back again: encode and decode; see [U] **26.2 Categorical string variables**. Also see [R] **destring** for information on mapping string variables to numeric.

15.4.3 Strings containing numeric data

If a string variable contains the character representation of a number—for instance, `myvar` contains
`"1"`, `"1.2"`, and `"-5.2"`—you can convert it directly into a numeric variable using the `real()`
function or the `destring` command; e.g., `generate newvar=real(myvar)`.

Similarly, if you want to convert a numeric variable to its string representation, you can use the
`string()` function: `generate as_str=string(numvar)`.

See [R] **functions**.

15.4.4 String storage types

Strings are stored in string variables with storage types `str1`, `str2`, ..., `str244`. The storage
type merely sets the maximum length of the string, not its actual length; thus, `"example"` has length
7 whether it is stored as a `str7`, a `str10`, or even a `str244`. On the other hand, an attempt to assign
the string `"example"` to a `str6` would result in `"exampl"`.

The maximum length of a string in Stata/SE is 244. String literals may exceed 244 characters, but
only the first 244 characters are significant.

The maximum length of a string in Intercooled Stata or Small Stata is 80 characters. Thus, string
literals may exceed 80 characters, but only the first 80 characters are significant.

15.5 Formats: controlling how data are displayed

Formats describe how a number or string is to be presented. For instance, how is the number
325.24 to be presented? As `325.2`, or `325.24`, or `325.240`, or `3.2524e+2`, or `3.25e+2`, or how?
The *display format* tells Stata exactly how you want this done. You do not have to specify display
formats, since Stata always makes reasonable assumptions on how to display a variable, but you
always have the option.

15.5.1 Numeric formats

A Stata numeric format is formed by

first type	%	to indicate the start of the format
then optionally type	-	if you want the result left-aligned
then optionally type	0	if you want to retain leading zeros (1)
then type	a number w	stating the width of the result
then type	.	
then type	a number d	stating the number of digits to follow the decimal point
then type		
either	e	for scientific notation; e.g., 1.00e+03
or	f	for fixed format; e.g., 1000.0
or	g	for general format; Stata chooses based on the number being displayed
then optionally type	c	to indicate comma format (not allowed with e)

(1) Specifying 0 to mean "include leading zeros" will be honored only with the f format.

For example,

%9.0g	general format, 9 columns wide	
	sqrt(2) =	1.414214
	1,000 =	1000
	10,000,000 =	1.00e+07
%9.0gc	general format, 9 columns wide, with commas	
	sqrt(2) =	1.414214
	1,000 =	1,000
	10,000,000 =	1.00e+07
%9.2f	fixed format, 9 columns wide, 2 decimal places	
	sqrt(2) =	1.41
	1,000 =	1000.00
	10,000,000 =	10000000.00
%9.2fc	fixed format, 9 columns wide, 2 decimal places, with commas	
	sqrt(2) =	1.41
	1,000 =	1,000.00
	10,000,000 =	10,000,000.00
%9.2e	exponential format, 9 columns wide	
	sqrt(2) =	1.41e+00
	1,000 =	1.00e+03
	10,000,000 =	1.00e+07

Stata has three numeric format types: e, f, and g. The formats are denoted by a leading percent sign (%) followed by the string *w.d*, where *w* and *d* stand for two integers. The first integer, *w*, specifies the width of the format. The second integer *d* specifies the number of digits that are to follow the decimal point. Logic requires that *d* be less that *w*. Finally, a character denoting the format type (e, f, or g) and to that may optionally be appended a c indicating commas are to be included in the result (c is not allowed with e.)

By default, every numeric variable is given a %*w*.0g format, where *w* is large enough to display the largest number of the variable's type. The %*w*.0g format is a set of formatting rules that present the values in as readable a fashion as possible without sacrificing precision. The g format changes the number of decimal places displayed whenever it improves the readability of the current value.

The default formats for each of the numeric variable types are

byte	%8.0g
int	%8.0g
long	%12.0g
float	%9.0g
double	%10.0g

You change the format of a variable using the format *varname* %*fmt* command.

In addition to %*w*.0g, allowed is %*w*.0gc to display numbers with commas. The number one thousand is displayed as 1000 in %9.0g format and as 1,000 in %9.0gc format.

In addition to %*w*.0g and %*w*.0gc, %*w*.dg and %*w*.dgc, *d* > 0, is allowed, such as %9.4g and %9.4gc. The 4 means to display approximately 4 significant digits. For instance, the number 3.14159265 in %9.4g format is displayed as 3.142, 31.4159265 as 31.42, 314.159265 as 314.2, and 3141.59265 as 3142. The format is not exactly a significant-digit format because 31415.9265 is displayed as 31416, not as 3.142e+04.

Under the f format, values are always displayed with the same number of decimal places, even if this results in a loss in the displayed precision. Thus, the f format is similar to the C f format. Stata's f format is also similar to the Fortran F format, but, unlike the Fortran F format, it switches to g whenever a number is too large to be displayed in the specified f format.

In addition to %*w*.df, %*w*.dfc is allowed to display numbers with commas.

The e format is similar to the C e and the Fortran E format. Every value is displayed as a leading digit (with a minus sign, if necessary), followed by a decimal point, the specified number of digits, the letter e, a plus sign or a minus sign, and the power of ten (modified by the preceding sign) that multiplies the displayed value. When the e format is specified, the width must exceed the number of digits that follow the decimal point by at least seven. This space is needed to accommodate the leading sign and digit, the decimal point, the e, and the signed power of ten.

▷ Example

Below we concoct a five-observation dataset with three variables: e_fmt, f_fmt, and g_fmt. All three variables have the same values stored in them; only the display format varies. describe shows the display format to the right of the variable type:

```
. use http://www.stata-press.com/data/r8/format
. describe
Contains data from http://www.stata-press.com/data/r8/format.dta
  obs:            5
  vars:           3                          12 Sep 2002 14:08
  size:          80 (99.9% of memory free)
```

variable name	storage type	display format	value label	variable label
e_fmt	float	%9.2e		
f_fmt	float	%10.2f		
g_fmt	float	%9.0g		

```
Sorted by:
```

The formats for each of these variables were set by typing

```
. format e_fmt %9.2e
. format f_fmt %10.2f
```

It was not necessary to set the format for the g_fmt variable, since Stata automatically assigned it the %9.0g format. Nevertheless, we could have typed format g_fmt %9.0g if we wished. Listing the data results in

```
. list
```

	e_fmt	f_fmt	g_fmt
1.	2.80e+00	2.80	2.801785
2.	3.96e+06	3962322.50	3962323
3.	4.85e+00	4.85	4.852834
4.	-5.60e-06	-0.00	-5.60e-06
5.	6.26e+00	6.26	6.264982

◁

❑ Technical Note

The discussion above is incomplete. There is one additional format available that will be of interest to numerical analysts. The %21x format displays base 10 numbers in a hexadecimal (base 16) format. The number is expressed in hexadecimal (base 16) digits; the number aX+b means $a \times 2^b$. For example,

```
. display %21x 1234.75
+1.34b0000000000X+00a
```

Thus, the base 10 number 1,234.75 has a base 16 representation of 1.34bX+0a, meaning

$$\left(1 + 3 \cdot 16^{-1} + 4 \cdot 16^{-2} + 11 \cdot 16^{-3}\right) \times 2^{10}$$

Remember, the hexadecimal—decimal equivalents are

hexadecimal	decimal
0	0
1	1
2	2
3	3
4	4
5	5
6	6
7	7
8	8
9	9
a	10
b	11
c	12
d	13
e	14
f	15

See [U] **15.2 Numbers**.

❏

15.5.2   European numeric formats

The three numeric formats e, f, and g will use ',' to indicate the decimal symbol if you specify their width and depth as w,d rather than $w.d$. For instance, the format %9,0g will display what Stata would usually display as 1.5 as 1,5.

If you use the European specification with fc or gc, the "comma" will be presented as a period. For instance, %9,0gc would display what Stata would usually display as 1,000.5 as 1.000,5.

If this way of presenting numbers appeals to you, also consider using Stata's set dp comma command. set dp comma tells Stata to interpret nearly all %$w.d$\{f|f|e\} formats as %w,d\{g|f|e\} formats. The fact is that most of Stata is written using period to represent the decimal symbol, and that means that even if you set the appropriate %w,d\{g|f|e\} format for your data, it will effect only displays of the data. For instance, if you type summarize to obtain summary statistics or regress to obtain regression results, the decimal will still be shown as a period.

set dp comma changes that and affects all of Stata. With set dp comma, it does not matter whether your data are formatted %$w.d$\{g|f|e\} or %w,d\{g|f|e\}. All results will be displayed using comma as the decimal character:

```
. use http://www.stata-press.com/data/r8/auto
(1978 Automobile Data)

. set dp comma

. summarize mpg weight foreign
```

Variable	Obs	Mean	Std. Dev.	Min	Max
mpg	74	21,2973	5,785503	12	41
weight	74	3019,459	777,1936	1760	4840
foreign	74	,2972973	,4601885	0	1

```
. regress mpg weight foreign
```

Source	SS	df	MS
Model	1619,2877	2	809,643849
Residual	824,171761	71	11,608053
Total	2443,45946	73	33,4720474

```
Number of obs =      74
F(  2,    71) =   69,75
Prob > F      =  0,0000
R-squared     =  0,6627
Adj R-squared =  0,6532
Root MSE      =  3,4071
```

mpg	Coef.	Std. Err.	t	P>|t|	[95% Conf. Interval]	
weight	-,0065879	,0006371	-10,34	0,000	-,0078583	-,0053175
foreign	-1,650029	1,075994	-1,53	0,130	-3,7955	,4954422
_cons	41,6797	2,165547	19,25	0,000	37,36172	45,99768

You can switch the decimal character back to period by typing `set dp period`.

❑ Technical Note

`set dp comma` makes drastic changes inside Stata, and we mention this because some older, user-written programs may not be up to dealing with those changes. If you are using an older user-written program, you might `set dp comma` and then find that the program does not work and instead presents some sort of syntax error.

If, using any program, you do get an unanticipated error, try setting `dp` back to `period`. See [R] **format** for more information.

Also understand that `set dp comma` affects how Stata outputs numbers, not how it inputs them. You must still use the period to indicate the decimal point on all input. Even with `set dp comma`, you type

```
. replace x=1.5 if x==2
```

❑

15.5.3 Date formats

Date formats are really a numeric format because Stata stores dates as the number of days from 01jan1960. See [U] **27 Commands for dealing with dates**.

`%d` is for displaying elapsed dates. The syntax of the `%d` format is

first type	%	to indicate the start of the format
then optionally type	–	if you want the result left-aligned
then type	d	
then optionally type	*other characters*	to indicate how the date is to be displayed

The %d format may be specified as simply %d, or the %d may be followed by up to 11 characters that specify how the date is to be presented. Allowable characters are

c and C	display the century without/with a leading 0
y and Y	display the two-digit year without/with a leading 0
m and M	display Month, first letter capitalized, in three-letter abbreviation (m) or spelled out (M)
l and L	display month, first letter not capitalized, in three-letter abbreviation (l) or spelled out (L)
n and N	display month number 1–12 without/with a leading 0
d and D	display day-within-month number 1–31 without/with a leading 0
j and J	display day-within-year number 1–366 without/with leading 0s
h	display the half of year number 1 or 2
q	display quarter of year number 1, 2, 3, or 4
w and W	display week-of-year number 1–52 without/with a leading 0
_	display a blank
.	display a period
,	display a comma
:	display a colon
-	display a dash
/	display a slash
'	display a close single quote
!c	display character c (code !! to display an exclamation point)

Specifying %d by itself is equivalent to specifying %dD1CY. The first day of January 1999 is displayed as 01jan1999 in this format. For examples of various date formats, see [U] **27 Commands for dealing with dates**.

15.5.4 Time-series formats

Time-series formats—also known as %t formats—are an extension of the date formats coded above. Stata's dates are coded 0 = 01jan1960 and 1 = 02jan1960. Stata also has dates where 0 represents the first week of 1960 (and 1 the second week), 0 represents the first month of 1960 (and 1 the second month), 0 represents the first quarter of 1960 (and 1 the second quarter), 0 represents the first half of 1960 (and 1 the second half), and 1960 represents the year 1960 (and 1961 the next year). %t formats are for displaying these quantities. The %t format is defined as

first type	%	to indicate the start of the format
then optionally type	-	if you want the result left-aligned
then type	t	
then type		a character to indicate how the date is encoded:
type	d	if 0 = 01jan1960, same as %d format
or type	w	if 0 = 1960w1
or type	m	if 0 = 1960m1
or type	q	if 0 = 1960q1
or type	h	if 0 = 1960h1
or type	y	if 1960 = 1960 (it records the year itself)
then optionally type	*other characters*	to indicate how the date is to be displayed

where the optional characters are the same as for the %d format given in the table of the previous section, [U] **15.5.3 Date formats**.

In addition to the above is a %tg format—the g stands for generic. The %tg format is provided merely for completeness; it is equivalent to %9.0g. %tg is provided for users who want to put some sort of %t format on a time variable that is encoded differently than Stata understands; whether they do this makes no difference.

The minimal %t formats are %td, %tw, and so on. The default formats for each are

format	default	0 is displayed as	2,000 is displayed as
%td	%tdD1CY	01jan1960	23jun1965
%tw	%twCY!ww	1960w1	1998w25
%tm	%tmCY!mn	1960m1	2126m9
%tq	%tqCY!qq	1960q1	2460q1
%th	%thCY!hh	1960h1	2960h1
%ty	%tyCY	.	2000
%tg	%9.0g	0	2000

There are no mistakes in the table above. For %ty encoded data, the year range is 100–9999, so year 0 displays as missing. More typically, years will be in the range 1900–2100.

For more examples of the %t format, see [U] **27 Commands for dealing with dates**.

15.5.5 String formats

The syntax for a string format is

first type	%	to indicate the start of the format
then optionally type	–	if you want the result left-aligned
then type	a number	indicating the width of the result
then type	s	

For instance, %10s represents a string format of width 10.

For str*w*, the default format is %*w*s or %9s, whichever is wider. For example, a str10 variable receives a %10s format. Strings are displayed right-justified in the field unless the minus sign is coded; %-10s would display the string left-aligned.

▷ Example

Our automobile data contains a string variable called make.

```
. use http://www.stata-press.com/data/r8/auto
(1978 Automobile Data)

. describe make
```

variable name	storage type	display format	value label	variable label
make	str18	%-18s		Make and Model

```
. list make in 63/67
```

	make
63.	Mazda GLC
64.	Peugeot 604
65.	Renault Le Car
66.	Subaru
67.	Toyota Celica

These values are left-aligned because `make` has a display format of %-18s. If we want to right-align the values, we could change the format:

```
. format %18s make
. list make in 63/67
```

	make
63.	Mazda GLC
64.	Peugeot 604
65.	Renault Le Car
66.	Subaru
67.	Toyota Celica

◁

15.6 Dataset, variable, and value labels

Labels are strings used to label things. Stata provides labels for datasets, variables, and values.

15.6.1 Dataset labels

Associated with every dataset is an 80-character *dataset label*. The dataset label is initially set to blanks. You can use the `label data "text"` command to define the dataset label.

▷ Example

You have just entered 1980 state data on marriage rates, divorce rates, and median ages. The `describe` command will describe the data in memory. The result of typing `describe` is

```
. describe
Contains data
  obs:            50
  vars:            4
  size:        1,200 (99.8% of memory free)
```

variable name	storage type	display format	value label	variable label
state	str8	%9s		
median_age	float	%9.0g		
marriage_rate	long	%12.0g		
divorce_rate	long	%12.0g		

```
Sorted by:
      Note:  dataset has changed since last saved
```

`describe` shows that there are 50 observations on four variables. The four variables are named `state`, `median_age`, `marriage_rate`, and `divorce_rate`. `state` is stored as a `str8`; `median_age` is stored as a `float`; and `marriage_rate` and `divorce_rate` are both stored as `long`s. Each variable's display format (see [U] **15.5 Formats: controlling how data are displayed**) is shown. Finally, the data are not in any particular sort order, and the dataset has changed since it was last saved on disk.

You can label the data by typing `label data "1980 state data"`. You type this and then type `describe` again:

```
. label data "1980 state data"
. describe
Contains data
  obs:           50                            1980 state data
  vars:           4
  size:        1,200  (99.7% of memory free)

                storage  display   value
variable name    type    format    label     variable label

state            str8    %9s
median_age       float   %9.0g
marriage_rate    long    %12.0g
divorce_rate     long    %12.0g

Sorted by:
     Note:  dataset has changed since last saved
```
                                                                    ◁

The dataset label is displayed by the `describe` and `use` commands.

15.6.2   Variable labels

In addition to the name, every variable has associated with it an 80-character *variable label*. The variable labels are initially set to blanks. You use the `label variable` *varname* "*text*" command to define a new variable label.

▷ Example

You have entered data on four variables: `state`, `median_age`, `marriage_rate`, and `divorce_rate`. `describe` portrays the data you entered:

```
. describe
Contains data from states.dta
  obs:           50                            1980 state data
  vars:           4
  size:        1,200  (99.7% of memory free)

                storage  display   value
variable name    type    format    label     variable label

state            str8    %9s
median_age       float   %9.0g
marriage_rate    long    %12.0g
divorce_rate     long    %12.0g

Sorted by:
     Note:  dataset has changed since last saved
```

You can associate labels with the variables by typing

```
. label variable median_age "Median Age"
. label variable marriage_rate "Marriages per 100,000"
. label variable divorce_rate "Divorces per 100,000"
```

From then on, the result of `describe` will be

```
. describe
Contains data
   obs:            50                        1980 state data
   vars:            4
   size:        1,200  (99.7% of memory free)

                storage  display   value
variable name    type    format    label      variable label

state            str8    %9s
median_age       float   %9.0g                Median Age
marriage_rate    long    %12.0g               Marriages per 100,000
divorce_rate     long    %12.0g               Divorces per 100,000

Sorted by:
     Note:  dataset has changed since last saved
```
◁

Whenever Stata produces output, it will use the variable labels rather than the variable names to label the results if there is room.

15.6.3 Value labels

Value labels define a correspondence or mapping between numeric data and the words used to describe what those numeric values represent. Mappings are named and defined by the `label define` *lblname # "string" # "string"*... command. The maximum length for the *lblname* is 32 characters. `#` must be an integer or an extended missing value (`.a`, `.b`, ..., `.z`). The maximum length of *string* is 244 characters in Stata/SE and 80 characters in Intercooled Stata and in Small Stata. Named mappings are associated with variables by the `label values` *varname lblname* command.

▷ Example

The definition makes value labels sound more complicated than they are in practice. You create a dataset on individuals in which you record a person's sex, coding 0 for males and 1 for females. If your dataset also contained an employee number and salary, it might resemble the following:

```
. describe
Contains data
   obs:             7                        2002 Employee data
   vars:            3
   size:          112  (99.9% of memory free)

                storage  display   value
variable name    type    format    label      variable label

empno            float   %9.0g                Employee number
sex              float   %9.0g                Sex
salary           float   %8.0fc               Annual salary, exclusive of
                                                bonus

Sorted by:
     Note:  dataset has changed since last saved
```

```
. list
```

	empno	sex	salary
1.	57213	0	24,000
2.	47229	1	27,000
3.	57323	0	24,000
4.	57401	0	24,500
5.	57802	1	27,000
6.	57805	1	24,000
7.	57824	0	22,500

You could create a mapping called sexlabel defining 0 as "Male" and 1 as "Female", and then associate that mapping with the variable sex, by typing

```
. label define sexlabel 0 "Male" 1 "Female"
. label values sex sexlabel
```

From then on, your data would appear as

```
. describe
Contains data
    obs:            7                      2002 Employee data
   vars:            3
   size:          112 (99.8% of memory free)
```

	storage	display	value	
variable name	type	format	label	variable label
empno	float	%9.0g		Employee number
sex	float	%9.0g	sexlabel	Sex
salary	float	%8.0fc		Annual salary, exclusive of bonus

```
Sorted by:
     Note:  dataset has changed since last saved
. list
```

	empno	sex	salary
1.	57213	Male	24,000
2.	47229	Female	27,000
3.	57323	Male	24,000
4.	57401	Male	24,500
5.	57802	Female	27,000
6.	57805	Female	24,000
7.	57824	Male	22,500

Notice not only that the value label is used to produce words when we list the data, but also that the association of the variable sex with the value label sexlabel is shown by the describe command.

◁

❏ Technical Note

Value labels and variables may share the same name. For instance, rather than calling the value label `sexlabel` in the example above, we could just as well have named it `sex`. We would then type `label values sex sex` to associate the value label named `sex` with the variable named `sex`.

❏

▷ Example

Stata's `encode` and `decode` commands provide a convenient way to go from string variables to numerically coded variables and back again. Let's pretend that in the example above, rather than coding 0 for males and 1 for females, you created a string variable recording either "male" or "female". Your data look like

```
. describe
Contains data
   obs:            7                     2002 Employee data
   vars:           3
   size:         126 (99.8% of memory free)

                 storage  display   value
variable name     type    format    label     variable label

empno            float    %9.0g               Employee number
sex              str6     %9s                 Sex
salary           float    %8.0fc              Annual salary, exclusive of
                                              bonus

Sorted by:
     Note:  dataset has changed since last saved
. list
```

	empno	sex	salary
1.	57213	male	24,000
2.	47229	female	27,000
3.	57323	male	24,000
4.	57401	male	24,500
5.	57802	female	27,000
6.	57805	female	24,000
7.	57824	male	22,500

You now want to create a numerically encoded variable, we will call it `gender`, from the string variable. (You want to do this, say, because you typed `anova salary sex` to perform a one-way ANOVA of salary on sex and you were told that there were "no observations". You then remembered that all of Stata's statistical commands treat string variables as if they contain nothing but missing values. The statistical commands work only with numerically coded data.)

(Continued on next page)

```
. encode sex, generate(gender)
. describe
Contains data
    obs:             7                              2002 Employee data
    vars:            4
    size:          154 (99.8% of memory free)
```

| | storage | display | value | |
variable name	type	format	label	variable label
empno	float	%9.0g		Employee number
sex	str6	%9s		Sex
salary	float	%8.0fc		Annual salary, exclusive of bonus
gender	long	%8.0g	gender	Sex

```
Sorted by:
     Note:   dataset has changed since last saved
```

encode adds a new long variable called gender to the data and defines a new value label called gender. The value label gender maps 1 to the string male and 2 to female, so if you were to list the data, you could not tell the difference between the gender and sex variables. However, they are different. Stata's statistical commands know how to deal with gender. sex they do not understand. See [R] **encode**.

❑

❑ Technical Note

Perhaps rather than employee data, your data are on persons undergoing sex-change operations. As such, there would be two sex variables in your data, sex before the operation and sex after the operation. Assume the variables are named presex and postsex. You can associate the *same* value label to each variable by typing

```
. label define sexlabel 0 "Male" 1 "Female"
. label values presex sexlabel
. label values postsex sexlabel
```

❑

❑ Technical Note

Stata's input commands (input and infile) have the ability to go from the words in a value label back to the numeric codes. Remember that encode and decode can translate a string to a numeric mapping and vice versa, so you can map strings to numeric codes either at the time of input or later.

For example,

```
. label define sexlabel 0 "Male" 1 "Female"
. input empno sex:sexlabel salary, label
          empno       sex      salary
  1. 57213 Male 24000
  2. 47229 Female 27000
  3. 57323 0 24000
  4. 57401 Male 24500
  5. 57802 Female 27000
  6. 57805 Female 24000
  7. 57824 Male 22500
  8. end
```

The `label define` command defines the value label `sexlabel`. `input empno sex:sexlabel salary, label` tells Stata to input three variables from the keyboard (`empno`, `sex`, and `salary`), to attach the value label `sexlabel` to the `sex` variable, and to look up any words that are typed in the value label to try to convert them to numbers. To prove it works, we `list` the data that we recently entered:

```
. list
```

	empno	sex	salary
1.	57213	Male	24000
2.	47229	Female	27000
3.	57323	Male	24000
4.	57401	Male	24500
5.	57802	Female	27000
6.	57805	Female	24000
7.	57824	Male	22500

Compare the information we typed for observation 3 with the result listed by Stata. We typed 57323 0 24000. Thus, the value of `sex` in the third observation is 0. When Stata listed the observation, it indicated the value is `Male` because we told Stata in our `label define` command that zero is equivalent to `Male`.

Let's now add one more observation to our data:

```
. input, label
        empno        sex     salary
8. 67223 FEmale 23000
'FEmale' cannot be read as a number
8. 67223 Female 23000
9. end
```

At first we typed 67223 FEmale 23000, and Stata responded with "'FEmale' cannot be read as a number". Remember that Stata always respects case, so `FEmale` is not at all the same thing as `Female`. Stata prompted us to enter the line again, and we did so, this time correctly.

❏

❏ Technical Note

Coupled with the `automatic` option, Stata can not only go from words to numbers, but can create the mapping as well. Let's input the data again, but this time, rather than type the data in at the keyboard, let's read the data from a file. Assume you have an ASCII file called `employee.raw` stored on your disk. It contains

```
57213 Male 24000
47229 Female 27000
57323 Male 24000
57401 Male 24500
57802 Female 27000
57805 Female 24000
57824 Male 22500
```

The `infile` command can read these data *and* create the mapping automatically:

```
. label list sexlabel
value label sexlabel not found
r(111);
```

```
. infile empno sex:sexlabel salary using employee, automatic
(7 observations read)
```

Our first command, `label list sexlabel`, is only to prove that we had not previously defined the
value label `sexlabel`. Stata `infiled` the data without complaint. We now have

```
. list
```

	empno	sex	salary
1.	57213	Male	24000
2.	47229	Female	27000
3.	57323	Male	24000
4.	57401	Male	24500
5.	57802	Female	27000
6.	57805	Female	24000
7.	57824	Male	22500

Of course, `sex` is just another numeric variable; it does not actually take on the values `Male` and
`Female`—it takes on numeric codes that have been automatically mapped to `Male` and `Female`. We
can find out what that mapping is by using the `label list` command:

```
. label list sexlabel
sexlabel:
            1 Male
            2 Female
```

We discover that Stata attached the codes 1 to `Male` and 2 to `Female`. Anytime we want to see what
our data really look like, ignoring the value labels, we can use the `nolabel` option:

```
. list, nolabel
```

	empno	sex	salary
1.	57213	1	24000
2.	47229	2	27000
3.	57323	1	24000
4.	57401	1	24500
5.	57802	2	27000
6.	57805	2	24000
7.	57824	1	22500

❑

15.7 Notes attached to data

A dataset may contain notes. These are nothing more than little bits of text that you define and
review with the `notes` command. Typing `note`, a colon, and the text defines a note:

```
. note:  Send copy to Bob once verified.
```

You can later display whatever notes you have previously defined by typing `notes`:

```
. notes
_dta:
  1.  Send copy to Bob once verified.
```

Notes are saved with the data, so once you save your dataset, you can replay this note in the future, too.

You can add more notes:

```
. note: Mary wants a copy, too.

. notes

_dta:
     1.  Send copy to Bob once verified.
     2.  Mary wants a copy, too.
```

The notes we have added so far are attached to the data generically, which is why Stata prefixes them with _dta when it lists them. You can attach notes to variables:

```
. note state: verify values for Nevada.

. note state: what about the two missing values?

. notes

_dta:
     1,  Send copy to Bob once verified.
     2.  Mary wants a copy, too.

state:
     1,  verify values for Nevada
     2.  what about the two missing values?
```

When you describe your data, you can see whether notes are attached to the dataset or to any of the variables:

```
. describe

Contains data from states.dta
   obs:          50                        1980 state data
   vars:          4
   size:       1,200 (99.3% of memory free)  (_dta has notes)

                storage  display     value
variable name    type    format      label     variable label

state            str8    %9s                *
median_age       float   %9.0g                 Median Age
marriage_rate    long    %12.0g                Marriages per 100,000
divorce_rate     long    %12.0g                Divorces per 100,000
                                          * indicated variables have notes

Sorted by:
    Note:  dataset has changed since last saved
```

See [R] **notes** for a complete description of this feature.

15.8 Characteristics

Characteristics are an arcane feature of Stata but are of great use to Stata programmers. In fact, the notes command described above was implemented using characteristics.

Most users do not care about this detail.

The dataset itself and each variable within the dataset have associated with them a set of characteristics. Characteristics are named and referred to as *varname*[*charname*], where *varname* is the name of a variable or _dta. The characteristics contain text. Characteristics are stored with the data in the Stata-format .dta dataset, so they are recalled whenever the data are loaded.

How are characteristics used? The [XT] **xt** commands need to know the name of the variable corresponding to time. These commands allow the variable name to be specified as an option but do not require it. When the user does not specify the variable, the commands somehow manage to remember it from last time, even when the last time was a different Stata session. They do this with characteristics. When the user does not specify the variable name, the commands check the characteristic _dta[tis] for the name of the variable. If the time variable's name is stored there, they continue; if not, they issue an error because they need to know it. When the user specifies the option identifying the time variable, these commands store that name in the characteristic _dta[tis] so that they will know it next time. This use of characteristics is hidden from the user—no mention is made of how the commands remember the identity of the time variable.

Occasionally, commands identify their use of characteristics explicitly. The **xi** command (see [R] **xi**) states that it drops the first level of a categorical variable but that, if you wish to control which level is dropped, you can set the variable's omit characteristic. In the documentation, an example is provided where the user types

> . char agegrp[omit] 3

to set the default omission group to 3. As with the [XT] **xt** commands, if the user saves the data after setting the characteristic, the preferred omission group will be remembered from one session to the next.

As a Stata user, you need only understand how to set and clear a characteristic for the few commands that explicitly reveal their use of characteristics. You set a variable *varname*'s characteristic *charname* to be *x* by typing

> . char *varname*[*charname*] *x*

You set the data's characteristic *charname* to be *x* by typing

> . char _dta[*charname*] *x*

You clear a characteristic by typing

> . char *varname*[*charname*]

where *varname* is either a variable name or _dta. You can clear a characteristic even if it has never been set.

The most important feature of characteristics is that Stata remembers them from one session to the next; they are saved with the data.

❑ Technical Note

Programmers will want to know more. A technical description is found in [P] **char**, but as an overview, you may reference *varname*'s *charname* characteristic by embedding its name in single quotes and typing '*varname*[*charname*]'; see [U] **21.3.12 Referencing characteristics**.

You can fetch the names of all characteristics associated with *varname* by typing

> . local *macname* : char *varname*[]

The maximum length of the contents of a characteristic is the same as for macros: 8,681 characters for Small Stata, 67,784 for Intercooled, and $33 * c(\text{max\_k\_theory}) + 200$ for Stata/SE, which for the default setting of 5,000 is 165,200. The association of names with characteristics is by convention. If you, as a programmer, wish to create new characteristics for use in your ado-files, do so, but include at least one capital letter in the characteristic name. The current convention is that all lowercase names are reserved for "official" Stata.

❑

16 Functions and expressions

Contents

16.1   Overview
16.2   Operators
    16.2.1   Arithmetic operators
    16.2.2   String operators
    16.2.3   Relational operators
    16.2.4   Logical operators
    16.2.5   Order of evaluation, all operators
16.3   Functions
16.4   System variables (_variables)
16.5   Accessing coefficients and standard errors
    16.5.1   Simple models
    16.5.2   ANOVA and MANOVA models
    16.5.3   Multiple-equation models
16.6   Accessing results from Stata commands
16.7   Explicit subscripting
    16.7.1   Generating lags and leads
    16.7.2   Subscripting within groups
16.8   Time-series operators
    16.8.1   Generating lags and leads
    16.8.2   Operators within groups
16.9   Label values
16.10   Precision and problems therein

If you have not read [U] **14 Language syntax**, please do so before reading this entry.

16.1   Overview

Examples of expressions include

```
2+2
miles/gallons
myv+2/oth
(myv+2)/oth
ln(income)
age<25 & income>50000
age<25 | income>50000
age==25
name=="M Brown"
fname + " " + lname
substr(name,1,10)
_pi
val[_n-1]
L.gnp
```

137

Expressions like those above are allowed anywhere *exp* appears in a syntax diagram. One example is [R] **generate**:

> generate *newvar* = *exp* [if *exp*] [in *range*]

The first *exp* specifies the contents of the new variable and the optional second expression restricts the subsample over which it is to be defined. Another is [R] **summarize**:

> summarize [*varlist*] [if *exp*] [in *range*]

The optional expression restricts the sample over which summary statistics are calculated.

Algebraic and string expressions are specified in a natural way using the standard rules of hierarchy. You may use parentheses freely to force a different order of evaluation.

▷ Example

myv+2/oth is interpreted as myv+(2/oth). If you wanted to change the order of the evaluation, you could type (myv+2)/oth.

◁

16.2   Operators

Stata has four different classes of operators: arithmetic, string, relational, and logical. Each type is discussed below.

16.2.1   Arithmetic operators

The *arithmetic operators* in Stata are + (addition), − (subtraction), * (multiplication), / (division), ^ (raise to a power), and the prefix ! (negation). Any arithmetic operation on a missing value or an impossible arithmetic operation (such as division by zero) yields a missing value.

▷ Example

The expression -(x+y^(x-y))/(x*y) denotes the formula

$$-\frac{x + y^{x-y}}{x \cdot y}$$

and evaluates to *missing* if x or y is missing or zero.

◁

16.2.2   String operators

The + sign is also used as a string operator for the *concatenation* of two strings. Stata determines whether + means addition or concatenation by context.

▷ Example

The expression "this"+"that" results in the string "thisthat", whereas the expression 2+3 results in the number 5. Stata issues the error message "type mismatch" if the arguments on either side of the + sign are not of the same type. Thus, the expression 2+"this" is an error, as is 2+"3".

The expressions on either side of the + can be arbitrarily complex:

```
substr(string(20+2),1,1) + upper(substr("rf",1+1,1))
```

The result of the above expression is the string "2F". See [R] **functions** below for a description of the substr(), string(), and upper() functions.

◁

16.2.3   Relational operators

The *relational operators* are > (greater than), < (less than), >= (greater than or equal), <= (less than or equal), == (equal), and != (not equal). Observe that the relational operator for equality is a pair of equal signs. This convention distinguishes relational equality from the =*exp* assignment phrase.

❏ Technical Note

You may use ~ anywhere ! would be appropriate to represent the logical operator "not". Thus, the not-equal operator may also be written as ~=.

❏

Relational expressions are either *true* or *false*. Relational operators may be used on either numeric or string subexpressions; thus, the expression 3>2 is *true*, as is "zebra">"cat". In the latter case, the relation merely indicates that "zebra" comes after the word "cat" in the dictionary. All uppercase letters precede all lowercase letters in Stata's book, so "cat">"Zebra" is also *true*.

Missing values may appear in relational expressions. If x were a numeric variable, the expression $x \geq .$ is *true* if x is missing and *false* otherwise. A missing value is greater than any nonmissing value; see [U] **15.2.1 Missing values**.

▷ Example

You have data on age and income and wish to list the subset of the data for persons aged 25 years or less. You could type

```
. list if age<=25
```

If you wanted to list the subset of data of persons aged exactly 25, you would type

```
. list if age==25
```

Note the double equal sign. It would be an error to type list if age=25.

◁

Although it is convenient to think of relational expressions as evaluating to *true* or *false*, they actually evaluate to numbers. A result of *true* is defined as 1 and *false* as 0.

▷ Example

The definition of *true* and *false* makes it easy to create indicator, or dummy, variables. For instance,

```
generate incgt10k=income>10000
```

creates a variable that takes on the value 0 when `income` is less than or equal to $10,000, and 1 when `income` is greater than $10,000. Since missing values are greater than all nonmissing values, the new variable `incgt10k` will also take on the value 1 when `income` is *missing*. It would be safer to type

```
generate incgt10k=income>10000 if income<.
```

Now, observations in which `income` is *missing* will also contain *missing* in `incgt10k`. See [U] **28 Commands for dealing with categorical variables** for more examples.

◁

❑ Technical Note

Although you will rarely wish to do so, since arithmetic and relational operators both evaluate to numbers, there is no reason you cannot mix the two types of operators in a single expression. For instance, (2==2)+1 evaluates to 2, since 2==2 evaluates to 1, and $1 + 1$ is 2.

Relational operators are evaluated after all arithmetic operations. Thus, the expression (3>2)+1 is equal to 2, whereas 3>2+1 is equal to 0. Evaluating relational operators last guarantees the *logical* (as opposed to the *numeric*) interpretation. It should make sense that 3>2+1 is *false*.

❑

16.2.4   Logical operators

The *logical operators* are & (and), | (or), and ! (not). The logical operators interpret any nonzero value (including *missing*) as *true* and zero as *false*.

▷ Example

If you have data on `age` and `income` and wish to `list` data for persons making more than $50,000 along with persons under the age of 25 making more than $30,000, you could type

```
list if income>50000 | income>30000 & age<25
```

The & takes precedence over the |. If you were unsure, however, you could have typed

```
list if income>50000 | (income>30000 & age<25)
```

In either case, the statement will also `list` all observations for which `income` is *missing*, since *missing* is greater than 50,000.

◁

❑ Technical Note

Like relational operators, logical operators return 1 for *true* and 0 for *false*. For example, the expression 5 & . evaluates to 1. Logical operations, except for ~, are performed after all arithmetic and relational operations; the expression 3>2 & 5>4 is interpreted as (3>2)&(5>4) and evaluates to 1.

❑

16.2.5   Order of evaluation, all operators

The order of evaluation (from first to last) of all operators is ! (or ~), ^, – (negation), /, *, – (subtraction), +, != (or ~=), >, <, <=, >=, ==, &, and |.

16.3   Functions

Stata provides mathematical functions, probability and density functions, matrix functions, string functions, functions for dealing with dates and time series, and a set of special functions for programmers. You can find all of these documented in [R] **functions**.

Functions are merely a set of rules; you supply the function with arguments and the function evaluates the arguments according to the rules that define the function. Since functions are essentially subroutines that evaluate arguments and cause no action on their own, functions must be used in conjunction with a Stata command. Functions are indicated by the function name, an open parenthesis, an expression or expressions separated by commas, and a close parenthesis.

For example,

```
. display sqrt(4)
2
```

or

```
. display sqrt(2+2)
2
```

demonstrates the simplest use of a function. In this case, we have used the mathematical function, sqrt(), which takes a single number (or expression) as its argument and returns its square root. Note that the function was used in conjunction with the Stata command display. If I had simply typed

```
. sqrt(4)
```

Stata would have returned the error message

```
unrecognized command:  sqrt
r(199);
```

Functions can operate on variables as well. For example, suppose you wanted to generate a random variable that has observations drawn from a log-normal distribution. You could type

```
. set obs 5
obs was 0, now 5
. generate y = uniform()
. replace y = invnorm(y)
(5 real changes made)
. replace y = exp(y)
(5 real changes made)
. list
```

	y
1.	.686471
2.	2.380994
3.	.2814537
4.	1.215575
5.	.2920268

You could have saved yourself quite a bit of typing by just typing

. generate y = exp(invnorm(uniform()))

Functions accept expressions as arguments.

All functions are defined over a specified domain and return values within a specified range. Whenever an argument is outside of a function's domain, the function will return a missing value or issue an error message, whichever is most appropriate. For example, if you supplied the log() function with an argument of zero, the log(0) would return a missing value because zero is outside of the natural logarithm function's domain. If you supplied the log() function with a string argument, Stata would issue a "type mismatch" error, because log() is a numerical function and is undefined for strings. If you supply an argument that evaluates to a value that is outside of the function's range, then the function will return a missing value. Whenever a function accepts a string as an argument, the string must be enclosed in double quotes unless you provide the name of a variable that has a string storage type.

16.4   System variables (_variables)

Expressions may also contain _*variables* (pronounced "underscore variables"). These are built-in system variables that are created and updated by Stata. They are called _*variables* because their names all begin with the underscore '_' character.

The _*variables* are

[*eqno*] _b[*varname*] (synonym: [*eqno*] _coef[*varname*]) contains the value (to machine precision) of the coefficient on *varname* from the most recently fitted model (such as ANOVA, regression, Cox, logit, probit, multinomial logit, and the like). See [U] **16.5 Accessing coefficients and standard errors** below for a complete description.

_cons is always equal to the number 1 when used directly and refers to the intercept term when used indirectly, as in _b[_cons].

_n contains the number of the current observation.

_N contains the total number of observations in the dataset.

_pi contains the value of π to machine precision.

_rc contains the value of the return code from the most recent capture command.

[*eqno*] _se[*varname*] contains the value (to machine precision) of the standard error of the coefficient on *varname* from the most recently fitted model (such as ANOVA, regression, Cox, logit, probit, multinomial logit, and the like). See [U] **16.5 Accessing coefficients and standard errors** below for a complete description.

16.5   Accessing coefficients and standard errors

After estimating a model, you can access the coefficients and standard errors and use them in subsequent expressions. You should also see [R] **predict** (and [U] **23 Estimation and post-estimation commands**) for an easier way to obtain predictions, residuals, and the like.

16.5.1   Simple models

Begin by considering estimation methods that yield a single estimated equation with a one-to-one correspondence between coefficients and variables such as cnreg, logit, ologit, oprobit, probit, regress, and tobit. _b[*varname*] (synonym _coef[*varname*]) contains the coefficient on *varname* and _se[*varname*] contains its standard error, and both are recorded to machine precision. Thus, _b[age] refers to the calculated coefficient on the age variable after typing, say, regress response age sex, and _se[age] refers to the standard error on the coefficient. _b[_cons] refers to the constant and _se[_cons] to its standard error. Thus, you might type

```
. regress response age sex
. generate asif = _b[_cons] + _b[age]*age
```

16.5.2   ANOVA and MANOVA models

In ANOVA there is no simple relationship between the coefficients and the variables. For continuous variables in the model, _b[*varname*] refers to the coefficient. This works just as it does in simple models. For categorical variables, you must specify the level as well as the variable. _b[drug[2]] refers to the coefficient on the second level of drug. For interactions, _b[drug[2]*disease[1]] refers to the coefficient on the second level of drug and the first level of disease. Standard errors are obtained similarly using _se[]. Thus, you might type

```
. anova outcome sex age drug sex*age drug*age, continuous(age)
. generate age_effect = _b[age]*age
. replace age_effect = age_effect + _b[sex[1]*age] if sex==1
. replace age_effect = age_effect + _b[sex[2]*age] if sex==2
```

The coefficients and standard errors in manova are accessed in the same manner as with anova, with the exception that manova is also a multiple-equation estimator. _b[] and _se[] must be preceded by an equation number in square brackets; see [U] **16.5.3 Multiple-equation models**.

16.5.3   Multiple-equation models

The syntax for referring to coefficients and standard errors in multiple-equation models is the same as in the simple-model case, except that _b[] and _se[] are preceded by an equation number in square brackets. There are, however, numerous alternatives in how you may type requests. The way that you are supposed to type requests is

$$[eqno]\,\_b[varname]$$
$$[eqno]\,\_se[varname]$$

but you may substitute _coef[] for _b[]. In fact, you may omit the _b[] altogether, and most Stata users do:

$$[eqno]\,[varname]$$

You may also omit the second pair of square brackets:

$$[eqno]\,varname$$

There are two ways to specify the equation number *eqno*: either as an absolute equation number or as an "indirect" equation number. In the absolute form, the number is preceded by a '#' sign. Thus, [#1]displ refers to the coefficient on displ in the first equation (and [#1]_se[displ] refers to its standard error). You can even use this form for simple models such as regress if you prefer. regress estimates a single equation, so [#1]displ refers to the coefficient on displ just as _b[displ] does. Similarly, [#1]_se[displ] and _se[displ] are equivalent. The logic works both ways—in the multiple equation context, _b[displ] refers to the coefficient on displ in the first equation and _se[displ] refers to its standard error. _b[*varname*] (_se[*varname*]) is just another way of saying [#1]*varname* ([#1]_se[*varname*]).

Equations may also be referenced indirectly. [res]displ refers to the coefficient on displ in the equation named res. Equations are often named after the corresponding dependent variable name if there is such a concept in the fitted model, so [res]displ might refer to the coefficient on displ in the equation for variable res.

In the case of multinomial logit (mlogit), however, equations are named after the levels of the single dependent categorical variable. In multinomial logit, there is one dependent variable and there is an equation corresponding to each of the outcomes (values taken on) recorded in that variable except for the one that is arbitrarily labeled the base. [res]displ would be interpreted as the coefficient on displ in the equation corresponding *to the outcome* res. If outcome res is the base outcome, Stata treats [res]displ as zero (and Stata does the same thing for [res]_se[displ]).

Continuing with the multinomial logit case, the outcome variable must be numeric, although it need not be an integer. [res]displ would only be understood if there were a value label associated with the numeric outcome variable and res was one of the labellings. If your data are not labeled, you may refer to the numeric value directly by omitting the '#' sign. [1]displ refers to the coefficient on displ in the equation corresponding to the outcome 1, which may be different from [#1]displ. [1.2]displ would be the coefficient on displ in the equation corresponding to outcome 1.2. [1.2]_cons refers to the constant in the equation corresponding to outcome 1.2. [1.2]_se[_cons] refers to the standard error on the constant.

Thus, you might type

```
. mlogit outcome displ weight
. gen cont_din1 = [1]displ*displ
```

For every observation in your data, cont_din1 would contain the coefficient on displ in the equation corresponding to the outcome 1 multiplied by displ, or the contribution of displ in determining outcome 1.

16.6   Accessing results from Stata commands

Most Stata commands—not just estimation commands—save results in a way that allows you to access them in subsequent expressions. You do that by referring to e(*name*), r(*name*), or s(*name*):

```
. summarize age
. gen agedev = age-r(mean)

. regress mpg weight
. display "The number of observations used is " e(N)
```

Most commands are categorized r-class, meaning they save results in r(). The returned results—such as r(mean)—are available immediately following the command, and if you are going to refer to them, you need to refer to them soon because the next command will probably replace what is in r().

e-class commands are Stata's estimation commands—commands that fit models. Results in e() stick around until the next model is fitted.

s-class commands are parsing commands—commands used by programmers to interpret commands you type. Very few commands save anything in s().

Every command of Stata is designated r-class, e-class, or s-class, or, if the command saves nothing, n-class. r stands for return as in returned results, e stands for estimation as in estimation results, s stands for string, and, admittedly, this last acronym is weak, n stands for null.

You can find out what is stored where by looking in the *Saved Results* section for the particular command in the *Reference* manual. If you know the class of a command—and it is easy enough to guess—you can also see what is stored by typing 'return list', 'ereturn list', or 'sreturn list':

See [R] **saved results** and [U] **21.8 Accessing results calculated by other programs**.

16.7   Explicit subscripting

Individual observations on variables can be referenced by subscripting the variables. Explicit subscripts are specified by following a variable name with square brackets that contain an expression. The result of the subscript expression is truncated to an integer, and the value of the variable for the indicated observation is returned. If the value of the subscript expression is less than 1 or greater than _N, a missing value is returned.

16.7.1   Generating lags and leads

When you type something like

```
. generate y = x
```

Stata interprets it as if you typed

```
. generate y = x[_n]
```

and what that means is that the first observation of y is to be assigned the value from the first observation of x, the second observation of y is to be assigned the value from the second observation on x, and so on. Were you instead to type

```
. generate y = x[1]
```

you would set each and every observation of y equal to the first observation on x. If you typed

```
. generate y = x[2]
```

you would set each and every observation of y equal to the second observation on x. What would happen if you typed

```
. generate y = x[0]
```

Nothing too bad would happen: Stata would merely copy missing value into every observation of y because observation 0 does not exist. Exactly the same thing would happen were you to type

```
. generate y = x[100]
```

and you had fewer than 100 observations in your data.

When you type the square brackets, you are specifying explicit subscripts. Explicit subscripting combined with the _*variable* _n can be used to create lagged values on a variable. The lagged value of a variable x can be obtained by typing

```
. generate xlag = x[_n-1]
```

If you are really interested in lags and leads, you probably have time-series data and would be better served by using the time-series operators, such as L.x. Time-series operators can be used with varlists and expressions and they are safer because they account for gaps in the data; see [U] **14.4.3 Time-series varlists** and [U] **16.8 Time-series operators**. Even so, it is important that you understand how the above works.

The built-in underscore variable ⊿n is understood by Stata to mean the observation number of the current observation. That is why

. generate y = x[_n]

results in observation 1 of x being copied to observation 1 of y, and similarly for the rest of the observations. We are considering

. generate xlag = x[_n-1]

and notice that ⊿n-1 evaluates to the observation number of the previous observation. So, for the first observation, ⊿n-1 = 0 and xlag[1] is set to missing. For the second observation, ⊿n-1 = 1 and xlag[2] is set to the value of x[1], and so on.

Similarly, the lead of x can be created by

. generate xlead = x[_n+1]

In this case, the last observation on the new variable xlead will be *missing* because ⊿n+1 will be greater than ⊿N (⊿N is the total number of observations in the dataset).

16.7.2   Subscripting within groups

When a command is preceded by the by *varlist*: prefix, subscript expressions and the underscore variables ⊿n and ⊿N are evaluated relative to the subset of the data currently being processed. For example, consider the following (admittedly not very interesting) data:

. list

	bvar	oldvar
1.	1	1.1
2.	1	2.1
3.	1	3.1
4.	2	4.1
5.	2	5.1

To see how ⊿n, ⊿N, and explicit subscripting work, let's create three new variables demonstrating each and then list their values:

. generate small_n = _n
. generate big_n = _N
. generate newvar = oldvar[1]
. list

	bvar	oldvar	small_n	big_n	newvar
1.	1	1.1	1	5	1.1
2.	1	2.1	2	5	1.1
3.	1	3.1	3	5	1.1
4.	2	4.1	4	5	1.1
5.	2	5.1	5	5	1.1

small_n (which is equal to _n) goes from 1 to 5, and big_n (which is equal to _N) is 5. This should not be surprising; there are 5 observations in the data, and _n is supposed to count observations, whereas _N is the total number. newvar, which we defined as oldvar[1], is 1.1. Indeed, we see that the first observation on oldvar is 1.1.

Now, let's repeat those same three steps, only this time precede each step with the prefix by bvar:. First, we will drop the old values of small_n, big_n, and newvar so that we start fresh:

```
. drop small_n big_n newvar
. by bvar, sort: generate small_n=_n
. by bvar: generate big_n =_N
. by bvar: generate newvar=oldvar[1]
. list
```

	bvar	oldvar	small_n	big_n	newvar
1.	1	1.1	1	3	1.1
2.	1	2.1	2	3	1.1
3.	1	3.1	3	3	1.1
4.	2	4.1	1	2	4.1
5.	2	5.1	2	2	4.1

The results are different. Remember that we claimed that _n and _N are evaluated relative to the subset of data in the by-group. Thus, small_n (_n) goes from 1 to 3 for bvar = 1 and from 1 to 2 for bvar = 2. big_n (_N) is 3 for the first group and 2 for the second. Finally, newvar (oldvar[1]) is 1.1 and 4.1.

▷ Example

You now know enough to do some amazing things.

Suppose you have data on individual states and you have another variable in your data, call it region, that divides the states into the four Census regions. You have a variable x in your data, and you want to make a new variable called avgx to include in your regressions. This new variable is to take on the average value of x for the region in which the state is located. Thus, for California you will have the observation on x and the observation on the average value in the region, avgx. Here's how:

```
. by region, sort: generate avgx=sum(x)/_n
. by region: replace avgx=avgx[_N]
```

First, by region, we generate avgx equal to the running sum of x divided by the number of observations so far. The , sort ensures that the data are in region order. We have, in effect, created the running average of x within region. It is the last observation of this running average, the overall average within the region, that interests us. So, by region, we replace every avgx observation in a region with the last observation within the region, avgx[_N].

Here is what we will see when we type these two commands:

```
. by region, sort: generate avgx=sum(x)/_n
. by region: replace avgx=avgx[_N]
(46 real changes made)
```

In our example, there are no missing observations on x. If there had been, we would have obtained the wrong answer. When we created the running average, we typed

```
. by region, sort: generate avgx=sum(x)/_n
```

The problem is not with the sum() function. When sum() encounters a missing, it adds zero to the sum. The problem is with _n. Let's assume that the second observation in the first region has recorded a missing for x. When Stata processes the third observation in that region, it will calculate the sum of two elements (remember that one is missing) and then divide the sum by 3 when it should be divided by 2. There is an easy solution:

```
. by region: generate avgx=sum(x)/sum(x<.)
```

Rather than divide by _n, we divide by the total number of nonmissing observations seen on x so far, namely, the sum(x<.).

If our goal were simply to obtain the mean, it could have been more easily accomplished by typing egen avgx=mean(x), by(region); see [R] **egen**. egen, however, is written in Stata, and the above is how egen's mean() function works. The general principles are worth understanding.

◁

▷ Example

You have some patient data recording vital signs at various times during an experiment. The variables include patient, an id-number or name of the patient; time, a variable recording the date or time or epoch of the vital sign reading; and vital, a vital sign. You probably have more than one vital sign, but one is enough to illustrate the concept. Each observation in your data represents a patient-time combination.

Let's assume you have 1,000 patients and, for every observation on the same patient, you want to create a new variable called orig that records the patient's initial value of this vital sign.

```
. sort patient time
. by patient: generate orig=vital[1]
```

Observe that vital[1] refers not to the first reading on the first patient, but to the first reading on the current patient, because we are performing the generate command by patient.

◁

▷ Example

Let's do one more example with this patient data. Suppose that we want to create a new dataset from our patient data that records not only the patient's identification, the time of the reading of the first vital sign, and the first vital sign reading itself, but also the time of the reading of the last vital sign and its value. We want one observation per patient. Here's how:

```
. sort patient time
. by patient: generate lasttime=time[_N]
. by patient: generate lastvital=vital[_N]
. by patient: drop if _n!=1
```

◁

16.8   Time-series operators

Time-series operators allow you to refer to the lag of gnp by typing L.gnp, the second lag by typing L2.gnp, etc. There are also operators for lead (F), difference D, and seasonal difference S.

Time-series operators can be used with varlists and with expressions. See [U] **14.4.3 Time-series varlists** if you have not read it already. This section has to do with using time-series operators in expressions such as with generate. You do not have to create new variables; you can use the time-series operated variables directly.

16.8.1   Generating lags and leads

In a time-series context, referring to L2.gnp is better than referring to gnp[_n-2] because there might be missing observations. Pretend that observation 4 contains data for $t = 25$ and observation 5 data for $t = 27$. L2.gnp will still produce correct answers; L2.gnp for observation 5 will be the value from observation 4 because the time-series operators look at t to find the relevant observation. The more mechanical gnp[_n-2] just goes two observations back, and that, in this case, would not produce the desired result.

Time-series operators can be used with varlists or with expressions, so you can type

        . regress val L.gnp r

or

        . generate gnplagged = L.gnp
        . regress val gnplagged

Before you can type either one, however, you must use the tsset command to tell Stata the identity of the time variable; see [TS] **tsset**. Once you have tsset the data, anyplace you see an *exp* in a syntax diagram, you may type time-series operated variables, so you can type

        . summarize r if F.gnp<gnp

or

        . generate grew = 1 if gnp>L.gnp & L.gnp<.
        . replace grew = 0 if grew>=. & L.gnp<.

or

        . generate grew = (gnp>L.gnp) if L.gnp<.

16.8.2   Operators within groups

Stata also understands panel or cross-sectional time-series data. For instance, if you type

        . tsset country time

then you are declaring that you have time-series data. The time variable is time, and you have time-series data for separate countries.

Once you have tsset both cross-sectional and time identifiers, you proceed just as you would if you had a simple time series.

        . generate grew = (gnp>L.gnp) if L.gnp<.

would produce correct results. The L. operator will not confuse the observation at the end of one panel with the beginning of the next.

16.9   Label values

(If you have not read [U] **15.6 Dataset, variable, and value labels**, please do so.) You may use labels in an expression in place of the numeric values with which they are associated. To use a label in this way, type the label in double quotes followed by a colon and the name of the value label.

▷ Example

If the value label `yesno` associates the label `yes` with 1 and `no` with 0, then `"yes":yesno` (said aloud as the value of `yes` under `yesno`) is evaluated as 1. If the double-quoted label is not defined in the indicated value label, or if the value label itself is not found, a missing value is returned. Thus, the expression `"maybe":yesno` is evaluated as *missing*.

```
. list
```

	name	answer
1.	Sribney	no
2.	Franks	no
3.	Hilbe	yes
4.	DeLeon	no
5.	Cain	no
6.	Willis	yes
7.	Pechacek	no
8.	cox	no
9.	Reimer	no
10.	Hardin	yes
11.	Lancaster	yes
12.	Johnson	no

```
. list if answer=="yes":yesno
```

	name	answer
3.	Hilbe	yes
6.	Willis	yes
10.	Hardin	yes
11.	Lancaster	yes

In the above example, the variable `answer` is not a string variable; it is a numeric variable that has the associated value label `yesno`. Since `yesno` associates `yes` with 1 and `no` with 0, we could have typed `list if answer==1` instead of what we did type. We could not have typed `list if answer=="yes"` because `answer` is not a string variable. If we had, we would have received the error message "type mismatch".

◁

16.10   Precision and problems therein

Examine the following short Stata session:

```
. drop _all
. input x y
            x           y
  1. 1 1.1
  2. 2 1.2
  3. 3 1.3
  4. end
```

```
. count if x==1
      1

. count if y==1.1
      0

. list
```

	x	y
1.	1	1.1
2.	2	1.2
3.	3	1.3

We created a dataset containing two variables, x and y. The first observation has x equal to 1 and y equal to 1.1. When we asked Stata to count the number of times that the variable x took on the value 1, we were told that it occurred once. Yet when we asked Stata to count the number of times y took on the value 1.1, we were told zero—meaning that it never occurred. What's gone wrong? When we list the data, we see that the first observation has y equal to 1.1.

Despite appearances, Stata has not made a mistake. Stata stores numbers internally in binary, and the number 1.1 has no exact binary representation—that is, there is no finite string of binary digits that is exactly equal to 1.1.

□ Technical Note

The number 1.1 in binary is 1.0001100110011 ..., where the period represents the binary point. The problem binary computers have with storing numbers like 1/10 is much like the problem we base-10 users have in precisely writing 1/11, which is 0.0909090909 ....

□

The number that appears as 1.1 in the listing above is actually 1.1000000238419, which is off by roughly 2 parts in 10^8. Unless we tell Stata otherwise, it stores all numbers as floats, which are also known as *single-precision* or *4-byte reals*. On the other hand, Stata performs all internal calculations in double, which is also known as *double-precision* or *8-byte reals*. This is what leads to the difficulty.

In the above example, we compared the number 1.1, stored as a float, with the number 1.1 stored as a double. The double-precision representation of 1.1 is more accurate than the single-precision representation, but what is important is that it is also different. Those two numbers are not equal.

There are a number of ways around this problem. The problem with 1.1 apparently not equaling 1.1 would never arise if the storage precision and the precision of the internal calculations were the same. Thus, you could store all your data as doubles. This takes more computer memory, however, and it is unlikely that (1) your data are really that accurate and (2) the extra digits would meaningfully affect any calculated result even if the data were that accurate.

□ Technical Note

This is unlikely to affect any calculated result because Stata performs all internal calculations in double precision. This is all rather ironic, since the problem would also not arise if we had designed Stata to use single precision for its internal calculations. Stata would be less accurate, but the problem would have been completely disguised from the user, making this entry unnecessary.

□

Another solution is to use the `float()` function. `float`(x) rounds x to its `float` representation. If we had typed `count if y==float(1.1)` in the above example, we would have been informed that there is one such value.

17 Matrix expressions

Contents
17.1 Overview
    17.1.1   Definition of a matrix
    17.1.2   matsize
17.2 Row and column names
    17.2.1   The purpose of row and column names
    17.2.2   Three-part names
    17.2.3   Setting row and column names
    17.2.4   Obtaining row and column names
17.3 Vectors and scalars
17.4 Inputting matrices by hand
17.5 Accessing matrices created by Stata commands
17.6 Creating matrices by accumulating data
17.7 Matrix operators
17.8 Matrix functions
17.9 Subscripting
17.10 Using matrices in scalar expressions

17.1 Overview

Matrices can be used interactively in Stata, and you might do so, for instance, in a teaching situation. The real power of matrices, however, is unleashed when they are used in Stata programs and ado-files to implement other statistical procedures. We do this ourselves, and you can, too.

17.1.1 Definition of a matrix

Stata's definition of a matrix includes a few details that go beyond the mathematics. To Stata, a matrix is a named entity containing an $r \times c$ ($0 < r \leq$ matsize, $0 < c \leq$ matsize) rectangular array of double-precision numbers (including missing values) that is bordered by a row and a column of names.

```
. matrix list A

A[3,2]
     c1  c2
r1    1   2
r2    3   4
r3    5   6
```

In this case, we have a 3×2 matrix named A containing elements 1, 2, 3, 4, 5, and 6. Row 1, column 2 (written $A_{1,2}$ in math and A[1,2] in Stata) contains 2. The columns are named c1 and c2 and the rows r1, r2, and r3. These are the default names Stata comes up with when it cannot do better. The names do not play a role in the mathematics, but they are of great help when it comes to labeling the output.

The names are operated on just as the numbers. For instance,

```
. matrix B=A'*A
. matrix list B
symmetric B[2,2]
      c1  c2
c1  35
c2  44  56
```

We defined $\mathbf{B} = \mathbf{A}'\mathbf{A}$. Note that the row and column names of $\mathbf{B}$ are the same. Multiplication is defined for any $a \times b$ and $b \times c$ matrices, the result being $a \times c$. Thus, the row and column names of the result are the row names of the first matrix and the column names of the second matrix. We formed $\mathbf{A}'\mathbf{A}$, using the transpose of $\mathbf{A}$ for the first matrix—which also interchanged the names—and so obtained the names shown.

17.1.2   matsize

Matrices are limited to being no larger than matsize $\times$ matsize. The default value of matsize is 400 for Stata/SE and 40 for Intercooled Stata, but you can reset this using the set matsize command; see [R] **matsize**.

The maximum value of matsize is 800 for Intercooled Stata, and thus matrices are not suitable for holding large amounts of data. This restriction does not prove a limitation because terms that appear in statistical formulas are of the form $(\mathbf{X}'\mathbf{W}\mathbf{Z})$ and Stata provides a command, matrix accum, for efficiently forming such matrices; see [U] **17.6 Creating matrices by accumulating data** below. The maximum value of matsize is 11,000 for Stata/SE, and thus performing matrix operations directly on large amounts of data is more feasible.

17.2   Row and column names

Matrix rows and columns always have names. Stata is smart about setting these names when the matrix is created, and the matrix commands and operators manipulate these names throughout calculations, with the result that the names typically are set correctly at the conclusion of matrix calculations.

For instance, consider the matrix calculation $\mathbf{b} = (\mathbf{X}'\mathbf{X})^{-1}\mathbf{X}'\mathbf{y}$ performed on real data:

```
. use http://www.stata-press.com/data/r8/auto
(1978 Automobile Data)
. matrix accum XprimeX = weight foreign
(obs=74)
. matrix vecaccum yprimeX = mpg weight foreign
. matrix b = syminv(XprimeX)*yprimeX'
. matrix list b
b[3,1]
                 mpg
 weight  -.00658789
foreign  -1.6500291
  _cons   41.679702
```

Note that these names were produced without us ever having given a special command to place the names on the result. When we formed matrix XprimeX, Stata produced the result

```
. matrix list XprimeX
symmetric XprimeX[3,3]
            weight    foreign      _cons
 weight  7.188e+08
foreign      50950         22
  _cons     223440         22         74
```

matrix accum forms $\mathbf{X}'\mathbf{X}$ matrices from data and it sets the row and column names to the variable names used. The names are correct in the sense that, for instance, the (1,1) element is the sum across the observations of squares of weight and the (2,1) element is the sum of the product of weight and foreign.

Similarly, matrix vecaccum forms $\mathbf{y}'\mathbf{X}$ matrices and it also sets the row and column names to the variable names used, so matrix vecaccum yprimeX = mpg weight foreign resulted in

```
. matrix list yprimeX

yprimeX[1,3]
        weight   foreign    _cons
  mpg  4493720       545     1576
```

The final step, matrix b = syminv(XprimeX)*yprimeX', manipulated the names, and, if you think carefully, you can derive for yourself the rules. syminv() (inversion) is much like transposition, so row and column names must be swapped. In this case, however, the matrix was symmetric, so that amounted to leaving the names as they were. Multiplication amounts to taking the column names of the first matrix and the row names of the second. The final result is

```
. matrix list b

b[3,1]
                  mpg
 weight     .00659790
 foreign  -1.6500291
  _cons    41.679702
```

and the interpretation is $mpg = -.00659\,weight - 1.65\,foreign + 41.68 + e$.

Researchers realized long ago that matrix notation simplifies the description of complex calculations. What they may not have realized is that, corresponding to each mathematical definition of a matrix operator, there is a definition of the operator's effect on the names that can be used to carry the names forward through long and complex matrix calculations.

17.2.1 The purpose of row and column names

Mostly, matrices in Stata are used in programming estimators and Stata uses row and column names to produce pretty output. For instance, say that we wrote code—interactively or in a program—that produced the following coefficient vector b and covariance matrix V:

```
. matrix list b

b[1,3]
          weight  displacement        _cons
 y1    -.00656711     .00528078    40.084522

. matrix list V

symmetric V[3,3]
                      weight  displacement         _cons
       weight       1.360e-06
 displacement      -.0000103      .00009741
        _cons      -.00207455     .01188356     4.0808455
```

We could now produce standard estimation output by coding two more lines:

(Continued on next page)

```
. ereturn post b V
. ereturn display
```

	Coef.	Std. Err.	z	P>|z|	[95% Conf.	Interval]
weight	-.0065671	.0011662	-5.63	0.000	-.0088529	-.0042813
displacement	.0052808	.0098696	0.54	0.593	-.0140632	.0246248
_cons	40.08452	2.02011	19.84	0.000	36.12518	44.04387

Stata's `ereturn` command knew to produce this output because of the row and column names on the coefficient vector and variance matrix. Moreover, in most cases, we do nothing special in our code that produces b and V to set the row and column names because, given how matrix names work, they work themselves out.

In addition, sometimes row and column names help us detect programming errors. Assume we wrote code to produce matrices b and V but made a mistake. Sometimes our mistake will result in the wrong row and column names. Rather than the b vector we previously showed you, we might produce

```
. matrix list b
b[1,3]
        weight          c2        _cons
y1  -.00656711       42.23    40.084522
```

Were we to `post` our estimation results now, Stata would refuse because it can tell by the names that there is a problem:

```
. ereturn post b V
name conflict
r(507);
```

Understand, however, that Stata follows the standard rules of matrix algebra; the names are just along for the ride. Matrices are summed by position, meaning a directive to form $C = A + B$ results in $C_{11} = A_{11} + B_{11}$ regardless of the names, and it is not an error to sum matrices with different names:

```
. matrix list a
symmetric a[3,3]
                c1          c2          c3
    mpg       14419
 weight     1221120   1.219e+08
  _cons         545       50950          22
. matrix list b
symmetric b[3,3]
                   c1          c2          c3
displacement  3211055
         mpg   227102       22249
       _cons    12153        1041          52
. matrix c = a + b
. matrix list c
symmetric c[3,3]
                   c1          c2          c3
displacement  3225474
         mpg  1448222   1.219e+08
       _cons    12698       51991          74
```

Matrix row and column names are used to label output; they do not affect how matrix algebra is performed.

17.2.2 Three-part names

Row and column names have three parts: *equation_name*:*ts_operator*.*subname*.

In the examples shown so far, the first two parts have been blank; the row and column names consisted of *subnames* only. This is typical. Run any single-equation model (such as those produced by regress, probit, logistic, etc.), and if you fetch the resulting matrices, you will find that they have row and column names of the *subname* form.

Those who work with time-series data will find matrices with row and column names of the form *ts_operator*.*subname*. For example,

```
. matrix list example1
symmetric example1[3,3]
                               L.
             rate           rate          _cons
  rate    3.0952534
L.rate     .0096504       .00007742
 _cons   -2.8413483      -.01821928     4.8578916
```

We obtained this matrix by running a linear regression on rate and L.rate and then fetching the covariance matrix. Think of the row and column name L.rate no differently than you think of rate or, in the previous examples, r1, r2, c1, c2, weight, and foreign.

Equation names are used to label partitioned matrices and, in estimation, occur in the context of multiple equations. Here is a matrix with *equation_names* and *subnames*,

```
. matrix list example2
symmetric example2[5,5]
                      mpg:          mpg:          mpg:          mpg:          mpg:
                   foreign         displ         _cons       foreign         _cons
   mpg:foreign    1.6483972
     mpg:displ     .004747       .00003876
     mpg:_cons   -1.4266352     -.00905773     2.4341021
weight:foreign  -51.208454     -4.665e-19      15.224135     24997.727
 weight:_cons    15.224135      2.077e-17     -15.224135    -7431.7565     7431.7565
```

and here is an example with all three parts filled in:

```
. matrix list example3
symmetric example3[5,5]
                      val:          val:          val:       weight:       weight:
                                       L.
                      rate          rate         _cons       foreign         _cons
      val:rate     2.2947268
    val:L.rate      .00385216      .0000309
     val:_cons    -1.4533912     -.0072726      2.2583357
weight:foreign  -163.86684       7.796e-17     49.384526     25351.696
 weight:_cons    49.384526     -1.566e-16    -49.384526     -7640.237      7640.237
```

val:L.rate is a column name, just as, in the previous section, c2 and foreign were names.

Let us pretend that this last matrix is the variance matrix produced by a program we wrote and that our program also produced a coefficient vector b:

```
. matrix list b
b[1,5]
                    val:          val:          val:       weight:       weight:
                                    L.
                    rate          rate         _cons       foreign         _cons
y1    4.5366753     -.00316923     20.68421    -1008.7968     3324.7059
```

Here is the result of posting and displaying the results:

```
. ereturn post b example3
. ereturn display
```

		Coef.	Std. Err.	z	P>|z|	[95% Conf.	Interval]
val							
rate							
	—	4.536675	1.514836	2.995	0.003	1.567652	7.505698
	L1	-.0031692	.0055591	-0.570	0.569	-.0140648	.0077264
_cons		20.68421	1.502776	13.764	0.000	17.73882	23.6296
weight							
foreign		-1008.797	159.2222	-6.336	0.000	-1320.866	-696.7271
_cons		3324.706	87.40845	38.036	0.000	3153.388	3496.023

The equation names are used to separate one equation from the next.

17.2.3 Setting row and column names

You reset row and column names using the `matrix rownames` and `matrix colnames` commands.

Before resetting the names, list the matrix to verify the names are not set correctly; often, they already are. When you enter a matrix by hand, however, the row names are unimaginatively set to r1, r2, ..., and the column names to c1, c2, ...

```
. matrix a = (1,2,3\4,5,6)
. matrix list a
a[2,3]
     c1  c2  c3
r1    1   2   3
r2    4   5   6
```

Regardless of the current row and column names, `matrix rownames` and `matrix colnames` will reset them:

```
. matrix colnames a = foreign alpha _cons
. matrix rownames a = one two
. matrix list a
a[2,3]
      foreign   alpha   _cons
one         1       2       3
two         4       5       6
```

You may set the *ts_operator* as well as the *subname*,

```
. matrix colnames a = foreign l.rate _cons
. matrix list a
a[2,3]
                    L.
      foreign     rate    _cons
one         1        2        3
two         4        5        6
```

and you may set equation names:

```
. matrix colnames a = this:foreign this:l.rate that:_cons
. matrix list a
a[2,3]
          this:    this:    that:
                    L.
       foreign    rate    _cons
  one        1       2        3
  two        4       5        6
```

See [P] **matrix rowname** for more information.

17.2.4   Obtaining row and column names

matrix list displays the matrix with its row and column names. In a programming context, you can fetch the row and column names into a macro using

$$
\begin{array}{l}
\texttt{local} \ldots \texttt{: rowfullnames } \textit{matname} \\
\texttt{local} \ldots \texttt{: colfullnames } \textit{matname} \\
\texttt{local} \ldots \texttt{: rownames } \textit{matname} \\
\texttt{local} \ldots \texttt{: colnames } \textit{matname} \\
\texttt{local} \ldots \texttt{: roweq } \textit{matname} \\
\texttt{local} \ldots \texttt{: coleq } \textit{matname}
\end{array}
$$

rowfullnames and colfullnames return the full names (*equation_name*: *ts_operator*. *subname*) listed one after the other.

rownames and colnames omit the equations and return *ts_operator*. *subname*, listed one after the other.

roweq and coleq return the equation names, listed one after the other.

See [P] **macro** and [P] **matrix define** for more information.

17.3   Vectors and scalars

Stata does not have vectors as such—they are considered special cases of matrices and are handled by the matrix command.

Stata does have scalars, although they are not strictly necessary because they, too, could be handled as special cases. See [P] **scalar** for a description of scalars.

17.4   Inputting matrices by hand

You input matrices using

matrix input *matname* = (...)

or

matrix *matname* = (...)

In either case, you enter the matrices rowwise. You separate one element from the next using commas (,) and one row from the next using backslashes (\). If you omit the word input, you are using the expression parser to input the matrix:

```
. matrix a = (1,2\3,4)
. matrix list a
a[2,2]
     c1  c2
r1    1   2
r2    3   4
```

This has the advantage that you can use expressions for any of the elements:

```
. matrix b = (1, 2+3/2 \ cos(_pi), _pi)
. matrix list b
b[2,2]
            c1          c2
r1           1         3.5
r2          -1   3.1415927
```

The disadvantage is that the matrix must be small, say, no more than 50 elements (regardless of the value of `matsize`).

`matrix input` has no such restriction, but you may not use subexpressions for the elements:

```
. matrix input c  = (1,2\3,4)
. matrix input d = (1, 2+3/2 \ cos(_pi), _pi)
invalid syntax
r(198);
```

Either way, after inputting the matrix, you will probably want to set the row and column names; see [U] **17.2.3 Setting row and column names** above.

17.5   Accessing matrices created by Stata commands

Some Stata commands—and all estimation commands—leave behind matrices that you can subsequently use. After executing an estimation command, type `ereturn list` to see what is available:

```
. use http://www.stata-press.com/data/r8/auto
(1978 Automobile Data)
. probit foreign mpg weight
 (output omitted )
. ereturn list
scalars:
                  e(N) =  74
               e(ll_0) =  -45.03320955699139
                 e(ll) =  -26.8441890057987
               e(df_m) =  2
               e(chi2) =  36.37804110238539
               e(r2_p) =  .403902380712477
macros:
            e(depvar) : "foreign"
               e(cmd) : "probit"
          e(crittype) : "log likelihood"
           e(predict) : "probit_p"
          e(chi2type) : "LR"
matrices:
                 e(b) :  1 x 3
                 e(V) :  3 x 3
functions:
            e(sample)
```

Most estimation commands leave behind e(b) (the coefficient vector) and e(V) (the variance–covariance matrix of the estimator):

```
. matrix list e(b)

e(b)[1,3]
            mpg      weight       _cons
y1  -.10395033  -.00233554    8.275464
```

You can refer to e(b) and e(V) in any matrix expression:

```
. matrix myb = e(b)

. matrix list myb

myb[1,3]
            mpg      weight       _cons
y1  -.10318674   -.0023264    8.234735

. matrix c = e(b)*syminv(e(V))*e(b)'

. matrix list c

symmetric c[1,1]
             y1
y1   77.848188
```

17.6   Creating matrices by accumulating data

In programming estimators, matrices of the form $X'X$, $X'Z$, $X'WX$, and $X'WZ$ often occur, where X and Z are data matrices. matrix accum, matrix glsaccum, matrix vecaccum, and matrix opaccum produce such matrices; see [P] **matrix accum**.

We recommend that you do not load the data into a matrix and use the expression parser directly to form such matrices, although see [P] **matrix mkmat** if that is your interest. If that is your interest, be sure to read the *Technical Note* at the end of [P] **matrix mkmat**. There is much to recommend learning how to use the matrix accum commands.

17.7   Matrix operators

You can create new matrices or replace existing matrices by typing

matrix *matname* = *matrix_expression*

For instance,

```
. matrix A = syminv(R*V*R')
. matrix IAR = I(rowsof(A)) - A*R
. matrix beta = b*IAR' + r*A'
. matrix C = -C'
. matrix D = (A, B \ B', A)
. matrix E = (A+B)*C'
. matrix S = (S+S')/2
```

The following operators are provided:

Operator	Symbol
Unary operators	
negation	−
transposition	'
Binary operators	
(lowest precedence)	
row join	\
column join	,
addition	+
subtraction	−
multiplication	*
division by scalar	/
Kronecker product	#
(highest precedence)	

Parentheses may be used to change the order of evaluation.

Note in particular that , and \ are operators; (1,2) creates a 1×2 matrix (vector) and (A,B) creates a rowsof(A) $\times$ colsof(A)+colsof(B) matrix, where rowsof(A) = rowsof(B). (1\2) creates a 2×1 matrix (vector) and (A\B) creates a rowsof(A)+rowsof(B) $\times$ colsof(A) matrix, where colsof(A) = colsof(B). Thus, expressions of the form

```
matrix R = (A,B)*Vinv*(A,B)'
```

are allowed.

17.8   Matrix functions

In addition to the functions listed below, see [P] **matrix svd** for singular value decomposition, [P] **matrix symeigen** for eigenvalues and eigenvectors of symmetric matrices, and [P] **matrix eigenvalues** for eigenvalues of nonsymmetric matrices. For a full description of the matrix functions, see [R] **functions**.

Matrix functions returning matrices:

cholesky(M)	I(n)	sweep(M,i)
corr(M)	inv(M)	syminv(M)
diag(v)	J(r,c,z)	vec(M)
e(*name*)	matuniform(r, c)	vecdiag(M)
get(*systemname*)	nullmat(*matname*)	
hadamard(M, N)	return(*name*)	

Matrix functions returning scalars:

colnumb(M,s)	el(s,i,j)	rownumb(M,s)
colsof(M)	issym(M)	rowsof(M)
det(M)	matmissing(M)	trace(M)
diag0cnt(M)	mreldif(X,Y)	

17.9   Subscripting

1. In matrix and scalar expressions, you may refer to *matname*$[r,c]$, where r and c are scalar expressions, to obtain a single element of *matname* as a scalar.

   Examples:
   ```
   matrix A = A / A[1,1]
   generate newvar = oldvar / A[2,2]
   ```

2. In matrix expressions, you may refer to *matname*$[s_r,s_c]$, where s_r and s_c are string expressions, to obtain a sub-matrix with a single element. The element returned is based on searching the row and column names.

   Examples:
   ```
   matrix B = V["price","price"]
   generate sdif = dif / sqrt(V["price","price"])
   ```

3. In matrix expressions, you may mix these two syntaxes and so refer to *matname*$[r,s_c]$ or to *matname*$[s_r,c]$.

   Examples:
   ```
   matrix b = b * R[1,"price"]
   ```

4. In matrix expressions, you may use *matname*$[r_1..r_2,c_1..c_2]$ to refer to submatrices; r_1, r_2, c_1, and c_2 may be scalar expressions. If r_2 evaluates to missing, it is taken as referring to the last row of *matname*; if c_2 evaluates to missing, it is taken as referring to the last column of *matname*. Thus, *matname*$[r_1...,c_1...]$ is allowed.

   Examples:
   ```
   matrix S = Z[1..4, 1..4]
   matrix R = Z[5..., 5...]
   ```

5. In matrix expressions, you may refer to *matname*$[s_{r1}..s_{r2},s_{c1}..s_{c2}]$ to refer to submatrices where s_{r1}, s_{r2}, s_{c1}, and s_{c2}, are string expressions. The matrix returned is based on looking up the row and column names.

   If the string evaluates to an equation name only, all the rows or columns for the equation are returned.

   Examples:
   ```
   matrix S = Z["price".."weight", "price".."weight"]
   matrix L = D["mpg:price".."mpg:weight", "mpg:price".."mpg:weight"]
   matrix T1 = C["mpg:", "mpg:"]
   matrix T2 = C["mpg:", "price:"]
   ```

6. In matrix expressions, any of the above syntaxes may be combined.

   Examples:
   ```
   matrix T1 = C["mpg:", "price:weight".."price:displ"]
   matrix T2 = C["mpg:", "price:weight"...]
   matrix T3 = C["mpg:price", 2..5]
   matrix T4 = C["mpg:price", 2]
   ```

7. When defining an element of a matrix, use

matrix *matname*[*i*,*j*] = *expression*

where *i* and *j* are scalar expressions. The matrix *matname* must already exist.

Example:
```
matrix A = J(2,2,0)
matrix A[1,2] = sqrt(2)
```

8. To replace a submatrix within a matrix, use the same syntax. If the expression on the right evaluates to a scalar or 1×1 matrix, the element is replaced. If it evaluates to a matrix, the submatrix with top-left element at (i, j) is replaced. The matrix *matname* must already exist.

Example:
```
matrix A = J(4,4,0)
matrix A[2,2] = C'*C
```

17.10   Using matrices in scalar expressions

Scalar expressions are documented as *exp* in the Stata manuals:

generate *newvar* = *exp* if *exp* ...
replace *newvar* = *exp* if *exp* ...
regress ... if *exp* ...
if *exp* {... }
while *exp* {... }

Most importantly, scalar expressions occur in `generate` and `replace`, in the `if` *exp* modifier allowed on the end of many commands, and in the `if` and `while` commands for program control.

It is rare that one needs to refer to a matrix in any of these situations except when using the `if` and `while` commands.

In any case, you may refer to matrices in any of these situations, but the expression cannot require evaluation of matrix expressions returning matrices. Thus, you could refer to `trace(A)` but not to `trace(A+B)`.

It can be difficult to predict when an evaluation of an expression requires evaluating a matrix; even experienced users can be surprised. If you get the error message "matrix operators that return matrices not allowed in this context", r(509), you have encountered such a situation.

The solution is to split the line in two. For instance, one would change

```
if trace(A+B)==0 {
        ...
}
```

to

```
matrix AplusB = A+B
if trace(AplusB)==0 {
        ...
}
```

or even to

```
matrix Trace = trace(A+B)
if Trace[1,1]==0 {
        ...
}
```

18 Printing and preserving output

Contents

18.1 Overview
    18.1.1   Starting and closing logs
    18.1.2   Appending to an existing log
    18.1.3   Temporarily suspending and resuming logging
18.2 Placing comments in logs
18.3 Logging only what you type
18.4 The log-button alternative
18.5 Printing logs

18.1 Overview

Stata will record your session into a file. By default, the resulting file—called a log file—contains what you type and what Stata produces in response, recorded in a format called SMCL, see [P] **smcl**. The file can be printed or converted to ASCII text for incorporation into documents you create with your word processor.

To start a log:
  Your session is now being recorded
  in file *filename*.`smcl`.
	`. log using` *filename*

To temporarily stop logging:
  Temporarily stop:
  Resume:
	
	`. log off`
	`. log on`

To stop logging and close the file:
  You can now print *filename*.`smcl` or type
  to create *filename*.`log` that you can
  load into your word processor.
	`. log close`
	`. translate` *filename*.`smcl` *filename*.`log`

Alternative ways to start logging:
  append to an existing log:
  replace an existing log:
	
	`. log using` *filename*`, append`
	`. log using` *filename*`, replace`

The above works, but in addition:
  To start a log:
  To temporarily stop logging:
  To resume:
  To stop logging and close the file:
  To print previous or current log:
	
	click the **Log** button
	click the **Log** button and choose **Suspend**
	click the **Log** button and choose **Resume**
	click the **Log** button and choose **Close**
	pull down **File**, select **View**, choose file,
	    then pull down **File** and select **Print Viewer**

In addition, `cmdlog` will produce logs containing solely what you typed—logs that, while not containing your results, are sufficient to recreate the session.

To start a command-only log:	`. cmdlog using` *filename*
To stop logging and close the file:	`. cmdlog close`
To recreate your session:	`. do` *filename*.`txt`

18.1.1 Starting and closing logs

With great foresight, you begin working in Stata and type `log using session` (or click the **Log** button) before starting your work:

```
. log using session

          log:  c:\example\session.smcl
     log type:  smcl
    opened on:  17 Dec 2002, 12:35:08
. use http://www.stata-press.com/data/r8/census
(1980 Census data by state)
. tabulate reg [freq=pop]

    Census
    region        Freq.     Percent        Cum.

        NE      49135283       21.75       21.75
   N Cntrl      58865670       26.06       47.81
     South      74734029       33.08       80.89
      West      43172490       19.11      100.00

     Total     225907472      100.00
. summarize medage
    Variable        Obs        Mean    Std. Dev.        Min         Max

      medage         50       29.54    1.693445       24.2        34.7
. log close
          log:  c:\example\session.smcl
     log type:  smcl
    closed on:  17 Dec 2002, 12:35:38
```

There is now a file named `session.smcl` on your disk. If you were to look at it in your word processor, you would see something like this:

```
{smcl}
{com}{sf}{ul off}{txt}{.-}
      log:  {res}c:\example\session.smcl
  {txt}log type:  {res}smcl
  {txt}opened on:  {res}17 Dec 2002, 12:35:08
{txt}
{com}. use http://www.stata-press.com/data/r8/census
{txt}(1980 Census data by state)

{com}. tabulate reg [freq=pop]

    {txt}Census {c |}
    region {c |}      Freq.      Percent        Cum.
{hline 12}{c +}{hline 35}
        NE {c |}{res}  49135283        21.75       21.75
{txt}    N Cntrl {c |}{res}  58865670        26.06        47.81
  (output omitted )
```

What you are seeing is SMCL; SMCL stands for Stata Markup and Control Language, and you can read about it, if you are interested, in [P] **smcl**.

Stata understands SMCL. Here is the result of typing the file using Stata's `type` command:

```
. type session.smcl
```

```
      log:  c:\example\session.smcl
 log type:  smcl
opened on:  17 Dec 2002, 12:35:08
. use http://www.stata-press.com/data/r8/census
(1980 Census data by state)
. tabulate reg [freq=pop]
   Census |
   region |      Freq.     Percent        Cum.
----------+-----------------------------------
       NE |   49135283       21.75       21.75
   N Cntrl |   58865670       26.06       47.81
    South |   74734029       33.08       80.89
     West |   43172490       19.11      100.00
----------+-----------------------------------
    Total |  225907472      100.00
. summarize medage

 Variable |       Obs        Mean    Std. Dev.       Min        Max
----------+---------------------------------------------------------
   medage |        50       29.54    1.693445       24.2       34.7
. log close
      log:  c:\example\session.smcl
 log type:  smcl
closed on:  17 Dec 2002, 12:35:38
```

```
. _
```

What you will see is a perfect copy of what you previously saw. If you use Stata to print the file, you will get a perfect printed copy, too.

SMCL files can be translated to ASCII text, which is a format more useful for inclusion into a word processing document. If you type 'translate *filename*.smcl *filename*.log', Stata will translate *filename*.smcl to ASCII and store the result in *filename*.log:

```
. translate session.smcl session.log
```

The resulting file session.log looks like this:

```
-----------------------------------------------------------------------------
      log:  c:\example\session.smcl
 log type:  smcl
opened on:  17 Dec 2002, 12:35:08
. use http://www.stata-press.com/data/r8/census
(1980 Census data by state)
. tabulate reg [freq=pop]
   Census |
   region |      Freq.     Percent        Cum.
----------+-----------------------------------
       NE |   49135283       21.75       21.75
   N Cntrl |   58865670       26.06       47.81
    South |   74734029       33.08       80.89
  (output omitted )
```

When you use translate to create *filename*.log from *filename*.smcl, *filename*.log must not already exist:

```
. translate session.smcl session.log
file session.log already exists
r(602);
```

If the file does already exist and you wish to overwrite the existing copy, then you can specify the `replace` option:

```
. translate session.smcl session.log, replace
```

See [R] **translate** for more information.

If you prefer, you can skip the SMCL and create ASCII text logs directly, either by specifying that you want the log in `text` format,

```
. log using session, text
```

or by specifying that the file to be created is to be a `.log` file:

```
. log using session.log
```

18.1.2 Appending to an existing log

You previously typed `log using session` and later typed `log close`. In some future Stata session, you type `log using session` again. Here is what happens:

```
. log using session
file c:\example\session.smcl already exists
r(602);
```

Stata never lets you accidentally write over a file. You have three choices: (1) choose a different name; (2) append onto the end of the existing file by typing `log using session, append`; or (3) replace the existing file by typing `log using session, replace`. In this last case, you are telling Stata that you know the file already exists and that it is okay to replace it.

18.1.3 Temporarily suspending and resuming logging

Once you are logging your session, you can turn logging on and off. When you turn logging off, Stata temporarily stops recording your session but leaves the log file open. When you turn logging back on, Stata continues to record your session, appending the additional record to the end of the file.

For instance, say the first time something interesting happens, you type `log using results` (or click on **Log** and open `results.smcl`). You then retype the command that produced the interesting result (or double-click on the command in the Review window, or use the editing key PrevLine to retrieve the command; see [U] **13 Keyboard use**). You now have a hard copy of the interesting result.

You are now reasonably sure that nothing interesting will occur, at least for a while. Rather than type `log close`, however, you type `log off`, or you click on **Log** and choose **Suspend**. From now on, nothing goes into the file. The next time something interesting happens, you type `log on` (or click on **Log** and choose **Resume**) and reissue the (interesting) command. After that, you type `log off`. You keep working like this—toggling the log on and off.

18.2 Placing comments in logs

Everything you type and everything Stata types in response goes into the log.

Stata treats lines starting with a '*' as comments and ignores them. Thus, if you are working interactively and wish to make a comment, you can type '*' followed by your comment:

```
. * check that all the spells are completed
. _
```

Stata ignores your comment but, if you have a log going, the comment now appears in the file.

❑ Technical Note

log can be combined with #review (see [U] **13 Keyboard use**) to bail you out when you have not adequately planned ahead. You have been working in front of your computer, and you now realize that you have done what you wanted to do. Unfortunately, you are not sure exactly what it is you have done. Did you make a mistake? Could you reproduce the result? Unfortunately, you have not been logging your output. #review will allow you to look over what you have done, and, combined with log, you can still make a record.

Type log using *filename*. Type #review 100. Stata will list the last 100 commands you gave, or however many it has stored. Since log is making a record, that list will also be stored in the file. Finally, type log close.

❑

18.3 Logging only what you type

Log files record everything that happens during a session, both what you type and what Stata produces in response.

Stata can also produce cmdlog files—files that contain only what you type. These files are perfect for later going back and creating a Stata do-file.

cmdlog creates cmdlog files and its basic syntax is

cmdlog using *filename* [, replace]	creates *filename*.txt
cmdlog off	temporarily suspends command logging
cmdlog on	resumes command logging
cmdlog close	closes the command log file

See [R] **log** for all the details.

Cmdlogs are plain ASCII text files. Were you to type

```
. cmdlog using session
(cmdlog c:\example\session.txt opened)
. use http://www.stata-press.com/data/r8/census
(Census Data)
. tabulate reg [freq=pop]
  (output omitted )
. summarize medage
  (output omitted )
. cmdlog close
(cmdlog c:\example\session.txt closed)
```

then file `mycmds.txt` would contain

```
use census
tabulate reg [freq=pop]
summarize medage
```

You can create both kinds of logs—full session logs and cmdlogs—simultaneously, if you wish.

18.4   The log-button alternative

The capabilities of the `log` command (but not the `cmdlog` command) are available in Stata's GUI interface; just click on the **Log** button.

You can use the Viewer to view logs, even logs that are in the process of being created. Just pull down **File** and select **View**. If you are currently logging, the filename to view will already be filled in with the current log file and all you need do is click **OK**. Periodically, you can click the **Refresh** button to bring the Viewer up to date.

You can also use the Viewer to view prior logs.

You can access the Viewer by pulling down **File** and choosing **View**, or you can use the `view` command:

```
. view myoldlog.smcl
```

18.5   Printing logs

You print logs from the Viewer. Pull down **File** and choose **View**, or type 'view *whatever*' from the command line to load the log into the viewer, and then pull down **File** and choose **Print Viewer**.

Alternatively, you can print logs by other means; see [R] **translate**.

19 Do-files

Contents

19.1   Description
    19.1.1   Version
    19.1.2   Comments and blank lines in do-files
    19.1.3   Long lines in do-files
    19.1.4   Error handling in do-files
    19.1.5   Logging the output of do-files
    19.1.6   Preventing –more– conditions
19.2   Calling other do-files
19.3   Ways to run a do-file (Stata for Windows)
19.4   Ways to run a do-file (Stata for Macintosh)
19.5   Ways to run a do-file (Stata for Unix)
19.6   Programming with do-files
    19.6.1   Argument passing
    19.6.2   Suppressing output

19.1 Description

Rather than typing commands at the keyboard, you can create a disk file containing commands and instruct Stata to execute the commands stored in that file. Such files are called *do-files*, since the command that causes them to be executed is do.

A do-file is a standard ASCII text file.

A do-file is executed by Stata when you type do *filename*.

Stata users can use any text editor to create do-files, or they can use the built-in do-file editor; see [GS] **15 Using the Do-file Editor**.

▷ Example

You can use do-files to create a batch-like environment in which you place all the commands you want to perform in a file and then instruct Stata to do that file. For instance, assume you use your text editor or word processor to create a file called myjob.do that contains the three lines:

```
──────────────────────────────────────────── top of myjob.do ────────────
use http://www.stata-press.com/data/r8/census5
tabulate region
summarize marriage_rate divorce_rate median_age if state!="Nevada"
──────────────────────────────────────────── end of myjob.do ────────────
```

You then enter Stata and instruct Stata to do the file:

```
. do myjob

. use http://www.stata-press.com/data/r8/census5
(1980 Census data by state)
```

```
. tabulate region

   Census
   region │     Freq.      Percent         Cum.
──────────┼─────────────────────────────────────
       NE │         9        18.00        18.00
   N Cntrl │        12        24.00        42.00
    South │        16        32.00        74.00
     West │        13        26.00       100.00
──────────┼─────────────────────────────────────
    Total │        50       100.00

. summarize marriage_rate divorce_rate median_age if state !="Nevada"

  Variable │       Obs        Mean    Std. Dev.        Min         Max
───────────┼──────────────────────────────────────────────────────────
marriage_r~e │        49    .0106791    .0021746    .0074654    .0172704
divorce_rate │        49    .0054268    .0015104    .0029436     .008752
 median_age │        49    29.52653    1.708286        24.2        34.7
```

You typed only do myjob to produce this output. Since you did not specify the file extension, Stata assumed you meant do myjob.do; see [U] **14.6 File-naming conventions**.

◁

19.1.1 Version

We recommend that the first line in your do-file declare the Stata release under which you wrote the do-file; myjob.do would better read as

── top of myjob.do ────────────

```
version 8
use http://www.stata-press.com/data/r8/census
tabulate region
summarize marriage_rate divorce_rate median_age if state!="Nevada"
```

── end of myjob.do ────────────

We admit that we do not always follow our own advice, as you will see many examples in this manual that do not include the version 8 line.

If you intend to keep the do-file, however, you should include this line since it ensures that your do-file will continue to work with future versions of Stata. Stata is under constant development, and sometimes things change in surprising ways.

For instance, in Stata 3.0, a new syntax for specifying the weights was introduced. If you had an old do-file written for Stata 2.1 that analyzed weighted data and did not have version 2.1 at the top, you would find that today's Stata would flag some of its lines as syntax errors. If you had the version 2.1 line, it would work just as it used to.

In Stata 4.0, we updated the random-number generator uniform() — the new one is better in that it has a longer period. If you wrote a do-file back in the days of Stata 3.1 that made a bootstrap calculation of variance and it did not include version 3.1 at the top, it would now produce different (but equivalent) results. If you had included the line, it would produce the same results as it used to.

When running an old do-file that includes a version statement, you need not worry about setting the version back. Stata automatically restores the previous value of version when the do-file completes.

19.1.2 Comments and blank lines in do-files

You may freely include blank lines in your do-file. In the previous example, the do-file could just as well have read

```
─────────────────────────────────────────── top of myjob.do ───────────
version 8

use http://www.stata-press.com/data/r8/census
tabulate region
summarize marriage_rate divorce_rate median_age if state!="Nevada"
─────────────────────────────────────────────── end of myjob.do ───────────
```

There are four ways to include comments in a do-file.

1. Begin the line with a '*'; Stata ignores such lines.

2. Place the comment in /* */ delimiters.

3. Place the comment after two forward slashes; i.e., //. Everything after the // to the end of the current line is considered a comment.

4. Place the comment after three forward slashes; i.e., ///. Everything after the /// to the end of the current line is considered a comment. However, when you use ///, the next line joins with the current line.

❑ Technical Note

The /* */, //, and /// comment indicators can be used in do-files and ado-files only; you may not use them interactively. You can, however, use the '*' comment indicator interactively.

❑

`myjob.do` then might read

```
─────────────────────────────────────────── top of myjob.do ───────────
* a sample analysis job
version 8

use http://www.stata-press.com/data/r8/census

/* obtain the summary statistics: */
tabulate region
summarize marriage_rate divorce_rate median_age if state!="Nevada"
─────────────────────────────────────────────── end of myjob.do ───────────
```

or, equivalently,

```
─────────────────────────────────────────── top of myjob.do ───────────
// a sample analysis job
version 8

use http://www.stata-press.com/data/r8/census

// obtain the summary statistics:
tabulate region
summarize marriage_rate divorce_rate median_age if state!="Nevada"
─────────────────────────────────────────────── end of myjob.do ───────────
```

Which style of comment indicator you use is up to you. One advantage of the /* */ method is that it can be put at the end of lines:

```
─────────────────────────────────────────── top of myjob.do ───────────
* a sample analysis job
version 8
```

———————————————————————————————— end of myjob.do ————

```
use http://www.stata-press.com/data/r8/census
tabulate region                    /* obtain summary statistics */
summarize marriage_rate divorce_rate median_age if state!="Nevada"
```

—— end of myjob.do ————

In fact, /* */ can be put anywhere, even in the middle of a line:

—— top of myjob.do ————

```
* a sample analysis job
version 8
use /* confirm this is latest */ http://www.stata-press.com/data/r8/census
tabulate region                    /* obtain summary statistics */
summarize marriage_rate divorce_rate median_age if state!="Nevada"
```

—— end of myjob.do ————

You can achieve the same results using the // and /// methods:

—— top of myjob.do ————

```
// a sample analysis job
version 8
use http://www.stata-press.com/data/r8/census
tabulate region                    // obtain summary statistics
summarize marriage_rate divorce_rate median_age if state!="Nevada"
```

—— end of myjob.do ————

or

—— top of myjob.do ————

```
// a sample analysis job
version 8
use /// confirm this is latest
http://www.stata-press.com/data/r8/census
tabulate region                    // obtain summary statistics
summarize marriage_rate divorce_rate median_age if state!="Nevada"
```

—— end of myjob.do ————

19.1.3   Long lines in do-files

When you use Stata interactively, you press *Enter* to end a line and tell Stata to execute it. If you need to type a line wider than the screen, you just do, letting it wrap or scroll.

You can follow the same procedure in do-files—if your editor or word processor will let you—but you can do better. You can change the end-of-line delimiter to ';' using #delimit, you can comment out the line break using /* */ comment delimiters, or you can use the /// comment indicator.

▷ Example

In the following fragment of a do-file, we temporarily change the end-of-line delimiter:

—— fragment of example.do ————

```
use mydata
#delimit ;
summarize weight price displ headroom rep78 length turn gear_ratio
        if substr(company,1,4)=="Ford" |
            substr(company,1,2)=="GM", detail ;
gen byte ford = substr(company,1,4)=="Ford" ;
#delimit cr
gen byte gm = substr(company,1,2)=="GM"
```

—— fragment of example.do ————

Once we change the line delimiter to semicolon, all lines, even short ones, must end in semicolons. Stata treats carriage returns as no different from blanks. We can change the delimiter back to carriage return by typing #delimit cr.

The #delimit command is allowed only in do-files—it is not allowed interactively. You need not remember to set the delimiter back to carriage return at the end of a do-file because Stata will reset it automatically.

◁

▷ Example

The other way around long lines is to comment out the carriage return using /* */ comment brackets or to use the /// comment indicator . Thus, our code fragment could also read

——————————————————————————————— fragment of example.do ———————————

```
use mydata
summarize weight price displ headroom rep78 length turn gear_ratio /*
      */  if substr(company,1,4)=="Ford" |    /*
      */      substr(company,1,2)=="GM", detail
gen byte ford = substr(company,1,4)=="Ford"
gen byte gm = substr(company,1,2)=="GM"
```

——————————————————————————————— fragment of example.do ———————————

or

——————————————————————————————— fragment of example.do ———————————

```
use mydata
summarize weight price displ headroom rep78 length turn gear_ratio ///
          if substr(company,1,4)=="Ford" |    ///
            substr(company,1,2)=="GM", detail
gen byte ford = substr(company,1,4)=="Ford"
gen byte gm = substr(company,1,2)=="GM"
```

——————————————————————————————— fragment of example.do ———————————

◁

19.1.4 Error handling in do-files

A do-file completes execution when (1) the end of the file is reached, (2) an exit is executed, or (3) an error (nonzero *return code*) occurs. If an error occurs, the remaining commands in the do-file are not executed.

If you press *Break* while executing a do-file, Stata responds as though an error occurred, stopping the do-file. This is because the return code is nonzero; see [U] **11 Error messages and return codes** for an explanation of return codes.

▷ Example

Here is what happens when we execute a do-file and then press *Break*:

```
. do myjob2
. version 8
. use census
(Census data)
```

```
. tabulate region

    Census  |
    region  |        Freq.        Percent        Cum.
—Break—
r(1);

end of do-file
—Break—
r(1);

. _
```

When we pressed *Break*, Stata responded by typing —Break— and showed a return code of 1. Stata seemingly repeated itself, typing first "end of do-file", and then —Break— and the return code of 1 again. Do not worry about the repeated messages. The first message indicates that Stata was stopping the `tabulate` because you pressed *Break*, and the second message indicates that Stata is stopping the do-file for the same reason.

◁

▷ Example

Let's try our example again, but this time, let's introduce an error. We change the file `myjob2.do` to read

```
───────────────────────────────────────────── top of myjob2.do ─────────────
use censas
tabulate region
summarize marriage_rate divorce_rate median_age if state!="Nevada"
───────────────────────────────────────────── end of myjob2.do ─────────────
```

Note our subtle typographical error. We typed `use censas` when we meant `use census`. We assume that there is no file called `censas.dta`, so now we have an error. Here is what happens when you instruct Stata to `do` the file:

```
. do myjob2

. version 8

. use censas
file censas.dta not found
r(601);

end of do-file
r(601);

. _
```

When Stata was told to `use censas`, it responded with "file censas.dta not found" and a return code of 601. Stata then typed "end of do-file" and repeated the return code of 601. The repeated message occurred for the same reason it did when we pressed *Break* in the previous example. The `use` resulted in a return code of 601, so the do-file itself resulted in the same return code. The important thing to understand is that Stata stopped executing the file because there was an error.

◁

❏ Technical Note

We can tell Stata to continue executing the file even if there are errors by typing do *filename*, `nostop`. Here is the result:

```
. do myjob2, nostop
. version 8
. use censas
file censas.dta not found
r(601);
. tabulate region
no variables defined
r(111);
summarize marriage_rate divorce_rate median_age if state!="Nevada"
no variables defined
r(111);
end of do-file

. _
```

None of the commands worked because the do-file's first command failed. That is why Stata ordinarily stops. However, if our file contained anything that could work, it would work.

❑

19.1.5 Logging the output of do-files

You log the output of do-files just as you would an interactive session; see [U] **18 Printing and preserving output**.

Many users include the commands to start and stop the logging in the do-file itself:

```
──────────────────────────────────── top of myjob3.do ──────────
version 8
log using myjob3, replace
* a sample analysis job
use http://www.stata-press.com/data/r8/census

tabulate region                  // obtain summary statistics
summarize marriage_rate divorce_rate median_age if state!="Nevada"
log close
────────────────────────────────────── end of myjob3.do ──────────
```

We chose to open with `log using myjob3, replace`, the important part being the `replace` option. Had we omitted the option, we could not easily rerun our do-file. If `myjob3.smcl` already existed and `log` was not told that it is okay to replace the file, the do-file would have stopped and instead reported "file myjob3.smcl already exists". We could get around that, of course, by erasing the log file before running the do-file.

19.1.6 Preventing —more— conditions

Assume you are running a do-file and logging the output so that you can look at it later. In that case, Stata's feature of pausing every time the screen is full is just an irritation: It means you have to sit and watch the do-file run so you can clear the —more—.

The way around this is to include the line `set more off` in your do-file. Setting more to `off`, as explained in [U] **10 —more— conditions**, prevents Stata from ever issuing a —more—.

19.2   Calling other do-files

Do-files may call other do-files. For instance, say you wrote `makedata.do` that infiles your data, generates a few variables, and saves `step1.dta`. Say you wrote `anlstep1.do` that performed a little analysis on `step1.dta`. You could then create a third do-file,

─── top of master.do ───────────
```
version 8
do makedata
do anlstep1
```
─── end of master.do ───────────

and so, in effect, combine the two do-files.

Do-files may call do-files, which, in turn, call do-files, and so on. Stata allows do-files to be nested 64 deep.

Be not confused: `master.do` above could call 1,000 do-files one after the other and still the maximum level of nesting would be only two.

19.3   Ways to run a do-file (Stata for Windows)

1. You can execute do-files by typing `do` followed by the filename as we did above.

2. Alternatively, you can execute do-files by pulling down **File** and choosing **Do...**.

3. You can use the do-file editor to compose, save, load, and execute do-files; see [GSW] **15 Using the Do-file Editor**. Click on the **Do-file Editor** button or type `doedit` in the Command window.

Regardless of the method chosen, if you wish to log the output, you can start the log before executing the do-file or you can include the `log using` and `log close` in your do-file.

4. You can double-click on the do-file to launch Stata and run the do-file. When the do-file completes, Stata will prompt you for the next command just as if you had started Stata the normal way. If you want Stata to exit instead, include `exit, STATA clear` as the last line of your do-file. If you want to log the output, you should include the `log using` and `log close` in your do-file.

5. You can run the do-file in batch mode. See [GSW] **A.9 Executing Stata in background (batch) mode** for details, but the short explanation is that you open a DOS window and type

       C:\DATA> c:\stata\wstata /s do myjob

   or

       C:\DATA> c:\stata\wstata /b do myjob

   to run in batch mode, assuming you have installed Stata in the folder `c:\stata`. /b and /s determine the kind of log produced, but put that aside for a second. What is important is that when you start Stata in these ways, Stata will run in the background. When the do-file completes, the Stata icon on the taskbar will flash. You can then click on it to close Stata. If you want to stop the do-file before it completes, click on the Stata icon on the taskbar, and Stata will ask you if you want to cancel the job.

   When you run Stata in these ways, Stata takes the following actions:

   a. Stata automatically opens a log. If you specified /s, Stata will open a SMCL log; if you specified /b, Stata will open an ASCII text log. If your do-file is named *xyz*.`do`, the log will be called *xyz*.`smcl` (/s) or *xyz*.`log` (/b) in the same directory.

   b. If your do-file explicitly opens another log, that is okay. You will end up with two copies of the output.

   c. Stata ignores —more— conditions and anything else that would cause the do-file to stop were it running interactively.

19.4   Ways to run a do-file (Stata for Macintosh)

1. You can execute do-files by typing do followed by the filename as we did above.

2. Alternatively, you can execute do-files by pulling down **File** and choosing **Do...**.

3. Alternatively, with Stata running, you can go to the desktop and double-click on the do-file.

4. You can use the do-file editor to compose, save, load, and execute do-files; see [GSM] **15 Using the Do-file Editor**. Click on the **Do-file Editor** button or type doedit in the Command window.

Regardless of the method chosen, if you wish to log the output, you can start the log before executing the do-file or you can include the log using and log close in your do-file.

5. With Stata not running, you can double-click on the do-file to launch Stata and run the do-file. When the do-file completes, Stata will prompt you for the next command just as if you had started Stata the normal way. If you want Stata to exit instead, include exit, STATA clear as the last line of your do-file. If you want to log the output, you should include the log using and log close in your do-file.

19.5   Ways to run a do-file (Stata for Unix)

1. You can execute do-files by typing do followed by the filename as we did above.

2. Alternatively, you can execute do-files by pulling down **File** and choosing **Do...**.

3. You can use the do-file editor to compose, save, load, and execute do-files; see [GSW] **15 Using the Do-file Editor**. Click on the **Do-file Editor** button or type doedit in the Command window.

Regardless of the method chosen, if you wish to log the output, you can start the log before executing the do-file or you can include the log using and log close in your do-file.

4. At the Unix prompt, you can type
     $ xstata do *filename*
or
     $ stata do *filename*
to bring up Stata and run the do-file. When the do-file completes, Stata will prompt you for the next command just as if you had started Stata the normal way. If you want Stata to exit instead, include exit, STATA clear as the last line of your do-file. If you want to log the output, you should include the log using and log close in your do-file.

5. At the Unix prompt, you can type
    $ stata -s do *filename* &
or
    $ stata -b do *filename* &
Do this and Stata runs the do-file in the background. Note that the above two examples both involve the use of stata, not xstata. Type stata even if you usually use the GUI version of Stata, xstata. The examples differ only in that one specifies the -s option and the other the -b option, and that determines the kind of log that will be produced. In the above examples, Stata takes the following actions:

  a. Stata automatically opens a log. If you specified -s, Stata will open a SMCL log; if you specified -b, Stata will open an ASCII text log. If your do-file is named *xyz*.do, the log will be called *xyz*.smcl (-s) or *xyz*.log (-b) in the current directory (the directory from which you issued the stata command).

  b. If your do-file explicitly opens another log, that is okay. You will end up with two copies of the output.

  c. Stata ignores —more— conditions and anything else that would cause the do-file to stop were it running interactively.

Thus, to reiterate, one way to run a do-file in the background and obtain an ASCII text log is by typing

    $ stata -b do myfile &

Another way uses standard redirection:

    $ stata < myfile.do > myfile.log &

The first way is slightly more efficient. Either way, Stata knows it is in the background and ignores —more— conditions and anything else that would cause the do-file to stop were it running interactively. However, if your do-file contains either the #delimit command or the comment characters (/* at the end of one line and */ at the beginning of the next), the second method will not work. We recommend that you use the first method: 'stata -b do myfile &'.

The choice between 'stata -b do myfile &' and 'stata -s do myfile &' is more personal. We prefer obtaining SMCL logs (-s) because they look better when printed, and, in any case, they can always be converted to ASCII text format using translate; see [R] **translate**.

19.6   Programming with do-files

This is an advanced topic, and we are going to refer to concepts not yet explained; see [U] **21 Programming Stata** for further information.

19.6.1   Argument passing

Do-files accept arguments just as Stata programs do; this is described in [U] **21 Programming Stata** and [U] **21.4 Program arguments**. In fact, the logic Stata follows when invoking a do-file is the same as when invoking a program: the local macros are saved and new ones are defined. Arguments are stored in the local macros '1', '2', and so on. When the do-file completes, the previous definitions are restored just as with programs.

Thus, if you wanted your do-file to

1. use a dataset of your choosing,

2. tabulate a variable named `region`, and

3. summarize variables `marriage_rate` and `divorce_rate`,

you could write the do-file

――――――――――――――――――――――――――――――― top of myxmpl.do ―――――――
```
use '1'
tabulate region
summarize marriage_rate divorce_rate
```
――――――――――――――――――――――――――― end of myxmpl.do ―――――――

and you could run this do-file by typing, for instance,

```
. do myxmpl census
  (output omitted )
```

The first command—`use '1'`—would be interpreted as `use census` because `census` was the first argument you typed after `do myxmpl`.

An even better version of the do-file would read

――――――――――――――――――――――――――――――― top of myxmpl.do ―――――――
```
args varname
use 'varname'
tabulate region
summarize marriage_rate divorce_rate
```
――――――――――――――――――――――――――― end of myxmpl.do ―――――――

The `args` command merely assigns a better name to the argument passed. `args varname` does not verify that what we type following `do myxmpl` is a variable name—we would have to use the `syntax` command if we wanted to do that—but substituting `'varname'` for `'1'` does make the code more readable.

If our program were to receive two arguments, we could refer to them as `'1'` and `'2'`, or we could put an `args varname other` at the top of our do-file and then refer to `'varname'` and `'other'`.

To learn more about argument passing, see [U] **21.4 Program arguments**.

19.6.2 Suppressing output

There is an alternative to typing do *filename*; it is run *filename*. `run` works in the same way as `do`, except that neither the instructions in the file nor any of the output caused by those instructions is shown on the screen or in the log file.

For instance, using the above `myxmpl.do`, typing `run myxmpl census` results in

```
. run myxmpl census

. _
```

All the instructions were executed, but none of the output was shown.

This is not useful in this case, but if the do-file contained only the definitions of Stata programs—see [U] **21 Programming Stata**—and you merely wanted to load the programs without seeing the code, `run` would be useful.

20 Ado-files

Contents

20.1   Description
20.2   What is an ado-file?
20.3   How can I tell if a command is built-in or an ado-file?
20.4   Can I look at an ado-file?
20.5   Where does Stata look for ado-files?
    20.5.1   Where are the official ado directories?
    20.5.2   Where is my personal ado directory?
20.6   How do I install an addition?
20.7   How do I add my own ado-files?
20.8   How do I install official updates?

20.1   Description

Stata is programmable, and even if you never write a Stata program, Stata's programmability is still important. Many of Stata's features are implemented as Stata programs, and new features are implemented every day, both by us and by others.

1. You can obtain additions from the *Stata Journal*. You subscribe to the printed journal, but the software additions are available for free over the Internet.

2. You can obtain additions from the Stata listserver. There is an active group of users advising each other on how to use Stata, and often, in the process, trading programs. Visit the Stata web site, *http://www.stata.com*, for instructions on how to subscribe; subscribing to the listserver is free.

3. A large and constantly growing number of Stata programs are available at the Boston College Statistical Software Components Archive. This is a distributed database. The Archive can be browsed and searched, and you can find links to the Archive from *http://www.stata.com*. Importantly, Stata knows how to access the Archive, and other places as well. You can search for additions using Stata's search, net command; see [R] **search**. The materials search, net finds can be immediately installed by Stata using the hyperlinks that will be displayed in the search, net in the Results window, or by using the net command. There is also a specialized command, ssc, that has a number of options available to aid you in locating and installing the user-written commands that are available from this site; see [R] **ssc**.

4. You may decide to write your own additions to Stata.

This chapter is written for people who want to consume ado-files. All users should read it. If you later decide you want to write ado-files, the technical description for programmers can be found in [U] **21.11 Ado-files**.

20.2   What is an ado-file?

An ado-file defines a Stata command, but not all Stata commands are defined by ado-files.

When you type summarize to obtain summary statistics, you are using a command built into Stata.

When you type ci to obtain confidence intervals, you are running an ado-file. The results of using a built-in command or a program are indistinguishable.

An ado-file is an ASCII text file that contains a Stata program. When you type a command that Stata does not know, it looks in certain places for an ado-file of that name. If Stata finds it, Stata loads and executes it, so it appears to you as if the ado-command is just another command built into Stata.

We just told you that Stata's ci command is implemented as an ado-file. That means that, somewhere, there is a file named ci.ado.

Ado-files tend to come with help files. When you type help ci (or pull down **Help** and work your way to ci's manual page), Stata looks for ci.hlp, just as it looks for ci.ado when you use the ci command. A help file is also an ASCII text file that tells Stata's help system what to display.

20.3   How can I tell if a command is built-in or an ado-file?

Use the which command to determine if a file is built in or implemented as an ado-file. For instance, logistic is an ado-file, and here is what happens when you type which logistic:

```
. which logistic
c:\stata\ado\base\l\logistic.ado
*! version 3.1.9  01oct2002
```

logit is a built-in command:

```
. which logit
built-in command:  logit
```

20.4   Can I look at an ado-file?

Certainly. When you type which followed by an ado-command, Stata reports where the file is stored:

```
. which logistic
c:\stata\ado\base\l\logistic.ado
*! version 3.1.9  01oct2002
```

Ado-files are just ASCII text files containing the Stata program, so you can type them (or even look at them in your editor or word processor):

```
. type c:\stata\ado\base\l\logistic.ado
*! version 3.1.9  01oct2002
program define logistic, eclass byable(recall)
        version 6.0
        local options '"Level(integer $S_level) COEF"'
  (output omitted)
end
```

You can also look at the corresponding help file in raw form if you wish. If there is a help file, it is stored in the same place as the ado-file:

```
. type c:\stata\ado\base\l\logistic.hlp, asis
{smcl}
{* 18nov2002}{...}
{hline}
help for {hi:logistic}{right:manual:  {hi:[R] logistic}    }
{right:dialogs:  {dialog logistic}  {dialog logit_p:predict}}
{hline}
  (output omitted)
```

20.5   Where does Stata look for ado-files?

Stata looks for ado-files in seven places, which can be categorized in three ways:

  I. the official ado-directories, meaning
      1. (UPDATES), the official updates directory, and
      2. (BASE), the official base directory;

 II. your personal ado directories, meaning
      3. (SITE), the directory for ado-files your site might have installed,
      4. (PLUS), the directory for ado-files you personally might have installed,
      5. (PERSONAL), the directory for ado-files you personally might have written, and
      6. (OLDPLACE), the directory where Stata users used to save their personally written ado-files; and

III. the current directory, meaning
      7. (.), the ado-files you have written just this instant or for just this project.

Where these directories are varies from computer to computer, but Stata's `sysdir` command will tell you where they are on your computer:

```
. sysdir
    STATA.  C.\STATA\
  UPDATES.  O.\STATA\ado\updates\
     BASE:  C:\STATA\ado\base\
     SITE:  C:\STATA\ado\site\
     PLUS:  C:\ado\plus\
 PERSONAL:  C:\ado\personal\
 OLDPLACE:  C:\ado\
```

20.5.1   Where are the official ado directories?

These are the directories listed as BASE and UPDATES by `sysdir`:

```
. sysdir
    STATA:  C:\STATA\
  UPDATES:  C:\STATA\ado\updates\
     BASE:  C:\STATA\ado\base\
     SITE:  C:\STATA\ado\site\
     PLUS:  C:\ado\plus\
 PERSONAL:  C:\ado\personal\
 OLDPLACE:  C:\ado\
```

1. BASE contains the ado-files we originally shipped to you.

2. UPDATES contains any updates you might have installed since then. You install these updates using the `update` command or by pulling down **Help** and choosing **Official Updates**; see [U] **20.8 How do I install official updates?**.

20.5.2   Where is my personal ado directory?

These are the directories listed as PERSONAL, PLUS, SITE, and OLDPLACE by `sysdir`:

```
. sysdir
    STATA:  C:\STATA\
  UPDATES:  C:\STATA\ado\updates\
     BASE:  C:\STATA\ado\base\
     SITE:  C:\STATA\ado\site\
     PLUS:  C:\ado\plus\
 PERSONAL:  C:\ado\personal\
 OLDPLACE:  C:\ado\
```

1. PERSONAL is for ado-files you personally have written. Store your private ado-files here; see [U] **20.7 How do I add my own ado-files?**.

2. PLUS is for ado-files you personally installed but did not write. Such ado-files are usually obtained from the SJ, but they are sometimes found in other places, too. You find and install such files using Stata's `net` command, or you can pull down **Help** and select **SJ and User-written Programs**; see [U] **20.6 How do I install an addition?**.

3. SITE is really the opposite of a personal ado directory—it is a public directory, and it corresponds to PLUS in a public sort of way. If you are on a networked computer, the site administrator can install ado-files here, and all Stata users will then be able to use them just as if each found and installed them in their PLUS directory for themselves. Site administrators find and install the ado-files just as you would, using Stata's `net` command, the difference being that they specify an option when they say to install something that tells Stata to write the files into SITE rather than PLUS. Site administrators should see [R] **net**.

4. OLDPLACE is for old Stata users. Prior to Stata 6, all "personal" ado-files, whether personally written or just personally installed, were written in the same directory—OLDPLACE. So that they do not have to go back and rearrange what they have already done, Stata still looks in OLDPLACE.

20.6   How do I install an addition?

Additions come in three flavors:

1. User-written additions, which you might find in the SJ, etc.
   Discussed here.

2. Ado-files you have written.
   Discussed in [U] **20.7 How do I add my own ado-files?**. If you have an ado-file obtained from the Stata listserver or a friend, treat it as belonging to this case.

3. Official updates provided by StataCorp.
   Discussed in [U] **20.8 How do I install official updates?**.

User-written additions which you might find in the *Stata Journal* (SJ), etc. are obtained over the Internet. To access the copy on the Internet,

1. pull down **Help** and choose **SJ and User-written Programs** and click on *http://www.stata.com*,

or

2. type `net from http://www.stata.com`.

What to do next should be obvious, but, in case it is not, see

Windows:	[GSW] **20 Using the Internet**
Macintosh:	[GSM] **20 Using the Internet**
Unix:	[GSU] **20 Using the Internet**

Also see [U] **32 Using the Internet to keep up to date** and [R] **net**.

20.7   How do I add my own ado-files?

You write a Stata program (see [U] **21 Programming Stata**), store it in a file ending in .ado, perhaps write a help file, and copy everything to the directory sysdir lists as PERSONAL:

```
. sysdir
    STATA:   C:\STATA\
  UPDATES:   C:\STATA\ado\updates\
     BASE:   C:\STATA\ado\base\
     SITE:   C:\STATA\ado\site\
     PLUS:   C:\ado\plus\
 PERSONAL:   C:\ado\personal\
 OLDPLACE:   C:\ado\
```

In this case, we would copy the files to c:\ado\personal.

While you are writing your ado-file, it is sometimes convenient to store the pieces in the current directory. Do that if you wish; you can move them to your personal ado directory when the program is debugged.

20.8   How do I install official updates?

Updates are available over the Internet:

1. pull down **Help** and select **Official Updates**, and then click on *http://www.stata.com*

or

2. type update from http://www.stata.com.

What to do next should be obvious, but, in case it is not, see

Windows:	[GSW] **20 Using the Internet**
Macintosh:	[GSM] **20 Using the Internet**
Unix:	[GSU] **20 Using the Internet**

Also see [U] **32 Using the Internet to keep up to date** and [R] **net**.

The official updates include bug fixes and new features. The official updates never change the syntax of an existing command, nor do they change the way Stata works. They add a new feature—which you can use or ignore—or they fix a bug.

Once you have installed the updates, you can enter Stata and type help whatsnew (or pull down **Help** and click **What's new**) to learn about what has changed.

21 Programming Stata

Contents

21.1   Description
21.2   Relationship between a program and a do-file
21.3   Macros
        21.3.1   Local macros
        21.3.2   Global macros
        21.3.3   The difference between local and global macros
        21.3.4   Macros and expressions
        21.3.5   Double quotes
        21.3.6   Extended macro functions
        21.3.7   Macro expansion operators and function
        21.3.8   Advanced local macro manipulation
        21.3.9   Advanced global macro manipulation
        21.3.10   Constructing Windows filenames using macros
        21.3.11   Accessing system values
        21.3.12   Referencing characteristics
21.4   Program arguments
        21.4.1   Renaming positional arguments
        21.4.2   Incrementing through positional arguments
        21.4.3   Using macro shift
        21.4.4   Parsing standard Stata syntax
        21.4.5   Parsing immediate commands
        21.4.6   Parsing nonstandard syntax
21.5   Scalars and matrices
21.6   Temporarily destroying the data in memory
21.7   Temporary objects
        21.7.1   Temporary variables
        21.7.2   Temporary scalars and matrices
        21.7.3   Temporary files
21.8   Accessing results calculated by other programs
21.9   Accessing results calculated by estimation commands
21.10   Saving results
        21.10.1   Saving results in r()
        21.10.2   Saving results in e()
        21.10.3   Saving results in s()
21.11   Ado-files
        21.11.1   Version
        21.11.2   Comments and long lines in ado-files
        21.11.3   Debugging ado-files
        21.11.4   Local subroutines
        21.11.5   Development of a sample ado-command
        21.11.6   Writing online help
21.12   A compendium of useful commands for programmers
21.13   References

This is an advanced topic. Some Stata users live productive lives without ever programming Stata. After all, you do not need to know how to program Stata to input data, create new variables, and fit models. On the other hand, programming Stata is not difficult—at least if the problem is not difficult—and Stata's programmability is one of its best features. The real power of Stata is not revealed until you program it.

If you are uncertain whether to read this chapter, we recommend you start reading and then bail out when it gets too arcane for you. You will learn things about Stata that you may find useful even if you never write a Stata program.

For those who want even more, we offer courses over the Internet on Stata programming; see [U] **2.7 NetCourses**.

21.1   Description

When you type a command that Stata does not recognize, the first thing Stata does is look in its memory for a program of that name. If Stata finds it, Stata executes the program.

There is no Stata command named `hello`,

```
. hello
unrecognized command
r(199);
```

but there could be if you defined a program named `hello`, and after that, the following might happen when you typed `hello`:

```
. hello
hi there
.  _
```

This would happen if, beforehand, you had typed

```
. program hello
  1. display "hi there"
  2. end
.  _
```

That is the overview of programming. A program is defined by

> `program` *progname*
> *Stata commands*
> end

and it is executed by typing *progname* to Stata's dot prompt.

21.2   Relationship between a program and a do-file

Programs and do-files are not much different to Stata, which is to say, Stata treats a do-file in the same way it treats a program. Below we will discuss argument passing, consuming results from Stata commands, and other topics, but realize that everything we say applies equally to do-files and programs.

The differences between a program and a do-file are

1. You invoke a do-file by typing do *filename*. You invoke a program by simply typing the program's name.

2. Programs must be defined (loaded) before they are used, whereas all that is required to run a do-file is that the file exist. There are ways to make programs load automatically, however, so this difference is not of great importance.

3. When you type do *filename*, Stata displays the commands it is executing and the results. When you type *progname*, Stata shows only the results. The display of the underlying commands is suppressed. This is an important difference in outlook: In a do-file, how it does something is as important as what it does. In a program, the how is no longer important. One might think of the program as a new feature of Stata.

Let us now mention some of the similarities:

1. Arguments are passed to programs and do-files in the same way.

2. Programs and do-files are implemented in terms of the same Stata commands, which is to say, any Stata commands.

3. Programs may call other programs. Do-files may call other do-files. Programs may call do-files (rarely happens) and do-files may call programs (often happens). Stata allows programs (and do-files) to be nested up to 64 deep.

Now, here is the interesting thing: *programs are typically defined in do-files* (or in a variant of do-files called ado-files; we will get to that later).

That is, you *can* define a program interactively, and that is useful for pedagogical purposes. However, in real applications, you compose programs in an editor or word processor and store the definition in a do-file.

You have already seen your first program:

```
program hello
        display "hi there"
end
```

You *could* type those commands interactively, but if the body of the program were more complicated, that would be inconvenient. So instead, imagine you typed the commands into a do-file:

```
───────────────────────────────────────────── top of hello.do ──────────
program hello
        display "hi there"
end
───────────────────────────────────────────── end of hello.do ──────────
```

Now, returning to Stata, you type

```
. do hello

. program hello
  1.        display "hi there"
  2. end

.
end of do-file
```

Do you see that typing do hello did nothing but load the program? Typing do hello is the same as typing out the program's definition because that is all the do-file contains. Understand that the do-file was executed, but all that the statements in the do-file did was define the program hello. Now that the program is loaded, we can use it interactively:

```
. hello
hi there
```

So, that is one way you could use do-files and programs together. If you wanted to create new commands for interactive use, you could

1. Write the command as a `program ... end` in a do-file.

2. `do` the do-file before you use the new command.

3. Use the new command during the rest of the session.

There are more convenient ways to do this that automatically load the definition-providing do-file, but put that aside. The above method would work.

Another way we could use do-files and programs together is to put the definition and the execution together into a do-file:

```
────────────────────────────────────────────────────────────── top of hello.do ──────────
program hello
        display "hi there"
end
hello
────────────────────────────────────────────────────────────── end of hello.do ──────────
```

Here is what would happen if we executed this do-file:

```
. do hello

. program hello
  1.          display "hi there"
  2. end

. hello
hi there

.
end of do-file
```

Do-files and programs are often used in such combinations. Why? Pretend program `hello` is long and complicated and you have a problem where you need to do it twice. That would be a good reason to write a program. Moreover, you may wish to carry forth this procedure as a step of your analysis and, being cautious, do not want to perform this analysis interactively. You never intended program `hello` to be used interactively—it was just something you needed in the midst of a do-file—so you defined the program and used it there.

Anyway, there are lots of variations on this theme, but understand that few people actually sit in front of Stata and *interactively* type `program` and then compose a program. They instead do that in front of their editor or word processor. They compose the program in a do-file and then execute the do-file.

There is one other (minor) thing to know: Once a program is defined, Stata will not allow you to redefine it:

```
. program hello
hello already defined
r(110);
```

Thus, in our most recent do-file that defines and executes `hello`, we could not rerun it in the same Stata session:

```
. do hello

. program hello
hello already defined
r(110);
end of do-file
r(110);
```

That problem is solved by typing `program drop hello` before redefining it. We could do that interactively or we could modify our do-file:

```
──────────────────────────────────────── top of hello.do ────────────
program drop hello
program hello
        display "hi there"
end
hello
──────────────────────────────────────── end of hello.do ────────────
```

There is a problem with this solution. We can now rerun our do-file, but the first time we tried to run it in a Stata session, it would fail:

```
. do hello

. program drop hello
hello not found
r(111);

end of do-file
r(111);
```

The way around this conundrum is

```
──────────────────────────────────────── top of hello.do ────────────
capture program drop hello
program hello
        display "hi there"
end
hello
──────────────────────────────────────── end of hello.do ────────────
```

`capture` in front of a command makes Stata indifferent to whether the command works; see [P] **capture**. In real do-files containing programs, you will often see `capture program drop` followed by `program`.

You will want to learn about the `program` command itself—see [P] **program**. It manipulates programs. `program` can define programs, drop programs, and show you a directory of programs you have defined.

A program can contain any Stata command, but certain Stata commands are of special interest to program writers; see [U] **21.12 A compendium of useful commands for programmers**.

21.3 Macros

Before we can undertake programming, we must discuss macros.

Macros are the variables of Stata programs.

A *macro* is a string of characters, called the *macroname*, that stands for another string of characters, called the *macro contents*.

Macros come in two flavors, local and global. We will start with local macros because they are the most commonly used, but nothing really distinguishes local from global at this stage.

21.3.1   Local macros

Local macro names are up to 31 (not 32) characters in length.

One sets the contents of a local macro with the `local` command. In fact, we can do this interactively. We will begin by experimenting with macros in this way to learn about them. If we type

```
. local shortcut "myvar thisvar thatvar"
```

then 'shortcut' is a synonym for "myvar thisvar thatvar". Note the single quotes around `shortcut`. We said that sentence exactly the way we meant to because

if you type	'shortcut'
which is to say	left-single-quote shortcut right-single-quote
Stata hears	myvar thisvar thatvar

To access the contents of the macro, we use a left single-quote (located at the upper left on most keyboards), the macro name, and a right single-quote (located under the " on the right side of most keyboards).

The single quotes bracketing the macroname `shortcut` are called the macro-substitution characters. `shortcut` means `shortcut`. 'shortcut' means `myvar thisvar thatvar`.

So, if you were to type

```
. list 'shortcut'
```

the effect is exactly as if you typed

```
. list myvar thisvar thatvar
```

Macros can be used literally anywhere. For instance, if we also defined

```
. local cmd "list"
```

then we could type

```
. 'cmd' 'shortcut'
```

to mean `list myvar thisvar thatvar`.

As another example, consider the definitions

```
. local prefix "my"
. local suffix "var"
```

Then

```
. 'cmd' 'prefix''suffix'
```

would mean `list myvar`.

21.3.2   Global macros

Let's put aside why Stata has two kinds of macros—local and global—and focus right now on how global macros work.

The first difference is that global macros are allowed to have names that are up to 32 (not 31) characters in length.

The second difference is that you set the contents of a global macro using the `global` rather than the `local` command:

```
. global shortcut "alpha beta"
```

You obtain the contents of a global macro by prefixing its name with a dollar sign: $shortcut is equivalent to "alpha beta".

In the previous section, we defined a local macro named shortcut. That is a different macro. 'shortcut' is still "myvar thisvar thatvar".

Local and global macros may have the same names, but even if they do, they are unrelated and are still distinguishable.

At this stage, the only thing to know about global macros is that they are just like local macros except that you set their contents with global rather than local and that you substitute their contents by prefixing them with a $ rather than enclosing them in ' '.

21.3.3   The difference between local and global macros

The difference between local and global macros is that local macros are private and global macros are public.

Pretend you have written a program

```
program myprog
        code using local macro alpha
end
```

The local macro alpha in myprog is private in that no other program can modify or even look at alpha's contents. To make this point absolutely clear, assume your program looks like

```
program myprog
        code using local macro alpha
        mysub
        more code using local macro alpha
end
program mysub
        code using local macro alpha
end
```

Note that myprog calls mysub and that both programs use a local macro named alpha. Even so, the local macros in each program are different. mysub's alpha macro may contain one thing, but that has nothing to do with what myprog's alpha contains. Even when mysub begins execution, its alpha macro is different from myprog's. It is not that mysub's inherits myprog's alpha macro contents but is then free to change it. It is that myprog's alpha and mysub's alpha are completely different things.

When you write a program using local macros, you need not worry that some other program has been written using local macros with the same names. Local macros are just that: local to your program.

Global macros, on the other hand, are known to all programs. If both myprog and mysub use global macro beta, they are using the same macro. Whatever are the contents of $beta when mysub is invoked, those are the contents when mysub begins execution, and, whatever are the contents of $beta when mysub completes, those are the contents when myprog regains control.

21.3.4 Macros and expressions

From now on we are going to use local and global macros according to whichever is convenient; understand that whatever is said about one applies to the other.

Consider the definitions

```
. local one 2+2
. local two = 2+2
```

(which we could just as well have illustrated using the global command). In any case, note the equal sign in the second macro definition and the lack of the equal sign in the first. Formally, the first should be

```
. local one "2+2"
```

but Stata does not mind if we omit the double quotes in the local (global) statement.

local one 2+2 (with or without double quotes) copies the string 2+2 into the macro named one.

local two = 2+2 evaluates the expression 2+2, producing 4, and stores 4 in the macro named two.

That is, you type

local *macname contents*

if you want to copy *contents* to *macname*, and you type

local *macname* = *expression*

if you want to evaluate *expression* and store the result in *macname*.

In the second form, *expression* can be numeric or string. 2+2 is a numeric expression. As an example of a string expression,

```
. local res = substr("this",1,2) + "at"
```

stores that in res.

Since the expression can be either numeric or string, what is the difference between the following statements?

```
. local a "example"
. local b = "example"
```

Both statements store example in their respective macros. The first does it by a simple copy operation, whereas the second evaluates the expression "example", which is a string expression because of the double quotes, and which, in this case, evaluates to itself.

There is, however, a difference. Stata's expression parser is limited to handling strings of 244 characters in Stata/SE or 80 characters in Intercooled Stata or Small Stata. If the string is longer than this, it is truncated.

The copy operation of the first syntax is not limited—it can copy up to the maximum length of a macro, which is currently 67,784 characters for Intercooled Stata and 3,400 for Small Stata. For Stata/SE, the limit is $33 * c(\text{max\_k\_theory}) + 200$ characters, which for the default setting of 5,000 is 165,200.

To a programmer, the length limit for string expressions may seem limiting, but it turns out it is not, due to another feature discussed in [U] **21.3.6 Extended macro functions** below.

Put that aside. There are some other issues of using macros and expressions that look a little strange to programmers from other languages the first time they see them. For instance, pretend the macro 'i' contains 5. How would you increment it so that it contained $5 + 1 = 6$? The command is

```
local i = 'i' + 1
```

Do you see why the single quotes are on the right but not the left? Remember, 'i' refers to the contents of the local macro named i, which, we just said, is 5. Thus, after expansion, the line reads

```
local i = 5 + 1
```

which is the desired result.

There is a way to increment local macros that is more familiar to programmers, in particular, C programmers

```
local ++i
```

which is equivalent in every way to

```
local i = 'i' + 1
```

Alternatively, you can decrement a local macro using

```
local --i
```

It is useful to learn both ways to increment a variable, however, in case you wish to increment a local macro by 5, say, which is done using the first construct

```
local i = 'i' + 5
```

Finally, note that (unexpectedly, but logical given the way Stata parses macro names)

```
local i++
```

will *not* increment the local macro i, but instead redefines the local macro i to contain ++. There is, however, a context in which i++ (and i--) do work as expected; see [U] **21.3.7 Macro expansion operators and function**.

Now consider another local macro 'answ', which might contain yes or might contain no. In a program that was supposed to do something different based on answ's content, you would code

```
if "'answ'" == "yes" {
        ...
}
else {
        ...
}
```

Note the odd-looking "'answ'", and now think about the line after substitution. The line reads either

```
if "yes" == "yes" {
```

or

```
if "no" == "yes" {
```

either of which is the desired result. Had we omitted the double quotes, the line would have read

```
if no == "yes" {
```

(assuming 'answ' contains no), and that is not at all the desired result. As the line reads now, no would not be a string, but would be interpreted as a variable in the data.

The key to all of this is to think of the line after substitution.

21.3.5   Double quotes

Double quotes are used to enclose strings: `"yes"`, `"no"`, `"my dir\my file"`, `"'answ'"` (meaning the contents of local macro `answ`, treated as a string), and so on. Double quotes are used with macros,

```
local a "example"
if "'answ'" == "yes" {
       . . .
}
```

and double quotes are used by lots of Stata commands,

```
. regress lnwage age ed if sex=="female"
. gen outa = outcome if drug=="A"
. use "person file"
```

In fact, Stata has two sets of double-quote characters, of which `""` is one. The other is `'"  "'`, and they work the same way as `""`:

```
. regress lnwage age ed if sex=='"female"'
. gen outa = outcome if drug=='"A"'
. use '"person file"'
```

No rational user would use `'"  "'` (called compound double quotes) instead of `""` (called simple double quotes), but smart programmers do use them:

```
local a '"example"'
if '"'answ'"' == '"yes"' {
       . . .
}
```

Why is `'"example"'` better than `"example"`, `'"'answ'"'` better than `'"answ"'`, and `'"yes"'` better than `"yes"`? The answer is that only `'"'answ'"'` is better than `"'answ'"`; `'"example"'` and `'"yes"'` are no better—and no worse—than `"example"` and `"yes"`.

`'"'answ'"'` is better than `"'answ'"` because the macro `answ` might itself contain (simple or compound) double quotes. The really great thing about compound double quotes is that they nest. Pretend `'answ'` contained the string "I "think" so". Then,

Stata would find	`if "'answ'"=="yes"`
confusing because it would expand to	`if "I "think" so"=="yes"`
Stata would not find	`if '"'answ'"'=='"yes"'`
confusing because it would expand to	`if '"I "think" so"'=='"yes"'`

Open and close double quote in the simple form look the same; open quote is `"` and so is close quote. Open and close double quote in the compound form are distinguishable; open quote is `'"` and close quote is `"'`, and so Stata can pair the close with the corresponding open double quote. `'"I "think" so"'` is easy for Stata to understand, whereas `"I "think" so"` is a hopeless mishmash. (If you disagree, consider what `"A"B"C"` might mean. Is it the quoted string A"B"C or is it quoted string A followed by B followed by quoted string C?)

Since Stata can distinguish open from close quotes, even nested compound double quotes are understandable: `'"I '"think"' so"'`. (What does `"A"B"C"` mean? Either it means `'"A'"B"'C"'` or it means `'"A"'B'"C"'`.)

Yes, compound double quotes make you think your vision is stuttering, especially when combined with the macro substitution `' '` characters. That is why we rarely use them, even when writing programs. You do not have to use exclusively one or the other style of quotes. It is perfectly acceptable to code

```
local a "example"
if `""`answ'""' == "yes" {
        ...
}
```

using compound double quotes where it might be necessary (`""`answ'""'`) and using simple double quotes in other places (such as `"yes"`). It is also acceptable to use simple double quotes around macros (e.g., `"`answ'"`) if you are certain that the macros themselves do not contain double quotes or if you do not care what happens if they do.

There are instances where careful programmers should use compound double quotes, however. Later you will learn that Stata's `syntax` command interprets standard Stata syntax, and so makes it easy to write programs that understand things like

```
. myprog mpg weight if index(make,"VW")!=0
```

`syntax` works—we are getting ahead of ourselves—by placing the if *exp* typed by the user in the local macro `if`. Thus, `'if'` will contain "if index(make,"VW")!=0" in this case. Now, say you are at a point in your program where you want to know whether the user specified an if *exp*. It would be natural to code

```
if `""`if'""' != "" {
                // the if exp was specified
        ...
}
else {
                // it was not
        ...
}
```

Note that we used compound double quotes around the macro `'if'`. The local macro `'if'` might contain double quotes, so we placed compound double quotes around it.

21.3.6 Extended macro functions

In addition to allowing =*exp*, `local` and `global` provide what are known as *extended functions*. The use of an extended function is denoted by a colon (:) following the macro name. For instance,

```
local lbl : variable label myvar
```

copies the variable label associated with `myvar` to the local macro `lbl`. Thus, if `myvar` had the variable label `My variable`, the macro `lbl` would now contain "`My variable`", also. If `myvar` had no variable label, the macro `lbl` would contain nothing. The full list of extended functions is

{ global | local } *macroname* : char { *varname*[ ] | *varname*[*charname*] } | { _dta[ ] |
                                                              _dta[*charname*] }
{ global | local } *macroname* : colfullnames *matrixname*
{ global | local } *macroname* : colnames *matrixname*
{ global | local } *macroname* : coleq *matrixname*
{ global | local } *macroname* : constraint { dir| # }
{ global | local } *macroname* : data label
{ global | local } *macroname* : dir ["] *dir*["] { files | dirs | other } ["] *pattern*["]
                                        [, nofail]
{ global | local } *macroname* : display *display_directive*
{ global | local } *macroname* : e(scalars | macros | matrices | functions)
{ global | local } *macroname* : environment *envvar*   Unix and Windows only

$\{$ global | local $\}$ *macroname* : format *varname*

$\{$ global | local $\}$ *macroname* : label *valuelabelname* # $[\#]$

$\{$ global | local $\}$ *macroname* : label (*varname*) # $[\#]$

$\{$ global | local $\}$ *macroname* : label *valuelabelname* maxlength

$\{$ global | local $\}$ *macroname* : label (*varname*) maxlength

$\{$ global | local $\}$ *macroname* : length $\{$local | global$\}$ *macname*

$\{$ global | local $\}$ *macroname* : list *list_directive*

$\{$ global | local $\}$ *macroname* : permname *suggested_name* $[$length(#)$]$

$\{$ global | local $\}$ *macroname* : piece # # of "*string*"

$\{$ global | local $\}$ *macroname* : rowfullnames *matrixname*

$\{$ global | local $\}$ *macroname* : r(scalars | macros | matrices | functions)

$\{$ global | local $\}$ *macroname* : rownames *matrixname*

$\{$ global | local $\}$ *macroname* : roweq *matrixname*

$\{$ global | local $\}$ *macroname* : s(macros)

$\{$ global | local $\}$ *macroname* : sortedby

$\{$ global | local $\}$ *macroname* : subinstr $\{$local | global$\}$ *macname* "*string*" "*string*" $[,$
all word count($\{$local | global$\}$ *macname*) $]$

$\{$ global | local $\}$ *macroname* : sysdir *dir*

$\{$ global | local $\}$ *macroname* : tempvar

$\{$ global | local $\}$ *macroname* : tempfile

$\{$ global | local $\}$ *macroname* : tsnorm *string* $[,$ varname$]$

$\{$ global | local $\}$ *macroname* : type *varname*

$\{$ global | local $\}$ *macroname* : value label *varname*

$\{$ global | local $\}$ *macroname* : variable label *varname*

$\{$ global | local $\}$ *macroname* : word count *string*

$\{$ global | local $\}$ *macroname* : word # of *string*

See [P] **macro** for details and examples.

colfullname, colnames, and coleq return lists of matrix column names; see [U] **17.2.4 Obtaining row and column names**.

constraint $\{$ dir | # $\}$ gives information on constraints. constraint # puts constraint # in *macroname*, or returns "" if constraint # is not defined.

constraint dir returns an unsorted numerical list of those constraints that are currently defined.

data label returns the dataset label; see [U] **15.6.1 Dataset labels**.

dir ["]*dir*["] $\{$files | dirs | other$\}$ ["]*pattern*["] $[,$ nofail] puts in *macroname* the specified files, directories, or entries that are neither files nor directories, from directory *dir* and matching pattern *pattern*, where the pattern matching is defined by Stata's match(s_1, s_2) function.

nofail specifies that if the directory contains too many filenames to fit into a macro, the filenames that fit should be returned rather than an error issued.

display fills in *macroname* with the *display_directive*. This is, in effect, Stata's display command— see [P] **display**—with its output redirected to a macro. display used in this way does not allow in *color*, _newline, _continue, or _request(*macname*).

e(scalars | macros | matrices | functions) returns the names of all the saved results in e() of the specified type.

environment, allowed under Unix or Windows, imports the contents of the Unix environment variable *envvar* into the Stata macro.

format returns the display format associated with *varname*.

label looks up # in *valuelabelname* and returns the label value of # if *valuelabelname* exists and # is recorded in it, and returns # otherwise. If a second number is specified, the result is trimmed to being no more than that many characters in length.

If (*varname*) (note parentheses) is specified rather than *valuelabelname*, the logic is the same except that the value label for variable *varname*, if any, is used.

maxlength specifies that, rather than looking up a number in a value label, label is to return the maximum length of the labels themselves. For instance, if value label yesno mapped 0 to no and 1 to yes, then its maxlength would be 3 because yes is the longest label and it has three characters.

length returns the length (in # of characters) of *macname*, or 0 if *macname* is undefined.

list fills in *macroname* with the *list_directive*, which specifies one of many available commands/operators for working with macros that contain lists.

For example, if mylist contains "a b a c a", then

```
        local result : list uniq mylist
```

will store "a b c" in the local macro result. See [P] **macro lists** for a complete list of the available list directives.

permname *suggested_name* [length(#)] return a valid new variable name based on *suggested_name*, where *suggested_name* must follow naming conventions, but may be either too long or correspond to an already existing variable.

length(#) specifies the maximum length of the returned variable name, 32 by default.

piece $#_i$ $#_l$ of "*string*" returns the ith piece of *string* with length not to exceed l. The pieces are formed at word breaks if that is possible. Say 'stub' contained a potentially long string and you needed to display it in 20 columns on multiple lines. "local x: piece 1 20 of "'stub'"" would produce the first line of text. "local x: piece 2 20 of "'stub'"" would produce the second line, and so on, until piece returned "" (nothing) in x. See [P] **macro**.

r(scalars|macros|matrices|functions) returns the names of all the saved results in r() of the specified type.

rowfullname, rownames, and roweq return lists of matrix row names; see [U] **17.2.4 Obtaining row and column names**.

s(macros) returns the names of all the saved results in s() of type macros (which is the only type that exists within s()).

sortedby fills in the *macroname* with the list of variables by which the data are sorted, or leaves the macro undefined if the dataset is not sorted.

subinstr {local|global} *macname* "*string*$_{from}$" "*string*$_{to}$" changes the first occurrence of "*string*$_{from}$" to "*string*$_{to}$" in *macname*.

Option all specifies that all occurrences are to be changed.

Option word specifies that substitutions are allowed only on space-separated tokens (words). In this case, tokens at the beginning or end also count as being space separated.

The count() option places the number of substitutions that occur into the named macro.

sysdir *dir* substitutes directory names for the codenames. Different people have Stata installed in different places. There are seven directories that, in system maintenance, are important; they are

Codename	directory
STATA	directory where Stata is installed
UPDATES	directory where Stata's official ado-file updates go
BASE	directory containing base-level, official ado-files
SITE	directory where personal, site-level ado-files go
PLUS	directory where user-written ado-files go
PERSONAL	directory where personal ado-files go
dirname	generic directory

For instance, coding "local x: sysdir PERSONAL" would place in 'x' the name of the directory containing the user's personal ado-files. 'x' will end in the appropriate directory separator for the operating system, so you could subsequently refer to 'x''fname' if 'fname' contained, say, "myfile.ado".

sysdir *dirname* returns *dirname*. This function is used to code local x : sysdir 'dir', where the macro dir may either contain the name of a directory specified by a user, or a keyword such as STATA, UPDATES, etc. The appropriate directory name will be returned.

tempvar returns a name that can be used as a temporary variable.

tempfile returns a name that can be used as a temporary file.

tsnorm *string* returns the canonical form of *string* when *string* is interpreted as a time-series operator. For instance, if string is 1d1, "L2D" is returned, and if *string* is l.1d1, "L3D" is returned. If *string* is nothing, "" is returned.

tsnorm with the varname option returns the canonical form when *string* is interpreted as a time-series operated variable. For instance, if *string* is 1d1.gnp, "L2D.gnp" is returned, and if string is l.1d1.gnp, "L3D.gnp" is returned. If string is just a variable name, the variable name is returned.

type returns the storage type for *varname*.

value label returns the name of the value label associated with the variable; see [U] **15.6.3 Value labels**.

variable label returns the variable label associated with the variable; see [U] **15.6.2 Variable labels**.

word count returns the number of words in the string.

word # of returns the #th word of the string.

❏ Technical Note

Users of previous versions of Stata should note that several old extended macro functions have been deprecated in favor of Stata's (new to version 8) c-class. For example, old constructs such as

```
local lsize : set linesize
```

followed by references to 'lsize' are now coded as direct references to c-class values, 'c(linesize)' in this case. c-class values cannot be set directly—they merely exist to reflect current settings of various Stata and system parameters. See [P] **creturn** for more information on c-class values.

❏

21.3.7   Macro expansion operators and function

We already mentioned incrementing macros in [U] **21.3.4 Macros and expressions**. The construct

```
        command that makes reference to 'i'
        local ++i
```

occurs so commonly in Stata programs that it is convenient (and faster when executed) to collapse both lines of code into one, and to increment (or decrement) i at that same time that you make reference to it.

Toward this end, Stata has the macro expansion operators ++ and --, which, when used within a reference to a local (not global) macro, increment (or decrement) the macro in addition to executing the command which makes reference to the macro.

For example, consider

```
. local i 4
. local j 'i++'
. display "i is " 'i' ", and j is " 'j'
i is 5, and j is 4
```

In the first line, Stata defines the local macro i to contain 4. In the second line, 'i++' makes reference to i, which Stata expands to 4, then increments the value of i to 5. Once i has been expanded and incremented, the command is then executed: a local macro j is defined to contain 4.

Alternately, we can place the ++ before the name of the local macro, thus changing the order in which the expansion and incrementing of the macro takes place.

```
. local i 4
. local j '++i'
. display "i is " 'i' ", and j is " 'j'
i is 5, and j is 5
```

By placing the ++ before i we tell Stata to *first* increment i, then expand it. Thus, after expansion i contains 5 and

```
        local j 5
```

is what executes as a result of the macro expansion.

The operator -- works the same way, except that the macros are decremented, and -- may also be placed before or after the name of the local macro.

```
. local i 4
. local j '--i'
. display "i is " 'i' ", and j is " 'j'
i is 3, and j is 3
```

The key to understanding macro expansion operators is to note that the macro operation occurs when macros are expanded (either immediately before or immediately after, as demonstrated above), and that the macro operation(s) are completed before the command that makes reference to the macro is executed.

For example, assuming that I do not have a variable u3 in my data,

```
. local i 3
. summarize u'i--'
variable u3 not found
r(111);
. display 'i'
2
```

This is because i was decremented before the summarize command was executed, and with no regard for whether that command produced an error.

As another example, compare

```
. local i 3
. if 0 display `i++'
. else display `i'
4
```

with

```
. local i 3
. if 0 {
.        display `i++'
. }
. else {
.        display `i'
3
. }
```

Both blocks of code look like they should produce identical results, yet in the first block, since "if 0 display `i++'" is all on the same line, the macro expansion took place before the truthfulness of the if condition was considered. When macro expansion took place, i was incremented.

In the second block of code, "display `i++'" is on a separate line from the if condition. When Stata encounters a false if condition, it is smart enough to skip over the statements in the if block, and instead merely searches for a line containing nothing but "}", which signals the end of the block. Since Stata skipped over the if block, i was not incremented.

When dealing with ++ (or --) and if, we certainly recommend the second syntax—it is much safer.

❏ Technical Note

Do not confuse

```
local ++i
```

mentioned in [U] **21.3.4 Macros and expressions** with the inline reference `++i'. Although both will perform the same function as far as i is concerned, the former is a command that can be issued by itself, yet the latter must be used within a context for which its expanded value makes sense.

Also note that the macros being incremented or decremented need not contain positive integers; any numeric value will do. However, attempting to use ++ or -- with non-numeric (or empty) macros will result in an error.

❏

Another useful macro expansion operator is the = operator which, when used within a macro context, allows the inline evaluation of Stata expressions. Said differently, typing

> *command that makes reference to* `= exp'

is equivalent to

> local *macroname* = *exp*
> *command that makes reference to* ` *macroname*'

although the former runs faster and is easier to type. When you use `=exp' within some larger command, *exp* is first evaluated by Stata's expression evaluator, and the results are inserted as a literal string into the larger command. Then the command is executed. For example,

```
summarize u4
summarize u'=2+2'
summarize u'=4*(cos(0)==1)'
```

all do the same thing. Note that *exp* can be any valid Stata expression, and thus may include references to macros, variables, matrices, scalars, etc.

Taking this idea of evaluating a Stata expression within a command a bit further, Stata has the : macro expansion operator, which is used to obtain inline results from Stata's expanded macro functions. Typing

> *command that makes reference to* ' : *extended macro function* '

is equivalent to

> local *macroname* : *extended macro function*
> *command that makes reference to* ' *macroname* '

but the former is faster and more concise. For example, using the auto data

```
. display "':format gear_ratio'"
%0.0f
```

or, better yet, if we wanted to set the display format of headroom to that of gear_ratio

```
. format ':format gear_ratio' headroom
```

which, after expansion, is the equivalent of

```
. format %6.2f headroom
```

There is another macro expansion operator, . (the dot operator), which is used in conjunction with Stata's class system; see [P] **class** for more information.

There is also one macro expansion function, macval(), which is for use only within a reference to a local macro; e.g., 'macval(name)'. macval() is used to confine macro expansion to the first level, thereby suppressing the expansion of any embedded references to local macros. For example,

```
. local middle S.
. local name Kathleen \'middle' Brynildsen
. display "'name'"
Kathleen S. Brynildsen
. display "'macval(name)'"
Kathleen 'middle' Brynildsen
```

Note the use of the backslash (\) in the definition of the local macro name. It is there to delay the expansion of middle and instead place the literal string 'middle' in name.

Those who use file to read from disk files will find macval() useful for processing what is read, since macval() will resist the temptation to expand anything that looks like a reference to a local macro; see [P] **file**.

21.3.8 Advanced local macro manipulation

This section is really an aside to help test your understanding of local macro substitution. The tricky examples illustrated below rarely, but sometimes, occur in real programs.

We have given much attention to the topic of incrementing local macros.

```
local ++i
```

Now pretend you had macros x1, x2, x3, and so on. Obviously, 'x1' refers to the contents of x1, 'x2' to the contents of x2, etc.

What does 'x'i'' refer to? Pretend 'i' contains 6. The rule is to expand the inside first:

> 'x'i''   expands to   'x6'
> 'x6'   expands to   the contents of local macro x6

So, there you have a vector of macros.

We have already shown adjoining expansions: 'alpha''beta' expands to myvar if 'alpha' contains my and 'beta' contains var.

Stata does not mind if you reference a nonexisting macro. A nonexisting macro is treated as a macro with no contents. If local macro gamma does not exist, then

> 'gamma'   expands to

which is to say, nothing. It is not an error. Thus, 'alpha''gamma''beta' still expands to myvar.

Correspondingly, you clear a local macro by setting its contents to nothing:

> local *macname*
>
> or   local *macname* ""
>
> or   local *macname* = ""

As demonstrated above, when references to local macros are nested, the rule is to expand the inside first. In cases where macro expansion operators ++ or -- are used, it is also important to know that macros are expanded from left to right, and in fact, the left-to-right rule takes precedence over the nesting rule.

To illustrate, consider the following

```
. local i5 6
. local j = 'i'--i5'++' / '--i'--i5''
. display "i5 is " 'i5' ", and j is " 'j'
i5 is 4, and j is 1.25
```

Let's look closely at the second line in the above sequence, local j = ...

1. Stata first attempts to expand the reference 'i'--i5'++' since it is the left-most reference. In doing so, however, it encounters the nested reference '--i5', which must be expanded before 'i'--i5'++' can be expanded. Upon expanding '--i5' to 5, i5 now contains 5, and the command in question now consists of (for the time being)

   ```
   local j = 'i5++' / '--i'--i5''
   ```

2. Stata now expands the left-most reference, which is 'i5++', and upon expanding this reference to 5, i5 now contains 6, and the command now consists of

   ```
   local j = 5 / '--i'--i5''
   ```

3. Stata now attempts to expand '--i'--i5'', and thus must first expand '--i5'. As a result, i5 now contains 5 and the command is now

   ```
   local j = 5 / '--i5'
   ```

4. Upon expanding '--i5' to 4, i5 now contains 4 and the command

   ```
   local j = 5 / 4
   ```

is ready to execute. Upon completion, the local macro j contains 1.25.

As mentioned previously, such contrivances rarely appear in Stata programs. Rather, the above is designed to test your knowledge of Stata macros.

21.3.9   Advanced global macro manipulation

We continue with our aside to test your understanding of macro substitution, this time with global macros.

In [U] **21.3.4 Macros and expressions**, we mentioned incrementing local macros:

```
local ++i
```

The corresponding command for global macros is

```
global i = $i + 1
```

although the construct never arises in practice. Global macros are rarely used and, when they are used, it is typically for communication between programs. You should never use a global macro where a local macro would suffice.

Things like xi are expanded sequentially. If $x contained this and $i 6, then xi expands to this6. If $x was undefined, then xi is just 6 because undefined global macros, like undefined local macros, are treated as containing nothing.

You can nest macro expansion by including braces, so assuming $i contains 6, ${x$i} expands to 0[xG], which expands to the contents of 0xG (which would be nothing if 0xG is undefined).

You can mix global and local macros. Assume local macro j contains 7. Then, ${x`j'} expands to the contents of $x7.

You also use braces to force the contents of global macros to run up against the succeeding text. For instance, assume the macro drive contains "b:". Were drive a local macro, you could type

`drive'myfile.dta

to obtain b:myfile.dta. Since drive is a global macro, however, you must type

${drive}myfile.dta

You could not type

$drive myfile.dta

because that would expand to b: myfile.dta. You could not type

$drivemyfile.dta

because that would expand to .dta.

Because Stata uses $ to mark global-macro expansion, printing a real $ is sometimes tricky. To display the string $22.15 using the display command, you can type display "\$22.15", although you can get away with display "$22.15" because Stata is rather smart. Stata would not be smart about display "$this" if you really wanted to display $this and not the contents of the macro this. You would have to type display "\$this". Another alternative would be to use the SMCL code for dollar sign when you wanted to display it: display "{c S|}this"; see [P] **smcl**.

Real dollar signs can also be placed into the contents of macros, thus postponing substitution. First, let's understand what happens when we do not postpone substitution; consider the following definitions:

```
global baseset "myvar thatvar"
global bigset "$baseset thisvar"
```

Then $bigset is equivalent to "myvar thatvar thisvar". Now, say we redefine the macro baseset:

```
global baseset "myvar thatvar othvar"
```

The definition of bigset has not changed—it is still equivalent to "myvar thatvar thisvar". It has not changed because bigset used the definition of baseset that was current at the time it was defined. bigset no longer knows that its contents are supposed to have any relation to baseset.

Instead, let us assume we had defined bigset as

```
global bigset "\$baseset thisvar"
```

at the outset. Then $bigset is equivalent to "$baseset thisvar", which in turn is equivalent to "myvar thatvar othvar thisvar". Since bigset explicitly depends upon baseset, anytime we change the definition of baseset, we will automatically change the definition of bigset as well.

21.3.10 Constructing Windows filenames using macros

Stata uses the \ character to tell its parser not to expand macros.

Windows uses the \ character as the directory path separator.

Mostly, there is no problem. However, if you are writing a program that contains a Windows path in macro path and a filename in fname, do *not* assemble the final result as

```
'path'\'fname'
```

because Stata will interpret the \ as an instruction to not expand 'fname'. Instead, assemble the final result as

```
'path'/'fname'
```

Stata understands / as a directory separator on all platforms.

21.3.11 Accessing system values

Stata programs often need access to system parameters and settings, such as the value of π, the current date and time, or the current working directory.

System values are accessed via Stata's c-class values. The syntax works much the same as if you were referencing a local macro. For example, a reference to the c-class value for π, 'c(pi)', will expand to a literal string containing 3.141592653589793, and could be used to do

```
. display sqrt(2*'c(pi)')
2.5066283
```

You could also access the current time

```
. display "'c(current_time)'"
11:34:57
```

C-class values are a feature new to Stata 8, and are designed to provide one, all-encompassing means to access system parameters and settings, including system directories, system limits, string limits, memory settings, properties of the data currently in memory, output settings, efficiency settings, network settings, debugging settings, etc.

See [P] **creturn** for a complete detailed list of what is available. Typing

```
. creturn list
```

will give you the whole list of current settings.

21.3.12 Referencing characteristics

Characteristics—see [U] **15.8 Characteristics**—are like macros associated with variables. They have names of the form *varname*[*charname*]—such as mpg[comment]—and you quote their names just as you do macro names to obtain their contents:

To substitute the value of *varname*[*charname*], type   `varname[charname]`
For example,   `mpg[comment]`

You set the contents using the char command:

char *varname*[*charname*] [["]*text*["]]

This is similar to the local and global commands, except that there is no =*exp* variation. You clear a characteristic by setting its contents to nothing just as you would with a macro:

Type   char *varname*[*charname*]
or   char *varname*[*charname*] " "

What is unique about characteristics is that they are saved with the data, meaning their contents survive from one session to the next, and they are associated with variables in the data, so if the user ever drops a variable, the associated characteristics disappear, too. (In addition, there is _dta[*charname*] that is associated with the data but not with any variable in particular.)

All the standard rules apply: characteristics may be referenced by quotation in any context, and all that happens is that the characteristic's contents are substituted for the quoted characteristic name. As with macros, referencing a nonexistent characteristic is not an error; it merely substitutes to nothing.

21.4 Program arguments

When you invoke a program or do-file, what you type following the program or do-file name are the arguments. For instance, if you have a program called xyz and type

. xyz mpg weight

then mpg and weight are the program's arguments, mpg being the first argument and weight being the second.

Program arguments are passed to programs via local macros:

Macro	Contents
`0`	what the user typed exactly as the user typed it, odd spacing, double quotes, and all
`1`	going back; the first argument (first word of `0`)
`2`	the second argument (second word of `0`)
`3`	the third argument (third word of `0`)
. . .	. . .
`*`	the arguments `1`, `2`, `3`, . . ., listed one after the other and with a single blank in between; similar to but different from `0` because odd spacing and double quotes are gone

That is, what the user types is passed to you in three different ways:

1. It is passed in '0' exactly as the user typed it, meaning quotes, odd spacing, and all.

2. It is passed in '1', '2', ... broken out into arguments on the basis of blanks (but with quotes used to force binding; we will get to that).

3. It is passed in '*' as "'1' '2' '3' ...", which is a sort of crudely cleaned up version of '0'.

It is not anticipated that you will use all three forms in one program.

We recommend that you ignore '*', at least for receiving arguments; it is included so that old Stata programs continue to work.

Operating directly with '0' takes considerable programming sophistication, although Stata's syntax command makes interpreting '0' according to standard Stata syntax easy. That will be covered in [U] **21.4.4 Parsing standard Stata syntax** below.

The easiest way to receive arguments, however, is to deal with the positional macros '1', '2', ...

At the start of this section, we imagined an xyz program invoked by typing xyz mpg weight. In that case, '1' would contain mpg, '2' would contain weight, and '3' would contain nothing.

Let's write a program to report the correlation between two variables. Of course, Stata already has a command that can do this—correlate—and, in fact, we will implement our program in terms of correlate. It is silly, but all we want to accomplish right now is to show how Stata passes arguments to a program.

Here is our program:

```
program xyz
        correlate '1' '2'
end
```

Once the program is defined, we can try it:

```
. use http://www.stata-press.com/data/r8/auto
(1978 Automobile Data)

. xyz mpg weight
(obs=74)
```

	mpg	weight
mpg	1.0000	
weight	-0.8072	1.0000

See how this works? We typed xyz mpg weight, which invoked our xyz program with '1' being mpg and '2' being weight. Our program gave the command correlate '1' '2', and that expanded to correlate mpg weight.

Stylistically, this is not a good example of the use of positional arguments, but realistically, there is nothing wrong with it. The stylistic problem is that if xyz is really to report the correlation between two variables, it ought to allow standard Stata syntax, and that is not really a difficult thing to do. Realistically, the program works.

Positional arguments, however, play an important role, even for programmers who care about style. When we write a subroutine—a program to be called by another program and not intended for direct human use—we often pass information using positional arguments.

Stata forms the positional arguments '1', '2', ... by taking what the user typed following the command (or do-file), parsing it on white space with double quotes used to force binding, and stripping the quotes. What that means is that the arguments are formed on the basis of words, but double-quoted strings are kept together as a single argument but with the quotes removed.

Let's create a program to illustrate these concepts. Although one does not normally define programs interactively, this program is short enough that we will:

```
. program listargs
  1.  display "The 1st argument you typed is:  '1'"
  2.  display "The 2nd argument you typed is:  '2'"
  3.  display "The 3rd argument you typed is:  '3'"
  4.  display "The 4th argument you typed is:  '4'"
  5.  end
```

The display command simply types the double-quoted string following it; see [P] **display**.

Let's try our program:

```
. listargs
The 1st argument you typed is:
The 2nd argument you typed is:
The 3rd argument you typed is:
The 4th argument you typed is:
```

We type listargs and the result shows us what we already know—we typed nothing after the word listargs. There are no arguments. Let's try it again, this time adding this is a test:

```
. listargs this is a test
The 1st argument you typed is:  this
The 2nd argument you typed is:  is
The 3rd argument you typed is:  a
The 4th argument you typed is:  test
```

We learn that the first argument is 'this', the second is 'is', and so on. Blanks always separate arguments. You can, however, override this feature by placing double quotes around what you type:

```
. listargs "this is a test"
The 1st argument you typed is:  this is a test
The 2nd argument you typed is:
The 3rd argument you typed is:
The 4th argument you typed is:
```

This time we typed only *one* argument, 'this is a test'. When we place double quotes around what we type, Stata interprets whatever we type inside the quotes to be a single argument. In this case, '1' contains 'this is a test' (note that the double quotes were removed).

We can use double quotes more than once:

```
. listargs "this is" "a test"
The 1st argument you typed is:  this is
The 2nd argument you typed is:  a test
The 3rd argument you typed is:
The 4th argument you typed is:
```

The first argument is 'this is' and the second argument is 'a test'.

21.4.1   Renaming positional arguments

Positional arguments can be renamed: in your code you do not have to refer to '1', '2', '3', ...; you can instead refer to more meaningful names such as n, a, and b; or numb, alpha, and beta; or whatever else you find convenient. You want to do this because programs coded in terms of '1', '2', ... are hard to read, and therefore are more likely to contain errors.

You obtain better-named positional arguments using the args command:

```
program progname
        args argnames
        . . .
end
```

For instance, if your program was to receive four positional arguments and you wanted to call them varname, n, oldval, and newval, you would code

```
program progname
        args varname n oldval newval
        . . .
end
```

varname, n, oldval, and newval become new local macros, and all args does is copy '1', '2', ... to them. It does not change '1', '2', etc.—you can still refer to the numbered macros if you wish—and it does not verify that your program receives the right number of arguments. If our example above were invoked with just two arguments, then 'oldval' and 'newval' would contain nothing. If it were invoked with five arguments, the fifth argument would still be out there, stored in local macro '5'.

Let's make a command to create a dataset containing n observations on x ranging from a to b. Such a command would be useful, for instance, if we wanted to graph some complicated mathematical function and experiment with different ranges. It is convenient if we can type the range of x over which we wish to make the graph rather than concocting the range by hand. (In fact, Stata already has such a command—range—but it will be instructive to write our own.)

Before writing this program, we had better know how to proceed, so here is how, in Stata, you could create a dataset containing n observations with x ranging from a to b:

1. drop _all to clear whatever data are in memory.

2. set obs n to make a dataset of n observations on no variables; if n were 100, we would type set obs 100.

3. gen x = (_n-1)/(n-1)*(b-a)+a because the built-in variable _n is 1 in the first observation, 2 in the second, and so on; see [U] **16.4 System variables (_variables)**.

So, the first version of our program might read

```
program rng // arguments are n a b
        drop _all
        set obs '1'
        generate x = (_n-1)/(_N-1)*('3'-'2')+'2'
end
```

The above is just a direct translation of what we just said. '1' corresponds to n, '2' corresponds to a, and '3' corresponds to b. This program, however, would be far more understandable if we changed it to read

```
program rng
        args n a b
        drop _all
        set obs 'n'
        generate x = (_n-1)/(_N-1)*('b'-'a')+'a'
end
```

21.4.2 Incrementing through positional arguments

Some programs contain k arguments, where k varies, but it does not much matter because the same thing is done to each argument. summarize is an example of a program like that: type summarize mpg to obtain summary statistics on mpg, and type summarize mpg weight to obtain first summary statistics on mpg and then summary statistics on weight.

```
program ...
        local i = 1
        while "'i'" != "" {
                logic stated in terms of ''i''
                local ++i
        }
end
```

Equivalently, if the logic that uses ''i'' contains only one reference to ''i'',

```
program ...
        local i = 1
        while "'i'" != "" {
                logic stated in terms of ''i++''
        }
end
```

Note the tricky construction ''i'', which then itself had to be placed in double quotes for the while loop; see [U] **21.3.8 Advanced local macro manipulation**. The outside while loop continues to process arguments, '1', '2', '3', etc., until ''i'' expands to '$k+1$', which expands to nothing.

Say you were writing a subroutine that was to receive k variables, but the code that processes each variable needs to know (while it is processing) how many variables were passed to the subroutine. You need first to count the variables (and so derive k), and then, knowing k, to pass through the list again.

```
program progname
                                        // count the number of arguments
        local k = 1
        while "'k'" != "" {
                local ++k
        }
        local --k                       // k contains one too many
                                        // now pass through again
        local i = 1
        while 'i' <= 'k' {
                code in terms of ''i'' and 'k'
                local ++i
        }
end
```

In the above example, we have used while, Stata's all-purpose looping command. Stata has two other looping commands, foreach and forvalues, and they sometimes produce code that is more readable and executes more quickly. We direct you to read [P] **foreach** and [P] **forvalues**, but emphasize that at this point that there is nothing they can do that while cannot do. Above, we coded

```
local i = 1
while 'i' <= 'k' {
        code in terms of ''i'' and 'k'
        local ++i
}
```

to produce logic that looped over the values 'i' = 1 to 'k'. We could have instead coded

```
forvalues i = 1(1)'k' {
        code in terms of ''i'' and 'k'
}
```

Similarly, at the beginning of this subsection, we said you could use the following code in terms of while to loop over the arguments received:

```
program ...
        local i = 1
        while "''i''" != "" {
                logic stated in terms of ''i''
                local ++i
        }
end
```

Equivalent to the above would be

```
program ...
        foreach x of local 0 {
                logic stated in terms of 'x'
        }
end
```

See [P] **foreach** and [P] **forvalues**.

You can combine args and incrementing through an unknown number of positional arguments. For instance, say you were writing a subroutine that was to receive (1) varname, the name of some variable, (2) n, which is some sort of count, and (3) at least one and maybe 20 variable names. Perhaps you are to sum the variables, divide by n, and store the result in the first variable. What the program does is irrelevant; here is how we could receive the arguments:

```
program progname
        args varname n
        local i 3
        while "''i''" != "" {
                logic stated in terms of ''i''
                local ++i
        }
end
```

21.4.3   Using macro shift

Another way to code the repeat-the-same-process problem for each argument is

```
program ...
        while "'1'" != "" {
                logic stated in terms of '1'
                macro shift
        }
end
```

macro shift shifts '1', '2', '3', ..., one to the left: what was '1' disappears, what was '2' becomes '1', what was '3' becomes '2', and so on.

The outside while loop continues the process until macro '1' contains nothing.

macro shift is an older construct that we no longer advocate using. Instead, we recommend that you use the techniques described in the previous subsection; i.e. references to ''i'' and foreach/forvalues.

There are two reasons we make this recommendation: (1) macro shift destroys the positional macros '1', '2', which must then be reset using tokenize should you wish to pass through the argument list again, and (more importantly) (2) if the number of arguments is large (which, with the release of Stata/SE, is more of a possibility), macro shift can be incredibly slow.

❏ Technical Note

There is one thing that macro shift can do that would be difficult by other means.

'*', the result of listing the contents of the numbered macros one after the other with a single blank in between, changes with macro shift. Say your program received a list of variables and that the first variable had the interpretation of the dependent variable and the rest were independent variables. You want to save the first variable name in 'lhsvar' and all the rest in 'rhsvars'. You could code

```
program progname
        local lhsvar "'1'"
        macro shift 1
        local rhsvars "'*'"
        ...
end
```

Getting ahead of ourselves, it sometimes happens that a single macro contains a list of variables and you want to split the contents of the macro in two. Perhaps 'varlist' is the result of a syntax command (see [U] **21.4.4 Parsing standard Stata syntax**) and you now wish to split 'varlist' into 'lhsvar' and 'rhsvars'. tokenize will reset the numbered macros:

```
program progname
        ...
        tokenize 'varlist'
        local lhsvar "'1'"
        macro shift 1
        local rhsvars "'*'"
        ...
end
```

❏

21.4.4   Parsing standard Stata syntax

Let us now switch to '0' from the positional arguments '1', '2', ...

You can parse '0' (what the user typed) according to standard Stata syntax with a single command. Remember that standard Stata syntax is

$$\left[\text{by } \textit{varlist}:\right] \; \textit{command} \; \left[\textit{varlist}\right] \; \left[=\textit{exp}\right] \; \left[\text{using } \textit{filename}\right] \; \left[\text{if } \textit{exp}\right] \; \left[\text{in } \textit{range}\right] \; \left[\textit{weight}\right]$$

$$\left[, \; \textit{options}\right]$$

See [U] **14 Language syntax**.

syntax parses standard syntax. You type out the syntax diagram in your program, and then syntax looks at '0' (it knows to look there) and compares what the user typed with what you are willing to accept. Then one of two things happens: either syntax stores the pieces in an easy-for-you-to-process way or, if what the user typed does not match what you specified, syntax issues the appropriate error message and stops your program.

Consider a program that is to take two or more variable names along with an optional if *exp* and in *range*. The program would read

```
program ...
        syntax varlist(min=2) [if] [in]
        ...
    end
```

You will have to read [P] **syntax** to learn how to specify the syntactical elements, but the command is certainly readable, and it will not be long until you are guessing correctly about how to fill it in. And yes, the square brackets really do mean optional, and you just use them with syntax in the natural way.

Understand that one command is the entire parsing process. In this case, if what the user typed matches "two-or-more variables and an optional if and in", syntax defines new local macros:

'varlist'	the two or more variable names
'if'	the if *exp* specified by the user (or nothing)
'in'	the in *range* specified by the user (or nothing)

To see that this works, experiment with the following program:

```
program tryit
        syntax varlist(min=2) [if] [in]
        display "varlist now contains |'varlist'|"
        display '"if now contains |'if'|"'
        display "in now contains |'in'|"
    end
```

Here is a little experiment:

```
. tryit mpg weight
varlist now contains |mpg weight|
if now contains ||
in now contains ||
. tryit mpg weight displ if foreign==1
varlist now contains |mpg weight displ|
if now contains |if foreign==1|
in now contains ||
. tryit mpg wei in 1/10
varlist now contains |mpg weight|
if now contains ||
in now contains |in 1/10|
. tryit mpg
too few variables specified
r(102);
```

Note that in our third try we abbreviated the weight variable as wei, yet, after parsing, syntax unabbreviated the variable for us.

If what this program were next going to do was step through the variables in the varlist, the positional macros '1', '2', ... could be reset by coding

```
        tokenize 'varlist'
```

see [P] **tokenize**. This step resets '1' to be the first word of 'varlist', '2' to be the second word, and so on (if there is a so on).

21.4.5 Parsing immediate commands

Immediate commands are described in [U] **22 Immediate commands**—they take numbers as arguments. By convention, when you name immediate commands, you should make the last letter *i*. Assume `mycmdi` takes as arguments two numbers, the first of which must be a positive integer, and also allows the options `alpha` and `beta`. The basic structure is

```
program mycmdi
        gettoken n 0 : 0, parse(" ,")          /* get first number */
        gettoken x 0 : 0, parse(" ,")          /* get second number */
        confirm integer number 'n'             /* verify first is integer */
        confirm number 'x'                     /* verify second is number */

        if 'n'<=0 { error 2001 }               /* check that n is positive */
        place any other checks here

        syntax [, Alpha Beta]                  /* parse remaining syntax */
        make calculation and display output
end
```

See [P] **gettoken**.

21.4.6 Parsing nonstandard syntax

If you wish to interpret nonstandard syntax and positional arguments are not adequate for you, you know that you face a formidable programming task. The key to the solution is the `gettoken` command.

`gettoken` has the ability to pull a single token from the front of a macro according to the parsing characters you specify and, optionally, to define another macro or redefine the initial macro to contain the remaining (unparsed) characters. That is,

Say '0' contains	"this is what the user typed"
After `gettoken`,	
new macro 'token' could contain	"this"
and '0' could still contain	"this is what the user typed"
or	
new macro 'token' could contain	"this"
and new macro 'rest' could contain	" is what the user typed"
and '0' could still contain	"this is what the user typed"
or	
new macro 'token' could contain	"this"
and '0' could contain	" is what the user typed"

A simplified syntax of `gettoken` is

> `gettoken` *emname1* [*emname2*] : *emname3* [, <u>p</u>arse(*pchars*) <u>q</u>uotes
>
> <u>m</u>atch(*lmacname*) bind]

where *emname1*, *emname2*, *emname3*, and *lmacname* are the names of local macros. (There is a way to work with global macros, but, in practice, that is seldom necessary; see [P] **gettoken**.)

`gettoken` pulls the first token from *emname3* and stores it in *emname1*, and if *emname2* is specified, stores the remaining characters from *emname3* in *emname2*. Any of *emname1*, *emname2*, and *emname3* may be the same macro. Typically, `gettoken` is coded

> `gettoken` *emname1*   : 0 [, *options*]
>
> `gettoken` *emname1* 0 : 0 [, *options*]

since '0' is the macro containing what the user typed. The first coding is used for token lookahead, should that be necessary, and the second is used for committing to taking the token.

gettoken's options are

parse("*string*")	for specifying parsing characters the default is parse(" "), meaning parse on white space it is common to specify parse('"" "'), meaning parse on white space and double quote ('"" "' is the string double-quote-space in compound double quotes)
quotes	to specify that outer double quotes are *not* to be stripped
match(*lmacname*)	to bind on parentheses and square brackets *lmacname* will be set to contain "(", "[", or nothing, depending on whether *emname1* was bound on parentheses, brackets, or match() turned out to be irrelevant *emname1* will have the outside parentheses or brackets removed

gettoken binds on double quotes whenever a (simple or compound) double quote is encountered at the beginning of *emname3*. Specifying parse('"" "') ensures that double-quoted strings are isolated.

quote specifies that double quotes are not to be removed from the source in defining the token. For instance, in parsing '"this is" a test', the next token is "this is" if quote is not specified and is '"this is"' if quote is specified.

match() specifies that parentheses and square brackets are to be matched in defining tokens. The outside level of parentheses or brackets is stripped. In parsing "(2+3)/2", the next token is "2+3" if match() is specified. In practice, match() might be used with expressions, but it is more likely to be used to isolate bound varlists and time-series varlists.

21.5   Scalars and matrices

In addition to macros, scalars and matrices are provided for programmers; see [U] **17 Matrix expressions**, [P] **scalar** and [P] **matrix**.

As far as scalar calculations go, you can use macros or scalars. Remember, macros can hold numbers. Stata's scalars are, however, slightly faster and are a little more accurate than macros. The speed issue is so slight as to be nearly immeasurable. As for accuracy, macros are accurate to a minimum of 12 decimal digits, and scalars are accurate to roughly 16 decimal digits. Which you use makes little difference except in iterative calculations.

21.6   Temporarily destroying the data in memory

It is sometimes necessary to modify the user's data to accomplish a particular task. A well-behaved program, however, ensures that the user's data are always restored. The preserve command makes this easy:

```
                code before the data need changing
                preserve
                code that changes data freely
```

When you give the preserve command, Stata makes a copy of the user's data on disk. When your program terminates—no matter how—Stata restores the data and erases the temporary file. preserve is described in [P] **preserve**.

21.7 Temporary objects

If you write a substantial program, it will invariably require the use of temporary variables in the data, or temporary scalars, or temporary matrices, or temporary files. By temporary, it is meant that these objects are necessary while the program is making its calculations and, once the program completes, they can be discarded.

Stata provides three commands to deal with this: `tempvar` creates names for variables in the dataset, `tempname` creates names for scalars and matrices, and `tempfile` creates names for files. All are described in [P] **macro**, and all have the same syntax:

> { tempvar | tempname | tempfile } *macname* [*macname* ...]

The commands create local macros containing names you may use.

21.7.1 Temporary variables

Say that, in the process of making a calculation, you need to add variables `sum_y` and `sum_z` to the data. One possible code fragment is

```
prior code
tempvar sum_y
gen 'sum_y' = etc.
tempvar sum_z
gen 'sum_z' = etc.
code continues
```

or you may obtain both temporary variable names in a single call:

```
prior code
tempvar sum_y sum_z
gen 'sum_y' = etc.
gen 'sum_z' = etc.
code continues
```

It is not necessary that you explicitly `drop 'sum_y'` and `'sum_z'` when you are finished, although you may if you wish. Stata will automatically drop any variables with names assigned by `tempvar`. After issuing the `tempvar` command, it is important to remember always to refer to the names with the enclosing quotes, which signifies macro expansion. Thus, after typing `tempvar sum_y`, the one case where you do not put single quotes around the name, refer thereafter to the variable `'sum_y'` with quotes. `tempvar` does not create a temporary variable, but instead creates a name that we may subsequently use as a temporary variable and stores that name in the local macro whose name you provide.

A full description of `tempvar` can be found in [P] **macro**.

21.7.2 Temporary scalars and matrices

`tempname` works just like `tempvar`. For instance, a piece of your code might read

```
tempname YXX XXinv
matrix accum 'YXX' = price weight mpg
matrix 'XXinv' = syminv('YXX'[2..., 2...])
tempname b
matrix 'b' = 'XXinv'*'YXX'[1..., 1]
```

The above code solves for the coefficients of a regression on `price` on `weight` and `mpg`; see [U] **17 Matrix expressions** and [P] **matrix** for more information on the matrix commands.

As with temporary variables, temporary scalars and matrices are automatically dropped at the conclusion of your program.

21.7.3   Temporary files

In cases where you ordinarily might think you need temporary files, you may not because of Stata's ability to preserve and automatically restore the data in memory; see [U] **21.6 Temporarily destroying the data in memory** above.

For more complicated programs, Stata does provide temporary files. A code fragment might read

```
preserve                         /* save original data */
tempfile males females
keep if sex==1
save "'males'"
restore, preserve                /* get back original data */
keep if sex==0
save "'females'"
```

As with temporary variables, scalars, and matrices, it is not necessary to delete the temporary files when you are through with them; Stata automatically erases them when your program ends.

21.8   Accessing results calculated by other programs

Stata commands that report results also save the results where they can be subsequently used by other commands or programs. This is documented in the *Saved Results* section of the particular command in the *Reference* manuals. Commands save results in one of three places:

1. r-class commands such as `summarize` save their results in `r()`; most commands are r-class.

2. e-class commands such as `regress` save their results in `e()`; e-class commands are Stata's model estimation commands.

3. s-class commands (there are no good examples) save their results in `s()`; this is a rarely used class that programmers sometimes find useful to help parse input.

Commands that do not save results are called n-class commands. More correctly, these commands require that you state where the result is to be saved, as in `generate` *newvar* = ...

▷ Example

You wish to write a program to calculate the standard error of the mean, which is given by the formula $\sqrt{s^2/n}$, where s^2 is the calculated variance. (You could obtain this statistic using the `ci` command, but we will pretend that is not true.) You look at [R] **summarize** and learn that the mean is stored in `r(mean)`, the variance in `r(Var)`, and the number of observations in `r(N)`. With that knowledge, you write the following program:

```
program meanse
        quietly summarize '1'
        display "     mean = " r(mean)
        display "SE of mean = " sqrt(r(Var)/r(N))
end
```

The result of executing this program is

```
. meanse mpg
      mean = 21.297297
SE of mean = .67255109
```

◁

If you run an r-class command and type `return list` or run an e-class command and type `ereturn list`, Stata will summarize what was saved:

```
. use http://www.stata-press.com/data/r8/auto
(1978 Automobile Data)

. regress mpg weight displ
  (output omitted )

. ereturn list

scalars:
                  e(N) =  74
               e(df_m) =  2
               e(df_r) =  71
                  e(F) =  66.78504752026517
                 e(r2) =  .6529306984682528
               e(rmse) =  3.45606176570828
                e(mss) =  1595.409691543724
                e(rss) =  848.0497679157352
               e(r2_a) =  .643154098425105
                 e(ll) =  -195.2397979466294
               e(ll_0) =  -234.3943376482347

macros:
             e(depvar) : "mpg"
                e(cmd) : "regress"
            e(predict) : "regres_p"
              e(model) : "ols"

matrices:
                 e(b) :  1 x 3
                 e(V) :  3 x 3

functions:
              e(sample)
```

```
. summarize mpg if foreign
```

Variable	Obs	Mean	Std. Dev.	Min	Max
mpg	22	24.77273	6.611187	14	41

```
. return list

scalars:
                  r(N) =  22
              r(sum_w) =  22
               r(mean) =  24.77272727272727
                r(Var) =  43.70779220779221
                 r(sd) =  6.611186898567625
                r(min) =  14
                r(max) =  41
                r(sum) =  545
```

In the example above, we ran `regress` followed by `summarize`. As a result, `e(N)` records the number of observations used by `regress` (equal to 74), and `r(N)` records the number of observations used by `summarize` (equal to 22). `r(N)` and `e(N)` are separate things.

Were we now to run another r-class command—say, `tabulate`—the contents of `r()` would change, but those in `e()` would remain unchanged. You might, therefore, think that were we then to run another e-class command, say, `probit`, the contents of `e()` would change, but `r()` would remain unchanged. While it is true that `e()` results remain in place until the next e-class command is executed, do not depend on `r()` remaining unchanged. If an e-class or n-class command were to use an r-class command as a subroutine, that would cause `r()` to change. Anyway, most commands are r-class, so the contents of `r()` change frequently.

❑ Technical Note

It is, therefore, of great importance that you access results stored in `r()` immediately after the command that sets them. If you need the mean and variance of the variable '1' for subsequent calculation, do *not* code

```
summarize '1'
...
... r(mean) ... r(Var) ...
```

Instead, code

```
summarize '1'
local mean = r(mean)
local var = r(Var)
...
... 'mean' ... 'var' ...
```

or

```
tempname mean var
summarize '1'
scalar 'mean' = r(mean)
scalar 'var' = r(Var)
...
... 'mean' ... 'var' ...
```

❑

Saved results, be they in `r()` or `e()`, come in three flavors: scalars, macros, and matrices. If you look back at the `ereturn list` and `return list` output, you will see that `regress` saves examples of all three, whereas `summarize` just saves scalars. (`regress` also saves the "function" `e(sample)`, as do all the other e-class commands; see [U] **23.5 Specifying the estimation subsample**.)

Regardless of the flavor of `e(`*name*`)` or `r(`*name*`)`, you can just refer to `e(`*name*`)` or `r(`*name*`)`. That was the rule we gave in [U] **16.6 Accessing results from Stata commands**, and that rule is sufficient for most users. There is, however, another way to refer to saved results. Rather than referring to `r(`*name*`)` and `e(`*name*`)`, you can embed the reference in macro substitution characters ' ' to produce `'r(`*name*`)'` and `'e(`*name*`)'`. The result is the same as macro substitution; the saved result is evaluated, and then the evaluation is substituted:

```
. display "You can refer to " e(cmd) " or to 'e(cmd)'"
You can refer to regress or to regress
```

This means, for instance, that typing `'e(cmd)'` is the same as typing `regress` because `e(cmd)` contains "regress":

```
. 'e(cmd)'
```

Source	SS	df	MS
Model	1595.40969	2	797.704846

```
Number of obs =      74
F( 2,    71) =   66.79
Prob > F      =  0.0000
```

(remaining output omitted)

In the `ereturn list`, `e(cmd)` was listed as being a macro, and when you place a macro's name in single quotes, the macro's contents are substituted, so this is hardly a surprise.

What is surprising is that you can do this with scalar and even matrix saved results. `e(N)` is a scalar equal to 74, and may be used as such in any expression such as "`display e(mss)/e(N)`" or "`local meanss = e(mss)/e(N)`". '`e(N)`' substitutes to the string "74", and may be used in any context whatsoever, such as "`local val'e(N)' = e(N)`" (which would create a macro named `val74`). The rules for referring to saved results are

1. You may refer to `r(`*name*`)` or `e(`*name*`)` without single quotes in any expression, and only in an expression. (Referring to s-class `s(`*name*`)` without single quotes is not allowed.)

    1.1 If *name* does not exist, missing value (.) is returned; it is not an error to refer to a nonexisting saved result.

    1.2 If *name* is a scalar, the full double-precision value of *name* is returned.

    1.3 If *name* is a macro, it is examined to determine whether its contents can be interpreted as a number. If so, the number is returned; otherwise, the first 80 characters of *name* are returned.

    1.4 If *name* is a matrix, the full *matrix* is returned.

2. You may refer to '`r(`*name*`)`', '`e(`*name*`)`', or '`s(`*name*`)`'—note the presence of quotes indicating macro substitution—in any context whatsoever.

    2.1 If *name* does not exist, nothing is substituted; it is not an error to refer to a nonexisting saved result. The resulting line is as if you had never typed '`r(`*name*`)`', '`e(`*name*`)`', or '`s(`*name*`)`'.

    2.2 If *name* is a scalar, a string representation of the number accurate to no less than 12 digits is substituted.

    2.3 If *name* is a macro, the full contents (up to 18,623 characters for Intercooled and 1,000 characters for Small Stata) are substituted.

    2.4 If *name* is a matrix, the word `matrix` is substituted.

In general, you should refer to scalar and matrix saved results without quotes—`r(`*name*`)` and `e(`*name*`)`—and to macro saved results with quotes—'`r(`*name*`)`', '`e(`*name*`)`', and '`s(`*name*`)`'—but it is sometimes convenient to switch. For instance, say returned result `r(example)` contains the number of time periods patients are observed, and assume that `r(example)` was saved as a macro and not as a scalar. One could still refer to `r(example)` without the quotes in an expression context and obtain the expected result. It would have made more sense for the programmer to have stored `r(example)` as a scalar, but really it would not matter, and you as a user would not even have to be cognizant of how the saved result was stored.

Switching the other way is sometimes useful, too. Say that returned result `r(N)` is a scalar and contains the number of observations used. You now want to use some other command that has an option `n(#)` that specifies the number of observations used. You could not type `n(r(N))` because the syntax diagram says that option `n()` expects its argument to be a literal number. Instead, you could type `n('r(N)')`.

21.9 Accessing results calculated by estimation commands

Estimation results are saved in `e()`, and you access them in the same way you access any saved result; see [U] **21.8 Accessing results calculated by other programs** above. In summary,

1. Estimation commands—`regress`, `logistic`, etc.—save results in `e()`.

2. Estimation commands save their name in e(cmd). For instance, regress saves "regress" and poisson saves "poisson" in e(cmd).

3. Estimation commands save the number of observations used in e(N), and they identify the estimation subsample by setting e(sample). You could type, for instance, "summarize if e(sample)" to obtain summary statistics on the observations used by the estimator.

4. Estimation commands save the entire coefficient vector and variance–covariance matrix of the estimators in e(b) and e(V). These are matrices, and they may be manipulated like any other matrix:

```
. matrix list e(b)

e(b)[1,3]
        weight        displ       _cons
y1  -.00656711     .00528078    40.084522

. mat y = e(b)*e(V)*e(b)'
. mat list y

symmetric y[1,1]
          y1
y1   6556.982
```

5. Estimation commands set _b[*name*] and _se[*name*] as convenient ways to use coefficients and their standard errors in expressions; see [U] **16.5 Accessing coefficients and standard errors**.

6. Estimation commands may set other e() scalars, macros, or matrices containing additional information. This is documented in the "Saved Results" section of the particular command in the command reference.

▷ Example

If you are writing a command for use after regress, early in your code you should include the following:

```
if "`e(cmd)'" != "regress" {
        error 301
}
```

This is how you verify that the estimation results that are stored have been set by regress and not by some other estimation command. Error 301 is Stata's "last estimates not found" error.

◁

21.10   Saving results

If your program calculates something, it should save the results of the calculation so that other programs can access them. This way your program can not only be used interactively, but also can be used as a subroutine for other commands.

Saving results is easy:

1. On the program line, specify the option rclass, eclass, or sclass according to whether you intend to return results in r(), e(), or s().

2. Code

```
return scalar name = exp      (same syntax as scalar without the return)
return local  name ...        (same syntax as local without the return)
return matrix name matname    (moves matname to r(name))
```

to save results in r().

3. Code

```
ereturn name = exp            (same syntax as scalar without the ereturn)
ereturn local  name ...       (same syntax as local without the ereturn)
ereturn matrix name matname   (moves matname to e(name))
```

to save results in e(). (You do not save the coefficient vector and variance matrix e(b) and e(V) in this way. Use ereturn post instead.)

4. Code

```
sreturn local  name       (same syntax as local without the sreturn)
```

to save results in s(). (The s-class has only macros.)

A program must be exclusively r-class, e-class, or s-class.

21.10.1   Saving results in r()

In [U] **21.8 Accessing results calculated by other programs**, we showed an example that reported the mean and standard error of the mean. A better version would save in r() the results of its calculations and would read

```
program meanse, rclass
        quietly summarize '1'
        local mean = r(mean)
        local sem  = sqrt(r(Var)/r(N))
        display "      mean = " 'mean'
        display "SE of mean = " 'sem'
        return scalar mean = 'mean'
        return scalar se = 'sem'
end
```

Running meanse now sets r(mean) and r(se):

```
. meanse mpg
      mean = 21.297297
SE of mean = .67255109

. return list

scalars:
        r(se)      =  .6725510870764975
        r(mean)    =  21.2972972972973
```

In this modification, we added option rclass to the program statement, and we added two return commands to the end of the program.

Although we placed the return statements at the end of the program, they may be placed at the point of calculation if that is more convenient. A more concise version of this program would read

```
program meanse, rclass
        quietly summarize '1'
        return scalar mean = r(mean)
        return scalar se = sqrt(r(Var)/r(N))
        display "     mean = " return(mean)
        display "SE of mean = " return(se)
end
```

The `return()` function is just like the `r()` function, except that `return()` refers to the results that this program *will* return rather than to the saved results that currently *are* returned (which in this case are due to `summarize`). That is, when you code the `return` command, the result is not immediately posted to `r()`. Rather, Stata holds onto the result in `return()` until your program concludes, and then it copies the contents of `return()` to `r()`. While your program is active, you may use the `return()` function to access results you have already "returned". (`return()` works just like `r()` works after your program returns, meaning you may code '`return()`' to perform macro substitution.)

21.10.2   Saving results in e()

Saving in `e()` is in some ways similar to saving in `r()`: you add the `eclass` option to the `program` statement, and then you use `ereturn` ... just as you used `return` ... to store results. There are, however, some significant differences:

1. Unlike `r()`, estimation results are saved in `e()` the instant you issue an `ereturn scalar`, `ereturn local`, or `ereturn matrix` command. This is because estimation results can consume considerable memory, and Stata does not want to have multiple copies of the results floating around. That means you must be more organized and post your results at the end of your program.

2. There will come a point in your code where you will have your estimates and be ready to begin posting. The first step is to clear the previous estimates, set the coefficient vector `e(b)` and corresponding variance matrix `e(V)`, and set the estimation-sample function `e(sample)`. How you do this depends on how you obtained your estimates:

   2.1 If you obtained your estimates using Stata's likelihood maximizer `ml`, this is automatically handled for you; skip to step 3.

   2.2 If you obtained estimates by "stealing" an already existing estimator, `e(b)`, `e(V)`, and `e(sample)` already exist, and you do not want to clear them; skip to step 3.

   2.3 If you write your own code from start to finish, you use the `ereturn post` command; see [P] **ereturn**. You will code something like "`ereturn post 'b' 'V', esample('touse')`", where '`b`' is the name of the coefficient vector, '`V`' is the name of the corresponding variance matrix, and '`touse`' is the name of a variable containing 1 if the observation was used and 0 if it was ignored. `ereturn post` clears the previous estimates and moves the coefficient vector, variance matrix, and variable into `e(b)`, `e(V)`, and `e(sample)`.

   2.4 A variation on (2.3) is when you use an already existing estimator to produce the estimates but do not want all the other `e()` results stored by the estimator. In that case, you code

```
tempvar touse
tempname b V
matrix 'b' = e(b)
matrix 'V' = e(V)
qui gen byte 'touse' = e(sample)
ereturn post 'b' 'V', esample('touse')
```

3. You now save anything else in e() that you wish by using the ereturn scalar, ereturn local, or ereturn matrix commands.

4. You code ereturn local cmd "*cmdname*". Stata does not consider estimation results complete until this is posted, and Stata considers the results to be complete when this is posted, so you must remember to do this and to do this last. If you set e(cmd) too early and the user pressed *Break*, Stata would consider your estimates complete when they are not.

Say you wish to write the estimation command with syntax

> myest *depvar* *var*$_1$ *var*$_2$ [if *exp*] [in *range*], *optset1* *optset2*

where *optset1* affects how results are displayed and *optset2* affects the estimation results themselves. One important characteristic of estimation commands is that, when typed without arguments, they redisplay the previous estimation results. The outline is

```
program myest, eclass
        local options "optset1"
        if replay() {
                if "`e(cmd)'"!="myest" {
                        error 301            /* last estimates not found     */
                }
                syntax [, 'options']
        }
        else {
                syntax varlist [if] [in] [, 'options' optset2]
                marksample touse
```
Code contains either this,
```
                tempnames b V
```
commands for performing estimation
assume produces 'b' and 'V'
```
                ereturn post 'b' 'V', esample('touse')
                ereturn local depvar "'depv'"
```
or this,
```
                ml model ... if 'touse' ...
```
and regardless, concludes,
perhaps other ereturn *commands appear here*
```
                ereturn local cmd "myest"
        }
                                            /* (re)display results ...      */
```
code typically reads
code to output header above coefficient table
```
        ereturn display                     /* displays coefficient table   */
```
or
```
        ml display                          /* displays header and coef. table */
end
```

Here is a list of the commonly saved e() results. Of course, you may create any e() results that you wish.

e(N) (scalar)
    Number of observations.

e(df_m) (scalar)
    Model degrees of freedom.

e(df_r) (scalar)
    "Denominator" degrees of freedom if estimates nonasymptotic.

e(r2_p) (scalar)
    Value of the pseudo-R^2 if it is calculated. (If a "real" R^2 is calculated as it would be in linear regression, it is stored in (scalar) e(r2).)

e(F) (scalar)
Test of the model against the constant-only model, if relevant, and if nonasymptotic results.

e(ll) (scalar)
Log-likelihood value, if relevant.

e(ll_0) (scalar)
Log-likelihood value for constant-only model, if relevant.

e(N_clust) (scalar)
Number of clusters, if any.

e(chi2) (scalar)
Test of the model against the constant-only model, if relevant, and if asymptotic results.

e(cmd) (macro)
Name of the estimation command.

e(depvar) (macro)
Name(s) of the dependent variable(s).

e(wtype) and e(wexp) (macros)
If weighted estimation was performed, e(wtype) contains the weight type (fweight, pweight, etc.) and e(wexp) contains the weighting expression.

e(clustvar) (macro)
Name of the cluster variable, if any.

e(vcetype) (macro)
Text to appear above standard errors in estimation output; typically "Robust" or "".

e(scorevars) (macro)
Name(s) of the variable(s) holding scores, if generated.

e(chi2type) (macro)
LR or Wald or other depending on how e(chi2) was performed.

e(crittype) (macro)
Type of optimization criterion used, such as log likelihood or deviance.

e(predict) (macro)
Name of the command that predict is to use; if this is blank, predict uses the default _predict.

e(b) and e(V) (matrices)
The coefficient vector and corresponding variance matrix. Saved when you coded ereturn post.

e(sample) (function)
This function was defined by ereturn post's esample() option if you specified it. You specified a variable containing 1 if you used an observation and 0 otherwise. ereturn post stole the variable and created e(sample) from it.

21.10.3 Saving results in s()

S is a strange class because, whereas the other classes allow scalars, macros, and matrices, s allows only macros.

S is seldom used.

S is for subroutines that you might write to assist in parsing the user's input prior to evaluating any user-supplied expressions.

Think of s as standing for strange, seldom used subroutines.

Here is the problem s solves: Say you create a nonstandard syntax for some command so that you have to parse through it yourself. The syntax is so complicated that you want to create subroutines to bite off pieces of it and then return information to your main routine. Assume that your syntax contains expressions that the user might type. Now, pretend that one of the expressions the user types is, say, r(mean)/sqrt(r(Var)) —perhaps the user is consuming results left behind by summarize.

If, in your parsing step, you call subroutines that return results in r(), you will wipe out r(mean) and r(Var) before you ever get around to seeing them, much less evaluating them.

So, you must be careful to leave r() intact until your parsing is complete; you must use no r-class commands, and any subroutines you write must not touch r().

The way to do this is to use s-class subroutines because s-class routines return results in s() rather than r(). S-class provides macros only because that is all you need to solve parsing problems.

To create an s-class routine, specify the sclass option on the program line and then use sreturn local to return results.

S-class results are posted to s() at the instant you issue the sreturn() command, so you must organize your results. In addition, s() is never automatically cleared, so occasionally coding sreturn clear at appropriate points in your code is a good idea.

Very few programs need s-class subroutines.

21.11   Ado-files

Ado-files were introduced in [U] **20 Ado-files**.

When a user types 'gobbledegook', Stata first asks itself if gobbledegook is one of its built-in commands. If so, the command is executed. Otherwise, it asks itself if gobbledegook is a defined program. If so, the program is executed. Otherwise, Stata looks in various directories for gobbledegook.ado. If there is no such file, the process ends with the "unrecognized command" error.

If Stata finds the file, it quietly issues to itself the command 'run gobbledegook.ado' (specifying the path explicitly). If that runs without error, Stata asks itself again if gobbledegook is a defined program. If not, Stata issues the "unrecognized command" error. (In this case, somebody wrote a bad ado-file.) If the program is defined, as it should be, Stata executes it.

Thus, you can arrange for programs you write to be loaded automatically. For instance, if you were to create hello.ado containing

```
──────────────────────────────────────── top of hello.ado ────────────
    program hello
            display "hi there"
    end
──────────────────────────────────────── end of hello.ado ────────────
```

and store the file in your current directory or your personal directory (we will tell you where that is momentarily), you could type hello and be greeted by a reassuring

```
    . hello
    hi there
```

You could, at that point, think of hello as just another part of Stata.

There are two places to put your personal ado-files. One is the current directory, and that is a good choice when the ado-file is unique to a project. You will want to use it only when you are in that directory. The other place is your *personal ado directory*, which is probably something like `c:\ado\personal` if you use Windows, `~/ado/personal` if you use Unix, and `~:ado:personal` if you use a Macintosh. We are guessing.

To find your personal ado directory, enter Stata and type

```
. personal
```

❏ Technical Note

Stata looks in various directories for ado-files, defined by the c-class value `c(adopath)`, which contains

$$UPDATES;BASE;SITE;.;PERSONAL;PLUS;OLDPLACE$$

The words in capital letters are codenames for directories, and the mapping from codenames to directories can be obtained by typing the `sysdir` command. Here is what `sysdir` shows on one particular Windows computer:

```
. sysdir
     STATA:  D:\STATA\
   UPDATES:  D:\STATA\ado\updates\
      BASE:  D:\STATA\ado\base\
      SITE:  D:\STATA\ado\site\
      PLUS:  C:\ado\plus\
  PERSONAL:  C:\ado\personal\
  OLDPLACE:  C:\ado\
```

Even if you use Windows, your mapping might be different because it all depends on where you installed Stata. That is the point of the codenames. They make it possible to refer to directories according to their logical purposes rather than their physical location.

The c-class value `c(adopath)` is the search path, so in looking for an ado-file, Stata first looks in UPDATES, then in BASE, and so on, until it finds the file. Actually, Stata not only looks in UPDATES, it also takes the first letter of the ado-file it is looking for and looks in the lettered subdirectory. Say Stata was looking for `gobbledegook.ado`. Stata would look up UPDATES (`D:\STATA\ado\updates` in our example) and, if the file were not found there, it would look in the g subdirectory of UPDATES (`D:\STATA\ado\updates\g`) before looking in BASE, whereupon it would follow the same rules.

Why the extra complication? We distribute literally hundreds of ado-files with Stata, and some operating systems have difficulty dealing with so many files in the same directory. All operating systems experience at least a performance degradation. To prevent this, the ado directory we ship is split 27 ways (letters *a–z* and underscore). Thus, the Stata command `ci`, which is implemented as an ado-file, can be found in the subdirectory c of BASE.

If you write ado-files, you can structure your personal ado directory this way, too, but there is no reason to do so until you have more than, say, 250 files in a single directory.

❏

❏ Technical Note

After finding and running *gobbledegook*.`ado`, Stata calculates the total size of all programs that it has automatically loaded. If this exceeds `adosize` (see [P] **sysdir**), Stata begins discarding the oldest automatically loaded programs until the total is less than `adosize`. Oldest here is measured by the time last used, not the time loaded. This discarding saves memory and does not affect you, since any program that was automatically loaded could be automatically loaded again if needed.

It does, however, affect performance. Loading the program takes time, and you will again have to wait if you use one of the previously loaded-and-discarded programs. Increasing `adosize` reduces this possibility, but at the cost of memory. The `set adosize` command allows you to change this parameter; see [P] **sysdir**. The default value of `adosize` is 400 for Stata/SE and Intercooled Stata, and 200 for Small Stata. A value of 400 for `adosize` means that up to 400K can be allocated to autoloaded programs. Experimentation has shown that this is a good number—increasing it does not improve performance much.

❑

21.11.1 Version

We recommend that the first line following `program` in your ado-file declare the Stata release under which you wrote the program; `hello.ado` would better read as

```
                                                                   top of hello.ado
    program hello
            version 8
            display "hi there"
    end
                                                           end of hello.ado
```

We introduced the concept of version in [U] **19.1.1 Version**. In regular do-files, we recommend the `version` line appear as the first line of the do-file. For ado-files, the line appears after the `program` because the loading of the ado-file is one step and the execution of the program is another. It is when Stata executes the program defined in the ado-file that we want to stipulate the interpretation of the commands.

The appearance of the `version` line is of more importance in ado-files than in do-files because (1) ado-files have longer lives than do-files, so it is more likely that you will use an ado-file with a later release, and (2) ado-files tend to use more of Stata's features, and so will increase the probability that, if a change to Stata is made, the change will affect them.

21.11.2 Comments and long lines in ado-files

Comments in ado-files are handled the same way as in do-files: enclose the text in /* comment */ brackets or
* begin the line with an asterisk; see [U] **19.1.2 Comments and blank lines in do-files**.

Logical lines longer than physical lines are also handled as they are in do-files: either you change the delimiter to semicolon;
or you comment out the newline using /// at the end of the previous physical line. You can also use // and /// to place comments at the end of lines; see [U] **19.1.2 Comments and blank lines in do-files**.

21.11.3 Debugging ado-files

Debugging ado-files is just a little tricky because it is Stata and not you that is in control of when the ado-file is loaded.

Assume you wanted to change `hello` to say "Hi, Mary". Assume that your editor is called `vi` and that you are in the habit of calling your editor from Stata, so you do this,

```
. !vi hello.ado
```

and make the obvious change to the program. Equivalently, you can pretend that you are using a windowed operating system and jump out of Stata—leaving it running—to modify the `hello.ado` file. Anyway, you change `hello.ado` to read

```
─────────────────────────────────────────────── top of hello.ado ───────────
program hello
        version 8
        display "hi, Mary"
end
─────────────────────────────────────────────── end of hello.ado ───────────
```

Back in Stata, you try it:

```
. hello
hi there
```

What just happened is that Stata ran the old copy of `hello`—the copy it still has in its memory. Stata wants to be fast about executing ado-files, so when it loads one, it keeps it around a while—waiting for memory to get short—before clearing it from its memory. Naturally, Stata can drop `hello` anytime because it can always reload it from disk.

So, you changed the copy on disk, but Stata still has the old copy loaded into memory. You type `discard` to tell Stata to forget these automatically loaded things and to force itself to get new copies of the ado-files from disk:

```
. discard
. hello
hi, Mary
```

Understand that you only had to type `discard` because you changed the ado-file while Stata was running. Had you exited Stata and returned later to use `hello`, the `discard` would not have been necessary because Stata forgets things between sessions anyway.

21.11.4   Local subroutines

A single ado-file can contain more than one `program`, but if it does, the other programs defined in the ado-file are assumed to be subroutines of the main program. For example,

```
─────────────────────────────────────────────── top of decoy.ado ───────────
program decoy
        ...
        duck ...
        ...
end
program  duck
        ...
end
─────────────────────────────────────────────── end of decoy.ado ───────────
```

`duck` is considered a local subroutine of `decoy`. Even after `decoy.ado` were loaded, if you were to type `duck`, you would be told "unrecognized command". To emphasize what local means, assume you have also written an ado-file named `duck.ado`:

```
─────────────────────────────────────────────── top of duck.ado ───────────
program duck
        ...
end
─────────────────────────────────────────────── end of duck.ado ───────────
```

Even so, when decoy called duck, it would be the program duck defined in decoy.ado that was called. To further emphasize what local means, assume that decoy.ado contains

```
───────────────────────────────────────────────── top of decoy.ado ───────────
program decoy
        ...
        manic ...
        ...
        duck ...
        ...
end
program duck
        ...
end
───────────────────────────────────────────────── end of decoy.ado ───────────
```

and that manic.ado contained

```
───────────────────────────────────────────────── top of manic.ado ───────────
program manic
        ...
        duck ...
        ...
end
───────────────────────────────────────────────── end of manic.ado ───────────
```

Here is what would happen when you executed decoy:

1. decoy in decoy.ado would begin execution. decoy calls manic.

2. manic in manic.ado would begin execution. manic calls duck.

3. duck in duck.ado (yes) would begin execution. duck would do whatever and return.

4. manic regains control and eventually returns.

5. decoy is back in control. decoy calls duck.

6. duck in decoy.ado would execute, complete, and return.

7. decoy would regain control and return.

Note that, when manic called duck, it was the global ado-file duck.ado that was executed, yet when decoy called duck, it was the local program duck that was executed.

Stata does not find this confusing and neither should you.

21.11.5 Development of a sample ado-command

Below we demonstrate how one creates a new Stata command. We will program an influence measure for use with linear regression. It is an interesting statistic in its own right, but even if you are not interested in linear regression and influence measures, the focus here is on programming, not on the particular statistic chosen.

Belsley, Kuh, and Welsch (1980, 24) present a measure of influence in linear regression defined as

$$\frac{\text{Var}\left(\widehat{y}_i^{(i)}\right)}{\text{Var}(\widehat{y}_i)}$$

which is the ratio of the variance of the ith fitted value based on regression estimates obtained by omitting the ith observation, to the variance of the ith fitted value estimated from the full dataset. This ratio is estimated using

$$\text{FVARATIO}_i \equiv \frac{n - k}{n - (k + 1)} \left\{ 1 - \frac{d_i^2}{1 - h_{ii}} \right\} (1 - h_{ii})^{-1}$$

where n is the sample size; k is the number of estimated coefficients; $d_i^2 = e_i^2 / e'e$ and e_i is the ith residual; and h_{ii} is the ith diagonal element of the hat matrix. The ingredients of this formula are all available through Stata, and so, after estimating the regression parameters, one can easily calculate FVARATIO_i. For instance, one might type

```
. regress mpg weight displ
. predict hii if e(sample), hat
. predict ei if e(sample), resid
. quietly count if e(sample)
. scalar nreg = r(N)
. gen eTe = sum(ei*ei)
. gen di2 = (ei*ei)/eTe[_N]
. gen FVi = (nreg - 3) / (nreg - 4) * (1 - di2/(1-hii)) / (1-hii)
```

The number 3 in the formula for `FVi` represents k, the number of estimated parameters (which is an intercept plus coefficients on `weight` and `displ`), and the number 4 represents $k + 1$.

❑ Technical Note

Do you understand why this works? `predict` can create h_{ii} and e_i, but the trick is in getting $e'e$—the sum of the squared e_i's. Stata's `sum()` function creates a running sum. The first observation of `eTe` thus contains e_1^2; the second, $e_1^2 + e_2^2$; the third, $e_1^2 + e_2^2 + e_3^2$; and so on. The last observation, then, contains $\sum_{i=1}^{N} e_i^2$, which is $e'e$. (We specified `if e(sample)` on our `predict` commands to restrict calculations to the estimation subsample, so `hii` and `eii` might have missing values, but that does not matter because `sum()` treats missing values as contributing zero to the sum.) We use Stata's explicit subscripting feature and then refer to `eTe[_N]`, the last observation. (See [U] **16.3 Functions** and [U] **16.7 Explicit subscripting**.) After that, we plug into the formula to obtain the result.   ❑

Assuming we often wanted this influence measure, it would be easier and less prone to error if we canned this calculation in a program. Our first draft of the program reflects exactly what we would have typed interactively:

———————————————————————————————— top of fvaratio.ado, version 1 ————————

```
program fvaratio
        version 8
        predict hii if e(sample), hat
        predict ei if e(sample), resid
        quietly count if e(sample)
        scalar nreg = r(N)
        gen eTe = sum(ei*ei)
        gen di2 = (ei*ei)/eTe[_N]
        gen FVi = (nreg - 3) / (nreg - 4) * (1 - di2/(1-hii)) / (1-hii)
        drop hii ei eTe di2
end
```

———————————————————————————————— end of fvaratio.ado, version 1 ————————

All we have done is enter what we would have typed into a file, bracketing it with `program fvaratio`—meaning that we decided to call our new command `fvaratio`—and `end`. Since our command is to be called `fvaratio`, the file must be named `fvaratio.ado`, and it must be stored in either the current directory or our personal ado directory (see [U] **20.5.2 Where is my personal ado directory?**).

That done, when we type fvaratio, Stata will be able to find it, load it, and execute it. In addition to copying the interactive lines into a program, we added the line 'drop hii ...' to eliminate the working variables we had to create along the way.

So, now we can interactively type

```
. regress mpg weight displ
. fvaratio
```

and add the new variable FVi to our data.

Our program is not general. It is suitable for use after fitting a regression model on two, and only two, independent variables because we coded a 3 in the formula for k. Stata statistical commands such as regress store information about the problem and answer in e(). Looking under "Saved Results" in [R] **regress**, we find that e(df_m) contains the model degrees of freedom, which is $k - 1$, assuming the model has an intercept. Also, the sample size of the dataset used in the regression is stored in e(N), eliminating our need to count the observations and define a scalar containing this count. Thus, the second draft of our program reads

```
────────────────────────────────────── top of fvaratio.ado, version 2 ──────
program fvaratio
        version 8
        predict hii if e(sample), hat
        predict ei if e(sample), resid
        gen eTe = sum(ei*ei)
        gen di2 = (ei*ei)/eTe[_N]
        gen FVi = (e(N)-(e(df_m)+1)) / (e(N)-(e(df_m)+2)) *     /// changed this
                (1 - di2/(1-hii)) / (1-hii)                     // version
        drop hii ei eTe di2
end
─────────────────────────────────────────── end of fvaratio.ado, version 2 ─────────
```

In the formula for FVi, we substituted (e(df_m)+1) for the literal number 3, (e(df_m)+2) for the literal number 4, and we use e(N) for the sample size.

Returning to the substance of our problem, regress also saves the residual sum of squares in e(rss), so the calculation of eTe is not really necessary:

```
────────────────────────────────────── top of fvaratio.ado, version 3 ──────
program fvaratio
        version 8
        predict hii if e(sample), hat
        predict ei if e(sample), resid
        gen di2 = (ei*ei)/e(rss)                      // changed this version
        gen FVi = (e(N)-(e(df_m)+1)) / (e(N)-(e(df_m)+2)) *     ///
                (1 - di2/(1-hii)) / (1-hii)
        drop hii ei di2
end
─────────────────────────────────────────── end of fvaratio.ado, version 3 ─────────
```

Our program is now shorter and faster, and it is completely general. This program is probably good enough for most users; if you were implementing this for solely your own occasional use, you could stop right here. The program does, however, have the following deficiencies:

1. When we use it with data with missing values, the answer is correct, but we see messages about the number of missing values generated. (These messages appear when the program is generating the working variables.)

2. We cannot control the name of the variable being produced—it is always called FVi. Moreover, when FVi already exists (say from a previous regression), we get an error message that FVi already exists. We then have to drop the old FVi and type fvaratio again.

3. If we have created any variables named `hii`, `ei`, or `di2`, we also get an error that the variable already exists and the program refuses to run.

Fixing these problems is not difficult. The fix for problem 1 is exceedingly easy; we embed the entire program in a `quietly` block:

```
───────────────────────────────────── top of fvaratio.ado, version 4 ─────────
program fvaratio
        version 8
        quietly {                                        // new this version
                predict hii if e(sample), hat
                predict ei if e(sample), resid
                gen di2 = (ei*ei)/e(rss)
                gen FVi = (e(N)-(e(df_m)+1)) / (e(N)-(e(df_m)+2)) *     ///
                        (1 - di2/(1-hii)) / (1-hii)
                drop hii ei di2
        }                                                // new this version
end
───────────────────────────────────── end of fvaratio.ado, version 4 ─────────
```

The output for the commands between the `quietly {` and `}` is now suppressed—the result is the same as if we had put `quietly` in front of each command.

Solving problem 2—that the resulting variable is always called `FVi`—requires use of the `syntax` command. Let's put that off and deal with problem 3—that the working variables have nice names like `hii`, `ei`, and `di2`, and so prevent users from using those names in their data.

One solution would be to change the nice names to unlikely names. We could change `hii` to `MyHiiVaR`—that would not guarantee the prevention of a conflict, but it would certainly make it unlikely. It would also make our program difficult to read, an important consideration should we want to change it in the future. There is a better solution. Stata's `tempvar` command (see [U] **21.7.1 Temporary variables**) places names into local macros that are guaranteed to be unique:

```
───────────────────────────────────── top of fvaratio.ado, version 5 ─────────
program fvaratio
        version 8
        tempvar hii ei di2                               // new this version
        quietly {
                predict `hii' if e(sample), hat    // changed, as are other lines
                predict `ei' if e(sample), resid
                gen `di2' = (`ei'*`ei')/e(rss)
                gen FVi = (e(N)-(e(df_m)+1)) / (e(N)-(e(df_m)+2)) *     ///
                        (1 - `di2'/(1-`hii')) / (1-`hii')
        }
end
───────────────────────────────────── end of fvaratio.ado, version 5 ─────────
```

At the top of our program, we declare the temporary variables. (We can do it outside or inside the `quietly`—it makes no difference—and we do not have to do it at the top or even all at once; we could declare them as we need them, but at the top is prettiest.) When we refer to a temporary variable, we do not refer directly to it (such as by typing `hii`), we refer to it indirectly by typing open and close single quotes around the name (`` `hii' ``). And at the end of our program, we no longer bother to `drop` the temporary variables—temporary variables are dropped automatically by Stata when a program concludes.

❏ Technical Note

Why do we type single quotes around the names? `tempvar` creates local macros containing the real temporary variable names. `hii` in our program is now a local macro, and `'hii'` refers to the contents of the local macro, which is the variable's actual name.

❏

We now have an excellent program—its only fault is that we cannot specify the name of the new variable to be created. Here is the solution to that problem:

```
─────────────────────────────────────────────────── top of fvaratio.ado, version 6 ───────────
    program fvaratio
            version 8
            syntax newvarname                                    // new this version
            tempvar hii ei di2
            quietly {
                    predict 'hii' if e(sample), hat
                    predict 'ei' if e(sample), resid
                    gen 'di2' = ('ei'*'ei')/e(rss)
                    gen 'typlist' 'varlist' = ///       changed this version
                        (e(N)-(e(df_m)+1)) / (e(N)-(e(df_m)+2)) *        ///
                        (1 - 'di2'/(1-'hii')) / (1 'hii')
            }
    end
─────────────────────────────────────────────────── end of fvaratio.ado, version 6 ───────────
```

It took a change to one line and the addition of another to obtain the solution. This magic all happens because of `syntax` (see [U] **21.4.4 Parsing standard Stata syntax** above).

'`syntax newvarname`' specifies that one new variable name must be specified (had we typed '`syntax [newvarname]`', the new varname would have been optional; had we typed '`syntax newvarlist`', the user would have been required to specify at least one new variable and allowed to specify more). In any case, `syntax` compares what the user types to what is allowed. If what the user types does not match what we have declared, `syntax` will issue the appropriate error message and stop our program. If it does match, our program will continue, and what the user typed will be broken out and stored in local macros for us. In the case of a `newvarname`, the new name typed by the user is placed in the local macro `varlist`, and the type of the variable (`float`, `double`, ...) is placed in `typlist` (even if the user did not specify a storage type, in which case, the type is the current default storage type).

This is now an excellent program. There are, however, two more improvements we could make. First, we have demonstrated that by the use of '`syntax newvarname`', we can not only allow the user to define the name of the created variable, but the storage type as well. However, when it comes to the creation of intermediate variables such as '`hii`' and '`di2`', it is good programming practice to keep as much precision as possible. We want our final answer to be precise as possible, regardless of where we ultimately decide to store it. Any calculation that uses a previously generated variable would benefit if the previously generated variable were stored in double precision. Below we modify our program appropriately:

(Continued on next page)

```
————————————————————————————————— top of fvaratio.ado, version 7 —————————
program fvaratio
        version 8
        syntax newvarname
        tempvar hii ei di2
        quietly {
                predict double `hii' if e(sample), hat        // changed, as are
                predict double `ei' if e(sample), resid       // other lines
                gen double `di2' = (`ei'*`ei')/e(rss)
                gen `typlist' `varlist' =  ///
                        (e(N)-(e(df_m)+1)) / (e(N)-(e(df_m)+2)) *     ///
                        (1 - `di2'/(1-`hii')) / (1-`hii')
        }
end
——————————————————————————————————————————— end of fvaratio.ado, version 7 ———————
```

As for the second improvement we could make, `fvaratio` is intended to be used sometime after `regress`. How do we know the user is not misusing our program and executing it after, say, `logistic`? `e(cmd)` will tell us the name of the last estimation command; see [U] **21.9 Accessing results calculated by estimation commands** and [U] **21.10.2 Saving results in e()** above. We should change our program to read

```
————————————————————————————————— top of fvaratio.ado, version 8 —————————
program fvaratio
        version 8
        if "`e(cmd)'"!="regress" {                           // new this version
                error 301
        }
        syntax newvarname
        tempvar hii ei di2
        quietly {
                predict double `hii' if e(sample), hat
                predict double `ei' if e(sample), resid
                gen double `di2' = (`ei'*`ei')/e(rss)
                gen `typlist' `varlist' = ///
                        (e(N)-(e(df_m)+1)) / (e(N)-(e(df_m)+2)) *     ///
                        (1 - `di2'/(1-`hii')) / (1-`hii')
        }
end
——————————————————————————————————————————— end of fvaratio.ado, version 8 ———————
```

The `error` command issues one of Stata's prerecorded error messages and stops our program. Error 301 is "last estimates not found"; see [P] **error**. (Try typing `error 301` at the command line.)

In any case, this is a perfect program.

❑ Technical Note

You do not have to go to all the trouble we did to program the FVARATIO measure of influence or any other statistic that appeals to you. Whereas version 1 was not really an acceptable solution—it was too specialized—version 2 was acceptable. Version 3 was better, and version 4 better yet, but the improvements were of less and less importance.

Putting aside the details of Stata's language, you should understand that final versions of programs do not just happen—they are the results of drafts that have been refined. How much refinement depends on how often and who will be using the program. In this sense, the "official" ado-files that come with Stata are poor examples. They have been subject to substantial refinement because they will be used by strangers with no knowledge of how the code works. When writing programs for yourself, you may want to stop refining at an earlier draft.

❑

21.11.6 Writing online help

When you write an ado-file, you should also write a help file to go with it. This file is a standard ASCII text file, named *command*`.hlp`, that you place in the same directory as your ado-file *command*`.ado`. This way, when the user (who may be you at a future date) types `help` followed by the name of your new command (or pulls down **Help**), they will see something better than "help for ... not found".

You can obtain examples of help files by examining the `.hlp` files in the official ado directory; type "`sysdir`" and look in the lettered subdirectories of the directory defined as BASE:

```
. sysdir
    STATA:  C:\STATA\
  UPDATES:  C:\STATA\ado\updates\
     BASE:  C:\STATA\ado\base\
     SITE:  C:\STATA\ado\site\
     PLUS:  C:\ado\plus\
 PERSONAL:  C:\ado\personal\
 OLDPLACE:  C:\ado\
```

In this case, you would find examples of `.hlp` files in the a, b, ... subdirectories of C:\STATA\ado\base.

Help files are physically written on the disk in ASCII text format, but their contents are SMCL—Stata Markup and Control Language. For the most part, you can ignore that. If the file contains a line that reads

```
Also see help for the finishup command
```

it will display in just that way. However, SMCL contains lots of special directives, so that if the line in the file were to read

```
Also see {hi:help} for the {help finishup} command
```

what would be displayed would be

Also see **help** for the **finishup** command

and moreover, **finishup** would appear as a hypertext link, meaning that if the user clicked on it, they would be taken to help on `finishup`.

You can read about the details of SMCL in [P] **smcl**. The following is a SMCL help file:

──────────── top of examplehelpfile.hlp ────────────
```
{smcl}
{* 19sep2002}{...}
{hline}
help for {hi:whatever}{right:manual:  {hi:[U] 21.11.6 Writing online help}}
{hline}

{title:Calculate whatever statistic}

{p 8 17 2}
{cmdab:wh:atever}
[{it:varlist}]
[{it:weight}]
[{cmd:if} {it:exp}]
[{cmd:in} {it:range}]
[{cmd:,}
    {cmdab:d:etail}
    {cmdab:mean:only}
    {cmdab:median:only}
    {cmdab:gen:erate(}{it:newvar}{cmd:)}
]
```

```
{p 4 4 2}
{cmd:by} {it:...}{cmd::} may be used with {cmd:whatever}; see help {help by}.
{p 4 4 2}
{cmd:fweight}s are allowed; see help {help weights}.


{title:Description}
{p 4 4 2}
{cmd:whatever} calculates the whatever statistic for the variables in
{it:varlist} when the data are not stratified.


{title:Options}
{p 4 8 2}
{cmd:detail} presents detailed output of the calculation.
{p 4 8 2}
{cmd:meanonly} and {cmd:medianonly} restrict the calculation to be based on
only the means or medians.  The default is to use a trimmed mean.
{p 8 8 2}
{cmd:meanonly} is based on the idea of Rogers (1998) and should only
be specified when the data are known to be approximately symmetrically
distributed.
{p 8 8 2}
{cmd:medianonly} is a variation on meanonly which has not appeared in the
literature.  Use with caution.
{p 4 8 2}
{cmd:generate(}{it:newvar}{cmd:)} creates {it:newvar} containing the whatever
values.


{title:Remarks}
{p 4 4 2}
For detailed information on the whatever statistic, see Rogers (1998).


{title:Examples}
{p 4 8 2}{cmd:. whatever mpg weight}
{p 4 8 2}{cmd:. whatever mpg weight, meanonly}


{title:Author}
{p 4 4 2}
{browse "http://dbss.eul.edu/~hrogers":H. Rogers}, Department of Biostatistical
Social Science, Astronomy Building, Univ. of Equatorial London.  Email
{browse "mailto:hrogers@dbss.uel.edu":hrogers@dbss.eul.edu}
if you observe any problems.


{title:Also see}
{p 4 13 2}
Manual:  {hi:[U] 21.11.6 Writing online help}
{p 4 13 2}
Online:  help for {help help}, {help summarize}
```
── end of examplehelpfile.hlp ───────────

If you were to pull down **Help**, select **Stata Command,** and type `examplehelpfile` and click okay, or if you were to type `help examplehelpfile`, this is what you would see:

---

help for **whatever**   manual: **[U] 21.11.6 Writing online help**

Calculate whatever statistic

       whatever [*varlist*] [*weight*] [**if** *exp*] [**in** *range*] [**, detail mean**only
            medianonly **gen**erate(*newvar*)]

by ...: may be used with **whatever**; see help **by**.

fweights are allowed; see help **weights**.

Description

    whatever calculates the whatever statistic for the variables in *varlist* when
    the data are not stratified.

Options

    detail presents detailed output of the calculation.

    meanonly and **medianonly** restrict the calculation to be based on only the means
      or medians. The default is to use a trimmed mean.

      meanonly is based on the idea of Rogers (1998) and should only be
      specified when the data are known to be approximately symmetrically
      distributed.

      medianonly is a variation on meanonly which has not appeared in the
      literature. Use with caution.

    generate(*newvar***)** creates *newvar* containing the whatever values.

Remarks

    For detailed information on the whatever statistic, see Rogers (1998).

Examples

    . **whatever mpg weight**

    . **whatever mpg weight, meanonly**

Author

    H. Rogers, Department of Biostatistical Social Science, Astronomy Building,
    Univ. of Equatorial London. Email **hrogers@dbss.eul.edu** if you observe any
    problems.

Also see

    Manual: **[U] 21.11.6 Writing online help**
    Online: help for **help**, **summarize**

Users will find it easier to understand your programs if you document them in the same way we document ours. We offer the following guidelines:

1. The help file must have {smcl} on the first line. This notifies Stata that the help file is in SMCL format.

2. The second line should contain {* *date*}{...}. The * indicates a comment, and the {...} will suppress the blank line. Whenever you edit the help file, update the date found in the comment line.

3. The third line should contain a row of dashes; {hline} will provide them.

4. The fourth line should contain

   `help for {hi:`*yourcmd*`}{right:`*ultimate source of information*`}`

   In the official Stata files, the ultimate source of information is the printed manual, so we reference it. When a new command has been published in the *Stata Journal*, we place the SJ issue and insert number there; for example, (`SJ1-1: st0001`). For your files, we suggest putting your name.

   The line should mention all commands documented here, so if there are multiple commands, list them separated by commas. Use more than one line if necessary.

5. The next line should contain a `{hline}` row of dashes.

6. Include one blank line. Include the title and one blank line after that:

   `{title:`*Your title*`}`

7. Include syntax diagrams. Note our use of paragraph mode `{p `$\#_1$ $\#_2$ $\#_3$`}`. In particular, in the example, we used `{ p 8 17 2}`. We typically use $\#_1 = 8$ unless the commands are long, in which case, we would use $\#_1 = 4$. $\#_2$ we set to whatever looks good for the particular syntax diagram. We set $\#_3 = 2$ to bring in the right margin. Our use of paragraph mode for syntax diagrams is discussed in [P] **smcl**.

8. Include the title "Description", meaning two blank lines, `{title:Description}`, and another blank line. In general, the format for a heading is

   ```
   ( blank line )
   ( blank line )
   {title:Heading of your choice}
   ( blank line )
   ```

9. Provide a short description of what the command does. Do not burden the user with details yet. Assume that the user is at the point of asking whether this is what he or she is looking for.

10. If your commands have options, include the title "Options" followed by a description of each option, preferably in the order they appear in the syntax diagram. Option paragraphs are reverse indented: `{p 4 8 2}`. The first line is indented four spaces and begins with the option's name; subsequent lines are indented eight spaces. One blank line separates paragraphs. If an option requires more than one paragraph, subsequent paragraphs are set using `{p 8 8 2}`.

11. Optionally include `{title:Remarks}` and whatever lengthy discussion you feel necessary. Stata's official online help files omit this because the discussions appear in the manual. Stata's official help files for features added between releases (obtained from the *Stata Journal*, the Stata web site, etc.), however, include this because the appropriate SJ issue may not be as accessible as the manuals. This choice is yours.

12. Include `{title:Examples}` and provide some examples. Nothing communicates better than providing something beyond theoretical discussion. Examples rarely need much explanation.

13. Optionally include `{title:Author}` and your identity. Exercise caution. If you include a telephone number, expect it to ring. An email address may be more appropriate.

14. Optionally include `{title:References}` and any printed references. Stata's official help files seldom do this because the references are in the printed manual.

15. Include `{title:Also see}` and provide cross references to related online help.

We also warn that it is easy to use too much `{hi:highlighting}`. Use sparingly. In text, use `{cmd:...}` to show what would be shown in typewriter typeface were the documentation printed in this manual.

❑ Technical Note

Sometimes it is more convenient to describe two or more related commands in the same `.hlp` file. Thus, `xyz.hlp` might document both the `xyz` and `abc` commands. To arrange that typing `help abc` displays `xyz.hlp`, create the file `abc.hlp`, containing

```
                                        ─────── top of abc.hlp ──────
    .h xyz
                                        ─────── end of abc.hlp ───────
```

When a `.hlp` file contains a single line of the form '`.h` *refname*', Stata interprets that as an instruction to display `help` for *refname*.

❑

❑ Technical Note

If you write a collection of programs, you need to somehow index the programs so that users (and you) can find the command they want. We do that with our `contents.hlp` entry. You should create a similar kind of entry. We suggest that you call your private entry `user.hlp` in your personal ado directory; see [U] **20.5.2 Where is my personal ado directory?**. This way, to review what you have added, you can type `help user`.

We suggest Unix users at large sites also add `site.hlp` to the SITE directory (typically `/usr/local/ado`, but type `sysdir` to be sure). Then you can type `help site` for a list of the commands available site-wide.

❑

21.12   A compendium of useful commands for programmers

You can use literally any Stata command in your programs and ado-files. In addition, some commands are intended solely for use by Stata programmers. You should see the section under the *Programming* heading in the subject table of contents at the beginning of this manual.

21.13   References

Belsley, D. A., E. Kuh, & R. E. Welsch. 1980. *Regression Diagnostics: Identifying Influential Data and Sources of Collinearity*. New York: John Wiley & Sons.

Gould, W. 2001. pr0001: Statistical software certification. *The Stata Journal* 1(1): 29–50.

22 Immediate commands

Contents

22.1   Overview
        22.1.1   Examples
        22.1.2   A list of the immediate commands
22.2   The display command

22.1 Overview

An *immediate* command is a command that obtains data not from the data stored in memory, but from numbers typed as arguments. Immediate commands, in effect, turn Stata into a glorified hand calculator.

There are many instances where you may not have the data, but you do know something about the data, and what you know is adequate to perform statistical tests. For instance, you do not have to have individual level data to obtain the standard error of the mean, and thereby a confidence interval, if you know the mean, standard deviation, and number of observations. There are other instances where you may actually have the data and you could enter the data and perform the test, but it would be easier if you could just ask for the statistic based on a summary. For instance, you flip a coin 10 times and it comes up heads twice. You could enter a 10 observation dataset with two 1s (standing for heads) and eight 0s (meaning tails).

Immediate commands are meant to solve those problems. Immediate commands have the following properties:

1. They never disturb the data in memory. You can perform an immediate calculation as an aside and your data remain unchanged.

2. The syntax for all is the same, the command name followed by numbers, which are the summary statistics from which the statistic is calculated. The numbers are almost always summary statistics, and the order in which they are specified is in some sense "natural".

3. Immediate commands all end in the letter *i*, although the converse is not true. In most cases, if there is an immediate command, there is a nonimmediate form also; that is, a form that works on the data in memory. For every statistical command in Stata, we have included an immediate form if it is reasonable to assume that you might know the requisite summary statistics without having the underlying data, and if typing those statistics is not absurdly burdensome.

4. Immediate commands are documented along with their nonimmediate cousins. Thus, if you want to obtain a confidence interval, whether it be from summary data with an immediate command or using the data in memory, use the table of contents or index to discover that [R] **ci** discusses confidence intervals. There, you learn that `ci` calculates confidence intervals using the data in memory, and that `cii` does the same with the data specified immediately following the command.

22.1.1   Examples

▷ Example

Taking the example of confidence intervals, professional papers often publish the mean, standard deviation, and number of observations for variables used in the analysis. Those statistics are sufficient for calculating a confidence interval. If we know that the mean mileage rating of cars in some sample is 24, the standard deviation is 6, and that there are 97 cars in the sample:

```
. cii 97 24 6
```

Variable	Obs	Mean	Std. Err.	[95% Conf. Interval]	
	97	24	.6092077	22.79073	25.20927

We learn that the mean's standard error is 0.61 and its 95% confidence interval is $[22.8, 25.2]$. To obtain this, we typed cii (the immediate form of the ci command) followed by the number of observations, the mean, and the standard deviation. We knew the order in which to specify the numbers because we had read [R] **ci**.

We could use the immediate form of the ttest command to test the hypothesis that the true mean is 22:

```
. ttesti 97 24 6 22
One-sample t test
```

	Obs	Mean	Std. Err.	Std. Dev.	[95% Conf. Interval]	
x	97	24	.6092077	6	22.79073	25.20927

```
Degrees of freedom: 96
                          Ho: mean(x) = 22
   Ha: mean < 22          Ha: mean != 22            Ha: mean > 22
      t =   3.2830            t =   3.2830             t =   3.2830
  P < t =   0.9993       P > |t| =   0.0014        P > t =   0.0007
```

The first three numbers were as we specified in the cii command. ttesti requires a fourth number, which is the constant against which the mean is being tested; see [R] **ttest**.

◁

▷ Example

We mentioned flipping a coin 10 times and having it come up heads twice. The 99% confidence interval can also be obtained from ci:

```
. cii 10 2, level(99)
```

				— Binomial Exact —	
Variable	Obs	Mean	Std. Err.	[99% Conf. Interval]	
	10	.2	.1264911	.0108398	.6482422

In the previous example, we specified cii with three numbers following it; in this example, we specify 2. Immediate commands often determine what to do by the number of arguments following the command. With two arguments, ci assumes that we are specifying the number of trials and successes from a binomial experiment; see [R] **ci**.

The immediate form of the bitest command performs exact hypothesis testing:

```
. bitesti 10 2 .5
           N   Observed k   Expected k    Assumed p   Observed p

          10            2            5      0.50000      0.20000
Pr(k >= 2)               = 0.989258   (one-sided test)
Pr(k <= 2)               = 0.054688   (one-sided test)
Pr(k <= 2 or k >= 8) = 0.109375   (two-sided test)
```

For a full explanation of this output, see [R] **bitest**.

◁

▷ Example

Stata's `tabulate` command makes tables and calculates various measures of association. The immediate form, `tabi`, does the same, but you specify the contents of the table following the command:

```
. tabi 5 10 \ 2 14
                    col
       row |       1          2  |     Total

         1 |       5         10  |        15
         2 |       2         14  |        16

     Total |       7         24  |        31
          Fisher's exact =                 0.220
  1-sided Fisher's exact =                 0.170
```

The `tabi` command is slightly different from most immediate commands because it uses '\' to indicate where one row ends and another begins.

◁

22.1.2 A list of the immediate commands

Command	Reference	Description
bitesti	[R] **bitest**	Binomial probability test
cci csi iri mcci	[ST] **epitab**	Tables for epidemiologists
cii	[R] **ci**	Confidence intervals for means, proportions, and counts
prtesti	[R] **prtest**	One and two-sample tests of proportions
sampsi	[R] **sampsi**	Sample size and power determination
sdtesti	[R] **sdtest**	Variance comparison tests
symmi	[R] **symmetry**	Symmetry and marginal homogeneity tests
tabi	[R] **tabulate**	One- and two-way tables of frequencies
ttesti	[R] **ttest**	Mean comparison tests

22.2   The display command

display is not really an immediate command, but it can be used as a hand-calculator.

```
. display 2+5
7
. display sqrt(2+sqrt(3^2-4*2*-2))/(2*3)
.44095855
```

See [R] **display**.

23 Estimation and post-estimation commands

Contents

23.1 All estimation commands work the same way
23.2 Standard syntax
23.3 Replaying prior results
23.4 Cataloging estimation results
23.5 Specifying the estimation subsample
23.6 Specifying the width of confidence intervals
23.7 Obtaining the variance–covariance matrix
23.8 Obtaining predicted values
    23.8.1 predict can be used after any estimation command
    23.8.2 Making in-sample predictions
    23.8.3 Making out-of-sample predictions
    23.8.4 Obtaining standard errors, tests, and confidence intervals for predictions
23.9 Accessing estimated coefficients
23.10 Performing hypothesis tests on the coefficients
    23.10.1 Linear tests
    23.10.2 test can be used after any estimation command
    23.10.3 Likelihood-ratio tests
    23.10.4 Nonlinear Wald tests
23.11 Obtaining linear combinations of coefficients
23.12 Obtaining nonlinear combinations of coefficients
23.13 Obtaining marginal effects
23.14 Obtaining robust variance estimates
23.15 Obtaining scores
23.16 Weighted estimation
    23.16.1 Frequency weights
    23.16.2 Analytic weights
    23.16.3 Sampling weights
    23.16.4 Importance weights
23.17 A list of post-estimation commands
23.18 References

23.1 All estimation commands work the same way

All Stata commands that fit statistical models—commands such as `regress`, `logit`, `sureg`, and so on—work the same way. Most single-equation commands have similar syntax

$$command\ varlist\ [weight]\ [\texttt{if}\ exp]\ [\texttt{in}\ range]\ [,\ options]$$

as do most multiple-equation commands:

$$command\ (varlist)\ (varlist)\ \ldots\ (varlist)\ [weight]\ [\texttt{if}\ exp]\ [\texttt{in}\ range]\ [,\ options]$$

Adopt a loose definition of single and multiple equation in interpreting this. For instance, `heckman` is a two-equation system mathematically speaking, yet we categorize it, syntactically, with single-equation commands because most researchers think of it as a linear regression with an adjustment for the censoring.

In any case, the important thing is that most estimation commands have one or the other of these two syntaxes.

In single-equation commands, the first variable in the *varlist* is the dependent variable and the remaining variables are the independent variables. There can be variations. For instance, anova allows you to specify *terms* in addition to variables for the independent variables, but the foregoing is generally correct.

All estimation commands—whether single or multiple equation—share the following features:

1. You can use the features of Stata's syntax to specify the estimation sample; you do not have to make a special dataset.

2. You can, at any time, review the last estimates by typing the estimation command without arguments. After fitting a regression model with regress, for instance, you can see the last estimates again by typing regress by itself. You do not have to do this immediately—any number of commands can occur in between the estimation and the replaying, and, in fact, you can even replay the last estimates after the data have changed or you have dropped the data altogether. Stata never forgets (unless you type discard; see [P] **discard**).

3. All estimation commands display confidence intervals for the coefficients and allow the level() option to indicate the width of the intervals. The default is level(95), meaning 95% confidence intervals. You can reset the default with set level; see [R] **level**.

4. You can use mfx to display model results in terms of marginal effects (dy/dx or even $df(y)/dx$), which can be displayed as either derivatives or elasticities; see [R] **mfx**.

5. You can obtain tables of adjusted means; see [R] **adjust**.

6. You can obtain the variance–covariance matrix of the estimators (VCE), presented as either a correlation matrix or a covariance matrix, using vce at any time after estimation. (You can also obtain the estimated coefficients and covariance matrix as vectors and matrices and manipulate them with Stata's matrix capabilities; see [U] **17.5 Accessing matrices created by Stata commands**.)

7. You can obtain predictions, residuals, influence statistics, and the like, either for the data on which you just estimated or for some other data, using predict. You can also obtain point estimates, standard errors, Wald tests, etc., for customized predictions using predictnl.

8. You can refer to coefficients and standard errors in expressions (such as with generate); see [U] **16.5 Accessing coefficients and standard errors**.

9. You can perform tests on the estimated parameters using test (Wald tests of linear hypotheses), testnl (Wald tests of nonlinear hypotheses), and lrtest (likelihood-ratio tests). You can also obtain point estimates and confidence intervals for linear combinations of the estimated parameters using lincom, or nonlinear combinations using nlcom.

10. You can hold estimates, perform other estimation commands, and then restore the prior estimates. This is of particular interest to programmers; see [P] **_estimates**.

11. You can store estimation results by name for later retrieval or for displaying/comparing multiple models using estimates; see [R] **estimates**.

12. You can obtain the joint parameter vector and variance–covariance matrix for coefficients from two different models using seemingly unrelated estimation (suest). This is especially useful for testing the equality, say, of coefficients across models; see [R] **suest**.

13. You can perform Hausman model specification tests using hausman; see [R] **hausman**.

Eventually, it will also be true that

14. You can obtain the Huber/White/robust alternate estimate of variance by specifying the `robust` option, and you can relax the assumption of independence of the observations by also specifying the `cluster()` option.

Right now there are a few estimation commands that do not allow `robust`.

23.2 Standard syntax

The syntax for `regress` is the same as that for all single-equation estimation commands. Most importantly, you can combine if *exp* and in *range* with the estimation command, and estimation commands also allow by *varlist:*, where it would be sensible.

▷ Example

You have data on 74 automobiles that record the mileage rating (`mpg`), weight (`weight`), and whether the car is domestic or foreign-produced (`foreign`). You can fit a linear regression model of `mpg` on `weight` and `weightsq`, using just the foreign-made automobiles, by typing

```
. use http://www.stata-press.com/data/r8/auto2
(1978 Automobile Data)
. regress mpg weight weightsq if foreign
```

Source	SS	df	MS		Number of obs =	22
					F(2, 19) =	8.31
Model	428.256889	2	214.128444		Prob > F =	0.0026
Residual	489.606747	19	25.7687762		R-squared =	0.4666
					Adj R-squared =	0.4104
Total	917.863636	21	43.7077922		Root MSE =	5.0763

mpg	Coef.	Std. Err.	t	P>|t|	[95% Conf. Interval]	
weight	-.0132182	.0275711	-0.48	0.637	-.0709252	.0444888
weightsq	5.50e-07	5.41e-06	0.10	0.920	-.0000108	.0000119
_cons	52.33775	34.1539	1.53	0.142	-19.14719	123.8227

You can run separate regressions for the domestic and foreign-produced automobiles with the by *varlist:* prefix:

```
. by foreign: regress mpg weight weightsq
```

```
-> foreign = Domestic
```

Source	SS	df	MS		Number of obs =	52
					F(2, 49) =	91.64
Model	905.395466	2	452.697733		Prob > F =	0.0000
Residual	242.046842	49	4.93973146		R-squared =	0.7891
					Adj R-squared =	0.7804
Total	1147.44231	51	22.4988688		Root MSE =	2.2226

mpg	Coef.	Std. Err.	t	P>|t|	[95% Conf. Interval]	
weight	-.0131718	.0032307	-4.08	0.000	-.0196642	-.0066794
weightsq	1.11e-06	4.95e-07	2.25	0.029	1.19e-07	2.11e-06
_cons	50.74551	5.162014	9.83	0.000	40.37205	61.11896

```
-> foreign = Foreign
      Source |       SS       df       MS              Number of obs =      22
-------------+------------------------------           F(  2,    19) =    8.31
       Model | 428.256889      2  214.128444           Prob > F      =  0.0026
    Residual | 489.606747     19  25.7687762           R-squared     =  0.4666
-------------+------------------------------           Adj R-squared =  0.4104
       Total | 917.863636     21  43.7077922           Root MSE      =  5.0763

         mpg |      Coef.   Std. Err.       t    P>|t|     [95% Conf. Interval]
-------------+----------------------------------------------------------------
      weight |  -.0132182   .0275711    -0.48   0.637    -.0709252    .0444888
     weightsq |   5.50e-07   5.41e-06     0.10   0.920    -.0000108    .0000119
        _cons |   52.33775    34.1539     1.53   0.142    -19.14719    123.8227
```

Although all estimation commands do allow if *exp* and in *range*, not all allow the by *varlist*: prefix. In the case of by(), the duration of Stata's memory is limited: it remembers only the *last* set of estimates. This means that if you were to use any of the other features described below, they would use the last regression estimated, which right now is mpg on weight and weightsq for the Foreign subsample.

◁

23.3   Replaying prior results

When you type an estimation command without arguments, it redisplays prior results.

▷ Example

To perform a regression of mpg on the variables weight and displacement, you type

```
. regress mpg weight displacement
      Source |       SS       df       MS              Number of obs =      74
-------------+------------------------------           F(  2,    71) =   66.79
       Model | 1595.40969      2  797.704846           Prob > F      =  0.0000
    Residual | 848.049768     71  11.9443629           R-squared     =  0.6529
-------------+------------------------------           Adj R-squared =  0.6432
       Total | 2443.45946     73  33.4720474           Root MSE      =  3.4561

         mpg |      Coef.   Std. Err.       t    P>|t|     [95% Conf. Interval]
-------------+----------------------------------------------------------------
      weight |  -.0065671   .0011662    -5.63   0.000    -.0088925   -.0042417
displacement |   .0052808   .0098696     0.54   0.594    -.0143986    .0249602
        _cons |   40.08452    2.02011    19.84   0.000     36.05654    44.11251
```

You now go on to do other things, summarizing data, listing observations, performing hypothesis tests, or anything else. If you decide that you want to see the last set of estimates again, type the estimation command without arguments.

(Continued on next page)

```
. regress
```

Source	SS	df	MS
Model	1595.40969	2	797.704846
Residual	848.049768	71	11.9443629
Total	2443.45946	73	33.4720474

Number of obs =	74
F(2, 71) =	66.79
Prob > F =	0.0000
R-squared =	0.6529
Adj R-squared =	0.6432
Root MSE =	3.4561

mpg	Coef.	Std. Err.	t	P>|t|	[95% Conf. Interval]
weight	-.0065671	.0011662	-5.63	0.000	-.0088925 -.0042417
displacement	.0052808	.0098696	0.54	0.594	-.0143986 .0249602
_cons	40.08452	2.02011	19.84	0.000	36.05654 44.11251

This feature works with *every* estimation command, so you could just as well have done it with, say, stcox or logit.

◁

23.4   Cataloging estimation results

Stata keeps only the results of the most recently fitted model in active memory. You may, however, use Stata's estimates command to store estimation results by name for later displaying, comparison, cross-model testing, etc. You may store up to 20 sets of estimation results.

▷ Example

Continuing with our automobile data, let's fit three models and store the results. We fit the models quietly so as to keep the output to a minimum.

```
. qui regress mpg weight displ
. estimates store reg_model
. qui logit for weight displ
. estimates store logit_model
. qui ologit rep78 price weight gear_ratio
. estimates store ologit_model
```

We can now obtain a directory (listing) of our stored results, select a set to replay, and display a table comparing the estimated parameters and other model summary statistics.

```
. estimates dir
```

model	command	depvar	npar	title
reg_model	regress	mpg	3	
logit_model	logit	foreign	3	
ologit_model	ologit	rep78	7	

(Continued on next page)

```
. estimates replay logit_model
```

Model **logit_model**

```
Logit estimates                                Number of obs   =         74
                                               LR chi2(2)      =      44.72
                                               Prob > chi2     =     0.0000
Log likelihood = -22.672104                    Pseudo R2       =     0.4965
```

foreign	Coef.	Std. Err.	z	P>|z|	[95% Conf. Interval]	
weight	.0019774	.0019023	1.04	0.299	-.001751	.0057059
displacement	-.0651206	.0291647	-2.23	0.026	-.1222824	-.0079588
_cons	3.369612	2.104804	1.60	0.109	-.7557268	7.494952

```
. estimates table _all, stats(aic bic)
```

Variable	reg_model	logit_mo~l	ologit_m~l
weight	-.00656711	.00197745	-.00105961
displacement	.00528078	-.06512057	
price			.00020787
gear_ratio			.82064832
_cut1			-3.3407627
_cut2			-1.5486078
_cut3			.88460924
_cut4			2.5450833
_cons	40.084522	3.3696124	
aic	396.4796	51.344208	183.22888
bic	403.39179	58.256404	198.86762

We can also select a set of results to be made active, for use with Stata's post-estimation commands.

```
. estimates restore logit_model
(results logit_model are active now)
. logit
Logit estimates                                Number of obs   =         74
                                               LR chi2(2)      =      44.72
                                               Prob > chi2     =     0.0000
Log likelihood = -22.672104                    Pseudo R2       =     0.4965
```

foreign	Coef.	Std. Err.	z	P>|z|	[95% Conf. Interval]	
weight	.0019774	.0019023	1.04	0.299	-.001751	.0057059
displacement	-.0651206	.0291647	-2.23	0.026	-.1222824	-.0079588
_cons	3.369612	2.104804	1.60	0.109	-.7557268	7.494952

◁

We invite you to read [R] **estimates**—there is a lot more you can do with estimates. In particular, estimates makes it quite easy to perform cross-model tests such as the Hausman specification test.

23.5   Specifying the estimation subsample

Once an estimation command has been run or previous estimates restored, Stata remembers the estimation subsample, and you can use the modifier `if e(sample)` on the end of Stata commands. The term estimation subsample refers to the set of observations used to produce the active estimation results. That might turn out to be all the observations (as it was in the above example) or a subset of the observations:

```
. generate excellent = rep78==5 if rep78 < .
(5 missing values generated)

. regress mpg weight excellent if foreign
```

Source	SS	df	MS		Number of obs =	21
					F(2, 18) =	10.21
Model	423.317154	2	211.658577		Prob > F =	0.0011
Residual	372.96856	18	20.7204756		R-squared =	0.5316
					Adj R-squared =	0.4796
Total	796.285714	20	39.8142857		Root MSE =	4.552

mpg	Coef.	Std. Err.	t	P>|t|	[95% Conf. Interval]	
weight	-.0131402	.0029684	-4.43	0.000	-.0193765	-.0069038
excellent	5.052676	2.13492	2.37	0.029	.5673764	9.537977
_cons	52.86088	6.540147	8.08	0.000	39.12054	66.60122

```
. summarize mpg weight excellent if e(sample)
```

Variable	Obs	Mean	Std. Dev.	Min	Max
mpg	21	25.28571	6.309856	17	41
weight	21	2263.333	364.7099	1760	3170
excellent	21	.4285714	.5070926	0	1

Note that 21 observations were used in the above regression, and we subsequently obtained the means for those same 21 observations by typing 'summarize ... if e(sample)'. There are two reasons observations were dropped: we specified `if foreign` when we ran the regression and there were observations for which `excellent` was missing. The reason does not matter; `e(sample)` is true if the observation was used and false otherwise.

You can use `if e(sample)` on the end of any Stata command that allows an `if` *exp*.

23.6   Specifying the width of confidence intervals

You specify the width of the confidence intervals for the coefficients using the `level()` option, and you can specify the width at estimation or when you play back the results.

▷ Example

To obtain narrower, 90% confidence intervals when we fit the model, we type

```
. regress mpg weight displ, level(90)
```

Source	SS	df	MS		Number of obs =	74
					F(2, 71) =	66.79
Model	1595.40969	2	797.704846		Prob > F =	0.0000
Residual	848.049768	71	11.9443629		R-squared =	0.6529
					Adj R-squared =	0.6432
Total	2443.45946	73	33.4720474		Root MSE =	3.4561

mpg	Coef.	Std. Err.	t	P>|t|	[90% Conf. Interval]
weight	-.0065671	.0011662	-5.63	0.000	-.0085108 -.0046234
displacement	.0052808	.0098696	0.54	0.594	-.0111679 .0217294
_cons	40.08452	2.02011	19.84	0.000	36.71781 43.45124

Were we to subsequently type `regress`, without arguments, 95% confidence intervals would be reported. Had we initially fit the model with 95% confidence intervals, we could later type 'regress, level(90)' to redisplay results with 90% confidence intervals.

In addition, we could type `set level 90` to make 90% intervals our default; see [R] **level**.   ◁

23.7   Obtaining the variance–covariance matrix

Typing `vce` displays the variance–covariance matrix of the estimators in active memory.

▷ Example

Continuing with our regression of `mpg` on variables `weight` and `displ` example, we have previously typed `regress mpg weight displ`. The full variance–covariance matrix of the estimators can be displayed at any time after estimation:

```
. vce
```

	weight	displa~t	_cons
weight	1.4e-06		
displacement	-.00001	.000097	
_cons	-.002075	.011884	4.08085

`vce` with the `corr` option will present this matrix as a correlation matrix:

```
. vce, corr
```

	weight	displa~t	_cons
weight	1.0000		
displacement	-0.8949	1.0000	
_cons	-0.8806	0.5960	1.0000

See [R] **vce**.

In addition, Stata's matrix commands understand that `e(V)` refers to the matrix:

```
. matrix list e(V)
symmetric e(V)[3,3]
                   weight  displacement          _cons
       weight   1.360e-06
 displacement   -.0000103    .00009741
       _cons   -.00207455    .01188356      4.0808455
. mat Vinv = syminv(e(V))
. mat list Vinv
symmetric Vinv[3,3]
                   weight  displacement          _cons
       weight    60175851
 displacement   4081161.2     292709.46
       _cons    18706.732     1222.3339      6.1953911
```

See [U] **17.5 Accessing matrices created by Stata commands**.   ◁

23.8   Obtaining predicted values

Our discussion below, while cast in terms of predicted values, applies equally to all the statistics generated by predict; see [R] **predict**.

When Stata fits a model, be it regression or whatever, it internally saves the results, which includes the estimated coefficients and the variable names. The predict command allows you to use that information.

▷ Example

Let's perform a linear regression of mpg on weight and weightsq:

```
. regress mpg weight weightsq
```

Source	SS	df	MS
Model	1642.52197	2	821.260986
Residual	800.937487	71	11.2808097
Total	2443.45946	73	33.4720474

Number of obs =	74
F(2, 71) =	72.80
Prob > F =	0.0000
R-squared =	0.6722
Adj R-squared =	0.6630
Root MSE =	3.3587

mpg	Coef.	Std. Err.	t	P>|t|	[95% Conf. Interval]
weight	-.0141581	.0038835	-3.65	0.001	-.0219016 -.0064145
weightsq	1.32e-06	6.26e-07	2.12	0.038	7.67e-08 2.57e-06
_cons	51.18308	5.767884	8.87	0.000	39.68225 62.68392

After the regression, predict is defined to be

$$-.0141581\text{weight} + 1.32 \cdot 10^{-6}\text{weightsq} + 51.18308$$

(Actually, it is more precise, because the coefficients are internally stored at much higher precision than shown in the output.) Thus, we can create a new variable—call it fitted—equal to the prediction by typing predict fitted, and then use scatter to make a display of the fitted and actual values separately for domestic and foreign automobiles:

```
. predict fitted
(option xb assumed; fitted values)
. scatter mpg fitted weight, by(foreign, total) c(. l) m(o i) sort
```

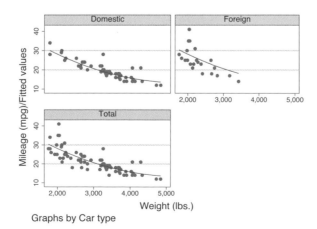

Graphs by Car type

`predict` can calculate much more than just predicted values. In the case of `predict` after linear regression, `predict` can calculate residuals, standardized residuals, studentized residuals, influence statistics, etc. In any case, what is to be calculated is specified via an option, so if we wanted the residuals stored in new variable `r`, we would type

```
. predict r, resid
```

What options may be specified following `predict` vary according to the estimation command previously used; the `predict` options are documented along with the estimation command. For instance, to discover all the things `predict` can do following `regress`, see [R] **regress**.

◁

23.8.1   predict can be used after any estimation command

The use of `predict` is not limited to linear regression.

▷ Example

You fit a logistic regression model of whether a car is manufactured outside the United States based on its weight and mileage rating using either the `logistic` or the `logit` command; see [R] **logistic** and [R] **logit**. We will use `logit`.

```
. use http://www.stata-press.com/data/r8/auto

. logit foreign weight mpg

Iteration 0:   log likelihood =  -45.03321
Iteration 1:   log likelihood = -29.898968
Iteration 2:   log likelihood = -27.495771
Iteration 3:   log likelihood = -27.184006
Iteration 4:   log likelihood = -27.175166
Iteration 5:   log likelihood = -27.175156
```

```
Logit estimates                                 Number of obs  =        74
                                                LR chi2(2)     =     35.72
                                                Prob > chi2    =    0.0000
Log likelihood = -27.175156                     Pseudo R2      =    0.3966
```

foreign	Coef.	Std. Err.	z	P>|z|	[95% Conf. Interval]	
weight	-.0039067	.0010116	-3.86	0.000	-.0058894	-.001924
mpg	-.1685869	.0919174	-1.83	0.067	-.3487418	.011568
_cons	13.70837	4.518707	3.03	0.002	4.851864	22.56487

After `logit`, `predict` without options calculates the probability of a positive outcome (we learned that by looking at [R] **logit**). To obtain the predicted probabilities that each car is manufactured outside the U.S.,

```
. predict probhat
(option p assumed; Pr(foreign))

. summarize probhat
```

Variable	Obs	Mean	Std. Dev.	Min	Max
probhat	74	.2972973	.3052979	.000729	.8980594

```
. list make mpg weight foreign probhat in 1/5
```

	make	mpg	weight	foreign	probhat
1.	AMC Concord	22	2,930	Domestic	.1904363
2.	AMC Pacer	17	3,350	Domestic	.0957767
3.	AMC Spirit	22	2,640	Domestic	.4220815
4.	Buick Century	20	3,250	Domestic	.0862625
5.	Buick Electra	15	4,080	Domestic	.0084948

◁

23.8.2 Making in-sample predictions

`predict` does not retrieve a vector of prerecorded values—it calculates the predictions based on the recorded coefficients and the data currently in memory. In the above examples, when we have typed things like

```
. predict phat
```

`predict` filled in the prediction everywhere that it could be calculated.

Sometimes we may have more data in memory than were used by the estimation command, either because we explicitly ignored some of the observations by specifying an `if` *exp* with the estimation command, or because there are missing values. In such cases, if we want to restrict the calculation to the estimation subsample, we would do that in the usual way by adding `if e(sample)` to the end of the command:

```
. predict phat if e(sample)
```

23.8.3 Making out-of-sample predictions

Because predict makes its calculations based on the recorded coefficients and the data in memory, `predict` can do more than calculate predicted values for the data on which the estimation took place—it can make out-of-sample predictions as well.

If we fit our model on a subset of the observations, we could then predict the outcome for all the observations:

```
. logit foreign weight mpg if rep78>3
. predict pall
```

If we do not specify `if e(sample)` at the end of the `predict` command, `predict` calculates the predictions for all observations possible.

In fact, because `predict` works from the active estimation results, we can use `predict` with *any* dataset that contains the necessary variables.

▷ Example

Continuing with our previous `logit` example, assume that you have a second dataset containing the `mpg` and `weight` of a different sample of cars. You have just fitted your model and now continue:

```
. use otherdat, clear
(Different cars)
. predict probhat                    Stata remembers previous model
(option p assumed; Pr(foreign))
```

```
. summarize probhat foreign
    Variable |     Obs        Mean   Std. Dev.       Min        Max
-------------+--------------------------------------------------------
     probhat |      12    .2505068    .3187104    .0084948    .8920776
     foreign |      12    .1666667    .3892495           0           1
```

◁

▷ Example

There are numerous ways to obtain out-of-sample predictions. Above, we estimated on one dataset and then used another. If our first dataset had contained both sets of cars, marked, say, by the variable difcars being 0 if from the first sample and 1 if from the second, we could type

```
. logit foreign weight mpg if difcars==0
same output as above appears

. predict probhat
(option p assumed; Pr(foreign))

. summarize probhat foreign if difcars==1
same output as directly above appears
```

If we just had a small number of additional cars, we could even input them after estimation. Assume that our data once again contain only the first sample of cars, and assume that we are interested in an additional sample of only 2 rather than 12 cars; we could type

```
. use http://www.stata-press.com/data/r8/auto

. keep make mpg weight foreign

. logit foreign weight mpg
same output as above appears

. input
                    make        mpg     weight    foreign
75. "Merc. Zephyr" 20 2830 0                we type in our new data
76. "VW Dasher" 23 2160 1
77. end
. predict probhat                  obtain all the predictions
(option p assumed; Pr(foreign))

. list in -2/1
```

```
          make    mpg   weight    foreign    probhat
     +------------------------------------------------+
75. | Merc. Zephyr    20    2830    Domestic   .3275397
76. |    VW Dasher    23    2160     Foreign   .8009743
     +------------------------------------------------+
```

◁

23.8.4 Obtaining standard errors, tests, and confidence intervals for predictions

When you use predict, you create, for each observation in the prediction sample, a statistic that is a function of the data and the estimated model parameters. You could have also generated your own customized predictions using generate. In either case, to get standard errors, Wald tests, and confidence intervals for your predictions, use predictnl. For example, if we wanted the standard errors for our predicted probabilities, we would type

```
. drop probhat
. predictnl probhat = predict(), se(phat_se)
. list in 1/5
```

	make	mpg	weight	foreign	probhat	phat_se
1.	AMC Concord	22	2,930	Domestic	.1904363	.0658386
2.	AMC Pacer	17	3,350	Domestic	.0957767	.0536296
3.	AMC Spirit	22	2,640	Domestic	.4220815	.0892845
4.	Buick Century	20	3,250	Domestic	.0862625	.0461927
5.	Buick Electra	15	4,080	Domestic	.0084948	.0093079

Comparing this output to our previous listing of the first five predicted probabilities, you will notice that the output is identical, except we now have an additional variable, phat_se, which contains the estimated standard error for each predicted probability.

Note that we first had to drop probhat since predictnl will regenerate it for us. Note also the use of predict() within predictnl—it specified that we wanted to generate a point estimate (and standard error) for the default prediction after logit; see [R] **predictnl** for more details.

23.9 Accessing estimated coefficients

You can access coefficients and standard errors after estimation by referring to _b[*name*] and _se[*name*]; see [U] **16.5 Accessing coefficients and standard errors**.

▷ Example

Let us return to linear regression. You are doing a study of earnings of men and women at a particular company. In addition to each person's earnings, you have information on their educational attainment and tenure with the company. You do the following:

```
. generate femten = female*tenure
. generate femed = female*ed
. regress lnearn ed tenure female femed femten
output appears
```

You now wish to predict everyone's income as if they were male and then compare these as-if earnings with the actual earnings:

```
. generate asif = _b[_cons] + _b[ed]*ed + _b[tenure]*tenure
```

◁

▷ Example

You are analyzing the mileage of automobiles and are using a slightly more sophisticated model than any we have used so far. As we have previously, you assume that mpg is a function of weight and weightsq, but you also add the interaction of foreign multiplied by weight (called fweight), the car's gear ratio (gear_ratio), and foreign multiplied by gear_ratio (fgear_ratio).

```
. use http://www.stata-press.com/data/r8/auto2
(1978 Automobile Data)
. generate fweight = foreign*weight
. generate fgear_ratio = foreign*gear_ratio
```

```
. regress mpg weight weightsq fweight gear_ratio fgear_ratio
```

Source	SS	df	MS		
Model	1737.05293	5	347.410585		
Residual	706.406534	68	10.3883314		
Total	2443.45946	73	33.4720474		

	Number of obs =	74
	F(5, 68) =	33.44
	Prob > F =	0.0000
	R-squared =	0.7109
	Adj R-squared =	0.6896
	Root MSE =	3.2231

mpg	Coef.	Std. Err.	t	P>|t|	[95% Conf. Interval]	
weight	-.0118517	.0045136	-2.63	0.011	-.0208584	-.002845
weightsq	9.81e-07	7.04e-07	1.39	0.168	-4.25e-07	2.39e-06
fweight	-.0032241	.0015577	-2.07	0.042	-.0063326	-.0001157
gear_ratio	1.159741	1.553418	0.75	0.458	-1.940057	4.259539
fgear_ratio	1.597462	1.205313	1.33	0.189	-.8077036	4.002627
_cons	44.61644	8.387943	5.32	0.000	27.87856	61.35432

Unless you are experienced in both regression technology and automobile technology, you may find it difficult to interpret this regression. Putting aside issues of statistical significance, you find that mileage decreases with a car's weight but increases with the square of weight; decreases even more rapidly with weight for foreign cars; increases with higher gear ratio; and increases even more rapidly with higher gear ratio in foreign cars.

Thus, do foreign cars yield better or worse gas mileage? Results are mixed. As the foreign cars' weight increases, they do more poorly in relation to domestic cars, but they do better at higher gear ratios. One way to compare the results is to predict what mileage foreign cars would have *if they were manufactured domestically*. The regression provides all the information necessary for making that calculation; mileage for domestic cars is estimated to be

$$-.012\, \text{weight} + 9.81 \cdot 10^{-7}\, \text{weightsq} + 1.160\, \text{gear\_ratio} + 44.6$$

We can use that equation to predict the mileage of foreign cars and then compare it with the true outcome. The _b[] function simplifies reference to the estimated coefficients. We can type

```
. gen asif=_b[weight]*weight + _b[weightsq]*weightsq +
      _b[gear_ratio]*gear_ratio + _b[_cons]
```

_b[weight] refers to the estimated coefficient on weight, _b[weightsq] to the estimated coefficient on weightsq, and so on.

We might now ask how the actual mileage of a Honda compares with the asif prediction:

```
. list make asif mpg if index(make,"Honda")
```

	make	asif	mpg
61.	Honda Accord	26.52597	25
62.	Honda Civic	30.62202	28

Notice the way we constructed our if clause to select Hondas. index() is the string function that returns the location in the first string where the second string is found or, if the second string does not occur in the first, zero. Thus, any recorded make that contains the string "Honda" anywhere in it would be listed; see [R] **functions**.

We find that both Honda models yield slightly lower gas mileage than the asif domestic-car-based prediction. (Please note that we do not endorse this model as a complete model of the determinants of mileage, nor do we single out the Honda for any special scorn. In fact, one should note that the observed values are within the root mean square error of the average prediction.)

We might wish to compare the overall average mpg and the asif prediction over all foreign cars in the data:

```
. summarize mpg asif if foreign
```

Variable	Obs	Mean	Std. Dev.	Min	Max
mpg	22	24.77273	6.611187	14	41
asif	22	26.67124	3.142912	19.70466	30.62202

We find that, on average, foreign cars yield slightly lower mileage than our asif prediction. This might lead us to ask if any foreign cars do better than the asif prediction:

```
. list make asif mpg if foreign & mpg>asif, sep(0)
```

	make	asif	mpg
55.	BMW 320i	24.31697	25
57.	Datsun 210	28.96818	35
63.	Mazda GLC	29.32015	30
66.	Subaru	28.85993	35
68.	Toyota Corolla	27.01144	31
71.	VW Diesel	28.90355	41

We find six such automobiles.

◁

23.10 Performing hypothesis tests on the coefficients

23.10.1 Linear tests

After estimation, test is used to perform tests of linear hypotheses based on the variance–covariance matrix of the estimators (Wald tests).

▷ Example

(test has numerous syntaxes and features, so do not use this example as an excuse for not reading [R] **test**.) Using the automobile data, you perform the following regression:

```
. http://www.stata-press.com/data/r8/auto
(1978 Automobile Data)
. generate weightsq=weight^2
. regress mpg weight weightsq foreign
```

Source	SS	df	MS		
Model	1689.15372	3	563.05124		
Residual	754.30574	70	10.7757963		
Total	2443.45946	73	33.4720474		

Number of obs =   74
F(3,   70) =   52.25
Prob > F   = 0.0000
R-squared   = 0.6913
Adj R-squared = 0.6781
Root MSE   = 3.2827

mpg	Coef.	Std. Err.	t	P>|t|	[95% Conf. Interval]
weight	-.0165729	.0039692	-4.18	0.000	-.0244892 -.0086567
weightsq	1.59e-06	6.25e-07	2.55	0.013	3.45e-07 2.84e-06
foreign	-2.2035	1.059246	-2.08	0.041	-4.3161 -.0909002
_cons	56.53884	6.197383	9.12	0.000	44.17855 68.89913

You can use the `test` command to calculate the joint significance of `weight` and `weightsq`:

```
. test weight weightsq
 ( 1)   weight = 0.0
 ( 2)   weightsq = 0.0
        F(  2,    70) =     60.83
              Prob > F =     0.0000
```

You are not limited to testing whether the coefficients are zero. You can test whether the coefficient on `foreign` is -2 by typing

```
. test foreign = -2
 ( 1)   foreign = -2.0
        F(  1,    70) =      0.04
              Prob > F =     0.8482
```

You can even test more complicated hypotheses since `test` has the ability to perform basic algebra. Here is an absurd hypothesis:

```
. test 2*(weight+weightsq)=-3*(foreign-(weight-weightsq))
 ( 1)  - weight + 5.0 weightsq + 3.0 foreign = 0
        F(  1,    70) =      4.31
              Prob > F =     0.0416
```

`test` simplified the algebra of our hypothesis and then presented the test results. We discover that the hypothesis may be absurd, but we cannot reject it at the 1% or even 4% level. You can also use `test`'s `accumulate` option to combine this test with another test:

```
. test foreign+weight=0, accum
 ( 1)  - weight + 5.0 weightsq + 3.0 foreign = 0
 ( 2)   weight + foreign = 0
        F(  2,    70) =      9.12
              Prob > F =     0.0003
```

There are limitations. `test` can test only linear hypotheses. If you attempt to test a nonlinear hypothesis, `test` will tell you that it is not possible:

```
. test weight/foreign=0
not possible with test
r(131);
```

Testing nonlinear hypotheses is discussed in [U] **23.10.4 Nonlinear Wald tests** below.

◁

23.10.2   test can be used after any estimation command

`test` bases its results on the estimated variance–covariance matrix of the estimators (i.e., performs a Wald test), and so can be used after any estimation command. In the case of maximum likelihood estimation, you will have to decide whether you want to perform tests based on the information matrix instead of constraining the equation, re-estimating it, and then calculating the likelihood-ratio test (see [U] **23.10.3 Likelihood-ratio tests**). Since `test` bases its results on the information matrix, its results have exactly the same standing as the asymptotic Z statistic presented in the coefficient table.

▷ Example

Let's examine the repair records of the cars in our automobile data as rated by *Consumers Reports*:

```
. tabulate rep78 foreign
```

Repair Record 1978	Car type Domestic	Foreign	Total
1	2	0	2
2	8	0	8
3	27	3	30
4	9	9	18
5	2	9	11
Total	48	21	69

The values are coded 1–5, corresponding to well below average to well above average. We will fit this variable using a maximum-likelihood ordered logit model (the `nolog` option suppresses the iteration log, saving us some paper):

```
. ologit rep78 price foreign weight weightsq displ, nolog
```

```
Ordered logit estimates                    Number of obs   =          69
                                           LR chi2(5)      =       33.12
                                           Prob > chi2     =      0.0000
Log likelihood = -77.133082                Pseudo R2       =      0.1767
```

rep78	Coef.	Std. Err.	z	P>|z|	[95% Conf. Interval]	
price	-.000034	.0001188	-0.29	0.775	-.0002669	.000199
foreign	2.685648	.9320398	2.88	0.004	.8588833	4.512412
weight	-.0037447	.0025609	-1.46	0.144	-.0087639	.0012745
weightsq	7.87e-07	4.50e-07	1.75	0.080	-9.43e-08	1.67e-06
displacement	-.0108919	.0076805	-1.42	0.156	-.0259455	.0041617
_cut1	-9.417196	4.298201	(Ancillary parameters)			
_cut2	-7.581864	4.23409				
_cut3	-4.82209	4.147679				
_cut4	-2.79344	4.156219				

We now wonder whether all of our variables other than `foreign` are jointly significant. We test the hypothesis just as we would after linear regression:

```
. test weight weightsq displ price
 ( 1)  weight = 0.0
 ( 2)  weightsq = 0.0
 ( 3)  displacement = 0.0
 ( 4)  price = 0.0
         chi2( 4) =      3.63
       Prob > chi2 =    0.4590
```

It is worthwhile comparing this with the results performed by a likelihood-ratio test; see [U] **23.10.3 Likelihood-ratio tests**. In this case, results differ little.

◁

23.10.3   Likelihood-ratio tests

After maximum likelihood estimation, you can obtain likelihood-ratio tests. This is done by fitting both the unconstrained and constrained models, storing the results using `estimates store`, and then running `lrtest`. See [R] **lrtest** for the full details.

▷ Example

In [U] **23.10.2 test can be used after any estimation command** above, we fitted an ordered logit on `rep78` and then tested the significance of all the explanatory variables except `foreign`.

To obtain the likelihood-ratio test, sometime after fitting the full model, we type `estimates store` *full_model_name*, where *full_model_name* is just a label that we assign to these results.

```
. ologit rep78 price for weight weightsq displ
(output omitted )
. estimates store myfullmodel
```

This command saves the current model results with the name `myfullmodel`.

Next, we fit the constrained model. After that, typing '`lrtest myfullmodel .`' compares the current model with the model we saved:

```
. ologit rep78 foreign

Iteration 0:   log likelihood = -93.692061
Iteration 1:   log likelihood = -79.696089
Iteration 2:   log likelihood = -79.044933
Iteration 3:   log likelihood = -79.029267
Iteration 4:   log likelihood = -79.029243

Ordered logit estimates                    Number of obs   =         69
                                           LR chi2(1)      =      29.33
                                           Prob > chi2     =     0.0000
Log likelihood = -79.029243                Pseudo R2       =     0.1565
```

rep78	Coef.	Std. Err.	z	P>|z|	[95% Conf. Interval]
foreign	2.98155	.6203637	4.81	0.000	1.76566 4.197441
_cut1	-3.158382	.7224269		(Ancillary parameters)	
_cut2	-1.362642	.3557343			
_cut3	1.232161	.3431227			
_cut4	3.246209	.5556646			

```
. lrtest myfullmodel .
likelihood-ratio test                      LR chi2(4)   =       3.79
(Assumption: . nested in full)             Prob > chi2  =     0.4348
```

When we tested the same constraint with `test` (which performed a Wald test), we obtained a χ^2 of 3.63 and a significance level of 0.4590. Note that we used . (the dot) to specify the results in active memory, although we could have also stored them with `estimates store` and referred to them by name instead. Also, the order in which you specify the two models to `lrtest` doesn't matter; `lrtest` is smart enough to know the full model from the constrained model.

◁

There are two other post-estimation commands that work in the same way as `lrtest`, meaning that they accept names of stored estimation results as their input: `hausman` for performing Hausman specification tests and `suest` for seemingly unrelated estimation. We do not cover these commands here; see [R] **hausman** and [R] **suest** for more details.

23.10.4 Nonlinear Wald tests

testnl can be used to test nonlinear hypotheses about the parameters of the active estimation results. testnl, like test, bases its results on the variance–covariance matrix of the estimators (i.e., performs a Wald test), and so can be used after any estimation command; see [R] **testnl**.

▷ Example

You fit the model

```
. regress price mpg weight foreign
  (output omitted )
```

and then type

```
. testnl (38*_b[mpg]^2 = _b[foreign]) (_b[mpg]/_b[weight]=4)
  (1)  38*_b[mpg]^2 = _b[foreign]
  (2)  _b[mpg]/_b[weight]=4
            F(2, 70) =        0.02
            Prob > F =        0.9806
```

We performed this test on linear regression estimates, but tests of this type could be performed after any estimation command.

◁

23.11 Obtaining linear combinations of coefficients

lincom computes point estimates, standard errors, t or z statistics, p-values, and confidence intervals for a linear combination of coefficients after any estimation command. Results can optionally be displayed as odds ratios, incidence rate ratios, or relative risk ratios.

▷ Example

We fit a linear regression:

```
. use http://www.stata-press.com/data/r8/regress
. regress y x1 x2 x3
```

Source	SS	df	MS		Number of obs	=	148
					F(3, 144)	=	96.12
Model	3259.3561	3	1086.45203		Prob > F	=	0.0000
Residual	1627.56282	144	11.3025196		R-squared	=	0.6670
					Adj R-squared	=	0.6600
Total	4886.91892	147	33.2443464		Root MSE	=	3.3619

y	Coef.	Std. Err.	t	P>|t|	[95% Conf. Interval]	
x1	1.457113	1.07461	1.36	0.177	-.6669339	3.581161
x2	2.221682	.8610358	2.58	0.011	.5197797	3.923583
x3	-.006139	.0005543	-11.08	0.000	-.0072345	-.0050435
_cons	36.10135	4.382693	8.24	0.000	27.43863	44.76407

Suppose that we want to see the difference of the coefficients of x2 and x1. We type

```
. lincom x2 - x1
 ( 1)  - x1 + x2 = 0.0
```

y	Coef.	Std. Err.	t	P>|t|	[95% Conf. Interval]
(1)	.7645682	.9950282	0.77	0.444	-1.20218 2.731316

◁

`lincom` is very handy for computing the odds ratio of one covariate group relative to another.

▷ Example

We estimate the parameters of a logistic model of low birth weight:

```
. use http://www.stata-press.com/data/r8/lbw3
(Hosmer & Lemeshow data)

. logit low age lwd black other smoke ptd ht ui

Iteration 0:   log likelihood =   -117.336
Iteration 1:   log likelihood = -99.431174
Iteration 2:   log likelihood = -98.785718
Iteration 3:   log likelihood =   -98.778
Iteration 4:   log likelihood = -98.777998
```

```
Logit estimates                                    Number of obs   =        189
                                                   LR chi2(8)      =      37.12
                                                   Prob > chi2     =     0.0000
Log likelihood = -98.777998                        Pseudo R2       =     0.1582
```

low	Coef.	Std. Err.	z	P>|z|	[95% Conf. Interval]
age	-.0464796	.0373888	-1.24	0.214	-.1197603 .0268011
lwd	.8420615	.4055338	2.08	0.038	.0472299 1.636893
black	1.073456	.5150752	2.08	0.037	.0639273 2.082985
other	.815367	.4452979	1.83	0.067	-.0574008 1.688135
smoke	.8071996	.404446	2.00	0.046	.0145001 1.599899
ptd	1.281678	.4621157	2.77	0.006	.3759478 2.187408
ht	1.435227	.6482699	2.21	0.027	.1646415 2.705813
ui	.6576256	.4666192	1.41	0.159	-.2569313 1.572182
_cons	-1.216781	.9556797	-1.27	0.203	-3.089878 .656317

If we want to get the odds ratio for black smokers relative to white nonsmokers (the reference group), we type

```
. lincom black + smoke, or
 ( 1)  black + smoke = 0.0
```

low	Odds Ratio	Std. Err.	z	P>|z|	[95% Conf. Interval]
(1)	6.557805	4.744692	2.60	0.009	1.588176 27.07811

`lincom` computed $\exp(\beta_{\text{black}} + \beta_{\text{smoke}}) = 6.56$.

◁

23.12 Obtaining nonlinear combinations of coefficients

lincom is limited to estimating linear combinations of coefficients, e.g., black + smoke, or exponentiated linear combinations as in the above. For general nonlinear combinations, use nlcom.

Continuing our previous example, suppose that we wanted the ratio of the coefficients (and standard errors, Wald test, confidence interval, etc.) of black and other:

```
. nlcom _b[black]/_b[other]
      _nl_1:  _b[black]/_b[other]
```

low	Coef.	Std. Err.	z	P>|z|	[95% Conf. Interval]	
_nl_1	1.316531	.7359262	1.79	0.074	-.1258574	2.75892

The Wald test given is that of the null hypothesis that the nonlinear combination is zero versus the two-sided alternative—probably not very informative in the case of a ratio. If we would instead like to test whether this ratio is one, we can rerun nlcom, this time subtracting one from our ratio estimate.

```
. nlcom _b[black]/_b[other] - 1
      _nl_1:  _b[black]/_b[other]  -  1
```

low	Coef.	Std. Err.	z	P>|z|	[95% Conf. Interval]	
_nl_1	.3165314	.7359262	0.43	0.667	-1.125857	1.75892

We can interpret this as not very much evidence that the "ratio minus 1" is different from zero, meaning we cannot reject the null hypothesis that the ratio equals one.

Note that when using nlcom, we needed to refer to the model coefficients by their "proper" names, e.g. _b[black], and not by the shorthand black such as when using lincom. If we had typed

```
. nlcom black/other
```

Stata would have reported an error. Consider this a limitation of Stata.

23.13 Obtaining marginal effects

Stata's mfx command will compute the marginal effects of the independent variables on predicted values.

▷ Example

Consider the logistic regression model we previously fitted on the automobile data:

(*Continued on next page*)

```
. use http://www.stata-press.com/data/r8/auto

. logit foreign weight mpg
Iteration 0:    log likelihood =   -45.03321
Iteration 1:    log likelihood =  -29.898968
Iteration 2:    log likelihood =  -27.495771
Iteration 3:    log likelihood =  -27.184006
Iteration 4:    log likelihood =  -27.175166
Iteration 5:    log likelihood =  -27.175156
```

Logit estimates				Number of obs	=	74
				LR chi2(2)	=	35.72
				Prob > chi2	=	0.0000
Log likelihood = -27.175156				Pseudo R2	=	0.3966

foreign	Coef.	Std. Err.	z	P>|z|	[95% Conf. Interval]	
weight	-.0039067	.0010116	-3.86	0.000	-.0058894	-.001924
mpg	-.1685869	.0919174	-1.83	0.067	-.3487418	.011568
_cons	13.70837	4.518707	3.03	0.002	4.851864	22.56487

Typing `mfx compute` will give the marginal effects for the default prediction, which, in the case of `logit`, is the predicted probability that the automobile is manufactured outside the U.S.

```
. mfx compute
Marginal effects after logit
      y  = Pr(foreign) (predict)
         =  .15733364
```

variable	dy/dx	Std. Err.	z	P>|z|	[95% C.I.]		X
weight	-.0005179	.00014	-3.73	0.000	-.00079	-.000246	3019.46
mpg	-.0223512	.0127	-1.76	0.079	-.04725	.002548	21.2973

Given the above output, we see that both `weight` and `mpg` have a negative effect on the predicted probability. For example, increased weight (from the mean weight of 3019.46 lbs.) decreases the likelihood that the automobile is `foreign` when controlling for gas mileage.

◁

`mfx` can also calculate elasticities, calculate at covariate values other than the covariate means (the default), and calculate marginal effects for predictions other than the default prediction; see [R] **mfx** for details.

23.14   Obtaining robust variance estimates

Estimates of variance refers to estimated standard errors or, more completely, the estimated variance–covariance matrix of the estimators of which the standard errors are a subset, being the square root of the diagonal elements. We will simply call this matrix the variance. All estimation commands produce an estimate of variance and, using that, produce confidence intervals and significance tests.

In addition to the conventional estimator of variance, there is another. This estimator has been called by various names because it has been derived independently in different ways by different authors. Two popular names associated with the calculation are Huber and White, but it is also known as the sandwich estimator of variance (because of how the calculation formula physically appears) and the robust estimator of variance (because of claims made about it). In addition, this estimator also has an independent and long tradition in the survey literature.

The conventional estimator of variance is derived by starting with a model. To fix ideas, let's assume that it is the regression model

$$y_i = \mathbf{x}_i \boldsymbol{\beta} + \epsilon_i, \qquad \epsilon_i \sim N(0, \sigma^2)$$

although it is not important for the discussion that we are using regression. Under the model-based approach, we assume that the model is true and thereby derive an estimator for $\boldsymbol{\beta}$ and its variance.

The estimator of the standard error of $\widehat{\boldsymbol{\beta}}$ we develop is based on the assumption that the model is true in every detail. To wit, the only reason y_i is not exactly equal to $\mathbf{x}_i \boldsymbol{\beta}$ (so that we would only need to solve an equation to obtain precisely that value of $\boldsymbol{\beta}$) is because the observed y_i has noise ϵ_i added to it, that noise is Gaussian, and it has constant variance. It is that noise that leads to the uncertainty about $\boldsymbol{\beta}$, and it is from the characteristics of that noise that we are able to calculate a sampling distribution for $\widehat{\boldsymbol{\beta}}$.

The key thought here is that the standard error of $\widehat{\boldsymbol{\beta}}$ arises because of ϵ and is valid only because the model is absolutely, without question, true; we just do not happen to know the particular values of $\boldsymbol{\beta}$ and σ^2 that make the model true. The implication is that, in an infinite-sized sample, the estimator $\widehat{\boldsymbol{\beta}}$ for $\boldsymbol{\beta}$ would converge to the true value of $\boldsymbol{\beta}$ and its variance would go to 0.

Now, here is another interpretation of the estimation problem: We are going to fit the model

$$y_i = \mathbf{x}_i \mathbf{b} + e_i$$

and, to obtain estimates of $\mathbf{b}$, we are going to use the calculation formula

$$\widehat{\mathbf{b}} = (\mathbf{X}'\mathbf{X})^{-1}\mathbf{X}'\mathbf{y}$$

Please note that we have made no claims that the model is true nor any claims about e_i or its distribution. We shifted our notation from $\boldsymbol{\beta}$ and ϵ_i to $\mathbf{b}$ and e_i to emphasize this. All we have stated are the physical actions we intend to carry out on the data. Interestingly, it is possible to calculate a standard error for $\widehat{\mathbf{b}}$ in this case! At least, it is possible if you will agree with us on what the standard error measures.

We are going to define the standard error as measuring the standard error of the calculated $\widehat{\mathbf{b}}$ were we to repeat the data collection followed by estimation over and over again.

Also note that this is a different concept of the standard error from the conventional, model-based ideas, but it is not unrelated. Both measure uncertainty about $\mathbf{b}$ (or $\boldsymbol{\beta}$). The regression model-based derivation states from where the variation arises, and so is able to make grander statements about the applicability of the measured standard error. The weaker second interpretation makes fewer assumptions and so produces a standard error suitable for one purpose.

There is a subtle difference in interpretation of these identically calculated point estimates. $\widehat{\boldsymbol{\beta}}$ is the estimate of $\boldsymbol{\beta}$ under the assumption that the model is true. $\widehat{\mathbf{b}}$ is the estimate of $\mathbf{b}$, which is merely what the estimator would converge to if we collected more and more data.

Is the estimate of $\mathbf{b}$ unbiased? If you mean, does $\mathbf{b} = \boldsymbol{\beta}$, that depends on whether the model is true. $\widehat{\mathbf{b}}$ is, however, an unbiased estimate of $\mathbf{b}$, which, admittedly, is not saying much.

What if $\mathbf{x}$ and e are correlated; don't we have a problem in that case? Answer: You may have an interpretation problem—$\mathbf{b}$ may not measure what you want to measure, namely, $\boldsymbol{\beta}$—but we measure $\widehat{\mathbf{b}}$ to be such and such and expect, were the experiment and estimation repeated, that you would observe results in the range we have reported.

So, we have two very different understandings of what the parameters mean and how the variance in their estimators arises. However, both interpretations must confront the issue of how to make valid statistical inference about the coefficient estimates when the data do not come from either a simple random sample, or the distribution of $(\mathbf{x}_i, \epsilon_i)$ is not independent and identically distributed. In essence, we need an estimator of the standard errors that is robust to this deviation from the standard case.

Hence, the name, the robust estimate of variance, and its associated authors are Huber (1967) and White (1980, 1982) (who developed it independently), although many others have extended its development, including Gail, Tan, and Piantadosi (1988), Kent (1982), Royall (1986), and Lin and Wei (1989). In the survey literature, this same estimator has been developed; see, for example, Kish and Frankel (1974), Fuller (1975), and Binder (1983).

Many of Stata's estimation commands can produce this alternative estimate of variance, and, if they can, they have a `robust` option. Without `robust`, you get one measure of variance:

```
. use http://www.stata-press.com/data/r8/auto7
(1978 Automobile Data)

. regress mpg weight foreign
```

Source	SS	df	MS		Number of obs =	74
					F(2, 71) =	69.75
Model	1619.2877	2	809.643849		Prob > F =	0.0000
Residual	824.171761	71	11.608053		R-squared =	0.6627
					Adj R-squared =	0.6532
Total	2443.45946	73	33.4720474		Root MSE =	3.4071

mpg	Coef.	Std. Err.	t	P>|t|	[95% Conf. Interval]	
weight	-.0065879	.0006371	-10.34	0.000	-.0078583	-.0053175
foreign	-1.650029	1.075994	-1.53	0.130	-3.7955	.4954422
_cons	41.6797	2.165547	19.25	0.000	37.36172	45.99768

With `robust`, you get another:

```
. regress mpg weight foreign, robust
```

Regression with robust standard errors

				Number of obs =	74
				F(2, 71) =	73.81
				Prob > F =	0.0000
				R-squared =	0.6627
				Root MSE =	3.4071

mpg	Coef.	Robust Std. Err.	t	P>|t|	[95% Conf. Interval]	
weight	-.0065879	.0005462	-12.06	0.000	-.007677	-.0054988
foreign	-1.650029	1.132566	-1.46	0.150	-3.908301	.6082424
_cons	41.6797	1.797553	23.19	0.000	38.09548	45.26392

Either way, the point estimates are the same. (See [R] **regress** for an example where specifying robust produces strikingly different standard errors.)

How do we interpret these results? Let's consider the model-based interpretation. Suppose that

$$y_i = \mathbf{x}_i\boldsymbol{\beta} + \epsilon_i,$$

where $(\mathbf{x}_i, \epsilon_i)$ are independently and identically distributed (i.i.d.) with variance σ^2. For the model-based interpretation, we also need to assume that $\mathbf{x}_i$ and ϵ_i are uncorrelated. With these assumptions, and a few technical regularity conditions, our first regression gives us consistent parameter estimates and standard errors that we can use for valid statistical inference about the coefficients. Now suppose that we weaken our assumptions so that $(\mathbf{x}_i, \epsilon_i)$ are independently and, but not necessarily, identically distributed. Our parameter estimates are still consistent, but the standard errors from the first regression can no longer be used to make valid inference. We need estimates of the standard errors that are robust to the fact that the error term is not identically distributed. The standard errors in our second regression are just what we need. We can use them to make valid statistical inference about our coefficients, even though our data are not identically distributed.

Now consider a nonmodel-based interpretation. If our data come from a survey design that ensures that $(\mathbf{x}_i, e_i)$ are i.i.d., then we can use the nonrobust standard errors for valid statistical inference about the population parameters $\mathbf{b}$. Note that for this interpretation, we do not need to assume that $\mathbf{x}_i$ and e_i are uncorrelated. If they are uncorrelated, then the population parameters $\mathbf{b}$ and the model parameters β are the same. However, if they are correlated, then the population parameters $\mathbf{b}$ that we are estimating are not the same as the model-based β. So, what we are estimating is different, but we still need standard errors that allow us to make valid statistical inference. So, if the process that we used to collect the data caused $(\mathbf{x}_i, e_i)$ to be independently but not identically distributed, then we need to use the robust standard errors to make valid statistical inference about the population parameters $\mathbf{b}$.

The robust estimator of variance has one feature that the conventional estimator does not have: the ability to relax the assumption of independence of the observations. That is, if you specify the cluster() option, it can produce "correct" standard errors (in the measurement sense), even if the observations are correlated.

In the case of the automobile data, it is difficult to believe that the models of the various manufacturers are truly independent. Manufacturers, after all, use common technology, engines, and drive trains across their model lines. The VW Dasher in the above regression has a measured residual of -2.80. Having been told that, do you really believe the residual for the VW Rabbit is as likely to be above 0 as below? (The residual is -2.32.) Similarly, the measured residual for the Chevrolet Malibu is 1.27. Does that provide information about the expected value of the residual of the Chevrolet Monte Carlo (which turns out to be 1.53)?

One wants to be careful about picking examples out of data; we have not told you about the Datsun 210 and 510 (residuals $+8.28$ and -1.01) or the Cadillac Eldorado and Seville (residuals -1.99 and $+7.58$), but you should, at least, be questioning the assumption of independence. It may be believable that the measured mpg given the weight of one manufacturer's vehicles is independent of other manufacturers' vehicles, but it is at least questionable whether a manufacturer's vehicles are independent of one another.

In commands with the robust option, another option—cluster()—relaxes the independence assumption and requires only that the observations be independent across the clusters:

```
. regress mpg weight foreign, robust cluster(manufacturer)
```

```
Regression with robust standard errors              Number of obs =      74
                                                    F(  2,    22) =   90.93
                                                    Prob > F      =  0.0000
                                                    R-squared     =  0.6627
Number of clusters (manufacturer) = 23              Root MSE      =  3.4071
```

mpg	Coef.	Robust Std. Err.	t	P>|t|	[95% Conf. Interval]	
weight	-.0065879	.0005339	-12.34	0.000	-.0076952	-.0054806
foreign	-1.650029	1.039033	-1.59	0.127	-3.804852	.5047939
_cons	41.6797	1.844559	22.60	0.000	37.85432	45.50508

It turns out that, in these data, whether we specify cluster() makes little difference. The VW and Chevrolet examples we quoted above were not representative; had they been, the confidence intervals would have widened. (In the above, manuf is a variable that takes on values such as "Chev.", "VW", etc., recording the manufacturer of the vehicle. We created this variable from variable make, which contains values such as "Chev. Malibu", "VW Rabbit", etc., by extracting the first word.)

As a demonstration of how well clustering can work, in [R] **regress** we fitted a random-effects model with regress, robust cluster() and then compare the results with ordinary least squares and the GLS random-effects estimator. Here we will simply summarize the results.

We start with a dataset on 4,782 women aged 16 to 46. Subjects appear an average of 7.14 times in this dataset, so there are a total of 34,139 observations. The model we use is log wage on age, age-squared, and grade of schooling completed. The focus of the example is the estimated coefficient on schooling. We obtain the following results:

Estimator	point estimate	confidence interval
(inappropriate) least squares	.081	[.079, .083]
robust, cluster	.081	[.077, .085]
GLS random effects	.080	[.076, .083]

We wish you to start by noticing how well robust with the cluster() option does as compared with the GLS random-effects model. We then run a Hausman specification test, obtaining $\chi^2(2) = 62$, which casts grave doubt on the assumptions justifying the use of the GLS estimator, and hence on the GLS results. At this point, we will simply quote our comments:

> Meanwhile, our robust regression results still stand as long as we are careful about the interpretation. The correct interpretation is that, were the data collection repeated (on women sampled the same way as in the original sample), and were we to re-estimate the model parameters, 95% of the time we would expect our obtained range to contain the true coefficient on grade.

> Even with robust regression, you must be careful about going beyond that statement. In this case, the Hausman test is probably picking up something that differs within and between persons, casting doubt on our robust regression model in terms of interpreting $[.077, .085]$ to contain the rate of return to additional schooling, economy wide, for all women, without exception.

The formula for the robust estimator of variance is

$$\widehat{\mathcal{V}} = \widehat{\mathbf{V}}\left(\sum_{j=1}^{N} \mathbf{u}'_j \mathbf{u}_j\right)\widehat{\mathbf{V}}$$

where $\widehat{\mathbf{V}} = (-\partial^2 \ln L/\partial\beta^2)^{-1}$ (the conventional estimator of variance) and $\mathbf{u}_j$ (a row vector) is the contribution from the jth observation to the scores $\partial \ln L/\partial\beta$.

In the above, observations are assumed to be independent. Assume, for a moment, that the observations denoted by j are not independent but that they can be divided into M groups G_1, G_2, ..., G_M that are independent. Then, the robust estimator of variance is

$$\widehat{\mathcal{V}} = \widehat{\mathbf{V}}\left(\sum_{k=1}^{M} \mathbf{u}_k^{(G)\prime}\mathbf{u}_k^{(G)}\right)\widehat{\mathbf{V}}$$

where $\mathbf{u}_k^{(G)}$ is the contribution of the kth group to the scores $\partial \ln L/\partial\beta$. That is, application of the robust variance formula merely involves using a different decomposition of $\partial \ln L/\partial\beta$, namely, $\mathbf{u}_k^{(G)}$, $k = 1, \ldots, M$ rather than $\mathbf{u}_j$, $j = 1, \ldots, N$. Moreover, if the log-likelihood function is additive in the observations denoted by j,

$$\ln L = \sum_{j=1}^{N} \ln L_j$$

then $\mathbf{u}_j = \partial \ln L_j / \partial \boldsymbol{\beta}$, and so

$$\mathbf{u}_k^{(G)} = \sum_{j \in G_k} \mathbf{u}_j$$

In other words, the group scores that enter the calculation are simply the sums of the individual scores within group. That is what the cluster() option does. (This point was first made in writing by Rogers (1993), although he considered the point an obvious generalization of Huber (1967) and the calculation—implemented by Rogers—had appeared in Stata a year earlier.)

❑ **Technical Note**

What is written above is asymptotically correct, but we have ignored a finite-sample adjustment to $\widehat{\mathcal{V}}$. For maximum likelihood estimators, when you specify robust but not cluster(), a better estimate of variance is $\widehat{\mathcal{V}}^* = \{N/(N-1)\}\widehat{\mathcal{V}}$. When you also specify the cluster() option, this becomes $\widehat{\mathcal{V}}^* = \{M/(M-1)\}\widehat{\mathcal{V}}$.

In the case of linear regression, the finite sample adjustment is $N/(N-k)$ without cluster()—where k is the number of regressors—and $\{M/(M-1)\}\{(N-1)/(N-k)\}$ with cluster(). In addition, two data-dependent modifications to the calculation for $\widehat{\mathcal{V}}^*$, suggested by MacKinnon and White (1985), are also provided by regress; see [R] **regress**.

❑

23.15 Obtaining scores

Many of the estimation commands that provide the robust option also provide the score() option. score() returns an important ingredient into the robust variance calculation that is sometimes useful in its own right. As explained in [U] **23.14 Obtaining robust variance estimates** above, ignoring the finite-sample corrections, the robust estimate of variance is

$$\widehat{\mathcal{V}} = \widehat{\mathbf{V}} \left(\sum_{j=1}^{N} \mathbf{u}_j' \mathbf{u}_j \right) \widehat{\mathbf{V}}$$

where $\widehat{\mathbf{V}} = (-\partial^2 \ln L / \partial \boldsymbol{\beta}^2)^{-1}$ (the conventional estimator of variance) and $\mathbf{u}_j$ (a row vector) is the contribution from the jth observations to the scores $\partial \ln L / \partial \boldsymbol{\beta}$. Let us consider likelihood functions that are additive in the observations,

$$\ln L = \sum_{j=1}^{N} \ln L_j$$

so that $\mathbf{u}_j = \partial \ln L_j / \partial \boldsymbol{\beta}$. In general, function L_j is a function of $\mathbf{x}_j$ and $\boldsymbol{\beta}$, $L_j(\boldsymbol{\beta}; \mathbf{x}_j)$. For many likelihood functions, however, it is only the linear form $\mathbf{x}_j \boldsymbol{\beta}$ that enters the function. In those cases,

$$\frac{\partial \ln L_j(\mathbf{x}_j \boldsymbol{\beta})}{\partial \boldsymbol{\beta}} = \frac{\partial \ln L_j(\mathbf{x}_j \boldsymbol{\beta})}{\partial (\mathbf{x}_j \boldsymbol{\beta})} \frac{\partial (\mathbf{x}_j \boldsymbol{\beta})}{\partial \boldsymbol{\beta}} = \frac{\partial \ln L_j(\mathbf{x}_j \boldsymbol{\beta})}{\partial (\mathbf{x}_j \boldsymbol{\beta})} \mathbf{x}_j$$

Writing $u_j = \partial \ln L_j(\mathbf{x}_j \beta)/\partial(\mathbf{x}_j \beta)$, this becomes simply $u_j \mathbf{x}_j$. Thus, the formula for the robust estimate of variance can be rewritten as

$$\widehat{V} = \widehat{V} \left(\sum_{j=1}^{N} u_j^2 \mathbf{x}_j' \mathbf{x}_j \right) \widehat{V}$$

We refer to u_j as the score (in the singular), and it is u_j that is returned when you specify option score(). u_j is like a residual in that

1. $\sum_j u_j = 0$, and

2. correlation of u_j and $\mathbf{x}_j$, calculated over $j = 1, \ldots, N$, is 0.

In fact, in the case of linear regression, u_j is the residual, normalized,

$$\frac{\partial \ln L_j}{\partial(\mathbf{x}_j \beta)} = \frac{\partial}{\partial(\mathbf{x}_j \beta)} \ln f \left\{ (y_j - \mathbf{x}_j \beta)/\sigma \right\}$$
$$= (y_j - \mathbf{x}_j \beta)/\sigma^2$$

where $f()$ is the normal density.

▷ Example

Command probit does provide both robust and score() options. The scores play an important role in calculating the robust estimate of variance, but you can specify score regardless of whether you specify robust:

```
. use http://www.stata-press.com/data/r8/auto
. probit foreign mpg weight, score(u)
Iteration 0:   log likelihood =  -45.03321
Iteration 1:   log likelihood = -29.244141
Iteration 2:   log likelihood = -27.041557
Iteration 3:   log likelihood =  -26.84658
Iteration 4:   log likelihood = -26.844189
Iteration 5:   log likelihood = -26.844189
```

Probit estimates

				Number of obs	=	74
				LR chi2(2)	=	36.38
				Prob > chi2	=	0.0000
Log likelihood = -26.844189				Pseudo R2	=	0.4039

foreign	Coef.	Std. Err.	z	P>|z|	[95% Conf. Interval]	
mpg	-.1039503	.0515689	-2.02	0.044	-.2050235	-.0028772
weight	-.0023355	.0005661	-4.13	0.000	-.003445	-.0012261
_cons	8.275464	2.554142	3.24	0.001	3.269438	13.28149

```
. summarize u
```

Variable	Obs	Mean	Std. Dev.	Min	Max
u	74	-3.93e-16	.5988325	-1.655439	1.660787

```
. correlate u mpg weight
(obs=74)
```

	u	mpg	weight
u	1.0000		
mpg	-0.0000	1.0000	
weight	-0.0000	-0.8072	1.0000

```
. list make foreign mpg weight u if abs(u)>1.65
```

	make	foreign	mpg	weight	u
24.	Ford Fiesta	Domestic	28	1,800	-1.6554395
64.	Peugeot 604	Foreign	14	3,420	1.6607871

The light, high-mileage Ford Fiesta is surprisingly domestic, while the heavy, low-mileage Peugeot 604 is surprisingly foreign.   ◁

❏ Technical Note

For some estimation commands, one score is not enough. Consider a likelihood that can be written as $L_j(\mathbf{x}_j\boldsymbol{\beta}_1, \mathbf{z}_j\boldsymbol{\beta}_2)$. Then, $\partial \ln L_j/\partial\boldsymbol{\beta}$ can be written $(\partial \ln L_j/\partial\boldsymbol{\beta}_1, \partial \ln L_j/\partial\boldsymbol{\beta}_2)$. Each of the components can in turn be written as $[\partial \ln L_j/\partial(\boldsymbol{\beta}_1\mathbf{x})]\mathbf{x} = u_1\mathbf{x}$ and $[\partial \ln L_j/\partial(\boldsymbol{\beta}_2\mathbf{z})]\mathbf{z} = u_2\mathbf{z}$. There are then two scores, u_1 and u_2, and, in general, there could be more.

Stata's streg, distribution(weibull) command is an example of this: it estimates $\boldsymbol{\beta}$ and a shape parameter $\ln p$, the latter of which can be thought of as a degenerate linear form $\ln p\mathbf{z}$ with $\mathbf{z} = 1$. streg's score() option requires that you specify two variable names, or you can specify score(stub*), which will generate two variables, stub1 and stub2; the first will be defined containing u_1—the score associated with $\boldsymbol{\beta}$—and the second will be defined containing u_2—the score associated with $\ln p$.

❏

❏ Technical Note

Using Stata's matrix commands—see [P] **matrix**—we can make the robust variance calculation for ourselves and then compare it with that made by Stata.

```
. quietly probit foreign mpg weight, score(u)

. matrix accum S =  mpg weight [iweight=u^2*74/73]
(obs=26.53642547)

. matrix rV = e(V)*S*e(V)

. matrix list rV

symmetric rV[3,3]
               mpg       weight        _cons
   mpg    .00352299
weight    .00002216    2.434e-07
 _cons  -.14090346   -.00117031    6.4474172

. quietly probit foreign mpg weight, robust

. matrix list e(V)

symmetric e(V)[3,3]
               mpg       weight        _cons
   mpg    .00352299
weight    .00002216    2.434e-07
 _cons  -.14090346   -.00117031    6.4474172
```

The results are the same.

There is an important lesson here for programmers. Given the scores, conventional variance estimates can be easily transformed to robust estimates. If one were writing a new estimation command, it would not be difficult to include a robust option.

It is, in fact, easy if we ignore clustering. With clustering, it is more work since the calculation involves forming sums within clusters. For programmers interested in implementing robust variance calculations, Stata provides an _robust command to ease the task. This is documented in [P] _robust.

To use _robust, you first produce conventional results (a vector of coefficients and covariance matrix) along with a variable containing the scores u_j (or variables if the likelihood function has more than one stub). You then call _robust, and it will transform your conventional variance estimate into the robust estimate. _robust will handle the work associated with clustering, the details of the finite-sample adjustment, and it will even label your output so that the word Robust appears above the standard error when the results are displayed.

Of course, this is all even easier if you write your commands using Stata's ml maximum likelihood optimization, in which case, you merely pass the robust and cluster() options on to ml. ml will then call _robust itself and do all the work for you.

❑

23.16   Weighted estimation

In [U] **14.1.6 weight**, we introduced the syntax for weights. Stata provides four kinds of weights: fweights, or frequency weights; pweights, or sampling weights; aweights, or analytic weights; and iweights, or importance weights. The syntax for using each is the same. Type

        . regress y x1 x2

and you obtain unweighted estimates; type

        . regress y x1 x2 [pweight=pop]

and you obtain (in this example) pweighted estimation.

Below, we explain in detail how each kind of weight is used in estimation.

23.16.1   Frequency weights

Frequency weights—fweights—are integers and nothing more than replication counts. The weight is statistically uninteresting, but, from a data processing perspective, it is of great importance. Consider the following data,

y	x1	x2
22	1	0
22	1	0
22	1	1
23	0	1
23	0	1
23	0	1

and the estimation command

        . regress y x1 x2

Exactly equivalent is the following, more compressed data

y	x1	x2	pop
22	1	0	2
22	1	1	1
23	0	1	3

and the corresponding estimation command

    . regress y x1 x2 [fweight=pop]

When you specify frequency weights, you are treating each observation as one or more real observations.

❑ Technical Note

One will occasionally run across a command that does not allow weights at all, especially among user-written commands. expand (see [R] **expand**) can be used with such commands to obtain frequency-weighted results. The expand command duplicates observations so that the data become self-weighting. For example, we want to run the command usercmd, which does something or other, and we would very much like to type usercmd y x1 x2 [fw=pop]. Unfortunately, usercmd does not allow weights. Instead, we type

    . expand pop
    . usercmd y x1 x2

to obtain our result. Moreover, there is an important principle here: The results of running any command with frequency weights should be exactly the same as running the command on the unweighted, expanded data. Unweighted, duplicated data and frequency-weighted data are merely two ways of recording identical information.

❑

23.16.2 Analytic weights

Analytic weights—analytic is a term made up by us—statistically arise in one particular problem: linear regression on data that are themselves observed means. That is, think of the model

$$y_i = \mathbf{x}_i\boldsymbol{\beta} + \epsilon_i, \qquad \epsilon_i \sim N(0, \sigma^2)$$

and now think about estimating this model on data $(\overline{y}_j, \overline{\mathbf{x}}_j)$ that are themselves observed averages. For instance, a piece of the underlying data for $(y_i, \mathbf{x}_i)$ might be $(3, 1)$, $(4, 2)$, and $(2, 2)$, but you do not know that. Instead, you have a single observation $\{(3 + 4 + 2)/3, (1 + 2 + 2)/3\} = (3, 1.67)$ and know only that the $(3, 1.67)$ arose as the average of 3 underlying observations. All of your data are like that.

regress with aweights is the solution to that problem:

    . regress y x [aweight=pop]

There is a history of misusing such weights. A researcher does not have cell-mean data, but instead has a probability-weighted random sample. Long before Stata existed, some researchers were using aweights to produce estimates from such samples. We will come back to this point in [U] **23.16.3 Sampling weights** below.

Anyway, the statistical problem to which aweights are the solution can be written as

$$y_i = \mathbf{x}_i\boldsymbol{\beta} + \epsilon_i, \qquad \epsilon_i \sim N(0, \sigma^2/w_i)$$

where the w_i are the analytic weights. The details of the solution, it turns out, are to make linear regression calculations using the weights as if they were fweights, but to first normalize them to sum to N before doing that.

Most commands that allow aweights handle them in this manner. That is, if you specify aweights, they are

1. normalized to sum to N, and then

2. inserted in the calculation formulas in the same way as fweights.

23.16.3 Sampling weights

Sampling weights—probability weights or pweights—refer to probability-weighted random samples. Actually, what you specify in [pweight=...] is a variable recording the number of subjects in the full population that the sampled observation in your data represents. That is, an observation that had probability 1/3 of being included in your sample has pweight 3.

We noted above that some researchers have used aweights with this kind of data. If they do, they are probably making a mistake. Consider the regression model

$$y_i = \mathbf{x}_i\boldsymbol{\beta} + \epsilon_i, \qquad \epsilon_i \sim N(0, \sigma^2)$$

Begin by considering the exact nature of the problem of fitting this model on cell-mean data—the problem for which aweights are the solution. That statistical problem is one of heteroskedasticity arising from the grouping. Note that the error term ϵ_i is homoskedastic (meaning it has constant variance σ^2). Pretend that the first observation in the data is the mean of three underlying observations. Then,

$$y_1 = \mathbf{x}_1\boldsymbol{\beta} + \epsilon_1, \qquad \epsilon_i \sim N(0, \sigma^2)$$
$$y_2 = \mathbf{x}_2\boldsymbol{\beta} + \epsilon_2, \qquad \epsilon_i \sim N(0, \sigma^2)$$
$$y_3 = \mathbf{x}_3\boldsymbol{\beta} + \epsilon_3, \qquad \epsilon_i \sim N(0, \sigma^2)$$

and taking the mean,

$$(y_1 + y_2 + y_3)/3 = \{(\mathbf{x}_1 + \mathbf{x}_2 + \mathbf{x}_3)/3\}\boldsymbol{\beta} + (\epsilon_1 + \epsilon_2 + \epsilon_3)/3$$

For another observation in the data—which may be the result of summing of a different number of observations—the variance will be different. Hence, the model for the data is

$$\overline{y}_j = \overline{x}_j\boldsymbol{\beta} + \overline{\epsilon}_j, \qquad \overline{\epsilon}_j \sim N(0, \sigma^2/N_j)$$

This makes intuitive sense. Consider two observations, one recording means over 2 subjects and the other means over 100,000 subjects. You would expect the variance of the residual to be less in the 100,000-subject observation, or, said differently, there is more information in the 100,000-subject observation than in the 2-subject observation.

Now, instead pretend you are fitting the same model, $y_i = \mathbf{x}_i\boldsymbol{\beta} + \epsilon_i$, $\epsilon_i \sim N(0, \sigma^2)$, on probability-weighted data. Each observation in your data is a single subject, it is just that the different subjects have differing chances of being included in your sample. Therefore, for each subject in your data, it is true that

$$y_i = \mathbf{x}_i\boldsymbol{\beta} + \epsilon_i, \qquad \epsilon_i \sim N(0, \sigma^2)$$

That is, there is no heteroskedasticity problem. The use of the aweighted estimator cannot be justified on these grounds.

As a matter of fact, based on the argument just given, you do not need to adjust for the weights at all, although the argument does not justify not making an adjustment. If you do not adjust, you are holding tightly to the assumed truth of your model. There are two issues when considering adjustment for sampling weights:

1. the efficiency of the point estimate $\widehat{\boldsymbol{\beta}}$ of $\boldsymbol{\beta}$; and

2. the reported standard errors (and, more generally, variance matrix of $\widehat{\boldsymbol{\beta}}$).

Efficiency argues in favor of adjustment, and that, by the way, is why many researchers have used aweights with pweighted data. The adjustment implied by pweights to the point estimates is the same as the adjustment implied by aweights.

It is with regard to the second issue that the use of aweights produces incorrect results because it interprets larger weights as designating more accurately measured points. In the case of pweights, however, the point is no more accurately measured—it is still just one observation with a single residual ϵ_j and variance σ^2. In [U] **23.14 Obtaining robust variance estimates** above, we introduced another estimator of variance that measures the variation that would be observed were the data collection followed by the estimation repeated. Those same formulas provide the solution to pweights, and they have the added advantage that they are not conditioned on the model being true. If one has any hopes of measuring the variation that would be observed were the data collection followed by estimation repeated, one must include the probability of the observations being sampled in the calculation.

In Stata, when you type

```
. regress y x1 x2 [pw=pop]
```

the results are the same as if you had typed

```
. regress y x1 x2 [pw=pop], robust
```

That is, specifying pweights implies the `robust` option, and hence the robust variance calculation (but weighted). In this example, we use `regress` simply for illustration. The same is true of `probit` and all of Stata's estimation commands. Estimation commands that do not have a `robust` option (there are a few) do not allow pweights.

pweights are adequate for handling random samples where the probability of being sampled varies. pweights may be all you need. If, however, the observations are not sampled independently but are sampled in groups—called clusters in the jargon—you should specify the estimator's `cluster()` option as well:

```
. regress y x1 x2 [pw=pop], cluster(block)
```

There are two ways of thinking about this:

1. The robust estimator answers the question of the variation that would be observed were the data collection followed by the estimation repeated, and, if that question is to be answered, the estimator must account for the clustered nature of how observations are selected. If observations 1 and 2 are in the same cluster, then one cannot select observation 1 without selecting observation 2 (and, by extension, one cannot select observations like 1 without selecting observations like 2).

2. If you prefer, you can think about potential correlations. Observations in the same cluster may not really be independent—that is an empirical question to be answered by the data. For instance, if the clusters are neighborhoods, it would not be surprising that the individual neighbors are similar in their income, their tastes, and their attitudes, and even more similar than two randomly drawn persons from the area at large with similar characteristics such as age and sex.

Either way of thinking leads to the same (robust) estimator of variance.

Sampling weights usually arise from complex sampling designs, and these designs often involve not only unequal probability sampling and cluster sampling, but also stratified sampling. There is a family of commands in Stata designed to work with the features of complex survey data, and those are the commands that begin with `svy`. To fit a linear regression model with stratification, for example, one would use the `svyregress` command.

Non-`svy` commands that allow pweights and clustering give essentially identical results to the `svy` commands. If the sampling design is simple enough that it can be accommodated by the non-`svy` command, that is a fine way to perform the analysis. The `svy` commands differ in that they have additional bells and whistles, and they do all the little details correctly for bona fide survey data. See [U] **30 Overview of survey estimation** for a brief discussion of some of the issues involved in the analysis of survey data and a list of all the differences between the svy and non-svy commands.

Not all model estimation commands in Stata allow pweights. In many of these cases, this is because they are computationally or statistically difficult to implement.

23.16.4   Importance weights

Stata's iweights—importance weights—are the emergency exit. These weights are for those who want to take control and create special effects. For example, programmers have used regress with iweights to compute iteratively reweighted least-squares solutions for various problems.

iweights are treated much like aweights except that they are not normalized. To wit, Stata's iweight rule is

1. the weights are not normalized; and

2. they are generally inserted into calculation formulas in the same way as fweights. There are exceptions; see the *Methods and Formulas* for the particular command.

iweights are used mostly by programmers who are often on the way to implementing one of the other kinds of weights.

23.17   A list of post-estimation commands

The following commands can be used after estimation:

[R]	**adjust**	Tables of adjusted means and proportions
[R]	**hausman**	Hausman specification test
[R]	**level**	Set default significance level
[R]	**lincom**	Obtain linear combinations of coefficients
[R]	**linktest**	Specification link test for single-equation models
[R]	**lrtest**	Likelihood-ratio test after estimation
[R]	**mfx**	Obtain marginal effects or elasticities after estimation
[R]	**nlcom**	Nonlinear combinations of estimators
[R]	**predict**	Obtain predictions, residuals, etc. after estimation
[R]	**predictnl**	Obtain nonlinear predictions, standard errors, etc. after estimation
[R]	**suest**	Seemingly unrelated estimation
[R]	**test**	Test linear hypotheses after estimation
[R]	**testnl**	Test nonlinear hypotheses after estimation
[R]	**vce**	Display covariance matrix of the estimators

Also see [U] **16.5 Accessing coefficients and standard errors** for accessing coefficients and standard errors.

23.18   References

Binder, D. A. 1983. On the variances of asymptotically normal estimators from complex surveys. *International Statistical Review* 51: 279–292.

Deaton, A. 1997. *The Analysis of Household Surveys: A Microeconometric Approach to Development Policy*. Baltimore, MD: Johns Hopkins University Press.

Fuller, W. A. 1975. Regression analysis for sample survey. *Sankhyā, Series C* 37: 117-132.

Gail, M. H., W. Y. Tan, and S. Piantadosi. 1988. Tests for no treatment effect in randomized clinical trials. *Biometrika* 75: 57–64.

Huber, P. J. 1967. The behavior of maximum likelihood estimates under non-standard conditions. In *Proceedings of the Fifth Berkeley Symposium on Mathematical Statistics and Probability*. Berkeley, CA: University of California Press, 1, 221–233.

Kent, J. T. 1982. Robust properties of likelihood ratio tests. *Biometrika* 69: 19–27.

Kish, L. and M. R. Frankel. 1974. Inference from complex samples. *Journal of the Royal Statistical Society* B 36: 1–37.

Lin, D. Y. and L. J. Wei. 1989. The robust inference for the Cox proportional hazards model. *Journal of the American Statistical Association* 84: 1074–1078.

Long, J. S. and J. Freese. 2000a. sg145: Scalar measures of fit for regression models. *Stata Technical Bulletin* 56: 34–40. Reprinted in *Stata Technical Bulletin Reprints*, vol. 10, pp. 197–205.

——. 2000b. sg152: Listing and interpreting transformed coefficients from certain regression models. *Stata Technical Bulletin* 57: 27–34. Reprinted in *Stata Technical Bulletin Reprints*, vol. 10, pp. 231–240.

MacKinnon, J. G. and H. White. 1985. Some heteroskedasticity consistent covariance matrix estimators with improved finite sample properties. *Journal of Econometrics* 29: 305–325.

Rogers, W. H. 1993. sg17: Regression standard errors in clustered samples. *Stata Technical Bulletin* 13: 19–23. Reprinted in *Stata Technical Bulletin Reprints*, vol. 3, 88–94.

Royall, R. M. 1986. Model robust confidence intervals using maximum likelihood estimators. *International Statistical Review* 54: 221–226.

Weesie, J. 2000. sg127: Summary statistics for estimation sample. *Stata Technical Bulletin* 53: 32–35. Reprinted in *Stata Technical Bulletin Reprints*, vol. 9, pp. 275–277.

White, H. 1980. A heteroskedasticity-consistent covariance matrix estimator and a direct test for heteroskedasticity. *Econometrica* 48: 817–830.

——. 1982. Maximum likelihood estimation of misspecified models. *Econometrica* 50: 1–25.

Advice on Stata

Chapters

24 Commands to input data 287

25 Commands for combining data 295

26 Commands for dealing with strings 297

27 Commands for dealing with dates 301

28 Commands for dealing with categorical variables 317

29 Overview of Stata estimation commands 327

30 Overview of survey estimation 343

31 Commands everyone should know 359

32 Using the Internet to keep up to date 361

24 Commands to input data

Contents

24.1   Six ways to input data
24.2   Eight rules for determining which input method to use
    24.2.1   If you wish to enter data interactively: Rule 1
    24.2.2   If the dataset is in binary format: Rule 2
    24.2.3   If the data are simple: Rule 3
    24.2.4   If the dataset is formatted and the formatting is significant: Rule 4
    24.2.5   If there are no string variables: Rule 5
    24.2.6   If all the string variables are enclosed in quotes: Rule 6
    24.2.7   If the undelimited strings have no blanks: Rule 7
    24.2.8   If you make it to here: Rule 8
24.3   If you run out of memory
24.4   Transfer programs
24.5   ODBC sources
24.6   References

24.1   Six ways to input data

The seven ways to input data into Stata are

[R] **edit** and [R] **input**	to enter data from the keyboard
[R] **insheet**	to read tab- or comma-separated data
[R] **infile (free format)**	to read unformatted data
[R] **infile (fixed format)** or [R] **infix (fixed format)**	to read formatted data
[R] **odbc**	to read from an ODBC source
[U] **24.4 Transfer programs**	to transfer data

Since dataset formats differ, you should familiarize yourself with each method.

Note that [R] **infile (fixed format)** and [R] **infix (fixed format)** are alternatives. These are two different commands that do the same thing. Read about both and then use whichever appeals to you.

After you have read this chapter, also see [R] **infile** for additional examples of the different commands to input data.

24.2   Eight rules for determining which input method to use

Below are eight rules that, when applied sequentially, will direct you to the appropriate method for entering your data. Following the eight rules is a description of each command as well as a reference to the corresponding entry in the *Reference* manuals. The rules are

1. If you have a small amount of data and simply wish to type the data directly into Stata at the keyboard, see [R] **input**—there are many examples and you should have little difficulty. Also see [R] **edit**.

2. If your dataset is in binary format or the "internal" format of some software package, you can

    a. Translate the data into ASCII (also known as character) format using the other software. For instance, you can save an Excel spreadsheet as tab-delimited or comma-separated text. Then, see [R] **insheet**.

    b. There are also software packages available that will automatically convert non-Stata format data files into Stata format files; see [U] **24.4 Transfer programs**.

    c. If the data are in a spreadsheet, you may also be able to copy-and-paste the data into Stata's Data Editor; see [R] **edit** for details.

    d. If the data are located in an ODBC source, which typically includes databases and spreadsheets; see [R] **odbc**. Currently `odbc` is only supported for Windows.

3. If the dataset has one observation per line and the data are tab- or comma-separated, see [R] **insheet**. This is the easiest way to read data.

4. If the dataset is formatted and that formatting information is required in order to interpret the data, see [R] **infile (fixed format)** or [R] **infix (fixed format)** (your choice).

5. If there are no string variables, see [R] **infile (free format)**.

6. If all the string variables in the data are enclosed in (single or double) quotes, see [R] **infile (free format)**.

7. If the undelimited string variables have no blanks, see [R] **infile (free format)**.

8. If you make it to here, see [R] **infile (fixed format)** or [R] **infix (fixed format)** (your choice).

Let us now back up and start again.

24.2.1   If you wish to enter data interactively: Rule 1

Rule 1 simply says that if you have a small amount of data, you can type the data directly into Stata; see [R] **input** or [R] **edit**. Otherwise, we assume your data are stored on disk.

24.2.2   If the dataset is in binary format: Rule 2

Stata can read ASCII datasets, which is technical jargon for datasets composed of characters—datasets that can be typed on your screen or printed on your printer. The alternative, binary datasets, cannot be read by Stata. Binary datasets are quite popular, and almost every software package has its own binary format. Stata `.dta` datasets are an example, although this is a binary format Stata can read. The Excel format is another binary format and is one that Stata cannot read.

Thus, rule 2: If your dataset is in binary format or the "internal" format of another software package, you must either translate it into ASCII or use some other program for conversion. If this dataset is located in a database or an ODBC source, see [U] **24.5 ODBC sources**.

Detecting whether data are stored in binary format can be tricky. For instance, many Windows users wish to read data that have been entered into a word processor—let's assume Word. Unwittingly, they have stored the dataset as a Word document. The dataset looks like ASCII to them: When they look at it in Word, they see readable characters. The dataset seems to even pass the printing test in that Word can print it. Nevertheless, the dataset is not ASCII; it is stored in an internal Word format and the data cannot really pass the printing test since only Word can print it. To read the dataset, Windows users must use it in Word and then store it as an MS-DOS text file, MS-DOS text being the term Word decided to use to mean ASCII.

So, how are you to know whether your dataset is binary? Here's a simple test: Regardless of the operating system you use, enter Stata and type `type` followed by the name of the file:

```
. type myfile.raw
output will appear
```

You do not have to print the entire file; press *Break* when you have seen enough.

Do you see things that look like hieroglyphics? If so, the dataset is binary. See [U] **24.4 Transfer programs** below.

If it looks like data, however, the file is (probably) ASCII.

Let us assume you have an ASCII dataset that you wish to read. The data's format will determine the command you need to use. The different formats are discussed in the following sections.

24.2.3   If the data are simple: Rule 3

The easiest way to read data is with `insheet`; see [R] **insheet**.

`insheet` is smart: it looks at the dataset, determines what it contains, and then reads it. That is, `insheet` is smart given certain restrictions, such as that the dataset has one observation per line and that the values are tab- or comma-separated. `insheet` can read this

```
────────────────────────────────────── top of data1.raw ──────────
      M,Joe Smith,288,14
      M,K Marx,238,12
      F,Farber,211,7
────────────────────────────────────── end of data1.raw ──────────
```

or this (which has variable names on the first line)

```
────────────────────────────────────── top of data2.raw ──────────
      sex, name, dept, division
      M,Joe Smith,288,14
      M,K Marx,238,12
      F,Farber,211,7
────────────────────────────────────── end of data2.raw ──────────
```

or this (which has one tab character separating the values)

```
────────────────────────────────────── top of data3.raw ──────────
      M       Joe Smith       288     14
      M       K Marx  238     12
      F       Farber  211     7
────────────────────────────────────── end of data3.raw ──────────
```

(which looks odd because of how tabs work; data3.raw could similarly have a variable header), but `insheet` cannot read

```
────────────────────────────────────── top of data4.raw ──────────
      M       Joe Smith       288     14
      M       K Marx          238     12
      F       Farber          211     7
────────────────────────────────────── end of data4.raw ──────────
```

which has spaces rather than tabs!

There is a way to tell `data3.raw` from `data4.raw`: Ask Stata to type the data and show the tabs:

```
. type data3.raw, showtabs
M<T>Joe Smith<T>288<T>14
M<T>K Marx<T>238<T>12
F<T>Farber<T>211<T>7
. type data4.raw, showtabs
M      Joe Smith       288      14
M      K Marx          238      12
F      Farber          211      7
```

24.2.4   If the dataset is formatted and the formatting is significant: Rule 4

Rule 4 says that if the dataset is formatted and that formatting information is required in order to interpret the data, see [R] **infile (fixed format)** or [R] **infix (fixed format)**, which being a matter of preference.

Using infix or infile with a data dictionary is something new users want to avoid if at all possible.

The purpose of this rule is only to take you to the most complicated of all cases if there is no alternative. Otherwise, let's wait and see if it is necessary. Do not misinterpret the rule and say, "Ah, my dataset is formatted, at last a solution."

Just because a dataset is formatted does not mean you have to exploit the formatting information. The following dataset is formatted,

```
—————————————————————————————————— top of data5.raw ——————————
   1    27.39     12
   2     1.00      4
   3   100.10    100
——————————————————————————————— end of data5.raw ——————————
```

in that the numbers line up in neat columns, but you do not need to know the information to read it. Alternatively, consider the same data run together:

```
—————————————————————————————————— top of data6.raw ——————————
   1 27.39 12
   2  1.00  4
   3100.10100
——————————————————————————————— end of data6.raw ——————————
```

This dataset is formatted, too, and we must know the formatting information in order to make sense of "3100.10100". We must know that variable 2 starts in column 4 and is 6 characters long to extract the 100.10. It is datasets like data6.raw that we are looking for at this stage—datasets that only make sense if we know the starting and ending columns of data elements. In order to read data such as data6.raw, we must use either infix or infile with a data dictionary.

It should be obvious why reading unformatted data is easier. If the formatting information is required to interpret the data, then you must communicate that information to Stata, which means you will have to type it. This is the hardest kind of data to read, but Stata can do it. See [R] **infile (fixed format)** or [R] **infix (fixed format)**.

Looking back at data4.raw,

```
—————————————————————————————————— top of data4.raw ——————————
M      Joe Smith       288      14
M      K Marx          238      12
F      Farber          211      7
——————————————————————————————— end of data4.raw ——————————
```

you may be uncertain whether you have to read it with a data dictionary. If you are uncertain, do not jump yet.

Finally, here is an obvious example of unformatted data:

```
──────────────────────────────────────────── top of data7.raw ────────────
  1 27.39              12
  2 1 4
  3 100.1 100
──────────────────────────────────────────── end of data7.raw ────────────
```

In this case, blanks separate one data element from the next and, in one case, lots of blanks, although there is no special meaning attached to more than one blank.

In the following sections, we will look at datasets that are unformatted or formatted in a way that do not require a data dictionary.

24.2.5 If there are no string variables. Rule 5

Rule 5 says that if there are no string variables, see |R| **infile (free format)**.

Although the dataset `data7.raw` is unformatted, it can still be read using `infile` without a dictionary. This is not the case with `data4.raw`, because this dataset contains undelimited string variables with embedded blanks.

❑ Technical Note

Some Stata users prefer to read data with a data dictionary even when we suggest differently, as above. They like the convenience of the data dictionary—one can sit in front of an editor and carefully compose the list of variables and attach variable labels rather than having to type the variable list (correctly) on the Stata command line. What they should understand, however, is that one can create a do-file containing the `infile` statement, and thus have all the advantages of a data dictionary without some of the (extremely technical) disadvantages of data dictionaries.

Nevertheless, we do tend to agree with such users—we, too, prefer data dictionaries. Our recommendations, however, are designed to work in all cases. If the dataset is unformatted and contains no string variables, it can always be read without a data dictionary, whereas only in some cases can it be read with a data dictionary.

The distinction is that `infile` without a data dictionary performs stream I/O, whereas with a data dictionary it performs record I/O. The difference is intentional—it guarantees that you will be able to read your data into Stata somehow. Some datasets require stream I/O, others require record I/O, and still others can be read either way. Recommendations 1–5 identify datasets that either require stream I/O or can be read either way.

❑

We are now left with datasets which contain at least one string variable.

24.2.6   If all the string variables are enclosed in quotes: Rule 6

Rule 6 reads: if all the string variables in the data are enclosed in (single or double) quotes, see [R] **infile (free format)**.

See [U] **26 Commands for dealing with strings** for a formal definition of strings, but as a quick guide, a string variable is a variable that takes on values such as "bob", "joe", etc., as opposed to numeric variables that take on values like 1, 27.5, and –17.393. Undelimited strings—strings not enclosed in quotes—can be difficult to read.

Here is an example including delimited string variables:

```
─────────────────────────────────────────── top of data8.raw ───────────
    "M" "Joe Smith" 288 14
    "M" "K Marx" 238 12
    "F" "Farber" 211 7
─────────────────────────────────────────── end of data8.raw ───────────
```

or

```
──────────────────────────────── top of data8.raw, alternative format ───────────
    "M" "Joe Smith" 288  14
    "M" "K Marx"    238  12
    "F" "Farber"    211   7
──────────────────────────────── end of data8.raw, alternative format ───────────
```

Both of these are merely variations on `data4.raw` except that the strings are enclosed in quotes. In this case, `infile` without a dictionary can be used to read the data.

Here is another version of `data4.raw` without delimiters or even formatting:

```
─────────────────────────────────────────── top of data9.raw ───────────
    M Joe Smith 288 14
    M K Marx 238 12
    F Farber 211 7
─────────────────────────────────────────── end of data9.raw ───────────
```

What makes these data difficult? Blanks sometimes separate values and sometimes are nothing more than a blank within a string. For instance, you cannot tell whether Farber has first initial F with missing sex or is instead female with a missing first initial.

Fortunately, such data rarely happens. Either the strings are delimited, as we showed in `data8.raw`, or the data is in columns, as in `data4.raw`.

24.2.7   If the undelimited strings have no blanks: Rule 7

There is a case in which uncolumnized, undelimited strings cause no confusion—when they contain no blanks. For instance, if our data contained only last names:

```
─────────────────────────────────────────── top of data10.raw ───────────
    Smith 288 14
    Marx 238 12
    Farber 211 7
─────────────────────────────────────────── end of data10.raw ───────────
```

Stata could read it without a data dictionary. Caution: the last names must contain no blanks—no Van Owen's or von Beethoven's.

Thus, rule 7: If the undelimited string variables have no blanks, see [R] **infile (free format)**.

This leaves us with our final rule:

24.2.8 If you make it to here: Rule 8

If you make it to here, see [R] **infile (fixed format)** or [R] **infix (fixed format)** (your choice).

Remember `data4.raw`?

```
                                                   ──────── top of data4.raw ────────
    M        Joe Smith       288     14
    M        K Marx          238     12
    F        Farber          211     7
                                                   ──────── end of data4.raw ────────
```

It must be read using either `infile` with a dictionary or `infix`.

24.3 If you run out of memory

You can increase the amount of memory allocated to Stata; see [U] **7 Setting the size of memory**.

You can also try to conserve memory.

When you read the data, did you specify variable types? Stata can store integers more compactly than floats, and small integers more compactly than large integers; see [U] **15 Data**.

If that is not sufficient, then you will have to resort to reading the data in pieces. Both `infile` and `infix` allow specifying an in *range* modifier, and, in this case, the range is interpreted as the observation range to read. Thus, `infile ... in 1/100` would read observations 1 through 100 of your data and stop.

`infile ... in 101/200` would read observations 101 through 200. The end of the range may be specified as larger than the actual number of observations in the data. If the dataset contained only 150 observations, `infile ... in 101/200` would read observations 101 through 150.

Another way of reading the data in pieces is to specify the `if` *exp* modifier. Say your data contained an equal number of males and females, coded as the variable `sex` (which you will read) being 0 or 1, respectively. You could type `infile ... if sex==0` to read the males. `infile` will read an observation, ask itself if `sex` is zero, and if not, throw the observation away. Obviously, you could read just the females by typing `infile ... if sex==1`.

If the dataset is really big, perhaps you only need a random sample of the data—it was never your intention to analyze the entire dataset. Since `infile` and `infix` allow `if` *exp*, you could type `infile ... if uniform()<.1`. `uniform()` is the uniformly distributed random number generator; see [R] **functions**. This method would read an approximate 10% sample of the data. If you are serious about random samples, do not forget to set the seed before using `uniform()`; see [R] **generate**.

The final approach is to read all the observations but only a subset of the variables. When reading data without a data dictionary, you can specify `_skip` for variables, indicating that the variable is to be skipped over. When reading with a data dictionary or using `infix`, you can specify the actual columns to read, skipping any columns you wish to ignore.

24.4 Transfer programs

To import data from, say, Excel, you can write the data out as a text file and then read it in according to the rules above, read it via an ODBC source, or purchase a program to translate the dataset from Excel's format to Stata's format.

One such program is Stat/Transfer, which is available for Windows, Macintosh OS X, and Unix. It reads and writes data in a variety of formats, including Microsoft Access, dBASE, Epi Info, Excel, GAUSS, LIMDEP, Lotus 1-2-3, MATLAB, ODBC, Paradox, Quattro Pro, S-Plus, SAS, SPSS, SYSTAT, and, of course, Stata.

Stat/Transfer, available from...	and manufactured by...
Stata Corporation	Circle Systems
4905 Lakeway Drive	1001 Fourth Avenue Plaza, Suite 3200
College Station, Texas 77845	Seattle, Washington 98154
Telephone: 979-696-4600	Telephone: 206-682-3783
Fax: 979-696-4601	Fax: 206-328-4788
Email: *stata@stata.com*	*sales@circlesys.com*

There are other transfer programs available, too. Our web site, *http://www.stata.com*, lists programs available from other sources.

Access and Excel are trademarks of Microsoft Corporation. dBASE is a trademark of dBASE Inc. Paradox and Quattro Pro are trademarks of Corel Corporation. Epi Info is a trademark of The Centers for Disease Control and Prevention. GAUSS is a trademark of Aptech Systems, Inc. LIMDEP is a trademark of Econometric Software, Inc. Lotus and 1-2-3 are trademarks of IBM-Lotus. MATLAB is a trademark of The Math Works, Inc. SAS is a trademark of the SAS Institute Inc. S-Plus is a trademark of Insightful Corporation. SPSS is a trademark of SPSS Inc. SYSTAT is a trademark of SYSTAT Software Inc.

24.5   ODBC sources

If your dataset is located in a network database, or shared spreadsheet, you may be able to import your data via ODBC. ODBC, an acronym for "Open Database Connectivity", is a standard for exchanging data between programs. Stata supports the ODBC standard for importing data via the odbc command and is capable of reading from any ODBC source on your computer.

This process requires a data source, such as a database located on a network. To use the odbc command to import data from a database requires that the database first be setup as an ODBC source on the same machine that Stata is running from. The database itself does not have to be on the same machine, just the definition of that database as the ODBC source. On a Windows machine, an ODBC source is added via a Control Panel called "Data Sources". Additionally, typing odbc list from Stata displays all the ODBC sources that are provided by the computer.

Assuming the database is functioning, and the appropriate data source has been set up on the same machine as Stata, a single call using odbc load is all that is needed to import data. For a more thorough description of this process; see [R] **odbc**.

24.6   References

Swagel, P. 1994. os14: A program to format raw data files. *Stata Technical Bulletin* 20: 10–12. Reprinted in *Stata Technical Bulletin Reprints*, vol. 4, pp. 80–82.

25 Commands for combining data

Pretend you have two datasets you wish to combine. Below, we will draw a dataset as a box where, in the box, the variables go across and the observations go down.

See [R] **append** if you want to combine datasets vertically:

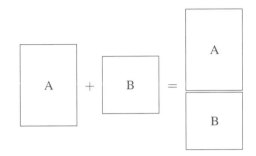

append adds observations to the existing variables. That is an oversimplification because append does not require that the datasets have exactly the same variables. **append** is appropriate, for instance, when you have data on hospital patients and then receive data on more patients.

See [R] **merge** if you want to combine datasets horizontally:

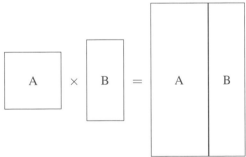

merge adds variables to the existing observations. That is an oversimplification because merge does not require that the datasets have exactly the same observations. merge is appropriate, for instance, when you have data on survey respondents and then receive data on part 2 of the questionnaire.

See [R] **joinby** when you want to combine datasets horizontally but form all pairwise combinations within group:

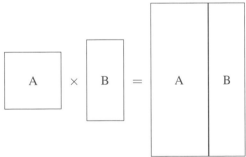

joinby is like merge, but forms all combinations of the observations where it makes sense. joinby would be appropriate, for instance, where A contained data on parents and B contained data on their children. joinby *familyid* would form a dataset of each parent joined with each of his or her children.

Also see [R] **cross** for a less frequently used command that forms every pairwise combination of two datasets.

26 Commands for dealing with strings

Contents

26.1   Description
26.2   Categorical string variables
26.3   Mistaken string variables
26.4   Complex strings

Please read [U] **15 Data** before reading this entry.

26.1   Description

The word *string* is shorthand for a string of characters. "Male" and "Female"; "yes" and "no"; and "R. Smith" and "P. Jones" are examples of strings. The alternative to strings is numbers—0, 1, 2, 5.7, and so on. Variables containing strings—called string variables—occur in data for a variety of reasons. Four of these reasons are listed below.

A variable might contain strings because it is an *identifying variable*. Employee names in a payroll file, patient names in a hospital file, and city names in a city data file are all examples of this. This is a proper use of string variables.

A variable might contain strings because it records categorical information. "Male" and "Female" and "yes and "no" are examples of such use, but this is not an appropriate use of string variables. It is not appropriate because the same information could be coded numerically, and, if it were, (1) it would take less memory to store the data, and (2) the data would be more useful. We will explain how to convert categorical strings to categorical numbers below.

In addition, a variable might contain strings because of a mistake. The variable contains things like 1, 5, 8.2, but due to an error in reading the data, the data were mistakenly put into a string variable. We will explain how to fix such mistakes.

Finally, a variable might contain strings because the data simply could not be coerced into being stored numerically. "15 Jan 1992", "1/15/92", and "1A73" are examples of such use. We will explain how to deal with such complexities.

26.2   Categorical string variables

A variable might contain strings because it records categorical information.

Suppose that you have read in a dataset that contains a variable called `sex`, recorded as "male" and "female", yet when you attempt to run an ANOVA, the following message is displayed:

```
. use http://www.stata-press.com/data/r8/hbp2
. anova hbp sex
no observations
r(2000);
```

There are no observations because `anova`, along with most of Stata's "analytic" commands, cannot deal with string variables. They want to see numbers, and when they do not, they treat the variable as if it contained numeric missing values. Despite this limitation, it is possible to obtain tables:

```
. encode sex, gen(gender)
. anova hbp gender
```

	Number of obs =		1128	R-squared	= 0.0123
	Root MSE	= .214223		Adj R-squared =	0.0114
Source	Partial SS	df	MS	F	Prob > F
Model	.644485682	1	.644485682	14.04	0.0002
gender	.644485682	1	.644485682	14.04	0.0002
Residual	51.6737767	1126	.045891454		
Total	52.3182624	1127	.046422593		

The magic here is to convert the string variable sex into a numeric variable called gender with an associated value label, a trick accomplished by encode; see [U] **15.6.3 Value labels** and [R] **encode**.

26.3   Mistaken string variables

A variable might contain strings because of a mistake.

Suppose that you have numeric data in a variable called x, but due to a mistake, x was made a string variable when you read the data. When you list the variable, it looks fine:

```
. list x
```

	x
1.	2
2.	2.5
3.	17

(*output omitted*)

Yet, when you attempt to obtain summary statistics on x,

```
. summarize x
```

Variable	Obs	Mean	Std. Dev.	Min	Max
x	0				

If this happens to you, type describe to confirm that x is stored as a string:

```
. describe
Contains data
  obs:            10
  vars:            3
  size:          160  (99.9% of memory free)
```

variable name	storage type	display format	value label	variable label
x	str4	%9s		
y	float	%9.0g		
z	float	%9.0g		

```
Sorted by:
```

x is stored as an str4.

The problem is that `summarize` does not know how to calculate the mean of string variables—how to calculate the mean of "Joe" plus "Bill" plus "Roger"—even when the string variable contains what could be numbers. By using the `destring` command, the variable mistakenly stored as a `str4` can be converted to a numeric variable.

```
. destring x, replace
. summarize x
```

Variable	Obs	Mean	Std. Dev.	Min	Max
newx	10	1.76	.8071899	.7	3

An alternative to using the `destring` command is to use `generate` with the `real()` function; see [R] **functions**.

26.4 Complex strings

A variable might contain strings because the data simply could not be coerced into being stored numerically.

A complex string is a string that contains more than one piece of information. The most common example of complex strings is dates: "15 Jan 1992" contains three pieces of information—a day, a month, and a year. If your complex strings are dates, see [U] **27 Commands for dealing with dates**.

Although Stata has functions for dealing with dates, you will have to deal with other complex strings yourself. For instance, assume that you have data including part numbers:

```
. list partno
```

	partno
1.	5A2713
2.	2B1311
3.	8D2712
(output omitted)	

The first digit of the part number is a division number and the character that follows identifies the plant at which the part was manufactured. The next three digits represent the major part number and the last digit is a modifier indicating the color. This complex variable can be decomposed using the `substr()` and `real()` functions described in [R] **functions**:

```
. gen byte div = real(substr(partno,1,1))
. gen str1 plant = substr(partno,2,1)
. gen int part = real(substr(partno,3,3))
. gen byte color = real(substr(partno,6,1))
```

We use the `substr()` function to extract pieces of the string, and use the `real()` function, when appropriate, to translate the piece into a number.

27 Commands for dealing with dates

Contents
27.1   Overview
27.2   Dates
     27.2.1   Inputting dates
     27.2.2   Conversion into elapsed dates
          27.2.2.1   The mdy() function
          27.2.2.2   The date() function
     27.2.3   Displaying dates
     27.2.4   Other date functions
     27.2.5   Specifying particular dates (date literals)
27.3   Time-series dates
     27.3.1   Inputting time variables
     27.3.2   Specifying particular dates (date literals)
     27.3.3   Time-series formats
     27.3.4   Translating between time units
     27.3.5   Extracting components of time
     27.3.6   Creating time variables
     27.3.7   Setting the time variable
     27.3.8   Selecting periods of time
     27.3.9   The %tg format

27.1   Overview

You can record dates however you want, but there is one format that Stata understands, called elapsed dates or %d dates. A %d date is the number of days from January 1, 1960. In this format,

0	corresponds to	01jan1960			
1	corresponds to	02jan1960	−1	corresponds to	31dec1959
2	corresponds to	03jan1960	−2	corresponds to	30dec1959
⋮			⋮		
31	corresponds to	01feb1960	−31	corresponds to	01dec1959
⋮					
15,000	corresponds to	25jan2001	−15,000	corresponds to	01dec1918
⋮			⋮		
2,936,549	corresponds to	31dec9999	−679,350	corresponds to	01jan0100

This format can be used with dates 01jan0100–31dec9999, although caution should be exercised in dealing with dates before Friday, 15oct1582, because that is when the Gregorian calendar went into effect.

Stata provides functions to convert dates into %d dates, formats to print %d dates in understandable forms, and other functions to manipulate %d dates.

Use of %d dates is described in [U] **27.2 Dates** below.

In addition to %d dates, Stata has five other date formats, called %t dates, that it understands:

Format	Description	Coding		
%td	daily (same as %d)	−1 = 31dec1959,	0 = 01jan1960,	1 = 02jan1960
%tw	weekly	−1 = 1959w52,	0 = 1960w1,	1 = 1960w2
%tm	monthly	−1 = 1959m12,	0 = 1960m1,	1 = 1960m2
%tq	quarterly	−1 = 1959q4,	0 = 1960q1,	1 = 1960q2
%th	half-yearly	−1 = 1959h2,	0 = 1960h1,	1 = 1960h2
%ty	yearly	1959 = 1959,	1960 = 1960,	1961 = 1961

Use of %t dates is described in [U] **27.3 Time-series dates** below.

27.2   Dates

In this section, we discuss %d dates, also called elapsed dates.

27.2.1   Inputting dates

The trick to inputting dates in Stata is to forget they are dates. Input them as strings and then later convert them into Stata's elapsed dates. You might have

———————————————————————————————————— top of bdays.raw ————————

```
Bill   21 Jan 1952   22
May    11 Jul 1948   18
Sam    12 Nov 1960   25
Kay     9 Aug 1975   16
```

———————————————————————————————————— end of bdays.raw ————————

and if you did, you could read these data by typing

```
. infix str name 1-5  str bday 7-17  x 20-21 using bdays
(4 observations read)
```

If you now listed the data, the data would look fine,

```
. list
```

	name	bday	x
1.	Bill	21 Jan 1952	22
2.	May	11 Jul 1948	18
3.	Sam	12 Nov 1960	25
4.	Kay	9 Aug 1975	16

but you would find there is not much you could do with bday because it is just a string variable. Turning it into a date Stata understands is easy,

```
. gen birthday = date(bday,"dmy")
. list
```

	name	bday	x	birthday
1.	Bill	21 Jan 1952	22	−2902
2.	May	11 Jul 1948	18	−4191
3.	Sam	12 Nov 1960	25	316
4.	Kay	9 Aug 1975	16	5699

and making the numeric birthday variable look like a date is equally easy:

```
. format birthday %d
. list
```

	name	bday	x	birthday
1.	Bill	21 Jan 1952	22	21jan1952
2.	May	11 Jul 1948	18	11jul1948
3.	Sam	12 Nov 1960	25	12nov1960
4.	Kay	9 Aug 1975	16	09aug1975

The convenient thing about the variable `birthday` is that it is numeric, which means you can make calculations on it. How old will each of these people be on January 1, 2000? It is easy to add such a variable:

```
. gen age2000 = (mdy(1,1,2000)-birthday)/365.25
. list
```

	name	bday	x	birthday	age2000
1.	Bill	21 Jan 1952	22	21jan1952	47.94524
2.	May	11 Jul 1948	18	11jul1948	51.47103
3.	Sam	12 Nov 1960	25	12nov1960	39.13101
4.	Kay	9 Aug 1975	16	09aug1975	24.39699

27.2.2   Conversion into elapsed dates

Two functions are provided—`mdy()` and `date()`—for converting variables into elapsed dates.

27.2.2.1   The mdy() function

`mdy()` takes three numeric arguments—a month, day, and year—and returns the corresponding elapsed date. For instance,

```
. list
```

	month	day	year
1.	7	11	1948
2.	1	21	1952
3.	11	2	1994
4.	8	12	93

```
. gen edate = mdy(month,day,year)
(1 missing value generated)
. list
```

	month	day	year	edate
1.	7	11	1948	-4191
2.	1	21	1952	-2902
3.	11	2	1994	12724
4.	8	12	93	.

Note that in the last observation, `mdy()` produced missing. It did this because the year was 93, and `mdy()` does not assume 93 means 1993.

27.2.2.2   The date() function

The second way to convert to elapsed dates is with the `date()` function. `date()` takes two string arguments. There is a variation on this—two strings arguments and a numeric argument—but let's postpone that. The first argument is the date to be converted. The second argument tells `date()` the order of the month, day, and year in the first argument. Typing `date(`*strvar*`,"mdy")` means that *strvar* contains the month, day, and year in that order. Typing `date(`*strvar*`,"dmy")` means *strvar* contains the day, month, and year. Knowing the order, `date()` allows *strvar* to be in almost any format. For instance,

```
. list
```

	mydate
1.	7-11-1948
2.	1/21/52
3.	11.2.1994
4.	Aug 12,1993
5.	Sept 11,2002
6.	November 13, 2005

```
. gen edate = date(mydate, "mdy")
(1 missing value generated)
. list
```

	mydate	edate
1.	7-11-1948	-4191
2.	1/21/52	.
3.	11.2.1994	12724
4.	Aug 12,1993	12277
5.	Sept 11,2002	15594
6.	November 13, 2005	16753

or, if you prefer,

```
. list
```

	mydate
1.	11-7-1948
2.	21/1/52
3.	2.11.1994
4.	12 Aug 1993
5.	11Sept2002
6.	13 November 2005

```
. gen edate = date(mydate, "mdy")
(1 missing value generated)
```

(Continued on next page)

```
. list
```

	mydate	edate
1.	11-7-1948	-4191
2.	21/1/52	.
3.	2.11.1994	12724
4.	12 Aug 1993	12277
5.	11Sept2002	15594
6.	13 November 2005	16753

date() can deal with virtually any date format: all it needs to know is the order of the month, day, and year, and that you indicate by the second argument using the letters m, d, and y. Second argument "mdy" means month–day–year order, "dmy" means day–month–year order, and so on.

Note, however, that like mdy(), date() refused to translate two-digit years: 1/21/52 and 21/1/52 both translated to missing. Unlike mdy(), date() would be willing to assume that 52 means 1952 or 2052 if you will tell it which. There are two ways to do this.

The first way involves specifying a default century, and you do that using date()'s second argument. Specify "md19y" or "dm20y", and date() will assume that two-digit years should be interpreted as being prefixed by 19 or 20; four digit years will still be correctly interpreted no matter which default you specify.

The second way involves specifying date()'s third argument. Specify date(...,...,2040), and date() will assume that two-digit years should be interpreted as the maximum year not greater than 2040. 52 would be interpreted as 1952, but 39 would be interpreted as 2039. 40 would be interpreted as 2040. You can specify whatever third argument works best for your data.

If you do neither, then two-digit years cannot be translated, and that is why we saw the missing values in the examples above; date() could not translate 1/21/52 (21/1/52 when we varied the order). We could have translated 1/21/52 had we typed

```
. gen edate = date(mydate, "md19y")
```

or

```
. gen edate = date(mydate, "mdy",2040)
```

Either method would translate 1/21/52 as 21jan1952, but they would differ on how they would treat dates with two-digit years 00, 01, . . . , 40. The first method would treat them as 1900, 1901, . . . , 1940, whereas the second would treat them as 2000, 2001, . . . , 2040.

To summarize: To get dates into Stata, either create three numeric variables containing the month, day, and year and use mdy() to convert them, or create a string variable containing the date in whatever format and use date() to translate it. If you are reading date data into Stata, the easiest way is to read the data into a string and then use date().

❏ Technical Note

How date() *works.* The date to be converted has three pieces of information, the month, day, and year, and the second argument specifies the order, of which there are six possibilities, "dmy", "mdy", "ymd", "ydm", "dym", and "myd", although the last three rarely occur. Knowing the order, date() examines the contents of the first argument and looks for transitions, meaning any separating character such as blanks, commas, dashes, and slashes, or changes from numeric to alpha or alpha to numeric. This allows dividing the source into its three components for translation. If the source divides into other than three components, or if it divides into three but they do not make sense, `date()` returns a missing value.

If you have two-digit years, `date()` will return missing unless you specify a default century on the second argument—e.g., `"md19y"`—or you specify a third argument.

`date()` can translate virtually any format except formats where all three elements run together and the months are indicated numerically, such as 012152 or 520121. It is the lack of blanks or other separating characters that confuses `date()`; `date` could translate 01 21 52 or 52 01 21. `date()` could also translate 21Jan52—note the absence of blanks—because the string Jan makes clear the separation.

Let us assume you have a date of the form 520121—the order is year, month, and day—stored in a numeric variable called `ymd`. Here is how you might translate it:

```
. gen year = int(ymd/10000)
. gen month = int((ymd-year*10000)/100)
. gen day = ymd - year*10000  - month*100
. gen edate = mdy(month, day, 1900 + year)
```

❑

27.2.3   Displaying dates

`%d` elapsed dates are convenient for computers and sometimes even for humans—you can, for instance, subtract them to obtain the number of days between dates. Nevertheless, they are unreadable. For instance, here are some birth dates:

```
. list
```

	bdate
1.	-2902
2.	-4191
3.	316
4.	5699

All you need do to make such dates readable is assign Stata's `%d` format to the variable:

```
. format bdate %d
. list bdate
```

	bdate
1.	21jan1952
2.	11jul1948
3.	12nov1960
4.	09aug1975

If you now saved the data, the date would forevermore be displayed in this format.

You may find the format 21jan1952 unappealing, and, if so, you can modify it. The `%d` format is equivalent to `%dD1CY`, meaning first display the day (D), then the month abbreviation in lowercase (1), then the century (C), and finally the year without the century (Y). This default was selected because, by using an abbreviated month, it makes clear the order of the day and month, and, by omitting the blanks, it is short.

Here is a variation on the format:

```
. format bdate %dD_m_CY
. list
```

	bdate
1.	21 Jan 1952
2.	11 Jul 1948
3.	12 Nov 1960
4.	09 Aug 1975

And here are two more variations:

```
. format bdate %dN/D/Y
. list
```

	bdate
1.	01/21/52
2.	07/11/48
3.	11/12/60
4.	08/09/75

```
. format bdate %dM_D,_CY
. list
```

	bdate
1.	January 21, 1952
2.	July 11, 1948
3.	November 12, 1960
4.	August 09, 1975

You can specify simply %d or you can follow the %d with up to 11 characters that tell Stata what to display. Here is what the characters mean:

(Continued on next page)

C	Display the century of the year with a leading 0; year 500 is 05, year 1994 is 19, year 2002 is 20.
c	Display the century of the year without a leading 0; year 500 is 5, year 1994 is 19.
Y	Display the year within century with a leading 0; 1908 is 08, 1994 is 94, 2002 is 02.
y	Display the year within century without a leading 0; 1908 is 8, 1994 is 94, 2002 is 2.
M	Display the month spelled out; January is January, February is February, . . ., December is December.
m	Display the 3-letter abbreviation of month; January is Jan, February is Feb, . . ., December is Dec.
L	Same as M except the month is presented in all lowercase; January is january, February is february, . . ., December is december.
l	Same as m except the 3-letter abbreviation is in all lowercase; January is jan, February is feb, . . ., December is dec.
N	Display the numeric month with a leading 0; January is 01, February is 02, . . ., December is 12.
n	Display the numeric month without a leading 0; January is 1, February is 2, . . ., December is 12.
D	Display the day with a leading 0; 1 is 01, 2 is 02, . . ., 31 is 31.
d	Display the day without a leading 0; 1 is 1, 2 is 2, . . ., 31 is 31.
J	display day-within-year with leading 0s (001 to 366)
j	display day-within-year without leading 0s (1 to 366)
h	display half of year (1 to 2)
q	display quarter of year (1, 2, 3, or 4)
W	display week of year with leading 0 (01 to 52)
w	display week of year without leading 0 (1 to 52)
_	Display a blank.
.	Display a period.
,	Display a comma.
:	Display a colon.
-	Display a dash.
/	Display a slash.
'	Display a close single quote.
!c	display character c (code !! to display exclamation point)

When using the detail characters, you need not specify all the components of the date. In our birth dates, if we wanted to see just the month and year, we might

```
. format bdate %dm,_CY

. list
```

	bdate
1.	Jan, 1952
2.	Jul, 1948
3.	Nov, 1960
4.	Aug, 1975

27.2.4 Other date functions

How you display the date does not matter; the variable itself always contains the number of days from January 1, 1960. Given a date variable, the following functions extract information from it:

year(*date*)	returns four-digit year; e.g., 1980, 2002
month(*date*)	returns month; 1, 2, . . . , 12
day(*date*)	returns day within month; 1, 2, . . . , 31
halfyear(*date*)	returns the half of year; 1 or 2
quarter(*date*)	returns quarter of year; 1, 2, 3, or 4
week(*date*)	returns week of year; 1, 2, . . . , 52
dow(*date*)	returns day of week; 0, 1, . . . , 6; 0 = Sunday
doy(*date*)	returns day of year; 1, 2, . . . , 366

For example,

```
. gen m = month(bdate)
. gen d = day(bdate)
. gen y = year(bdate)
. gen week_d = dow(bdate)
. list
```

	bdate	m	d	y	week_d
1.	Jan, 1952	1	21	1952	1
2.	Jul, 1948	7	11	1948	0
3.	Nov, 1960	11	12	1960	6
4.	Aug, 1975	8	9	1975	6

dow() returns 0–6, 0 meaning Sunday, 1 Monday, . . . , 6 Saturday. Thus, the person born on January 21, 1952 was born on a Monday.

27.2.5 Specifying particular dates (date literals)

If you work with dates, you will want to type dates in expressions. For instance, in a previous example when we needed to calculate the age of persons as of 1jan2000, we typed

```
. gen age2000 = (mdy(1,1,2000)-birthday)/365.25
```

We used mdy() to obtain 1jan2000 as an elapsed date. Alternatively, we could have used Stata's d(*constant*) function

```
. gen age2000 = (d(1jan2000)-birthday)/365.25
```

d() is unusual in that you cannot type any expression inside the parentheses; instead, you must type a day followed by a month followed by a four-digit year. Do that, however, and d() returns the corresponding %d date value.

You may type the day, month, and year however you wish, but you must specify them in the order day, month, and year. Stata will understand d(1/1/2000) or d(1-1-2000) or d(1.1.2000) or d(1 jan 2000) or d(1 January 2000), etc., but it would not understand d(jan.1,2000):

```
. gen age2000 = (d(jan.1,2000)-birthday)/365.25
d(jan.1,2000) invalid
r(198);
```

In addition, if you type the date in a style where two numbers appear one after the other, you must put some form of punctuation other than a space between the two numbers: You would think Stata would understand d(1 1 2000), but it does not because spaces disappear in expressions:

```
. gen age2000 = (d(1 1 2000)-birthday)/365.25
d(112000) invalid
r(198);
```

When you spell out the month, you can include spaces or not:

```
. gen age2000 = (d(1 jan 2000)-birthday)/365.25
```

Finally, d() is allowed only in expressions. There may be occasions when you need to specify the numeric value of a date in the option of some command, such as

```
. ksm value time, xline(d(15apr1998))
xline() invalid
r(198);
```

Unfortunately, Stata does not understand xline(d(15apr1998)). In such cases, you can use display to obtain the numeric equivalent:

```
. display d(15apr1998)
13984

. ksm value time, xline(13984)
  (output omitted)
```

27.3   Time-series dates

In addition to %d date formats, Stata has five other date formats called %t or time-series dates. These dates work like %d in that 1jan1960 is mapped to 0, but the meaning of 1 varies:

Format	Description	Coding		
%td	daily (same as %d)	−1 = 31dec1959,	0 = 01jan1960,	1 = 02jan1960
%tw	weekly	−1 = 1959w52,	0 = 1960w1,	1 = 1960w2
%tm	monthly	−1 = 1959m12,	0 = 1960m1,	1 = 1960m2
%tq	quarterly	−1 = 1959q4,	0 = 1960q1,	1 = 1960q2
%th	half-yearly	−1 = 1959h2,	0 = 1960h1,	1 = 1960h2
%ty	yearly	1959 = 1959,	1960 = 1960,	1961 = 1961

To best understand these formats, think of what happens when variable d contains a date and you add 1 to it. If d is in %td format, you obtain the next day. If d is in %tw format, you obtain the next week. If d is in %tm format, you obtain the next month. If d is in %tq format, you obtain the next quarter. If d is in %th format, you obtain the next half-year. If d is in %ty format, you obtain the next year.

Or, think of it like this: subtract two dates and you obtain the number of days, weeks, months, quarters, half-years, or years between them.

%td daily format.
    This format is equivalent to %d. In any context, whether a variable is %td or %d makes no difference. Examples of dates in this format are 12jun1998, 22dec2002, etc.

%tw weekly format.
    The year is divided into 52 weeks: week 1 is the first 7 days of year, week 2 the second 7 days, and so on. Since years have just over 52 weeks in them, the 52nd week is defined as having 8 or 9 days. Examples of dates in this format are 1998w24, 2002w52, etc.

%tm monthly format.

The year is divided into the 12 calendar months. Examples of dates in this format are 1998m6, 2002m12, etc.

%tq quarterly format.

The year is divided into 4 quarters based on months; quarter 1 is January through March; quarter 2 is April through June; quarter 3 is July through September; and quarter 4 is October through December. Examples of dates in this format are 1998q2, 2002q4, etc.

%th half-yearly format.

The year is divided into 2 halves based on months; half 1 is January through June and half 2 is July through December. Examples of dates in this format are 1998h1 and 2002h2.

%ty yearly format.

The year is not divided at all. Examples of dates in this format are 1998 and 2002.

27.3.1 Inputting time variables

Our advice for inputting %d variables, summarized in [U] **27.2.1 Inputting dates** and [U] **27.2.2 Conversion into elapsed dates**, was

1. use the `mdy()` function if you have three integers recording the month, day, and year; or

2. input the date as a string and then use the `date()` function to convert it.

We offer the same advice for %t variables; just the names of the functions change:

format	integer conversion	string conversion
%td	mdy(*month*,*day*,*year*)	date(*string*, "md[##]y" or "dm[##]Y" ... , [*topyear*])
%tw	yw(*year*, *week*)	weekly(*string*, "w[##]y" or "[##]yw", [*topyear*])
%tm	ym(*year*, *month*)	monthly(*string*, "m[##]y" or "[##]ym", [*topyear*])
%tq	yq(*year*, *quarter*)	quarterly(*string*, "q[##]y" or "[##]yq", [*topyear*])
%th	yh(*year*, *halfyear*)	halfyearly(*string*, "q[##]y" or "[##]yq", [*topyear*])
%ty	*year*	yearly(*string*, "[##]y", [*topyear*])

For instance, just as `mdy(5,30,1998)` = 14,029 (30may1998),

   `yw(1998,22)` = 1,997 (1998w22)
   `ym(1998,5)` = 460 (1998m5)
   `yq(1998,2)` = 153 (1998q2)
   `yh(1998,1)` = 76 (1998h1)
   `1998` = 1998

A dataset containing numeric variables `year` and `quarter` could be translated into a Stata %tq variable by typing

```
. gen date = yq(year,quarter)
(1 missing value generated)
. format date %tq
```

```
. list
```

	year	quarter	date
1.	1998	1	1998q1
2.	1999	5	.
3.	2005	3	2005q3

Note that the mistaken year 1999 quarter 5 translated to a missing value. Had our years all been in the 20th century and recorded in two digits (e.g., 95, 98, etc.), we would have typed 'generate date = yq(1900+year,quarter)'.

The string-conversion functions work just like the date() function: the second argument specifies the order in which the components of the date are expected to occur in the string; the y of the second argument may be prefixed with 19 or 20 as one way of handling two-digit years; a third argument specifying the maximum year may be specified as another way of handling two-digit years; and if you do not prefix y and do not specify a third argument, years in the string must be four digits.

For example, monthly(s,"my") could translate s containing "jan 1999" or "January, 1999" or "jan1999" or "1/1999" or "1-1999", but would return missing for "jan99". "jan99" could be decoded by specifying monthly(s,"m19y") (in which case it would be interpreted as 1jan1999) or by specifying, say, monthly(s,"my",2040) (in which case it would also be interpreted as 1jan1999 because 1999 is the maximum year not greater than 2040).

Below we translate a string variable sdate containing quarterly dates to a Stata date:

```
. use a different dataset

. gen date = quarterly(sdate,"yq",2040)
(1 missing value generated)

. format date %tq

. list
```

	sdate	date
1.	1995q2	1995q2
2.	1996 3	1996q3
3.	1996 5	.
4.	1997 quarter 1	1997q1
5.	98q.4	1998q4
6.	2001-3	2001q3
7.	2002q2	2002q2

27.3.2 Specifying particular dates (date literals)

Just as you can use the d(constant) function to type a date in an expression, such as

```
. gen age2000 = (d(1jan2000)-birthday)/365.25
```

Stata provides one-letter functions for typing weekly, monthly, quarterly, half-yearly, and yearly dates:

format	function	argument	examples
%td	d()	type day, month, year	d(15feb1998), d(15-5-2002)
%tw	w()	type year, week	w(1998w7), w(2002-25)
%tm	m()	type year, month	m(1998m2), m(2002-5)
%tq	q()	type year, quarter	q(1998q1), q(2002-2)
%th	h()	type year, half	h(1998h1), h(2002-1)
%ty	y()	type year	y(1998), y(2002)

For instance, if variable `qtr` contained a `%tq` date, you could type

```
. list if qtr>=q(1998q1)
```
   (*output omitted*)

The `y()` function is included largely for completeness. For the `%ty` format, the year maps to the year, so 1960 means 1960, and there is little reason to type `y(1960)`. Programmers, however, sometimes find `y()` useful in terms of syntax checking. `y()`—just as all the single-letter functions—produces an error when given an invalid date, which, in this case, means year < 100 or year > 9999.

27.3.3   Time-series formats

Just as with the `%d` format, the `%td`, `%tw`, `%tm`, `%tq`, `%th`, and `%ty` formats may be modified so that you can display the date in the form you wish. This is done using the same letter codes used with the `%d` format; see [U] **27.2.3 Displaying dates**. The default formats for each of the types are

format	means
%td	%tdDlCY
%tw	%twCY!ww
%tm	%tmCY!mn
%tq	%tqCY!qq
%th	%thCY!hh
%ty	%tyCY

Think of the `%t` format as

   `%t`⟨*character stating how data encoded*⟩⟨*optional characters saying how displayed*⟩

If you had variable `qtr` containing `%tq` dates and you wanted the dates displayed as, for instance, 1980 Q.1, you could type

```
. format qtr %tqCY_!Q.q
```

27.3.4   Translating between time units

A time unit, such as `%tq` quarterly, can be translated to any other time unit, such as `%th` half-yearly. Stata provides functions to translate any time unit to and from `%td` daily units. The trick is to combine these functions:

Input	Output %td daily	%tw weekly	%tm monthly	%tq quarterly	%th half-yearly	%ty yearly
%td		wofd(*d*)	mofd(*d*)	qofd(*d*)	hofd(*d*)	yofd(*d*)
%tw	dofw(*w*)		mofd(dofw(*w*))	qofd(dofw(*w*))	hofd(dofw(*w*))	yofd(dofw(*w*))
%tm	dofm(*m*)	wofd(dofm(*m*))		qofd(dofm(*m*))	hofd(dofm(*m*))	yofd(dofm(*m*))
%tq	dofq(*q*)	wofd(dofq(*q*))	mofd(dofq(*q*))		hofd(dofq(*q*))	yofd(dofq(*q*))
%th	dofh(*h*)	wofd(dofh(*h*))	mofd(dofh(*h*))	qofd(dofh(*h*))		yofd(dofh(*h*))
%ty	dofy(*y*)	wofd(dofy(*y*))	mofd(dofy(*y*))	qofd(dofy(*y*))	hofd(dofy(*y*))	

The functions that translate *from* `%td` dates—`wofd(`*d*`)`, `mofd(`*d*`)`, `qofd(`*d*`)`, `hofd(`*d*`)`, and `yofd(`*d*`)`—return the date containing *d*. Thus, `qofd()` of 6apr1998 is 1998q2.

The functions that translate *to* `%td` dates—`dofw()`, `dofm()`, `dofq()`, `dofh()`, and `dofy()`—return the `%td` date of the beginning of the period. Thus, `dofq()` of 1998q2 is 01apr1998.

27.3.5   Extracting components of time

With %d dates, functions `month()`, `day()`, and `year()` extract the month, day, and year. In fact, if you look back at [U] **27.2.4 Other date functions**, you will find that there are functions to extract the week (`week()`), quarter (`quarter()`), and so on.

To extract values from a %t date, combine these extraction functions with the `dof*()` functions. For instance, if variable `qtr` contains a %tq date and you want new variable `q` to contain the quarter, type

        . gen q = quarter(dofq(qtr))

If you wanted to create a variable equal to 1 for the first two quarters of each year and 0 otherwise, you could type

        . gen first = quarter(dofq(qtr))<=2

or

        . gen first = halfyear(dofq(qtr))==1

27.3.6   Creating time variables

If you have data for which you know that the first observation is for the first quarter of 1990, the second for the second quarter of 1990, and so on, but the dataset contains no variable recording that fact, you can generate one by typing

        . generate time = q(1990q1)+_n-1
        . format time %tq

Remember that _n is Stata's built-in observation counter variable; _n = 1 in the first observation, 2 in the second, and so on. The single-letter functions make it easy to type a date.

27.3.7   Setting the time variable

If you find the %t format useful, it is likely you are performing time-series analysis and will find Stata's other time-series features, such as time-series varlists, useful as well. If so, you need to set the time variable to turn on those other features:

        . tsset time

The time variable you set does not need to be a %t variable, but results will be more readable if it is; see [TS] **tsset**.

27.3.8   Selecting periods of time

Once you have `tsset` a time variable, two new functions become available to you: `tin()` and `twithin()`. These functions are useful for selecting contiguous subsets of data:

        . regress y x l.x if tin(01jan1998,31dec2000)

        . list if tin(01jan1998,)

$\texttt{tin}(t_0, t_1)$ and $\texttt{twithin}(t_0, t_1)$ select observations in the range t_0 to t_1 and differ only in whether observations for which $t = t_0$ and $t = t_1$ are included ($\texttt{tin()}$ includes them and $\texttt{twithin()}$ excludes them).

The time variable t is assumed to be the time variable you previously $\texttt{tsset}$.

The arguments you specify should be typed in the same way you would type them using the single-letter $\texttt{d()}$, $\texttt{w()}$, $\texttt{m()}$, $\texttt{q()}$, $\texttt{h()}$, or $\texttt{y()}$ functions; see [U] **27.3.2 Specifying particular dates (date literals)**. That means that what you type depends on the time variable's $\%\texttt{t}$ format.

If the time variable t does not have a $\%\texttt{t}$ variable, you type numbers.

```
. list if tin(5,20)
```

means the same as

```
. list if t>=5 & t<=20
```

where t is the name of the time variable you previously $\texttt{tsset}$.

If the time variable t has a $\%\texttt{td}$ or $\%\texttt{d}$ format, then you can type things like

```
. list if tin(5jun1995,20jun1995)
```

and that means the same as

```
. list if t>=d(5jun1995) & t<=d(20jun1995)
```

If the time variable t has a $\%\texttt{tq}$ format, then you can type things like

```
. list if tin(1998q1,1998q4)
```

which means the same as

```
. list if t>=q(1998q1) & t<=d(1998q4)
```

$\texttt{tin()}$ and $\texttt{twithin()}$ work by accessing the time variable previously $\texttt{tsset}$ and then examining its display format to determine how the arguments you type should appear.

In typing $\texttt{tin()}$ and $\texttt{twithin()}$, you may omit either or both arguments, but you must type the comma unless you omit both arguments. The following are all valid:

```
. list if tin(1998q1,)        (meaning 1998q1 and thereafter)
. list if tin(,1998q4)        (meaning up to and including 1998q4)
. list if tin(,)              (meaning all the data)
. list if tin()               (also meaning all the data)
```

There is no reason to type if $\texttt{tin(,)}$ or if $\texttt{tin()}$, but you can.

27.3.9   The %tg format

In addition to the $\%\texttt{t}$ formats documented above, there is one more: $\%\texttt{tg}$. The g stands for generic. The $\%\texttt{tg}$ format is like not putting a $\%\texttt{t}$ format on your variable at all—it is equivalent to $\%9.0\texttt{g}$. It is included for those who have a time variable, and wish to emphasize that it is a time variable, but the variable is in units that Stata does not understand.

When you have such a variable, it does not matter whether you place a $\%\texttt{tg}$ format on it.

28 Commands for dealing with categorical variables

Contents
28.1   Continuous, categorical, and indicator variables
    28.1.1   Converting continuous to indicator variables
    28.1.2   Converting continuous to categorical variables
    28.1.3   Converting categorical to indicator variables
28.2   Using indicator variables in estimation
    28.2.1   Testing the significance of indicator variables
    28.2.2   Importance of omitting one of the indicators

28.1   Continuous, categorical, and indicator variables

Categorical and indicator variables occur so often in statistical data that it is worth emphasizing certain elements of Stata that might escape your attention. Although to Stata a variable is a variable, it is helpful to distinguish among three conceptual types: *continuous*, *categorical*, and *indicator* variables.

A *continuous variable* measures something. Such a variable might measure, for instance, a person's age, height, or weight; a city's population or land area; or a company's revenues or costs.

A *categorical variable* identifies a group to which the thing belongs. For example, one can categorize persons according to their race or ethnicity, cities according to their geographic location, or companies according to their industry. Often, but not always, categorical variables are stored as strings.

An *indicator variable* denotes whether something is true. For example, is a person a veteran, does a city have a mass transit system, or is a company profitable?

Indicator variables are a special case of categorical variables. Consider a variable that records a person's sex. Examined one way, it is a categorical variable. A categorical variable identifies the group to which a thing belongs, and in this case, the thing is a person and the basis for categorization is anatomy. Looked at another way, however, it is an indicator variable. It indicates whether a person is, say, female.

In fact, we can use the same logic on any categorical variable that divides the data into two groups. It is a categorical variable since it identifies whether an observation is a member of this or that group; it is an indicator variable since it denotes the truth value of the statement "the observation is in this group".

All indicator variables are categorical variables, but the opposite is not true. A categorical variable might divide the data into more than two groups. For clarity, let's reserve the term *categorical variable* for variables that divide the data into more than two groups, and let's use the term *indicator variable* for categorical variables that divide the data into exactly two groups.

Stata has the capability to convert continuous variables to categorical and indicator variables and categorical variables to indicator variables.

28.1.1   Converting continuous to indicator variables

Stata treats logical expressions as taking on the values *true* or *false*, which it identifies with the numbers 1 and 0; see [U] **16 Functions and expressions**. For instance, if you have a continuous variable measuring a person's age and you wish to create an indicator variable denoting persons aged 21 and over, you could type

```
. generate age21p = age>=21
```

The variable `age21p` takes on the value 1 for persons aged 21 and over and 0 for persons under 21.

Since `age21p` can take on only 0 or 1, it would be more economical to store the variable as a `byte`. Thus, it would be better to type

```
. generate byte age21p = age>=21
```

This solution has a problem. The value of `age21` is set to 1 for all persons whose `age` is missing because Stata defines missing to be larger than all other numbers. In our data, we have no such missing ages, but it still would have been safer to type

```
. generate byte age21p = age>=21 if age<.
```

That way, persons whose age is missing would also have a missing `age21p`.

❏ Technical Note

Put aside missing values and consider the following alternative to `generate age21p = age>=21` that may have occurred to you:

```
. generate age21p = 1 if age>=21
```

That does not produce the desired result. This statement makes `age21p` 1 (*true*) for all persons aged 21 and above, but makes `age21p` missing for everyone else.

If you followed this second approach, you would have to combine it with

```
. replace age21p = 0 if age<21
```

to make the result identical to that produced by the single statement `gen age21p = age>=21`.

❏

28.1.2   Converting continuous to categorical variables

Suppose that you wish to categorize persons into four groups based on their age. You want a variable to denote whether a person is 21 or under, between 22 and 38, between 39 and 64, or 65 and above. Although most people would label these categories 1, 2, 3, and 4, there is really no reason to restrict ourselves to such a meaningless numbering scheme. Let's call this new variable `agecat` and make it so that it takes on the topmost value for each group. Thus, persons in the first group will be identified with an `agecat` of 21, persons in the second with 38, persons in the third with 64, and persons in the last (drawing a number out of the air) with 75. Here is one way that will work, but it is not the best method:

```
. generate byte agecat=21 if age<=21
(176 missing values generated)

. replace agecat=38 if age>21 & age<=38
(148 real changes made)
```

```
. replace agecat=64 if age>38 & age<=64
(24 real changes made)
. replace agecat=75 if age>64 & age<.
(4 real changes made)
```

We mechanically created the categorical variable according to the definition by using the generate and replace commands. The only thing that deserves comment is the opening generate. We (wisely) told Stata to generate the new variable agecat as a byte, thus conserving memory.

We can create the same result with one command using the recode() function:

```
. generate byte agecat=recode(age,21,38,64,75)
```

recode() takes three or more arguments. It examines the first argument (in this case, age) against the remaining arguments in the list. It returns the first element in the list that is greater than or equal to the first argument, or, failing that, the last argument in the list. Thus, for each observation, recode() asked if age was less than or equal to 21. If so, the value is 21. If not, is it less than or equal to 38? If so, the value is 38. If not, is it less than or equal to 64? If so, the value is 64. If not, the value is 75.

One typically makes tables of categorical variables, so we will tabulate the result:

```
. tabulate agecat
```

agecat	Freq.	Percent	Cum.
21	28	13.73	13.73
38	148	72.55	86.27
64	24	11.76	98.04
75	4	1.96	100.00
Total	204	100.00	

There is another way to convert continuous variables into categorical variables, and it is even more automated: autocode(). autocode() works something like recode(), except that all you tell the function is the range and the total number of cells that you want that range broken into:

```
. generate agecat=autocode(age,4,18,65)
. tabulate agecat
```

agecat	Freq.	Percent	Cum.
29.75	96	47.06	47.06
41.5	92	45.10	92.16
53.25	8	3.92	96.08
65	8	3.92	100.00
Total	204	100.00	

In one instruction, we told Stata to break age into four evenly spaced categories from 18 to 65. When we tabulate agecat, we see the result. In particular, we see that the break points of the four categories are 29.75, 41.5, 53.25, and 65. The first category contains everyone aged 29.75 years or less; the second category contains persons over 29.75 who are 41.5 years old or less; the third category contains persons over 41.5 who are 53.25 years old or less; and the last category contains all persons over 53.25.

❑ Technical Note

We chose the range 18 to 65 arbitrarily. Although you cannot tell from the table above, there are persons in this dataset who are under 18 and there are persons over 65. Those persons are counted in the first and last cells, but we have not divided the age range in the data evenly. We could split the full age range into four categories by obtaining the overall minimum and maximum ages (by typing `summarize`) and substituting the overall minimum and maximum for the 18 and 65 in the `autocode()` function:

```
. summarize age

    Variable |      Obs        Mean   Std. Dev.       Min        Max
-------------+--------------------------------------------------------
         age |      204    29.64706    9.805645         2         66
. generate agecat2=autocode(age,4,2,66)
```

Alternatively, we could `sort` the data into ascending order of `age`, and tell Stata to construct four categories over the range `age[1]` (the minimum) to `age[_N]` (the maximum):

```
. sort age
. generate agecat2=autocode(age,4,age[1],age[_N])
. tabulate agecat2

     agecat2 |      Freq.     Percent        Cum.
-------------+-----------------------------------
          18 |         20        9.80        9.80
          34 |        148       72.55       82.35
          50 |         28       13.73       96.08
          66 |          8        3.92      100.00
-------------+-----------------------------------
       Total |        204      100.00
```

❑

28.1.3 Converting categorical to indicator variables

The easiest way to convert categorical to indicator variables is to use the `xi` command, which will construct indicator variables on the fly. Here we use `xi` with the `logistic` command; `grp` is a variable taking on values 1, 2, and 3:

```
. xi: logistic outcome i.grp age
i.grp             _Igrp_1-3           (naturally coded; _Igrp_1 omitted)

Logit estimates                               Number of obs   =        189
                                              LR chi2(3)      =       6.54
                                              Prob > chi2     =     0.0879
Log likelihood = -114.06375                   Pseudo R2       =     0.0279

------------------------------------------------------------------------------
     outcome | Odds Ratio   Std. Err.       z    P>|z|     [95% Conf. Interval]
-------------+----------------------------------------------------------------
     _Igrp_2 |   2.106974    .9932407     1.58   0.114     .8363679    5.307878
     _Igrp_3 |   1.767748    .6229325     1.62   0.106     .8860686    3.526738
         age |   .9612592    .0311206    -1.22   0.222     .9021588    1.024231
------------------------------------------------------------------------------
```

See [R] **xi**. `xi` is an important command; do not ignore it.

There will be circumstances, however, where we will want to convert categorical to indicator variables permanently, so let's consider how to do that.

We should ask ourselves how this variable is stored. Is it a set of numbers, with different numbers reflecting the different categories, or is it a string? Things will be easier if it is numeric, so if it is not, use encode to convert it; see [U] **26.2 Categorical string variables**. Making categorical variables numeric is not really necessary, but it is a good thing to do because numeric variables can be stored more compactly than string variables. More importantly, all of Stata's statistical commands know how to deal with numeric variables; some do not know what to make out of a string.

Let's suppose you have a categorical variable that divides your data into four groups. To make matters concrete, we will assume that an observation in your data is a state and that the categorical variable denotes the geographical region for each state. Each state is in one of the four Census regions known as the Northeast, North Central, South, and West.

Typing one command will create four new variables, the first indicating whether it is true that the state is in the North Central, the second whether it is true the state is in the Northeast, and so on. Such variables are sometimes called *dummy* variables, and you can use them in regressions to control for the effects of, for instance, geographic region.

Here is the dataset before we type this miraculous command:

```
. use http://www.stata-press.com/data/r8/states3
(State data)

. describe

Contains data from http://www.stata-press.com/data/r8/states3.dta
  obs:            50                          State data
  vars:            6                          17 Sep 2002 11:06
  size:         1,700 (99.8% of memory free)  (_dta has notes)

              storage  display    value
variable name   type   format     label      variable label

state          str8    %9s
reg            int     %8.0g      reg         Census Region
median_age     float   %9.0g                  Median Age
marriage_rate  long    %12.0g                 Marriages per 100,000
divorce_rate   long    %12.0g                 Divorces per 100,000
region         str8    %9s                    Census Region

Sorted by:  reg
. label list reg
reg:
           1 N. Centr
           2 N. East
           3 South
           4 West
```

reg is the categorical variable and, in our example, it is numeric, although that is not important for what we are about to do. The regions are numbered 1 to 4, and a value label, also named reg, maps those numbers into the words N. Centr, N. East, South, and West.

We can make the four indicator variables from this categorical variable by typing

```
. tabulate reg, generate(reg)
   Census
   Region |      Freq.     Percent        Cum.

  N. Centr |         12       24.00       24.00
  N. East  |          9       18.00       42.00
     South |         16       32.00       74.00
      West |         13       26.00      100.00

     Total |         50      100.00
```

```
. describe

Contains data from http://www.stata-press.com/data/r8/states3.dta
  obs:            50                       State data
  vars:           10                       17 Sep 2002 11:06
  size:        1,900 (99.8% of memory free)   (_dta has notes)
```

variable name	storage type	display format	value label	variable label
state	str8	%9s		
reg	int	%8.0g	reg	Census Region
median_age	float	%9.0g		Median Age
marriage_rate	long	%12.0g		Marriages per 100,000
divorce_rate	long	%12.0g		Divorces per 100,000
region	str8	%9s		Census Region
reg1	byte	%8.0g		reg==N. Centr
reg2	byte	%8.0g		reg==N. East
reg3	byte	%8.0g		reg==South
reg4	byte	%8.0g		reg==West

```
Sorted by:  reg
    Note:   dataset has changed since last saved
```

Typing `tabulate reg, generate(reg)` produced a table of the number of states in each region (which is, after all, what `tabulate` does) *and*, because we specified the `generate()` option, it silently created four new variables—one for each line of the table.

Describing the data, we see that there are four new variables called `reg1`, `reg2`, `reg3`, and `reg4`. They are called this because we said `generate(reg)`. If we had said `tabulate reg, gen(junk)`, they would have been called `junk1`, `junk2`, `junk3`, and `junk4`.

Each of the new variables is stored as a `byte`, and each has been automatically labeled for us. The variable `reg1` takes on the value 1 if the state is in the North Central and 0 otherwise. (It also takes on the value missing if `reg` is missing for the observation, which never occurs in our data.)

Just to be clear about the relationship of `reg1` to `reg`, here is a tabulation:

```
. tabulate reg reg1
```

Census Region	reg==N. Centr 0	1	Total
N. Centr	0	12	12
N. East	9	0	9
South	16	0	16
West	13	0	13
Total	38	12	50

If `reg1` is 1, the region is North Central and vice versa.

28.2   Using indicator variables in estimation

Indicator variables allow you to control for the effects of a variable in a regression. Using the indicator variables we generated in the previous example, we can control for region in the following regression:

$$y_j = \beta_0 + \beta_1 age_j + \beta_2 \delta_{2j} + \beta_3 \delta_{3j} + \beta_4 \delta_{4j} + \epsilon_j$$

where y_j represents the marriage rate in state j; age_j represents the median age of the state's population; and δ_{ij} is 1 if state j is in region i and 0 otherwise. We also eliminate the state of Nevada from our regression.

```
. regress marriage_rate median_age reg1 reg3 reg4 if state!="NEVADA"
```

Source	SS	df	MS		Number of obs =	49
					F(4, 44) =	8.71
Model	.000232847	4	.000058212		Prob > F =	0.0000
Residual	.000294193	44	6.6862e-06		R-squared =	0.4418
					Adj R-squared =	0.3910
Total	.000527039	48	.00001098		Root MSE =	.00259

marriage_r~e	Coef.	Std. Err.	t	P>|t|	[95% Conf. Interval]	
median_age	-.0008815	.0002723	-3.237	0.002	-.0014303	-.0003326
reg1	.0003526	.0012315	0.286	0.776	-.0021293	.0028345
reg3	.002699	.0011637	2.319	0.025	.0003537	.0050442
reg4	.0022186	.0014201	1.562	0.125	-.0006434	.0050807
_cons	.039585	.0085496	4.630	0.000	.0223544	.0568155

We see from the results above that the marriage rate, after controlling for age, is significantly higher in region 3, the South.

28.2.1   Testing the significance of indicator variables

After seeing these results, you might wonder if region, taken as a whole, significantly contributes toward the explanatory power of the regression. We can find out using the `test` command:

```
. test reg1=0

 ( 1)  reg1 = 0.0

       F(  1,    44) =    0.08
            Prob > F =    0.7760

. test reg3=0, accumulate

 ( 1)  reg1 = 0.0
 ( 2)  reg3 = 0.0

       F(  2,    44) =    4.07
            Prob > F =    0.0239

. test reg4=0, accumulate

 ( 1)  reg1 = 0.0
 ( 2)  reg3 = 0.0
 ( 3)  reg4 = 0.0

       F(  3,    44) =    2.86
            Prob > F =    0.0478
```

We typed three commands. The first, `test reg1=0`, tested the coefficient on the variable `reg1` against 0. The resulting F test showed the same significance level as the corresponding t test presented in the regression output.

We next typed `test reg3=0, accumulate`. This command tests whether `reg3` is zero and accumulates that test with any previous tests. Thus, we are now testing the hypothesis that `reg1` and `reg3` are jointly zero. The F statistic is 4.07, and the result is significant at the 2.4% level; thus, at any significance level above 2.4%, it appears that `reg1` and `reg3` are not both zero.

We finally typed `test reg4=0, accumulate`. As before, this command tests the newly introduced constraint that `reg4` is zero and accumulates that with the previous tests. We are now testing whether `reg1`, `reg3`, and `reg4` are jointly zero. The F statistic is 2.86, and its significance level is roughly 4.8%; thus, at the 5% level, we can reject the hypothesis that, taken together, region has no effect on the marriage rate after controlling for age.

Stata's `test` command has a shorthand for tests of two or more variables simultaneously equal to zero. Type `test` followed by the names of the variables:

```
. test reg1 reg3 reg4
 ( 1)   reg1 = 0.0
 ( 2)   reg3 = 0.0
 ( 3)   reg4 = 0.0
        F(  3,    44) =     2.86
              Prob > F =     0.0478
```

❑ Technical Note

Sometimes tests of this kind are embedded in an ANOVA or ANCOVA model. In the language of ANOVA, we are doing a one-way layout after controlling for the effect of age. Hence, we did not have to fit the model with `regress`: we could have used Stata's `anova` command. We would have typed `anova marriage_rate median_age reg, continuous(median_age)` to obtain the ANOVA table and the desired test directly. We could have seen the underlying regression after estimation of the model by then typing `regress` without arguments. Typing any estimation command without arguments is taken as a request to reshow the last estimation results. Since all of Stata's estimation commands are tightly coupled, after estimation of an ANOVA or ANCOVA model, you can ask `regress` to show you the underlying regression coefficients.

❑

28.2.2   Importance of omitting one of the indicators

Some people prefer to fit models with dummy variables in the context of ANOVA, and others prefer the regression context. The choice is yours.

If you opt for the regression method, however, remember to leave one of the indicator variables out of the regression so that the coefficients have the interpretation of changes from a base group. If you fail to follow that advice, your regression will still be correct, but it is important that you understand what it is you are estimating and testing.

In the example above, for instance, we omitted `reg2`, the dummy for the Northeast. Let's rerun that regression and include the `reg2` dummy:

(Continued on next page)

```
. regress marriage_rate median_age reg1-reg4 if state!="NEVADA"
```

Source	SS	df	MS
Model	.000232847	4	.000058212
Residual	.000294193	44	6.6862e-06
Total	.000527039	48	.00001098

```
                                           Number of obs =      49
                                           F(  4,    44) =    8.71
                                           Prob > F      =  0.0000
                                           R-squared     =  0.3910
                                           Adj R-squared =  0.3910
                                           Root MSE      =  .00259
```

marriage_r~e	Coef.	Std. Err.	t	P>|t|	[95% Conf. Interval]	
median_age	-.0008815	.0002723	-3.237	0.002	-.0014303	-.0003326
reg1	.0399376	.0080754	4.946	0.000	.0236628	.0562124
reg2	.039585	.0085496	4.630	0.000	.0223544	.0568155
reg3	.0422839	.0080922	5.225	0.000	.0259752	.0585926
reg4	.0418036	.0076958	5.432	0.000	.0262938	.0573135
_cons	(dropped)					

If you compare the top half of the regression output with that of the previous example, you will find that they are identical. You will also find that the estimated coefficient, standard error, and related statistics for the variable **median_age** are identical, yet the estimates for each of the **reg** variables are different.

You will also note that the models have been fitted on different variables. In the first case, dummies for regions 1, 3, and 4 were included along with a constant. In the last case, dummies for regions 1, 2, 3, and 4 were included, and the constant was mysteriously dropped by **regress**. There are the same number of variables in each regression, but their identities have changed. In one case, there is a constant; in the other, there is a dummy for region 2.

Let's first explain why the constant was dropped; it was dropped because it was unnecessary. You can think of this model as making four different predictions for a given median age, one for each region. Each prediction is given by $-.0008815$ multiplied by the median age plus a constant for the region. The constant for the region is the estimated coefficient for the region from the table above. We can write this mathematically as

$$-.0008815age + c_i$$

For instance, for region 2, $c_2 = .039585$.

An overall constant is unnecessary in the sense that it is arbitrary, and zero is a good choice if one is choosing an arbitrary number. Let's choose another arbitrary number, say, 1. Then, each prediction would be given by $-.0008815$ multiplied by the median age plus a (different) constant for the region plus 1:

$$-.0008815age + c_i' + 1$$

The estimated constants for each of the regions would then have to change by exactly 1, so that

$$c_i' + 1 = c_i$$

The constant was dropped because including all four dummy variables in the regression made it unnecessary. In our first model we turned the problem around. We included a constant but left out the dummy for region 2.

Let's spend a moment proving that the results are identical. The model we just fitted indicates that the marriage rate in region 2 is given by

$$-.0008815age + .039585$$

Now, turn back to the previous model. The marriage rate in region 2 is equal to the same thing! The constant in the original regression is identical to the coefficient on `reg2` in the second regression.

Let's look at region 1. The model we just fitted indicates that the marriage rate is given by

$$-.0008815age + .0399376$$

In the first model, it is

$$-.0008815age + .0003526 + .039585 = -.0008815age + .0399376$$

which is again the same thing! If you perform the calculations for region 3 and region 4, you will discover two more equivalencies.

The models are identical in the sense that they make the same predictions. They differ, however, in other ways. In the first model, the coefficients on `reg1`, `reg3`, and `reg4` measure the difference between that region and region 2. In the second model, the coefficients measure the region's level directly.

Notice that the t statistic on the coefficient for `reg2` in the second model is equal to the t statistic on the constant in the first—namely, 4.630—as it should be. Both test the same hypothesis—that the marriage rate is zero in Northeastern states with a median age of zero. (Yes, there are no such states, and yes, the hypothesis sounds silly. Perhaps you prefer the hypothesis stated as "the constant for the Northeast is zero".)

The comparison of the t statistics for region 1, however, does not yield such equivalent results. In the first model, the t statistic is 0.286; in the second model, the statistic is 4.946. They are different because they test different hypotheses. In the first case, the statistic tests whether the constant for region 1 is the same as that for region 2; in the second case, it tests the hypothesis that the constant for region 1 is zero.

The differences in meaning of these statistical tests carry over to any `test` commands that you type. In the technical note above, we tested that `reg1`, `reg3`, and `reg4` were all simultaneously zero. We obtained an F statistic of 2.86. If we were to type the same statements after estimating our second model, we would obtain a whopping F statistic of 13.54! What we are testing, however, is different. If we wanted to perform the same test—namely, that all the regions have the same intercept—we would type `test reg1=reg2`; followed by `test reg3=reg2, accumulate`; followed by `test reg4=reg2, accumulate`. That test gives identical results.

29 Overview of Stata estimation commands

Contents

29.1   Introduction
29.2   Linear regression with simple error structures
29.3   ANOVA, ANCOVA, MANOVA, and MANCOVA
29.4   Generalized linear models
29.5   Binary outcome qualitative dependent variable models
29.6   Conditional logistic regression
29.7   Multiple outcome qualitative dependent variable models
29.8   Simple count dependent variable models
29.9   Linear regression with heteroskedastic errors
29.10   Stochastic frontier models
29.11   Linear regression with systems of equations (correlated errors)
29.12   Models with endogenous sample selection
29.13   Models with time-series data
29.14   Panel-data models
      29.14.1   Linear regression with panel data
      29.14.2   Censored linear regression with panel data
      29.14.3   Generalized linear models with panel data
      29.14.4   Qualitative dependent variable models with panel data
      29.14.5   Count dependent variable models with panel data
      29.14.6   Random-coefficient models with panel data
29.15   Survival-time (failure-time) models
29.16   Commands for estimation with survey data
29.17   Multivariate analysis
29.18   Pharmacokinetic data
29.19   Cluster analysis
29.20   Not elsewhere classified
29.21   Have we forgotten anything?
29.22   References

29.1 Introduction

By estimation commands, we mean commands that fit models such as linear regression, probit, and the like. Stata has many such commands, so many that it is easy to overlook a few. Some of these commands differ greatly from each other, others are gentle variations on a theme, and still others are outright equivalent.

If you have not yet read [U] **23 Estimation and post-estimation commands**, do so soon. Estimation commands share features that we shall not deal with here. We especially direct your attention to [U] **23.14 Obtaining robust variance estimates**, which discusses an alternative calculation for the estimated variance matrix (and hence standard errors) that many of Stata's estimation commands provide, and we direct your attention to [U] **23.10 Performing hypothesis tests on the coefficients**.

Here, however, we will put aside all of that—and all issues of syntax—and deal solely with matching commands to their statistical concepts. We will also put aside cross-referencing when it is obvious. We will not say "the `regress` command—see [R] **regress**—allows ...", nor will we even say "the `tobit` command—see [R] **tobit**—is related . . . ". To find the details on a particular command, look up its name in the index.

29.2 Linear regression with simple error structures

Let us begin by considering models of the form

$$y_j = \mathbf{x}_j \beta + \epsilon_j$$

for a continuous y variable. In this category, we restrict ourselves to estimation when σ_ϵ^2 is constant across observations j. The model is called the linear regression model, and the estimator is often called the (ordinary) least squares estimator.

`regress` is Stata's linear regression command. (`regress` will produce the robust estimate of variance as well as the conventional estimate, and `regress` has a collection of commands that can be run after it to explore the nature of the fit.)

In addition, the following commands will do linear regressions as does `regress`, but offer special features:

1. `ivreg` will fit instrumental variables models.

2. `areg` fits models $y_j = \mathbf{x}_j \beta + \mathbf{d}_j \gamma + \epsilon_j$, where $\mathbf{d}_j$ is a mutually exclusive and exhaustive dummy variable set. Through numerical trickery, `areg` obtains estimates of β (and associated statistics) without ever forming $\mathbf{d}_j$, meaning it also does not report the estimated γ. If your interest is in fitting fixed-effects models, Stata has a better command—`xtreg`—discussed in [U] **29.14.1 Linear regression with panel data** below. Most users who find `areg` appealing are probably seeking `xtreg` because it provides more useful summary and test statistics. `areg` literally duplicates the output `regress` would produce were you to generate all the dummy variables. This means, for instance, that the reported R^2 includes the effect of γ.

3. `boxcox` obtains maximum likelihood estimates of the coefficients and the Box–Cox transform parameter(s) in a model of the form

$$y_i^{(\theta)} = \beta_0 + \beta_1 x_{i1}^{(\lambda)} + \beta_2 x_{i2}^{(\lambda)} + \cdots + \beta_k x_{ik}^{(\lambda)} + \gamma_1 z_{i1} + \gamma_2 z_{i2} + \cdots + \gamma_l z_{il} + \epsilon_i$$

where $\epsilon \sim N(0, \sigma^2)$. Here the *depvar* y is subject to a Box–Cox transform with parameter θ. Each of the *indepvars* $x_1, x_2, \ldots, x_k$ is transformed by a Box–Cox transform with parameter λ. The $z_1, z_2, \ldots, z_l$ are independent variables that are not transformed. In addition to the general form specified above, `boxcox` can fit three other versions of this model defined by the restrictions $\lambda = \theta$, $\lambda = 1$, and $\theta = 1$.

4. `tobit` allows estimation of linear regression models when y_i has been subject to left censoring, right censoring, or both. For instance, say y_i is not observed if $y_i < 1000$, but for those observations, it is known that $y_i < 1000$. `tobit` fits such models.

5. `cnreg` (censored-normal regression) is a generalization of `tobit`. The lower and upper censoring points, rather than being constants, are allowed to vary observation-by-observation. Any model `tobit` can fit, `cnreg` can fit.

6. `intreg` (interval regression) is a generalization of `cnreg`. In addition to allowing open-ended intervals, `intreg` allows closed intervals, too. Rather than observing y_j, it is assumed that y_{0j} and y_{1j} are observed, where $y_{0j} \le y_j \le y_{1j}$. Survey data might report that a subject's monthly income was in the range \$1,500 to \$2,500. `intreg` allows such data to be used to fit a regression model. `intreg` allows $y_{0j} = y_{1j}$, and so can reproduce results reported by `regress`. `intreg` allows y_{0j} to be $-\infty$ and y_{1j} to be $+\infty$, and so can reproduce results reported by `cnreg` and `tobit`.

7. `truncreg` fits the regression model when the sample is drawn from a restricted part of the population, and so is `tobit`-like, except that in this case, neither are the independent variables observed. Under the normality assumption for the whole population, the error terms in the truncated regression model have a truncated-normal distribution.

8. `cnsreg` allows placing linear constraints on the coefficients.

9. `eivreg` adjusts estimates for errors in variables.

10. `nl` provides the nonlinear least-squares estimator of $y_j = f(\mathbf{x}_j, \beta) + \epsilon_j$.

11. `rreg` fits robust regression models, a term not to be confused with regression with robust standard errors. Robust standard errors are discussed in [U] **23.14 Obtaining robust variance estimates**. Robust regression concerns point estimates more than standard errors, and it implements a data-dependent method for downweighting outliers.

12. `qreg` produces quantile-regression estimates, a variation which is not linear regression at all but is an estimator of $y_j = \mathbf{x}_j \beta + \epsilon_j$. In the basic form of this model, sometimes called median regression, $\mathbf{x}_j \beta$ measures not the predicted mean of y_j conditional on $\mathbf{x}_j$, but its median. As such, `qreg` is of most interest when ϵ_j does not have constant variance. `qreg` allows you to specify the quantile, so you can produce linear estimates for the predicted 1st, 2nd, ..., 99th percentile.

Another command, `bsqreg`, is identical to `qreg` but presents bootstrapped standard errors.

The `sqreg` command estimates multiple quantiles simultaneously; standard errors are obtained via the bootstrap.

The `iqreg` command estimates the difference between two quantiles; standard errors are obtained via the bootstrap.

13. `vwls` (variance-weighted least squares) produces estimates of $y_j = \mathbf{x}_j \beta + \epsilon_j$, where the variance of ϵ_j is calculated from group data or is known *a priori*. As such, `vwls` is of most interest to categorical-data analysts and physical scientists.

29.3   ANOVA, ANCOVA, MANOVA, and MANCOVA

ANOVA and ANCOVA are certainly related to linear regression, but we classify them separately. The related Stata commands are `anova`, `oneway`, and `loneway`. The `manova` command provides MANOVA and MANCOVA (multivariate ANOVA and ANCOVA).

`anova` fits ANOVA and ANCOVA models, one-way and up—including two-way factorial, three-way factorial, etc.—and it fits nested and mixed-design models and repeated-measures models. It is probably what you are looking for.

`oneway` fits one-way ANOVA models. It is quicker at producing estimates than `anova`, although `anova` is so fast that this probably does not matter. The important difference is that `oneway` can report multiple-comparison tests.

`loneway` is an alternative to `oneway`. The results are numerically the same, but `loneway` can deal with more levels (limited only by dataset size; `oneway` is limited to 376 levels and `anova` to 798, but for `anova` to reach 798 requires a lot of memory), and `loneway` reports some additional statistics such as the intraclass correlation.

`manova` fits MANOVA and MANCOVA models, one-way and up–including two-way factorial, three-way factorial, etc.–and it fits nested and mixed-design models.

29.4 Generalized linear models

The generalized linear model is

$$g\{E(y_j)\} = \mathbf{x}_j\boldsymbol{\beta}, \qquad y_j \sim F$$

where $g()$ is called the link function and F is a member of the exponential family, both of which you specify prior to estimation. `glm` fits this model.

The GLM framework encompasses a surprising array of models known by other names, including linear regression, Poisson regression, exponential regression, and others. Stata provides dedicated estimation commands for many of these. Stata has, for instance, `regress` for linear regression, `poisson` for Poisson regression, and `ereg` and `streg` for exponential regression, and that is not all of the overlap.

`glm` will by default use maximum likelihood estimation, and alternately estimate via iterated reweighted least squares (IRLS) when the `irls` option is specified. For each family F there is a corresponding link function $g()$, called the canonical link, for which IRLS estimation produces results identical to maximum likelihood estimation. You can, however, match families and link functions as you wish, and, when you match a family to a link function other than the canonical link, you obtain a different but valid estimator of the standard errors of the regression coefficients. The estimator you obtain is asymptotically equivalent to the maximum likelihood estimator, which, in small samples, will produce slightly different results.

For example, the canonical link for the binomial family is logit. `glm, irls` with that combination produces results identical to the maximum-likelihood `logit` (and `logistic`) command. The binomial family with the probit link produces the probit model, but probit is not the canonical link in this case. Hence, `glm, irls` produces standard error estimates that differ slightly from those produced by Stata's maximum-likelihood `probit` command.

Many researchers feel that the maximum-likelihood standard errors are preferable to IRLS estimates (when they are not identical), but they would have a difficult time justifying that feeling. Maximum likelihood probit is an estimator with (solely) asymptotic properties; `glm, irls` with the binomial family and probit link is an estimator with (solely) asymptotic properties, and in finite samples, the standard errors differ a little.

Still, we recommend that you use Stata's dedicated estimators whenever possible. IRLS—the theory—and `glm, irls` —the command—are all-encompassing in their generality, and that means they rarely use quite the right jargon or provide things in quite the way you wish they would. The narrower commands, such as `logit`, `probit`, `poisson`, etc., focus on the issue at hand and are invariably more convenient.

`glm` is useful when you want to match a family to a link function that is not provided elsewhere.

`glm` also offers a number of estimators of the variance–covariance matrix that are consistent even when the errors are heteroskedastic and/or autocorrelated. Another advantage of a `glm` version of a model over a model-specific version is that many of these VCE estimators are available only for the `glm` implementation. In addition, one may also obtain the ML–based estimates of the VCE from `glm`.

29.5 Binary outcome qualitative dependent variable models

There are lots of ways to write these models; one way is

$$\Pr(y_j \neq 0) = F(\mathbf{x}_j\boldsymbol{\beta})$$

where F is some cumulative distribution. Two popular choices for $F()$ are the normal and logistic, and the models are called the probit and logit (or logistic regression) models. A third is the complementary log–log function; maximum likelihood estimates are obtained by Stata's `cloglog` command.

The two parent commands for the maximum likelihood estimator of probit and logit are `probit` and `logit`, although `logit` has a sibling, `logistic`, that provides the same estimates but displays results in a slightly different way.

Do not read anything into the names logit and logistic, although, even with that warning, we know you will. Logit and logistic have two completely interchanged definitions in two scientific camps. In the medical sciences, logit means the minimum χ^2 estimator and logistic means maximum likelihood. In the social sciences, it is the other way around. From our experience, it appears that neither reads the other's literature, since both talk (and write books) asserting that logit means one thing and logistic the other. Our solution is to provide both `logit` and `logistic` that do the same thing so that each camp can latch on to the maximum likelihood command under the name each expects.

There are two slight differences between `logit` and `logistic`. `logit` reports estimates in the coefficient metric, whereas `logistic` reports exponentiated coefficients—odds ratios. This is in accordance with the expectations of each camp and makes no substantive difference. The other difference is that `logistic` has a family of post-`logistic` commands that you can run to explore the nature of the fit. Actually, that is not exactly true because all the commands for use after `logistic` can also be used after `logit`. A note is even made of that fact in the `logit` documentation.

If you have not already selected one of `logit` or `logistic` as your favorite, we recommend you try `logistic`. Logistic regression (logit) models are more easily interpreted in the odds-ratio metric.

In addition to `logit` and `logistic`, Stata provides `glogit`, `blogit`, and `binreg` commands.

`blogit` is the maximum likelihood estimator (same as `logit` or `logistic`), but applied on data organized in a different way. Rather than individual observations, your data are organized so that each observation records the number of observed successes and failures.

`glogit` is the weighted-regression, grouped-data estimator.

`binreg` can be used to model either individual-level or grouped data in an application of the generalized linear model. The family is assumed to be binomial, and each link provides a distinct parameter interpretation. In addition, `binreg` offers several options for setting the link function according to the desired biostatistical interpretation. The available links and interpretation options are

Option	Implied link	Parameter
or	logit	Odds ratios = $\exp(\beta)$
rr	log	Risk ratios = $\exp(\beta)$
hr	log complement	Health ratios = $\exp(\beta)$
rd	identity	Risk differences = β

Related to logit, the skewed logit estimator scobit adds a power to the logit link function and is estimated by Stata's `scobit` command.

Turning to probit, you have two choices: `probit` and `dprobit`. Both are maximum likelihood, and it makes no substantive difference which you use. They differ only in how they report results. `probit` reports coefficients. `dprobit` reports changes in probabilities. Many researchers find changes in probabilities easier to interpret.

As in the logit case, Stata also provides `bprobit` and `gprobit`. `bprobit` is maximum likelihood—equivalent to `probit` or `dprobit`—but works with data organized in the different way outlined above. `gprobit` is the weighted-regression, grouped-data estimator.

Continuing with probit, `hetprob` fits heteroskedastic probit models. In these models, the variance of the error term is parameterized.

In addition, Stata's `biprobit` command will fit bivariate probit models, meaning two correlated outcomes. `biprobit` will also fit partial-observability models in which only the outcomes $(0, 0)$ and $(1, 1)$ are observed.

29.6 Conditional logistic regression

`clogit` is Stata's conditional logistic regression estimator. In this model, observations are assumed to be partitioned into groups and a predetermined number of events occur in each group. The model measures the risk of the event according to the observation's covariates x_j. The model is used in matched case–control studies (`clogit` allows $1 : 1$, $1 : k$, and $m : k$ matching) and is also used in natural experiments whenever observations can be grouped into pools in which a fixed number of events occur.

29.7 Multiple outcome qualitative dependent variable models

For more than two outcomes, Stata provides ordered logit, ordered probit, rank ordered logit, multinomial logistic regression, McFadden's choice model (conditional fixed-effects logistic regression), and nested logistic regression.

`oprobit` and `ologit` provide maximum-likelihood ordered probit and logit. These are generalizations of probit and logit models known as the proportional odds model, and are used when the outcomes have a natural ordering from low to high. The idea is there is an unmeasured $z_j = x_j \beta$, and the probability that the kth outcome is observed is $\Pr(c_{k-1} < z_j < c_k)$, where $c_0 = -\infty$, $c_k = +\infty$, and $c_1, \ldots, c_{k-1}$ along with β are estimated from the data.

`rologit` fits the rank-ordered logit model for rankings. This model is also known as the Plackett–Luce model, as the exploded logit model, and as choice-based conjoint analysis.

`mlogit` fits maximum-likelihood multinomial logistic models, also known as polytomous logistic regression. It is intended for use when the outcomes have no natural ordering and all that is known are the characteristics of the outcome chosen (and, perhaps, the chooser).

`clogit` fits McFadden's choice model, also known as conditional logistic regression. In the context denoted by the name McFadden's choice model, the model is used when the outcomes have no natural ordering, just as multinomial logistic regression, but the characteristics of the outcomes chosen and not chosen are known (along with, perhaps, the characteristics of the chooser).

In the context denoted by the name conditional logistic regression—mentioned above—subjects are members of pools and one or more are chosen, typically to be infected by some disease or to have some other unfortunate event befall them. Thus, the characteristics of the chosen and not chosen are known, and the issue of the characteristics of the chooser never arises. Said either way, it is the same model.

In their choice-model interpretations, `mlogit` and `clogit` assume that the odds-ratios are independent of any other, unspecified, alternatives. Since this assumption is frequently rejected by the data, the nested logit model is a very useful generalization. `nlogit` fits a nested logit model using full maximum likelihood. The model may contain one or more levels.

29.8 Simple count dependent variable models

These models concern dependent variables that count the number of occurrences of an event. In this category, we include Poisson and negative-binomial regression. For the Poisson model,

$$E(\text{count}) = E_j \exp(\mathbf{x}_j \boldsymbol{\beta})$$

where E_j is the exposure time. poisson fits this model.

Negative-binomial regression refers to estimating with data that are a mixture of Poisson counts. One derivation of the negative-binomial model is that individual units follow a Poisson regression model, but there is an omitted variable that follows a gamma distribution with variance α. Negative-binomial regression estimates β and α. nbreg fits such models. A variation on this, unique to Stata, allows you to model α. gnbreg fits those models.

Zero inflation refers to count models in which the number of 0 counts is more than would be expected in the regular model, and that is due to there being a probit or logit process that must first generate a positive outcome before the counting process can begin.

Stata's zip command fits zero inflated Poisson models.

Stata's zinb command fits zero-inflated negative-binomial models.

29.9 Linear regression with heteroskedastic errors

We now consider the model $y_j = \mathbf{x}_j \boldsymbol{\beta} + \epsilon_j$, where the variance of ϵ_j is nonconstant.

First, regress can fit such models if you specify the robust option. What we call robust is also known as the White correction for heteroskedasticity.

For scientists who have data where the variance of ϵ_j is known *a priori*, vwls is the command. vwls produces estimates for the model given each observation's variance, which is recorded in a variable in the data.

Finally, as mentioned above, qreg performs quantile regression, and it is in the presence of heteroskedasticity that this is most of interest. Median regression (one of qreg's capabilities) is an estimator of $y_j = \mathbf{x}_j \boldsymbol{\beta} + \epsilon_j$ when ϵ_j is heteroskedastic. Even more usefully, one can fit models of other quantiles and so model the heteroskedasticity. Also see the sqreg and iqreg commands; sqreg estimates multiple quantiles simultaneously. iqreg estimates differences in quantiles.

29.10 Stochastic frontier models

frontier fits stochastic production or cost frontier models on cross-sectional data. The model can be expressed as

$$y_i = \mathbf{x}_i \boldsymbol{\beta} + v_i - s u_i$$

where

$$s = \begin{cases} 1, & \text{for production functions} \\ -1, & \text{for cost functions} \end{cases}$$

u_i is a nonnegative disturbance, standing for technical inefficiency in the production function or cost inefficiency in the cost function. While the idiosyncratic error term v_i is assumed to have a normal distribution, the inefficiency term is assumed to be one of the three distributions: half-normal, exponential, or truncated-normal. In addition, when the nonnegative component of the disturbance is assumed to be either half-normal or exponential, `frontier` can fit models in which the error components are heteroskedastic conditional on a set of covariates. When the nonnegative component of the disturbance is assumed to be from a truncated-normal distribution, `frontier` can also fit a conditional mean model, where the mean of the truncated-normal distribution is modeled as a linear function of a set of covariates.

For panel data stochastic frontier models, see [U] **29.14.1 Linear regression with panel data**.

29.11   Linear regression with systems of equations (correlated errors)

If by correlated errors, you mean that observations are grouped, and that within group, the observations might be correlated but, across groups, they are uncorrelated, realize that `regress` with the `robust` and `cluster()` options can produce "correct" estimates, which is to say, inefficient estimates with correct standard errors and lots of robustness; see [U] **23.14 Obtaining robust variance estimates**. Obviously, if you know the correlation structure (and are not mistaken), you can do better, so `xtreg` and `xtgls` are also of interest in this case; we discuss them in [U] **29.14.1 Linear regression with panel data** below.

Turning to simultaneous multiple-equation models, Stata can produce three-stage least squares (3SLS) and two-stage least squares (2SLS) estimates using the `reg3` and `ivreg` commands. Two-stage models can be estimated by either `reg3` or `ivreg`. Three-stage models require use of `reg3`. The `reg3` command can produce constrained and unconstrained estimates.

In the case where we have correlated errors across equations but no endogenous right-hand side variables,

$$y_{1j} = \mathbf{x}_{1j}\boldsymbol{\beta} + \epsilon_{1j}$$
$$y_{2j} = \mathbf{x}_{2j}\boldsymbol{\beta} + \epsilon_{2j}$$
$$\vdots$$
$$y_{mj} = \mathbf{x}_{mj}\boldsymbol{\beta} + \epsilon_{mj}$$

where $\epsilon_{k\cdot}$ and $\epsilon_{l\cdot}$ are correlated with correlation ρ_{kl}, a quantity to be estimated from the data. This is called Zellner's seemingly unrelated regressions, and `sureg` fits such models. In the case where $\mathbf{x}_{1j} = \mathbf{x}_{2j} = \cdots = \mathbf{x}_{mj}$, the model is known as multivariate regression, and the corresponding command is `mvreg`.

Estimation in the presence of autocorrelated errors is discussed in [U] **29.13 Models with time-series data**.

29.12   Models with endogenous sample selection

What has become known as the Heckman model refers to linear regression in the presence of sample selection: $y_j = \mathbf{x}_j\boldsymbol{\beta} + \epsilon_j$ is not observed unless some event occurs which itself has probability $p_j = F(\mathbf{z}_j\boldsymbol{\gamma} + \nu_j)$, where ϵ and ν might be correlated and $\mathbf{z}_j$ and $\mathbf{x}_j$ may contain variables in common.

`heckman` fits such models by maximum likelihood or Heckman's original two-step procedure.

This model has recently been generalized to replacing the linear regression equation with another probit equation, and that model is fitted by heckprob.

Another important case of endogenous sample selection is the treatment effects model. The treatment effects model considers the effect of an endogenously chosen binary treatment on another endogenous, continuous variable, conditional on two sets of independent variables. treatreg fits a treatment effects model using either a two-step consistent estimator or full maximum likelihood.

29.13   Models with time-series data

ARIMA refers to models with autoregressive integrated moving average processes, and Stata's arima command fits models with ARIMA disturbances via the Kalman filter and maximum likelihood. These models may be fitted with or without confounding covariates.

Stata's prais command performs regression with AR(1) disturbances using the Prais–Winsten or Cochrane–Orcutt transformation. Both two-step and iterative solutions are available, as well as a version of the Hildreth–Lu search procedure. The Prais–Winsten estimates for the model are an improvement over the Cochrane–Orcutt estimates in that the first observation is preserved in the estimation. This is particularly important with trended data in small samples.

prais automatically produces the Durbin–Watson d-statistic, which can also be obtained after regress using dwstat.

newey produces linear regression estimates with the Newey–West variance estimates that are robust to heteroskedasticity and autocorrelation of specified order.

Stata provides estimators for regression models with autoregressive conditional heteroskedastic (ARCH) disturbances:

$$y_t = \mathbf{x}_t \boldsymbol{\beta} + \mu_t$$

where μ_t is distributed $N(0, \sigma_t^2)$ and σ_t^2 is given by some function of the lagged disturbances.

Stata's arch, aparch, and egarch commands provide different parameterizations of the conditional heteroskedasticity. All three of these commands also allow ARMA disturbances and/or multiplicative heteroskedasticity.

Stata provides var and svar for fitting vector autoregression (VAR) and structural vector autoregression (SVAR) models. See [TS] **var** for information on Stata's suite of commands for forecasting, specification testing, and inference on VAR and SVAR models. See [TS] **varirf** for information on Stata's suite of commands for estimating, analyzing, and presenting impulse–response functions and forecast error variance decompositions. There is also a set of commands for performing Granger causality tests, lag-order selection, and residual analysis.

29.14   Panel-data models

29.14.1   Linear regression with panel data

This section could just as well be called linear regression with complex error structures. The letters xt are the prefix for the commands in this class.

xtreg fits models of the form

$$y_{it} = \mathbf{x}_{it} \boldsymbol{\beta} + \nu_i + \epsilon_{it}$$

xtreg can produce the between regression (random-effects) estimator, the within regression (fixed-effects) estimator, or the GLS random-effects (matrix-weighted average of between and within results) estimator. In addition, it can produce the maximum-likelihood random-effects estimator.

xtregar can produce the within estimator and a GLS random-effects estimator when the ϵ_{it} are assumed to follow an AR(1) process.

xtivreg contains the between-2SLS estimator, the within-2SLS estimator, the first-differenced-2SLS estimator, and two GLS random-effects-2SLS estimators to handle cases in which some of the covariates are endogenous.

xtabond is for use with dynamic panel-data models (models in which there are lagged dependent variables) and can produce the one-step, one-step robust, and the two-step Arellano–Bond estimator. xtabond can handle predetermined covariates and it reports both the Sargan and autocorrelation tests derived by Arellano and Bond.

xtgls produces generalized least squares estimates for models of the form

$$y_{it} = \mathbf{x}_{it}\boldsymbol{\beta} + \epsilon_{it}$$

where you may specify the variance structure of ϵ_{it}. If you specify that ϵ_{it} is independent for all i and t, xtgls produces the same results as regress up to a small-sample degrees-of-freedom correction applied by regress but not by xtgls.

You may choose among three variance structures concerning i and three concerning t, producing a total of nine different models. Assumptions concerning i deal with heteroskedasticity and cross-sectional correlation. Assumptions concerning t deal with autocorrelation and, more specifically, AR(1) serial correlation.

Alternative methods report the OLS coefficients and a version of the GLS variance–covariance estimator. xtpcse produces panel-corrected standard error (PCSE) estimates for linear cross-sectional time-series models, where the parameters are estimated by OLS or Prais–Winsten regression. When computing the standard errors and the variance–covariance estimates, the disturbances are, by default, assumed to be heteroskedastic and contemporaneously correlated across panels.

In the jargon of GLS, the random-effects model fitted by xtreg has exchangeable correlation within i—xtgls does not model this particular correlation structure. xtgee, however, does.

xtgee will fit population-averaged models, and it will optionally provide robust estimates of variance. Moreover, xtgee will allow other correlation structures. One that is of particular interest to those with lots of data goes by the name unstructured. The within-panel correlations are simply estimated in an unconstrained way. In [U] **29.14.3 Generalized linear models with panel data**, we have more to say about this estimator since it is not restricted to just linear regression models.

xthtaylor uses instrumental variables estimators to estimate the parameters of panel data random-effects models of the form

$$y_{it} = \mathbf{X}_{1it}\boldsymbol{\beta}_1 + \mathbf{X}_{2it}\boldsymbol{\beta}_2 + \mathbf{Z}_{1i}\boldsymbol{\delta}_1 + \mathbf{Z}_{2i}\boldsymbol{\delta}_2 + u_i + e_{it}$$

The individual effects u_i are correlated with the explanatory variables $\mathbf{X}_{2it}$ and $\mathbf{Z}_{2i}$, but are uncorrelated with $\mathbf{X}_{1it}$ and $\mathbf{Z}_{1i}$, where $\mathbf{Z}_1$ and $\mathbf{Z}_2$ are constant within panel.

xtfrontier fits stochastic production or cost frontier models for panel data. You may choose from a time-invariant model or a time-varying decay model. In both models, the nonnegative inefficiency term is assumed to have a truncated-normal distribution. In the time-invariant model, the inefficiency term is constant within panels. In the time-varying decay model, the inefficiency term is modeled as a truncated-normal random variable multiplied by a specific function of time. In both models, the idiosyncratic error term is assumed to have a normal distribution. The only panel-specific effect is the random inefficiency term.

29.14.2   Censored linear regression with panel data

xttobit fits random-effects tobit models and generalizes that to observation-specific censoring.

xtintreg performs random-effects interval regression and generalizes that to observation-specific censoring. Interval regression, in addition to allowing open-ended intervals, also allows closed intervals.

29.14.3   Generalized linear models with panel data

In [U] **29.4 Generalized linear models** above, we discussed the model

$$g\{E(y_j)\} = \mathbf{x}_j\boldsymbol{\beta}, \qquad y_j \sim F$$

where $g()$ is the link function and F is a member of the exponential family, both of which you specify prior to estimation. This model can be further generalized to work with cross-sectional time-series data, so let us rewrite it:

$$g\{E(y_{it})\} = \mathbf{x}_{it}\boldsymbol{\beta}, \qquad y_{it} \sim F \text{ with parameters } \theta_{it}$$

We refer to this as the GEE method for panel data models, where GEE stands for Generalized Estimating Equations. xtgee fits this model and allows specifying the correlation structure of the errors.

If you specify that errors are independent within i, xtgee is equivalent to glm. Thus, since glm can reproduce (to name a few), the estimates produced by regress, logit, and poisson, so can xtgee.

If you specify errors are exchangeable within i, xtgee fits equal-correlation models. This means that with the identity link and Gaussian family, xtgee can reproduce the models fitted by xtreg. The only difference is that xtgee can provide standard errors that are robust to the correlations not being exchangeable.

xtgee provides other correlation structures, including multiplicative; AR(m); stationary(m); nonstationary(m); unstructured; and fixed (meaning user-specified). Unstructured should be of particular interest to those with large datasets, even if you ultimately plan to impose a structure such as exchangeability (equal correlation). If relaxing the equal-correlation assumption in a large dataset causes your results to change importantly, there is an issue before you worthy of some thought.

xtgee provides 175 models from which to choose.

29.14.4   Qualitative dependent variable models with panel data

xtprobit fits random-effects probit regression via maximum likelihood. It will also fit population-averaged models via GEE. This last is nothing more than xtgee with the binomial family, probit link, and exchangeable error structure.

xtlogit fits random-effects logistic regression models via maximum likelihood. It will also fit conditional fixed-effects models via maximum likelihood. Finally, as with xtprobit, it will fit population-averaged models via GEE.

xtcloglog estimates random-effects complementary log-log regression via maximum likelihood. It will also fit population-averaged models via GEE.

clogit is also of interest since it provides the conditional fixed-effects logistic estimator.

29.14.5   Count dependent variable models with panel data

xtpoisson fits two different random-effects Poisson regression models via maximum likelihood. The two distributions for the random effect are gamma and normal. It will also fit conditional fixed-effects models. It will also fit population-averaged models via GEE. This last is nothing more than xtgee with the Poisson family, log link, and exchangeable error structure.

xtnbreg fits random-effects negative-binomial regression models via maximum likelihood (the distribution of the random effects is assumed to be beta). It will also fit conditional fixed-effects models and population-averaged models via GEE.

29.14.6   Random-coefficient models with panel data

xtrchh will fit the Hildreth–Houck random-coefficients model. In this model, rather than just the intercept being constant within group and varying across groups, all the coefficients vary across groups.

29.15   Survival-time (failure-time) models

Commands are provided to fit Cox proportional hazards models, as well as several parametric survival models including exponential, Weibull, Gompertz, log-normal, log-logistic, and generalized gamma (see [ST] **stcox** and [ST] **streg**). The commands all allow for right-censoring, left-truncation, gaps in histories, and time-varying regressors. The commands are appropriate for use with single- or multiple-failure per subject data. Conventional and robust standard errors are available, with and without clustering.

Both the Cox model and the parametric models (as fitted using Stata) allow for two additional generalizations. First, the models may be modified to allow for latent random effects, or *frailties*. Second, the models may be stratified in the sense that the baseline hazard function may vary completely over a set of strata. The parametric models also allow the modeling of ancillary parameters.

stcox and streg require that the data be stset so that the proper response variables may be established. After stsetting the data, the response is taken as understood and one need only supply the regressors (and other options) to stcox and streg.

29.16   Commands for estimation with survey data

Many of Stata's estimation commands allow sampling weights and, if they do, provide a cluster() option as well; see [U] **23.16 Weighted estimation**. That still leaves the issue of stratification, and a parallel set of commands beginning with the letters svy are provided. The list currently includes

1. svyregress for linear regression,
2. svyivreg for instrumental-variables regression,
3. svyintreg for censored and interval regression,
4. svylogit for logistic regression,
5. svyprobit for probit,
6. svymlogit for multinomial logistic regression,
7. svyologit for ordered logistic regression,
8. svyoprobit for ordered probit,

9. svypoisson for Poisson regression,

10. svynbreg for negative binomial regression,

11. svygnbreg for generalized negative binomial regression,

12. svyheckman for linear regression with sample selection, and

13. svyheckprob for probit with sample selection.

See [U] **30 Overview of survey estimation**.

29.17   Multivariate analysis

In this category, we include canonical correlation, principal components, and factor analysis. See [U] **29.11 Linear regression with systems of equations (correlated errors)** above for multivariate regression. See [U] **29.3 ANOVA, ANCOVA, MANOVA, and MANCOVA** for multivariate analysis of variance.

Canonical correlation attempts to describe the relationship between two sets of variables. Given $\mathbf{x}_j = (x_{1j}, x_{2j}, \ldots, x_{Kj})$ and $\mathbf{y}_j = (y_{1j}, y_{2j}, \ldots, y_{Lj})$, the goal is to find linear combinations

$$\widehat{x}_{1j} = \widehat{h}_{11}x_{1j} + \widehat{h}_{12}x_{2j} + \cdots + \widehat{h}_{1K}x_{Kj}$$
$$\widehat{y}_{1j} = \widehat{g}_{11}y_{1j} + \widehat{g}_{12}y_{2j} + \cdots + \widehat{g}_{1L}y_{Lj}$$

such that the correlation of $\widehat{x}_1$ and $\widehat{y}_1$ is maximized. That is called the first canonical correlation. The second canonical correlation is defined similarly, with the added proviso that $\widehat{x}_2$ and $\widehat{y}_2$ are orthogonal to $\widehat{x}_1$ and $\widehat{y}_1$, and the third canonical correlation with the proviso that $\widehat{x}_3$ and $\widehat{y}_3$ are orthogonal to $\widehat{x}_1$, $\widehat{y}_1$, $\widehat{x}_2$, and $\widehat{y}_2$, and so on. canon estimates canonical correlations and their corresponding loadings.

Principal components concerns finding

$$\widehat{x}_{1j} = \widehat{b}_{11}x_{1j} + \widehat{b}_{12}x_{2j} + \cdots + \widehat{b}_{1K}x_{Kj}$$
$$\widehat{x}_{2j} = \widehat{b}_{21}x_{1j} + \widehat{b}_{22}x_{2j} + \cdots + \widehat{b}_{2K}x_{Kj}$$
$$\vdots$$
$$\widehat{x}_{Kj} = \widehat{b}_{K1}x_{1j} + \widehat{b}_{K2}x_{2j} + \cdots + \widehat{b}_{KK}x_{Kj}$$

such that $\widehat{x}_1$ has maximum variance, $\widehat{x}_2$ has maximum variance subject to being orthogonal to $\widehat{x}_1$, $\widehat{x}_3$ has maximum variance subject to being orthogonal to $\widehat{x}_1$ and $\widehat{x}_2$, and so on. pca extracts principal components and reports eigenvalues and loadings.

Factor analysis is concerned with finding a small number of common factors $\widehat{z}_k$, $k = 1, \ldots, q$ that linearly reconstruct the original variables $\mathbf{y}_i$, $i = 1, \ldots, L$.

$$\widehat{y}_{1j} = \widehat{z}_{1j}\widehat{b}_{11} + \widehat{z}_{2j}\widehat{b}_{12} + \cdots + \widehat{z}_{qj}\widehat{b}_{1q} + e_{1j}$$
$$\widehat{y}_{2j} = \widehat{z}_{1j}\widehat{b}_{21} + \widehat{z}_{2j}\widehat{b}_{22} + \cdots + \widehat{z}_{qj}\widehat{b}_{2q} + e_{2j}$$
$$\vdots$$
$$\widehat{y}_{Lj} = \widehat{z}_{1j}\widehat{b}_{L1} + \widehat{z}_{2j}\widehat{b}_{L2} + \cdots + \widehat{z}_{qj}\widehat{b}_{Lq} + e_{Lj}$$

Note that everything on the right-hand side is fitted so the model has an infinite number of solutions. Various constraints are introduced along with a definition of "reconstruct" to make the model determinate. Reconstruction, for instance, is typically defined in terms of prediction of the covariance of the original variables. factor fits such models and provides principal factors, principal component factors, iterated principal components, and maximum likelihood solutions.

29.18   Pharmacokinetic data

The are four estimation commands for the analysis of pharmacokinetic data. See [R] **pk** for an overview of the pk system.

1. pkexamine calculates pharmacokinetic measures from time-and-concentration subject-level data. pkexamine computes and displays the maximum measured concentration, the time at the maximum measured concentration, the time of the last measurement, the elimination time, the half-life, and the area under the concentration-time curve (AUC).

2. pksumm obtains the first four moments from the empirical distribution of each pharmacokinetic measurement and tests the null hypothesis that the distribution of that measurement is normally distributed.

3. pkcross analyzes data from a crossover design experiment. When analyzing pharmaceutical trial data, if the treatment, carryover, and sequence variables are known, the omnibus test for separability of the treatment and carryover effects is calculated.

4. pkequiv performs bioequivalence testing for two treatments. By default, pkequiv calculates a standard confidence interval symmetric about the difference between the two treatment means. pkequiv also calculates confidence intervals symmetric about zero, and intervals based on Fieller's theorem. Additionally, pkequiv can perform interval hypothesis tests for bioequivalence.

29.19   Cluster analysis

Strictly speaking, cluster analysis does not fall into the category of statistical estimation. Rather, it is a set of techniques for exploratory data analysis.

Stata's cluster analysis routines give you a choice of several hierarchical and partition clustering methods. Post-clustering summarization methods as well as cluster management tools are also provided. We briefly describe the analytic commands here. See [CL] **cluster** for a more detailed introduction to these commands and the post-clustering user utilities and programmer utilities that work with them. Stata's cluster environment has many different similarity and dissimilarity measures for continuous and binary data. See [CL] **cluster** for the details of these measures.

Stata's clustering methods fall into two general types: partition and hierarchical. Partition methods break the observations into a distinct number of nonoverlapping groups. Stata has implemented two partition methods, kmeans and kmedians. See [CL] **cluster kmeans** and [CL] **cluster kmedians** for details on these methods.

The partition clustering methods will generally be quicker and will allow larger datasets than the hierarchical clustering methods outlined below. However, if you wish to examine clustering to various numbers of clusters, you will need to execute cluster numerous times with the partition methods. Clustering to various numbers of groups using a partition method will typically not produce clusters that are hierarchically related. If this is important for your application, consider using one of the hierarchical methods.

Hierarchical clustering methods are generally of two types: agglomerative or divisive. Hierarchical clustering creates (by either dividing or combining) hierarchically related sets of clusters. Stata has seven agglomerative hierarchical methods. See [CL] **cluster averagelinkage**, [CL] **cluster centroidlinkage**, [CL] **cluster completelinkage**, [CL] **cluster medianlinkage**, [CL] **cluster singlelinkage**, [CL] **cluster wardslinkage**, and [CL] **cluster waveragelinkage** for details on these methods.

29.20   Not elsewhere classified

There are three other commands that are not really estimation commands but estimation-command modifiers: sw, fracpoly, and mfp.

sw, typed in front of an estimation command as a separate word, provides stepwise estimation. You can use the sw prefix with some, but not all, estimation commands. In [R] **sw** is a table of which estimation commands are currently supported, but do not take it too literally. It was accurate as of the day Stata 8 was released, but, if you install the official updates, sw may now work with other commands, too. If you want to use sw with some estimation command, our advice is to try it. Either it will work or you will get the message that the estimation command is not supported by sw.

fracpoly and mfp are commands to assist performing specification searches.

29.21   Have we forgotten anything?

We have discussed all the estimation commands included in Stata 8.0 the day it was released; by now, there may be more. To obtain an up-to-date list, type search estimation.

And, of course, you can always write your own; see [R] **ml**

29.22   References

Gould, W. W. 2000. sg124: Interpreting logistic regression in all its forms. *Stata Technical Bulletin* 53: 19–29. Reprinted in *Stata Technical Bulletin Reprints*, vol. 9, pp. 257–270.

30 Overview of survey estimation

Contents

30.1   Introduction
30.2   Accounting for the sample design in survey analyses
        30.2.1   Multistage sample designs
        30.2.2   Finite population corrections
        30.2.3   Design effects: deff and deft
30.3   Example of the effects of weights, clustering, and stratification
        30.3.1   Halfway isn't enough: the importance of stratification and clustering
30.4   Linear regression and other models
30.5   Pseudo-likelihoods
30.6   Building your own survey estimator using ml
30.7   Hypothesis testing using test and testnl
30.8   Estimation of linear and nonlinear combinations of parameters
30.9   Two-way contingency tables
30.10   Differences between the svy commands and other commands
30.11   References

30.1   Introduction

There is a family of commands in Stata designed especially for data from sample surveys. All the commands for analyzing these data begin with svy, the only exceptions being lincom, ml, nlcom, predictnl, test, and testnl.

(*Continued on next page*)

The svy commands are

svyregress	[SVY] **svy estimators**	Linear regression for survey data
svyivreg	[SVY] **svy estimators**	Instrumental variables regression for survey data
svyintreg	[SVY] **svy estimators**	Censored and interval regression for survey data
svylogit	[SVY] **svy estimators**	Logistic regression for survey data
svyprobit	[SVY] **svy estimators**	Probit models for survey data
svymlogit	[SVY] **svy estimators**	Multinomial logistic regression for survey data
svyologit	[SVY] **svy estimators**	Ordered logistic regression for survey data
svyoprobit	[SVY] **svy estimators**	Ordered probit models for survey data
svypoisson	[SVY] **svy estimators**	Poisson regression for survey data
svynbreg	[SVY] **svy estimators**	Negative binomial regression for survey data
svygnbreg	[SVY] **svy estimators**	Generalized negative binomial regression for survey data
svyheckman	[SVY] **svy estimators**	Heckman selection model for survey data
svyheckprob	[SVY] **svy estimators**	Probit estimation with selection for survey data
lincom	[SVY] **lincom for svy**	Estimate linear combinations of parameters (e.g., differences of means, regression coefficients)
ml	[SVY] **ml for svy**	Pseudo-maximum-likelihood estimation for survey data
nlcom	[SVY] **nonlinear for svy**	Nonlinear combinations of parameters
predictnl	[SVY] **nonlinear for svy**	Nonlinear predictions for survey data
testnl	[SVY] **nonlinear for svy**	Nonlinear hypothesis tests for survey data
svydes	[SVY] **svydes**	Describe strata and PSUs of survey data
svymean	[SVY] **svymean**	Estimation of population and subpopulation means
svyprop	[SVY] **svymean**	Estimation of population and subpopulation proportions
svyratio	[SVY] **svymean**	Estimation of population and subpopulation ratios
svytotal	[SVY] **svymean**	Estimation of population and subpopulation totals
svyset	[SVY] **svyset**	Set variables for survey data
svytab	[SVY] **svytab**	Two-way tables for survey data
test	[SVY] **test for svy**	Linear hypothesis tests for survey data

Before using any of the svy commands, first take a quick look at [SVY] **svyset**. The svyset command allows one to specify a variable containing the sampling weights, strata and PSU identifier variables, and finite-population correction variable (if any). Once set, the svy commands will automatically use these design specifications until they are cleared or changed.

After the design variables are set, the svy estimation commands can be used in a manner essentially identical to that of the corresponding nonsurvey command. The svyregress command is the parallel of regress; svyivreg corresponds to ivreg; svyintreg corresponds to intreg (and can fit basic tobit models as well as fancier generalizations); svylogit corresponds to logit; svyprobit to probit; svymlogit to mlogit; svyologit to ologit; svyoprobit to oprobit; svypoisson to poisson; svynbreg to nbreg; svygnbreg to gnbreg; svyheckman to heckman; and svyheckprob to heckprob.

The [SVY] **svymean** entry describes the commands for the estimation of means, totals, ratios, and proportions: `svymean`, `svytotal`, `svyratio`, and `svyprop`.

The `svytab` command can be used to produce tests of independence for two-way contingency tables with survey data. `svytab` will also produce estimates of proportions with standard errors and confidence intervals.

All of the `svy` commands that compute standard errors use the "linearization" variance estimator—so called because it is based on a first-order Taylor series linear approximation. See the *Methods and Formulas* sections of [SVY] **svymean** and [SVY] **svy estimators** for full details.

The `_robust` command is a programmer's command that can be used to compute the linearization variance estimator after user-programmed estimators. If estimation is carried out using maximum pseudo-likelihood with `ml`, then `_robust` is not necessary since the variance calculations required for survey data may be obtained by simply specifying the appropriate options to [R] **ml**; see [R] **ml** for more details. Even if you are not interested in programming your own estimator, the [P] **_robust** entry may be worth reading since it contains an elementary introduction to the linearization variance estimator.

The post-estimation commands `test`, `testnl`, `lincom`, `nlcom`, and `predictnl` can be used after svy estimation commands. `test` will compute *p*-values for multidimensional hypothesis tests after any `svy` estimation command (namely, the commands described in [SVY] **svy estimators** and [SVY] **svymean**). A generalization of `test` is `testnl`, which allows the testing of hypotheses that are nonlinear in the estimated parameters.

The `lincom` command has many capabilities. It can be used after `svymean, by()` to do the equivalent of a two-sample *t* test for survey data. It can be used after `svylogit` to produce odds ratios for any linear combination of coefficients; after `svymlogit`, it can calculate relative risk ratios; and after `svypoisson`, incidence rate ratios. For general nonlinear combinations of coefficients and their standard errors, confidence intervals, etc., use `nlcom`.

A generalization of both `lincom` and `nlcom` is `predictnl`, which performs pointwise inference over the observations in the data.

The `svydes` command describes the strata and primary sampling units for a survey dataset, and is useful in tracking down the cause of the error message "stratum with only one PSU detected, r(460)".

Most of the `svy` commands appeared in the *Stata Technical Bulletin* (STB) before they appeared in a new release of Stata, and this pattern will likely be followed in the future, with new additions to the `svy` commands appearing in the *Stata Journal* between releases. See [U] **2.5 The Stata Journal and the Stata Technical Bulletin** for more information on the *Stata Journal* and the STB.

30.2 Accounting for the sample design in survey analyses

Why should one use these `svy` commands rather than, say, the `ci` command for means, or `regress` for linear regression? To answer this question, we need to discuss some of the characteristics of survey design and survey data collection because these characteristics affect how we must do our analysis if we want to "get it right".

Survey data generally have three important characteristics:

1. sampling weights—also called probability weights,

2. clustering, and

3. stratification.

These factors arise from the design of the data collection procedure. Here's a brief description of how these design features affect the analysis of the data:

1. *Sampling weights.* In sample surveys, observations are selected through a random process, but different observations may have different probabilities of selection. Weights are equal to (or proportional to) the inverse of the probability of being sampled. Various post-sampling adjustments to the weights are sometimes done as well. A weight of w_j for the jth observation means, roughly speaking, that the jth observation represents w_j elements in the population from which the sample was drawn.

   Including sampling weights in the analysis gives estimators that are approximately unbiased for whatever we are attempting to estimate in the full population. If we omit the weights in the analysis, our estimates may be very biased. Weights also affect the standard errors of our estimates.

2. *Clustering.* Observations are not sampled independently in almost all survey designs. Groups (for example, counties, city blocks, or households) may be sampled as a group—what we term a "cluster".

   There may also be further subsampling within the clusters. For example, counties may be sampled, then city blocks within counties, then households within city blocks, and then finally persons within households. The units at the first level of sampling are called "primary sampling units"—in this example, counties are the primary sampling units (PSUs). The PSUs play a special role in the analysis of the data, as we will discuss in the following subsection.

   Because of the sampling design, observations in the same cluster are not independent. If we use estimators that assume independence (e.g., `regress` or `regress, robust`), the standard errors may be too small—the difference can be as much as a factor of 2 or more. Accounting for clustering is necessary for "honest" estimates of standard errors, valid p-values, and confidence intervals whose true coverage is close to 95% (or whatever level you use).

3. *Stratification.* In surveys, different groups of clusters are often sampled separately. These groups are called strata. For example, the 254 counties of a state might be divided into two strata, say, urban counties and rural counties. Then ten counties might be sampled from the urban stratum, and fifteen from the rural stratum.

   Sampling is done independently across strata, with the stratum divisions fixed in advance. Thus, strata are statistically independent and can be analyzed as such. In many cases, this produces smaller (and honestly so) estimates of standard errors.

To put it succinctly: It is important to use sampling weights in order to get the point estimates right. We must consider the clustering and stratification of the survey design to get the standard errors right. If our analysis ignores the clustering in our design, we would likely produce standard errors that are much smaller than they should be. Stratification and weighting can also have a substantial effect on standard errors.

30.2.1 Multistage sample designs

Many surveys involve multistage sampling; i.e., cluster sampling with two or more levels of clustering. Earlier we mentioned an example of multistage sampling, with selection stages at the county, city block, and household levels.

The variance estimators in the current `svy` commands are suitable for use with multistage sample data. Note, however, that the variance estimates are based only on computations at the primary-sampling-unit level; i.e., the first stage of sampling. They do not require information about the secondary (and beyond) sampling units.

These variance estimators make minimal assumptions about the nature of the sample. They allow any amount of correlation within the primary sampling units. Thus, elements within a primary sampling unit do not have to be independent; that is, there can be secondary clustering. Hence, the variance estimators of the svy commands can be used for multistage designs. They produce variance estimates that generally will be either approximately unbiased or biased toward more conservative estimates; i.e., larger standard errors.

There are other variance estimation methods (not yet implemented in Stata) that explicitly account for secondary sampling. However, these methods require more assumptions than the svy variance estimator. If these assumptions are correct, then these other methods may yield more efficient variance estimates. However, the svy estimator, because it makes fewer assumptions, can be more robust than these methods.

30.2.2 Finite population corrections

Some surveys are based on simple random sampling *without* replacement of individual persons. Without-replacement sampling generally increases the precision of our estimators. If the number of persons sampled is large relative to the total number of persons in that stratum, then the increase in precision can be substantial.

For these cases, we can reflect this increase in precision by using a "finite population correction" term in our variance estimator. The svy commands will compute this correction when an fpc variable is set using svyset.

The same applies to PSUs, but only when all the elements of PSUs are, by design, included in the study. For example, if our PSUs were households and we included every member of the household in our study, then a finite population correction term would be appropriate when the households are sampled using simple random sampling without replacement within each stratum.

If, however, we subsampled within the household, the finite population correction calculation of the svy commands should not be used. Let's look at a simple example to illustrate why. Suppose that our survey design is a two stage cluster design and the population is made up of N PSUs and each PSU contains M secondary sampling units (SSUs). To keep it simple, assume there is one stratum and the sampling units are taken at random without replacement at both stages. Let y_{ij} be the measurements taken on each person sampled, $i = 1, \ldots, n$ and $j = 1, \ldots, m$; thus, we sampled n out of N PSUs and m out of M SSUs from each PSU; thus, $f_1 = n/N$ and $f_2 = m/M$. If $\bar{y}_i$ and $\bar{\bar{y}}$ represent the mean for PSU i and the overall mean, respectively, then we might estimate the variance of the sample mean by one of the following.

$$v_1 = s_1^2/n$$
$$v_2 = (1 - f_1)s_1^2/n$$
$$v_3 = (1 - f_1)s_1^2/n + (1 - f_2)s_2^2/nm$$

Here v_1 is the estimator used by svymean when the fpc option is not svyset, v_2 is the estimator used by svymean when the fpc option is svyset, v_3 is the unbiased estimate of the variance of the mean for this design as given by Theorem 10.2 of Cochran (1977), and the between/within PSU sample variances are

$$s_1^2 = \sum_{i=1}^{n} \frac{(\overline{y}_i - \overline{\overline{y}})^2}{n-1}$$

$$s_2^2 = \sum_{i=1}^{n} \sum_{j=1}^{m} \frac{(\overline{y}_{ij} - \overline{y}_i)^2}{n(m-1)}$$

Now, v_3 is unbiased. Thus,

$$E(v_3) = (1-f_1)S_1^2/n + (1-f_2)S_2^2/nm$$

where S_1^2 and S_2^2 are the population counterparts of s_1^2 and s_2^2, respectively. The bias of v_1 is

$$E(v_1) - E(v_3) = f_1 S_1^2/n$$

and thus v_1 is a conservative variance estimator; however, the bias of v_2 is

$$E(v_2) - E(v_3) = -f_1(1-f_2)S_2^2/mn$$

so v_2 will tend to be smaller than it should. Since this result easily generalizes to other multistage sampling designs, it is clear that the fpc option should not be svyset when analyzing survey data from multistage sampling designs.

30.2.3 Design effects: deff and deft

In addition to handling stratification and clustering effects, the svy commands have another special feature that other Stata estimation commands do not have: they calculate the design effects deff and deft.

The design effect deff is equal to the design-based variance estimate divided by an estimate of the variance we would have obtained if we had carried out a similar survey using simple random sampling. This ratio is a measure of how the survey design affects variance estimates. The related measure deft is approximately equal to the square root of deff.

See [SVY] **svymean** for detailed descriptions of deff and deft.

30.3 Example of the effects of weights, clustering, and stratification

Below we present an example of the effects of weights, clustering, and stratification. This is a typical case, but it is still, nevertheless, dangerous to draw general rules from any single example. One could find particular analyses from other surveys that are counterexamples for each of the trends for standard errors exhibited here.

We use data from the Second National Health and Nutrition Examination Survey (NHANES II) (McDowell et al. 1981) as our example. This is a national survey, and the dataset has sampling weights, strata, and clustering. In this example, we will consider the estimation of the mean serum zinc level of all adults in the U.S.

First, consider a proper design-based analysis, which accounts for weighting, clustering, and stratification. Before we issue our svy estimation command, we set the weight, strata, and PSU identifier variables:

```
. use http://www.stata-press.com/data/r8/nhanes2f

. svyset [pweight=finalwgt], strata(stratid) psu(psuid)
pweight is finalwgt
strata is stratid
psu is psuid
```

We now estimate the mean using the proper design-based analysis:

```
. svymean zinc

Survey mean estimation

pweight:  finalwgt                    Number of obs    =      9189
Strata:   stratid                     Number of strata =        31
PSU:      psuid                       Number of PSUs   =        62
                                      Population size  = 1.042e+08
```

Mean	Estimate	Std. Err.	[95% Conf. Interval]		Deff
zinc	87.18207	.4944827	86.17356	88.19057	10.3472

If we ignore the survey design and use `ci` to estimate the mean, we get

```
. ci zinc
```

Variable	Obs	Mean	Std. Err.	[95% Conf. Interval]	
zinc	9189	86.51518	.1510744	86.21904	86.81132

Note that the point estimate from the unweighted analysis is smaller by more than one standard error than the proper design-based estimate. In addition, note that design-based analysis produced a standard error that is 3.27 times larger than the standard error produced by our incorrect `ci` analysis.

30.3.1   Halfway isn't enough: the importance of stratification and clustering

When some people analyze survey data, they say, "I know I have to use my survey weights, but I'll just ignore the stratification and clustering information." If we follow this strategy, we will obtain the proper design-based point estimates, but our standard errors, confidence intervals, and test statistics will usually be wrong.

To illustrate this, suppose we used the `svymean` procedure with `pweight`s only.

```
. svyset [pweight=finalwgt], clear
pweight is finalwgt

. svymean zinc

Survey mean estimation

pweight:  finalwgt                    Number of obs    =      9189
Strata:   <one>                       Number of strata =         1
PSU:      <observations>              Number of PSUs   =      9189
                                      Population size  = 1.042e+08
```

Mean	Estimate	Std. Err.	[95% Conf. Interval]		Deff
zinc	87.18207	.1828747	86.82359	87.54054	1.415232

This gives us the same point estimate as our design-based analysis, but the reported standard error is less than one-half of the design-based standard error.

In addition, we emphasize that it is important to account for stratification and for weighting and clustering in the analysis of survey data. For example, if we only accounted for clustering and weights, and ignored stratification in NHANES II, we would obtain the following analysis:

```
. svyset, psu(psuid)
pweight is finalwgt
psu is psuid

. svymean zinc

Survey mean estimation

pweight:  finalwgt              Number of obs   =      9189
Strata:   <one>                 Number of strata =        1
PSU:      psuid                 Number of PSUs  =        62
                                Population size = 1.042e+08
```

Mean	Estimate	Std. Err.	[95% Conf. Interval]		Deff
zinc	87.18207	.4425806	86.29707	88.06706	8.289062

Here our standard error is still about 10% different from what we obtained in our proper design-based analysis.

❏ Technical Note

Although `ci` with `aweights` gives the same point estimates as `svymean` with `pweights`, it gives different standard errors.

```
. ci zinc [aweight=finalwgt]
```

Variable	Obs	Mean	Std. Err.	[95% Conf. Interval]	
zinc	9189	87.18207	.1537233	86.88074	87.4834

`svymean` with `pweights` gave a slightly different standard error of 0.183.

`ci` with `aweights` uses a different formula for the standard error. `svymean` with `pweights` handles sampling weights properly. Indeed, any command in Stata that allows `pweights` handles `pweights` properly in this regard. But, as we illustrated above, it is also important to include clustering and stratification in the analysis.

❏

30.4   Linear regression and other models

Let's look at a regression. We model zinc based on age, sex, race, and rural or urban residence. We compare a proper design-based analysis with an ordinary regression (which assumes i.i.d. error).

Here is our design-based analysis:

(Continued on next page)

```
. svyset [pweight=finalwgt], strata(stratid) psu(psuid)
pweight is finalwgt
strata is stratid
psu is psuid
. svyregress zinc age age2 weight female black orace rural
Survey linear regression
```

pweight:	finalwgt			Number of obs	=	9189
Strata:	stratid			Number of strata	=	31
PSU:	psuid			Number of PSUs	=	62
				Population size	=	1.042e+08
				F(7, 25)	=	62.50
				Prob > F	=	0.0000
				R-squared	=	0.0698

zinc	Coef.	Std. Err.	t	P>|t|	[95% Conf.	Interval]
age	-.1701161	.0844192	-2.02	0.053	-.3422901	.002058
age2	.0008744	.0008655	1.01	0.320	-.0008907	.0026396
weight	.0535225	.0139115	3.85	0.001	.0251499	.0818951
female	-6.134161	.4403625	-13.93	0.000	-7.032286	-5.236035
black	-2.881813	1.075958	-2.68	0.012	-5.076244	-.687381
orace	-4.118051	1.621171	-2.54	0.016	-7.424340	.8117590
rural	-.5386327	.6171836	-0.87	0.390	-1.797387	.7201216
_cons	92.47495	2.228263	41.50	0.000	87.93038	97.01952

If we had improperly ignored our survey weights, stratification, and clustering (i.e., if we used the usual Stata `regress` command), we would have obtained the following results:

```
. regress zinc age age2 weight female black orace rural
```

Source	SS	df	MS		Number of obs =	9189
					F(7, 9181) =	79.72
Model	110417.827	7	15773.9753		Prob > F =	0.0000
Residual	1816535.3	9181	197.85811		R-squared =	0.0573
					Adj R-squared =	0.0566
Total	1926953.13	9188	209.724982		Root MSE =	14.066

zinc	Coef.	Std. Err.	t	P>|t|	[95% Conf.	Interval]
age	-.090298	.0638452	-1.41	0.157	-.2154488	.0348528
age2	-.0000324	.0006788	-0.05	0.962	-.0013631	.0012983
weight	.0606481	.0105986	5.72	0.000	.0398725	.0814237
female	-5.021949	.3194705	-15.72	0.000	-5.648182	-4.395716
black	-2.311753	.5073536	-4.56	0.000	-3.306279	-1.317227
orace	-3.390879	1.060981	-3.20	0.001	-5.470637	-1.311121
rural	-.0966462	.3098948	-0.31	0.755	-.7041089	.5108165
_cons	89.49465	1.477528	60.57	0.000	86.59836	92.39093

The point estimates differ by 3–100%, and the standard errors for the proper designed-based analysis are 30–110% larger. The differences are not as dramatic as we saw with the estimation of the mean, but they are still very substantial.

30.5   Pseudo-likelihoods

Many of the svy estimators are based on maximum likelihood estimators; namely, svyheck-man, svyheckprob, svyintreg, svylogit, svyprobit, svymlogit, svyologit, svyoprobit, svypoisson, svynbreg, and svygnbreg. It should be noted that the "likelihood" for weighted or clustered designs is not a true likelihood; that is, the "likelihood" is not the distribution of the sample. When there is clustering, individual observations are no longer independent, and the "likelihood" does not reflect this. When there are sampling weights, the "likelihood" does not fully account for the "randomness" of the weighted sampling.

The "likelihood" for the svy maximum likelihood estimators is used only for the computation of the point estimates. It should not be used for variance estimation using standard formulas, and the standard likelihood-ratio test should not be used after the svy maximum likelihood estimators. This is why the svy maximum-likelihood commands do not display or save the log likelihood. Instead of likelihood-ratio tests, test should be used to calculate adjusted Wald tests.

30.6   Building your own survey estimator using ml

Despite the name "pseudo-likelihood", the maximization of a weighted likelihood (weights being the sampling weights) will produce correct, consistent point estimates of the parameters of interest in the case of survey data. Correct variance estimation, however, requires a post-estimation application of linearization techniques to the usual, non-survey, Hessian–based variance estimator.

When the maximum likelihood is carried out using ml, the weighting during estimation and post-estimation linearization is performed automatically provided that the user specifies the appropriate survey options to ml; see [R] **ml** for details.

30.7   Hypothesis testing using test and testnl

After any of the svy estimation commands, you can use test for hypothesis testing. The test command computes an adjusted Wald F test when used after a svy estimation command. This adjustment to the Wald test statistic is a simple multiplicative adjustment that accounts for the sample design. See [SVY] **test for svy** for formulas, references, and a description of the use of test.

Consider our previous example:

(Continued on next page)

```
. svyregress zinc age age2 weight female black orace rural

Survey linear regression

pweight:  finalwgt                              Number of obs    =      9189
Strata:   stratid                               Number of strata =        31
PSU:      psuid                                 Number of PSUs   =        62
                                                Population size  = 1.042e+08
                                                F(  7,    25)    =     62.50
                                                Prob > F         =    0.0000
                                                R-squared        =    0.0698
```

zinc	Coef.	Std. Err.	t	P>|t|	[95% Conf. Interval]	
age	-.1701161	.0844192	-2.02	0.053	-.3422901	.002058
age2	.0008744	.0008655	1.01	0.320	-.0008907	.0026396
weight	.0535225	.0139115	3.85	0.001	.0251499	.0818951
female	-6.134161	.4403625	-13.93	0.000	-7.032286	-5.236035
black	-2.881813	1.075958	-2.68	0.012	-5.076244	-.687381
orace	-4.118051	1.021101	-2.54	0.016	-7.424349	-.8117528
rural	-.5386327	.6171836	-0.87	0.390	-1.707797	.7201216
_cons	92.47495	2.220067	41.50	0.000	87.93038	97.01952

Since we fit an age term (age) and an age-squared term (age2), it would be reasonable to carry out a joint test for these coefficients:

```
. test age age2

Adjusted Wald test

 ( 1)   age = 0
 ( 2)   age2 = 0

       F(  2,    30) =    26.91
            Prob > F =   0.0000
```

Tested jointly, we see that the age terms have a p-value of less than 0.0001.

In the above example, test was used to test the two simultaneous linear hypotheses $H_o: \beta_{age} = 0$ and $\beta_{age2} = 0$. Suppose that we instead wanted a test of $H_o: -\beta_{age}/(2\beta_{age2}) = 70$, i.e., that the peak zinc level occurs at age 70. Such nonlinear tests are carried out using testnl:

```
. testnl -_b[age]/(2*_b[age2]) = 70

  (1)  -_b[age]/(2*_b[age2]) = 70

            F(1, 31) =        0.31
            Prob > F =      0.5811
```

and we see that there is insufficient evidence to reject the null hypothesis in this case. See [SVY] **non-linear for svy** for more information on nonlinear estimation and inference for survey data.

test and testnl can be used after any of the model estimation commands described in [SVY] **svy estimators**. They can also be used after svymean, svytotal, and svyratio; see [SVY] **test for svy**.

30.8   Estimation of linear and nonlinear combinations of parameters

In addition to test and testnl, there are the post-estimation commands called lincom and nlcom. lincom will display an estimate of a linear combination of parameters, along with its standard error, a confidence interval, and a test that the linear combination is zero. nlcom will do likewise, for nonlinear combinations of parameters.

Most commonly, lincom is used to compute the differences of two subpopulation means. For example, suppose we wish to estimate the difference of zinc levels in white males versus black males. First, we estimate the subpopulation means:

```
. generate male = (sex==1)

. svymean zinc, by(race) subpop(male)

Survey mean estimation

pweight:  finalwgt                        Number of obs    =       9189
Strata:   stratid                         Number of strata =         31
PSU:      psuid                           Number of PSUs   =         62
Subpop.:  male==1                         Population size  = 1.042e+08
```

Mean	Subpop.	Estimate	Std. Err.	[95% Conf. Interval]		Deff
zinc						
	White	91.16652	.5417394	90.06163	92.2714	4.814037
	Black	88.269	1.208336	85.80458	90.73342	2.474595
	Other	85.54716	2.608974	80.22612	90.8682	3.147332

Then we run lincom:

```
. lincom [zinc]White - [zinc]Black

 ( 1)  [zinc]White - [zinc]Black = 0
```

Mean	Estimate	Std. Err.	t	P>|t|	[95% Conf. Interval]	
(1)	2.897512	1.103773	2.63	0.013	.6463533	5.148671

Note that the t statistic and its p-value give a survey analysis equivalent of a two-sample t test.

Suppose that we instead wanted to estimate the ratio of the means for white males and black males

```
. nlcom ratio: [zinc]White / [zinc]Black

     ratio:  [zinc]White / [zinc]Black
```

Mean	Coef.	Std. Err.	t	P>|t|	[95% Conf. Interval]	
ratio	1.032826	.012908	80.01	0.000	1.0065	1.059152

lincom and nlcom can also be used after any of the model estimation commands described in [SVY] **svy estimators**. lincom can, for example, display results as odds ratios after svylogit, and can be used to compute odds ratios for one covariate group relative to another. nlcom can display odds ratios as well, and additionally allows more general nonlinear combinations of the parameters. See [SVY] **lincom for svy** and [SVY] **nonlinear for svy** for full details.

Finally, note that test, testnl, lincom, and nlcom all operate on the estimated parameters only. To obtain estimates and inference for functions of the parameters and of the data, such as for an exponentiated linear predictor or a predicted probability of success from a logit model, use predictnl; see [R] **predictnl**.

30.9   Two-way contingency tables

The tabulate command with iweights will produce tabulations for weighted data. It will not, however, compute tests of independence for two-way tables. For this, use the svytab command. This command produces tests of independence appropriate for complex survey data.

Here is an example of its use:

```
. svytab race diabetes, row

pweight:  finalwgt                   Number of obs      =       10335
Strata:   stratid                    Number of strata   =          31
PSU:      psuid                      Number of PSUs     =          62
                                     Population size    = 1.170e+08


  1=white,    diabetes, 1=yes,
  2=black,          0=no
  3=other        0       1   Total

    White     .968    .032       1
    Black     .941    .060       1
    Other    .9797   .0303       1

    Total    .9657   .0343       1

  Key:  row proportions

  Pearson:
     Uncorrected   chi2(2)         =      21.2661
     Design-based  F(1.52, 47.26)  =      14.9435      P = 0.0000
```

Actually, svytab has several capabilities that tabulate does not have. Not only can it display proportions (or percentages), but it can also compute standard errors and confidence intervals for these proportions:

(Continued on next page)

```
. svytab race diabetes, row se ci format(%7.4f)
```

pweight:	finalwgt	Number of obs	= 10335
Strata:	stratid	Number of strata	= 31
PSU:	psuid	Number of PSUs	= 62
		Population size	= 1.170e+08

1=white, 2=black, 3=other	diabetes, 1=yes, 0=no 0	1	Total
White	0.9680 (0.0020) [0.9637,0.9718]	0.0320 (0.0020) [0.0282,0.0363]	1.0000
Black	0.9410 (0.0061) [0.9271,0.9523]	0.0590 (0.0061) [0.0477,0.0729]	1.0000
Other	0.9797 (0.0076) [0.9566,0.9906]	0.0203 (0.0076) [0.0094,0.0434]	1.0000
Total	0.9657 (0.0018) [0.9618,0.9692]	0.0343 (0.0018) [0.0308,0.0382]	1.0000

```
  Key:  row proportions
        (standard errors of row proportions)
        [95% confidence intervals for row proportions]

  Pearson:
    Uncorrected   chi2(2)          =   21.2661
    Design-based  F(1.52, 47.26) =   14.9435     P = 0.0000
```

The test of independence that is displayed by default is based on the usual Pearson χ^2 statistic for two-way tables. To account for the survey design, the statistic is turned into an F statistic with noninteger degrees of freedom using a second-order Rao and Scott (1984) correction.

svytab will actually compute a total of eight statistics for the test of independence. It will compute a Rao-and-Scott corrected Pearson statistic and a Rao-and-Scott corrected likelihood-ratio statistic, using either of two variants of the correction, yielding four of the statistics. It will compute a "Pearson" Wald statistic and a log-linear Wald statistic, both either adjusted or unadjusted, yielding the other four statistics.

This dizzying array of statistics is not intended to dazzle the user. The two Wald statistics in their unadjusted form have been in use for many years, so we felt compelled to implement them in Stata. In many situations, however, they possess poor statistical properties. The adjusted variants of these Wald statistics have better statistical properties, but based on simulations (Sribney 1998), they do not appear to be as good as the best Rao-and-Scott corrected statistic, which is the default. Hence, we advise researchers to use the default statistic in all situations, and conversely, we recommend that the other statistics only be used for comparative or pedagogical purposes.

A summary of the properties of these statistics is given in the [SVY] **svytab** entry, but anyone wishing to see the results of the simulations should look at the article by Sribney (1998).

30.10   Differences between the svy commands and other commands

As we mentioned earlier, all of the svy model estimation commands (the commands documented in [SVY] **svy estimators**) have corresponding non-svy commands: svyregress corresponds to regress, svylogit to logit, etc. The corresponding non-svy commands all allow pweights and have a cluster() option that corresponds to the psu() option of the svy commands. When pweights, cluster(), or the robust option is specified, the non-svy commands use the same robust (linearization) variance estimator as the svy commands.

The point estimates from the svy commands are exactly the same as the weighted point estimates from the non-svy commands.

Despite their similarities, the svy commands and corresponding non-svy commands have a number of differences:

1. All of the svy commands handle stratified sampling, but none of the non-svy commands do. Since stratification usually makes standard errors smaller, ignoring stratification is usually conservative. So, using the non-svy commands for stratified sampling is not a terrible thing to do. However, to get the smallest possible "honest" standard error estimates for stratified sampling, use the svy commands.

2. All of the svy commands use t-statistics with $n - L$ degrees of freedom for testing the significance of coefficients, where n is the total number of sampled PSUs (clusters) and L is the number of strata. Some of the non-svy commands use t-statistics, but most use z-statistics. If the non-svy command uses z-statistics for its standard variance estimator, then it also uses z-statistics with the robust (linearization) variance estimator. Strictly speaking, t-statistics are always appropriate with the robust (linearization) variance estimator; see [P] **_robust** for the theoretical rationale. But, using z rather than t-statistics only yields a nontrivial difference when there is a small number of clusters (< 50). If a non-svy command uses t-statistics and the cluster() option is specified, then the degrees of freedom used are the same as that of an svy command.

3. svy commands produce adjusted Wald tests for the model test, and test can be used to produce adjusted Wald tests for other hypotheses. Non-svy commands can only produce unadjusted Wald tests. The adjustment can be important when the degrees of freedom $n - L$ are small relative to the dimension of the test. (If the dimension is one, then the adjusted and unadjusted Wald tests are identical.) This fact along with point (2) makes it important to use the svy command if the number of sampled PSUs (clusters) is small (< 50).

4. svyregress differs slightly from regress and svyivreg differs slightly from ivreg in that they use different multipliers for the variance estimator. regress and ivreg use a multiplier of $\{(N - 1)/(N - k)\}\{n/(n - 1)\}$, where N is the number of observations, n is the number of clusters (PSUs), and k is the number of regressors including the constant. svyregress and svyivreg use a multiplier of merely $n/(n - 1)$. Thus, they produce slightly different standard errors. The $(N - 1)/(N - k)$ is ad hoc and has no rigorous theoretical justification; hence, the purist svy commands do not use it. The svy commands tacitly assume that $N \gg k$. If $(N - 1)/(N - k)$ is not close to 1, then you may be well advised to use regress or ivreg, so that some punishment is inflicted on your variance estimates. Note that maximum likelihood estimators in Stata (e.g., logit) do no such adjustment, but rely on the sensibilities of the analyst to ensure that N is reasonably larger than k. Thus, the svy maximum-likelihood estimators (e.g., svylogit) produce exactly the same standard errors as the corresponding non-svy commands (e.g., logit), but p-values are slightly different because of point (2).

5. svy commands can produce proper estimates for subpopulations using the subpop() option. Use of an if restriction with svy or non-svy commands can yield incorrect standard error estimates for subpopulations. Often, an if restriction will yield exactly the same standard error as subpop(); most other times, the two standard errors will be slightly different; but, in some cases—usually for thinly sampled subpopulations—the standard errors can be appreciably different. Hence, only svy commands with the subpop() option should be used to obtain estimates for thinly sampled subpopulations. See [SVY] **svymean** for more information.

6. svy commands handle zero sampling weights properly. Non-svy commands ignore any observation with a weight of zero. Usually, this will yield exactly the same standard errors, but sometimes they will differ. Sampling weights of zero can arise from various post-sampling adjustment procedures. If the sum of weights for one or more PSUs is zero, svy and non-svy commands will produce different standard errors, but usually this difference is very small.

7. You can svyset iweights and let these weights be negative. Negative sampling weights can arise from various post-sampling adjustment procedures. If you want to use negative sampling weights, then you must svyset iweights instead of pweights; no other commands will allow negative sampling weights.

8. Only the svy commands will compute finite population corrections (FPC). Finite population corrections are only justified for some special sampling designs. Omitting the FPC is conservative, so failing to specify an FPC cannot be harshly condemned. See [U] **30.2.2 Finite population corrections** earlier in this chapter for details.

9. Only the svy commands will compute the design effects deff and deft.

30.11   References

Cochran, W. G. 1977. *Sampling Techniques*. 3d ed. New York: John Wiley & Sons.

McDowell, A. A., A. Engel, J. T. Massey, and K. Maurer. 1981. Plan and operation of the Second National Health and Nutrition Examination Survey, 1976–1980. *Vital and Health Statistics* 15(1). National Center for Health Statistics, Hyattsville, MD.

Rao, J. N. K. and A. J. Scott. 1984. On chi-squared tests for multiway contingency tables with cell proportions estimated from survey data. *Annals of Statistics* 12: 46–60.

Sribney, W. M. 1998. svy7: Two-way contingency tables for survey or clustered data. *Stata Technical Bulletin* 45: 33–49. Reprinted in *Stata Technical Bulletin Reprints*, vol. 8, pp. 297–322.

31 Commands everyone should know

Contents

31.1   Forty-one commands
31.2   The by construct

31.1 Forty-one commands

Putting aside the statistical commands that might particularly interest you, here are 41 commands everyone should know:

Getting online help
    help, net search, search   [U] **8 Stata's online help and search facilities**

Keeping Stata up to date
    ado, net, update   [U] **32 Using the Internet to keep up to date**

Operating system interface
    pwd, cd   [R] **cd**

Using and saving data from disk
    use, save   [R] **save**
    append, merge   [U] **25 Commands for combining data**
    compress   [R] **compress**

Inputting data into Stata   [U] **24 Commands to input data**
    input   [R] **input**
    edit   [R] **edit**
    infile   [R] **infile (free format)**; [R] **infile (fixed format)**
    infix   [R] **infix (fixed format)**
    insheet   [R] **insheet**

Basic data reporting
    describe   [R] **describe**
    codebook   [R] **codebook**
    list   [R] **list**
    browse   [R] **edit**
    count   [R] **count**
    inspect   [R] **inspect**
    table   [R] **table**
    tabulate   [R] **tabulate**

(Continued on next page)

Data manipulation	[U] **16 Functions and expressions**
generate, replace	[R] **generate**
egen	[R] **egen**
rename	[R] **rename**
drop, keep	[R] **drop**
sort	[R] **sort**
encode, decode	[R] **encode**
order	[R] **order**
by	[U] **14.5 by varlist: construct**
reshape	[R] **reshape**
Keeping track of your work	
log	[U] **18 Printing and preserving output**
notes	[R] **notes**
Convenience	
display	[R] **display**

31.2   The by construct

If you do not understand the by *varlist*: construct, _n, and _N, and their interaction, and if you process data where observations are related, you are missing out on something. See

[U] **16.7 Explicit subscripting**
[U] **14.5 by varlist: construct**

For instance, say you have a dataset with multiple observations per person, and you want the average value of each person's blood pressure (bp) for the day. You could

```
. egen avgbp = mean(bp), by(person)
```

but you should understand that you could also

```
. by person, sort: gen avgbp = sum(bp)/_N
. by person: replace avgbp = bp[_N]
```

Yes, typing two commands is more work than typing just one, but understanding the two-command construct is the key to generating more complicated things that no one ever thought about adding to egen.

For instance, say your dataset also contains time recording when each observation was made. If you want to add the total time the person is under observation (last time minus first time) to each observation, type

```
. by person (time), sort: gen ttl = time[_N]-time[1]
```

Or, suppose you want to add how long it has been since the person was last observed to each observation:

```
. by person (time), sort: gen howlong = time - time[_n-1]
```

If instead you wanted how long it would be until the next observation, type

```
. by person (time), sort: gen whennext = time[_n+1] - time
```

by *varlist*:, _n, and _N are often the solution to difficult calculations.

32 Using the Internet to keep up to date

Contents

32.1 Overview
32.2 Sharing datasets (and other files)
32.3 Official updates
    32.3.1 Example
    32.3.2 Updating ado-files
    32.3.3 Frequently asked questions about updating the ado-files
    32.3.4 Updating the executable
    32.3.5 Frequently asked questions about updating the executable
    32.3.6 Updating both ado-files and the executable
32.4 Downloading and managing additions by users
    32.4.1 Downloading files
    32.4.2 Managing files
    32.4.3 Finding files to download
32.5 Making your own download site

32.1 Overview

Stata has the ability to read files over the Internet. Just to prove that to yourself, type the following:

```
. use http://www.stata.com/manual/chapter32, clear
```

You have just reached out and gotten a dataset from our web site. The dataset is not in HTML format, nor does this have anything to do with your browser. We just copied the Stata data file chapter32.dta onto our server, and now people all over the world can use it. If you have a web page, you can do the same thing. It is a very convenient way to share datasets with colleagues.

Now type the following:

```
. update from http://www.stata.com
```

We promise nothing bad will happen. update will read a short file from www.stata.com that will allow Stata to report whether your copy of Stata is up to date. Is your copy up to date? Now you know. If it is not, we will show you how to update it—it is no more difficult than typing update.

Now type the following:

```
. net from http://www.stata.com
```

That will go to www.stata.com and tell you what is available from our user-download site. The material there is not official, but it is useful. More usefully, type

```
. search kernel regression, net
```

or equivalently,

```
. net search kernel regression
```

That will search the entire web for additions to Stata having to do with kernel regression, whether it be from the *Stata Journal*, *Stata Technical Bulletin*, Statalist, archive sites, or user private sites.

To summarize, Stata has the ability to read files over the Internet:

1. You can share datasets, do-files, etc., with colleagues all over the world. This requires no special expertise, but you do need to have a web page.

2. You can update Stata; it is free, easy, and nearly instant.

3. You can find and add new features to Stata; it is also free, easy, and nearly instant.

Finally, you can create a site to distribute new features for Stata.

32.2   Sharing datasets (and other files)

There is just nothing to it: you copy the file as-is (in binary) onto the server and then let your colleagues know the file is there. This works for .dta files, .do files, .ado files, and, in fact, all files.

On the receiving end, you can use the file (if it is a .dta dataset) or you can copy it:

```
. use http://www.stata.com/manual/chapter32, clear
. copy http://www.stata.com/manual/chapter32.dta mycopy.dta
```

Stata includes a copy-file command and it works over the Internet just as use does; see [R] **copy**.

❏ Technical Note

If you are concerned about transmission errors, you can create a checksum file before you copy the file onto the server. In placing chapter32.dta on our site, we started with chapter32.dta in our working directory and typed

```
. checksum chapter32.dta, save
```

This created the new file chapter32.sum. We then placed both files on our server. We did not have to create this second file, but, since we did, when you use the data, Stata will be able to detect transmission errors and warn you if there are problems.

How would Stata know? chapter32.sum is a very short file containing the result of a mathematical calculation made on the contents of chapter32.dta. When your Stata receives chapter32.dta, it repeats the calculation and then compares that result with what is recorded in chapter32.sum. If the results are different, then there must have been a transmission error.

Whether you create a checksum file is optional.

See [R] **checksum**.

❏

32.3   Official updates

Although we follow no formal schedule for the release of updates, the fact is that we update Stata about once every two weeks. You do not have to update every that often, although we recommend that you do. There are two ways to check whether your copy of Stata is up to date:

type

```
. update query
```

or

Pull down **Help** and select **Official Updates**
Click on *http://www.stata.com*

After that, you will either

type:	or:
. update ado	click on *update ado-files*

or

type:	or:
. update executable	click on *update executable*

or

type:	or:
. update all	click on *update ado-files and executable*

and which, if any, of those things need doing will be obvious.

After you have updated your Stata, to find out what has changed

type:	or:
. help whatsnew	Pull down **Help** and select **What's New?**

32.3.1 Example

When you type update from http://www.stata.com or when you pull down **Help**, select **Official Updates**, and click on *http://www.stata.com*, Stata presents a report:

```
. update from http://www.stata.com
(contacting http://www.stata.com)

Stata executable
    folder:                C:\STATA\
    name of file:          wstata.exe
    currently installed:   04 Nov 2002
    latest available:      04 Nov 2002

Ado-file updates
    folder:                C:\STATA\ado\updates\
    names of files:        (various)
    currently installed:   01 Mar 2003
    latest available:      01 Mar 2003

Recommendation
    Do nothing; all files up-to-date.
```

There are two components of official Stata: the binary Stata executable and the ado-files that we shipped with it. Ado-files are just programs written in Stata. For instance, when you use generate, you are using a command that was compiled into the Stata executable. When you use stcox, you are using a command that was implemented as an ado-file.

Both components of our Stata are up to date.

(Continued on next page)

32.3.2 Updating ado-files

When you obtain the above report, you might see

```
. update from http://www.stata.com
(contacting http://www.stata.com)
Stata executable
      folder:               C:\STATA\
      name of file:         wstata.exe
      currently installed:  04 Nov 2002
      latest available:     04 Nov 2002
Ado-file updates
      folder:               C:\STATA\ado\updates\
      names of files:       (various)
      currently installed:  04 Nov 2002
      latest available:     01 May 2003
Recommendation
      Type -update ado-
```

If you go with the point-and-click alternative, at the bottom of the screen you will see

```
Recommendation
      update ado-files
```

where *update ado-files* is in blue and is therefore clickable.

Anyway, what you are to do next is type `update ado` or click on *update ado-files*. Either way, you will see something like the following:

```
. update ado
(contacting http://www.stata.com)
Ado-file update log
      1.   verifying C:\STATA\ado\updates\ is writeable
      2.   obtaining list of files to be updated
      3.   downloading relevant files to temporary area
           downloading filename.ado
           downloading filename.hlp
           ...
           downloading filename.ado
      4.   examining files
      5.   installing files
      6.   setting last date updated
Updates successfully installed.

Recommendation
      Type -help whatsnew- to learn about the new features
```

That is all there is to it, but do type `help whatsnew` to learn about the new features. (If you go the point-and-click path, click on *whatsnew*.)

Here is what happens if you type `update ado` and you are already up to date:

```
. update ado
(contacting http://www.stata.com)
ado-files already up to date
```

32.3.3   Frequently asked questions about updating the ado-files

1. Could something go wrong and make my Stata become unusable?

    No. The updates are copied to a temporary place on your computer, Stata examines them to make sure they are complete before copying them to the official place. Thus, either the updates are installed or they are not.

2. I do not believe you. Pretend that something you did not anticipate goes wrong, such as the power fails at the instant Stata is doing the local disk to local disk copy.

    If the improbable should happen, you can erase the update directory and then your Stata is back to being just as it was shipped. Updates go into a different directory than the originals and the originals are never erased.

    Stata tells you where it is installing your updates. You can also find out by typing sysdir. The directory you want is the one listed opposite UPDATES.

    (By the way, power failure should not cause a problem; the marker that the update is applied is set last, so you could also just type update ado again and Stata would refetch the partially installed update.)

3. How much is downloaded?

    A typical update is 100k to 300k. Ado-files are small; the biggest file that is copied is probably the database for search.

4. I am using Unix or a networked version of Stata. When I try to update ado, I am told that the directory is not writeable. Can I copy the updates into another directory and then copy them to the official directory myself?

    Yes, assuming you are a system administrator. Type 'update ado, into(*dirname*)'. Stata will download the updates just as it would ordinarily, but will place them in the directory you specify. We recommend that *dirname* be a new, empty directory, because later you will need to copy the entire contents of the directory to the official place. The official place is the directory listed next to UPDATES if you type sysdir. When you copy the files, copy over any existing files. Previously existing files in the official update directory are just previous updates. Also remember to make the files globally readable if necessary. See [R] **update**.

32.3.4   Updating the executable

Ado-file updates are released every other week; updates for the executable are rarer than that. If the executable needs updating, Stata will mention it when you type update:

```
. update from http://www.stata.com
(contacting http://www.stata.com)
Stata executable
     folder:              C:\STATA\
     name of file:        wstata.exe
     currently installed: 04 Nov 2002
     latest available:    12 Feb 2003
Ado-file updates
     folder:              C:\STATA\ado\updates\
     names of files:      (various)
     currently installed: 01 Nov 2002
     latest available:    01 Nov 2002
Recommendation
     Type -update executable-
```

Here is what happens when you type `update executable`:

```
. update executable
(contacting http://www.stata.com)

Executable update log
    1.  verifying C:\STATA\ is writeable
    2.  downloading new executable

New executable successfully downloaded

Instructions
    1.  Type -update swap-
```

Just follow the instructions, which will vary depending on your computer. In this case, `update swap` is a command that automatically copies the newly downloaded executable over the current one. It then briefly restarts Stata to begin using the new executable.

32.3.5   Frequently asked questions about updating the executable

1. If I understand this, `update executable` does not really install the update; it just copies one file onto my computer, and that one file happens to be the new executable, right?

   > Probably. There can be more than one file such as a DLL. All the files are copied to the same place. In the case where no DLLs are downloaded, this statement is true. On most systems, using the `update swap` command after downloading a new executable will perform the copy and actually install the executable.

2. How big is the downloaded file?

   > 2 to 5 megabytes, depending on operating system.

3. What happens if I type `update executable` and my executable is already up to date?

   > Nothing. You are told "executable already up to date".

4. I am using Unix or a networked version of Stata. When I try to `update executable`, I am told that the directory is not writeable. Can I download the updated executable to another directory and then copy it to the official directory myself?

   > Yes, assuming you are a system administrator.
   > Type '`update executable, into(`*dirname*`)`'. We recommend that *dirname* be a new, empty directory, because there may be more than one file and later you will need to copy all of them to the official place. The official place is the directory listed next to STATA if you type `sysdir`. When you copy the files, copy over any existing files; we recommend that you make a backup of the originals first. See [R] **update**.

32.3.6   Updating both ado-files and the executable

When you type `update`, you may be told you need to update both ado-files and executable:

```
. update from http://www.stata.com
(contacting http://www.stata.com)

Stata executable
    folder:             C:\STATA\
    name of file:       wstata.exe
    currently installed: 04 Nov 2002
    latest available:   22 Apr 2003
```

```
Ado-file updates
    folder:                C:\STATA\ado\updates\
    names of files:        (various)
    currently installed:   01 Mar 2003
    latest available:      01 May 2003
Recommendation
    Type -update all-
```

Typing update all is the same as typing update ado and then typing update executable. You could type the separate commands if you preferred. The order does not matter.

Note, you could skip the update from step. You could just type update all and follow the instructions. If nothing needed updating, you would see

```
. update all

> update ado
(contacting http://www.stata.com)
ado-files already up to date

> update executable
(contacting http://www.stata.com)
executable already up to date

. _
```

32.4 Downloading and managing additions by users

Try the following:

type

. net from http://www.stata.com

or

Pull down **Help** and select **SJ and User-written Programs**
Click on *http://www.stata.com*

32.4.1 Downloading files

We are not the only ones developing additions to Stata. Stata is supported by a large and highly competent user community. An important part of this is the *Stata Journal* (SJ) and the *Stata Technical Bulletin* (STB). The *Stata Journal* is a refereed, quarterly journal containing articles of interest to Stata users. For more details and subscription information, visit the *Stata Journal* web site at *http://www.stata-journal.com/*.

The *Stata Journal* is a printed and electronic journal with corresponding software. If you want the journal, you must subscribe, but the software is available for free from our web site at *http://www.stata-journal.com*.

The predecessor to the *Stata Journal* is the *Stata Technical Bulletin* (STB). The STB was also a printed and electronic journal with corresponding software. Individual STB journals may still be purchased. The STB software is available for free from our web site at *http://www.stata.com*.

Below are instructions for installing the *Stata Journal* and the *Stata Technical Bulletin* software from our web site.

Installing the Stata Journal software

1. Pull down **Help** and select **SJ and User-written Programs**.

2. Click on *http://www.stata-journal.com/software*.

3. Click on *sj2-2*.

4. Click on *st0001_1*.

5. Click on *click here to install*.

or type

1. Type: `. net from http://www.stata-journal.com/software`

2. Type: `. net cd sj2-2`

3. Type: `. net describe st0001_1`

4. Type: `. net install st0001_1`

The above could be shortened to

```
. net from http://www.stata-journal.com/software/sj2-2
. net describe st0001_1
. net install st0001_1
```

Alternatively, you could type

```
. net sj 2-2
. net describe st0001_1
. net install st0001_1
```

Installing the STB software

1. Pull down **Help** and select **SJ and User-written Programs**.

2. Click on *www.stata.com*.

3. Click on *stb*.

4. Click on *stb58*.

5. Click on *sg84_3*.

6. Click on *click here to install*.

or type

1. Type: `. net from http://www.stata.com`

2. Type: `. net cd stb`

3. Type: `. net cd stb58`

4. Type: `. net describe sg84_3`

5. Type: `. net install sg84_3`

The above could be shortened to

```
. net from http://www.stata.com/stb/stb58
. net describe sg84_3
. net install sg84_3
```

32.4.2 Managing files

You now have the `concord` command, because we just downloaded and installed it. Convince yourself of this by typing

. help concord

and you might try it out, too. Let's now list the additions you have installed—that is probably just `concord`—and then get rid of `concord`.

In command mode, you can type

. ado dir
[1] package sg84_3 from http://www.stata.com/stb/stb58
       STB-58 sg84_3. Concordance correlation coefficient: minor corrections

If you had more additions installed, they would be listed. Now knowing that you have *sg84_3* installed, you can obtain a more thorough description by typing

. ado describe sg84_3

---

[1] package **sg84_3** from http://www.stata.com/stb/stb58

TITLE
       STB-58 sg84_3. Concordance correlation coefficient: minor corrections
DESCRIPTION/AUTHOR(S)
       STB insert by Thomas J. Steichen, RJRT
                     Nicholas J. Cox, University of Durham, UK
       Support: steicht@rjrt.com, n.j.cox@durham.ac.uk
       After installation, see help **concord**
INSTALLATION FILES
       c/concord.ado
       c/concord.hlp
INSTALLED ON
       5 Oct 2002

You can erase *sg84_3* by typing

. ado uninstall sg84_3
package **sg84_3** from http://www.stata.com/stb/stb58
       STB-58 sg84_3. Concordance correlation coefficient: minor corrections
(package uninstalled)

You can do all of this from the point-and-click interface, too. Pull down **Help** and select **SJ and User-written Programs** and then click on *List*. From there, you can click on *sg84_3* to see the detailed description of the package and from there you can click on *click here to uninstall* if you want to erase it.

For more information on the `ado` command and the corresponding menu, see [R] **net**.

32.4.3 Finding files to download

There are two ways to find useful files to download. One is simply to thumb through sites. That is inefficient but entertaining. If you want to do that,

1. Pull down **Help** and choose **SJ and User-written Programs**.

2. Click on *http://www.stata.com*.

3. Click on *links*.

What you are doing is starting at our download site and then working out from there. We maintain a list of other sites and those sites will have more links. You can do this from command mode, too:

```
. net from http://www.stata.com
. net cd links
```

The efficient way to find files—at least if you know what you are looking for—is to search. There are two ways to do that. If you suspect what you are looking for might already be in Stata (or published in the SJ), use Stata's `search` command:

```
. search concordance correlation
```

Equivalently, you could pull down **Help** and select **Search**. Either way, you will learn about *sg84_3* and you can even click to install it.

If you want to search for additions over the net, which is to say, the SJ and archive sites and user sites, type

```
. net search concordance correlation
```

or pull down **Help**, select **Search**, and this time click *Search net resources*, rather than the default "*Search documentation and FAQs*".

32.5   Making your own download site

There are two reasons you may wish to create your own download site:

1. You have datasets and the like, you want to share them with colleagues, and you want to make it easier for colleagues to download the files.

2. You have written Stata programs, etc., that you wish to share with the Stata user community.

Making a download site is easy; the full instructions are found in [R] **net**.

At the beginning of this chapter, we pretended that you had a dataset you wanted to share with colleagues. We said you just had to copy the dataset onto your server and then let your colleagues know the dataset is there.

Let's now pretend that you had two datasets, `ds1.dta` and `ds2.dta`, and you wanted your colleagues to be able to learn about and fetch the datasets using the `net` command or by pulling down **Help** and selecting **SJ and User-written Programs**.

First, you would copy the datasets to your homepage just as before. Then you would create three more files, one to describe your site named `stata.toc` and two more to describe each "package" you want to provide:

```
———————————————————————————————————————— top of stata.toc ————————
v 3
d My name and affiliation (or whatever other title I choose)
d Datasets for the PAR study
p ds1 The base dataset
p ds2 The detail dataset
———————————————————————————————————————— end of stata.toc ————————
```

```
———————————————————————————————————————— top of ds1.pkg ————————
v 3
d ds1.  The base dataset
d My name or whatever else I wanted to put
d This dataset contains the baseline values for ...
p ds1.dta
———————————————————————————————————————— end of ds1.pkg ————————
```

```
————————————————————————————————— top of ds2.pkg ————————
v 3
d ds1.  The detail dataset
d My name or whatever else I wanted to put
d This dataset contains the follow-up information ...
p ds2.dta
————————————————————————————————— end of ds2.pkg ————————
```

Here is what users would see when they went to your site:

```
. net from http://www.myuni.edu/hande/~aparker

http://www.myuni.edu/hande/~aparker
My name and whatever else I wanted to put

Datasets for the PAR study
PACKAGES you could -net describe-:
    ds1                 The base dataset
    ds2                 The detail dataset

. net describe ds1

package ds1 from http://www.myuni.edu/hande/~aparker

TITLE
      ds1.  The base dataset
DESCRIPTION/AUTHOR(S)
      My name and whatever else I wanted to put
      This dataset contains the baseline values for ...
ANCILLARY FILES                                (type net get ds1)
      ds1.dta

. net get ds1
checking ds1 consistency and verifying not already installed...

copying into current directory...
      copying  ds1.dta
ancillary files successfully copied.

. —
```

See [R] **net**.

Combined Author Index

This is the combined author index for this manual and for the *Stata Base Reference Manual*, the *Stata Cluster Analysis Reference Manual*, the *Stata Cross-Sectional Time-Series Reference Manual*, the *Stata Programming Reference Manual*, the *Stata Survey Data Reference Manual*, the *Stata Survival Analysis & Epidemiological Tables Reference Manual*, and the *Stata Time-Series Reference Manual*.

A

Aalen, O. O., [ST] **sts**

Abraham, B., [TS] **tssmooth**, [TS] **tssmooth dexponential**, [TS] **tssmooth exponential**, [TS] **tssmooth hwinters**, [TS] **tssmooth shwinters**

Abraira-García, L., [ST] **epitab**

Abramowitz, M., [R] **functions**, [R] **orthog**, [U] **16 Functions and expressions**

Abrams, K. R., [R] **meta**

Abramson, J. H., [R] **kappa**, [R] **meta**

Abramson, Z. H., [R] **kappa**, [R] **meta**

Afifi, A. A., [R] **anova**, [R] **sw**

Agresti, A., [R] **tabulate**

Aigner, D. J., [R] **frontier**, [XT] **xtfrontier**

Aisbett, C. W., [ST] **stcox**, [ST] **streg**

Aitchison, J., [R] **ologit**, [R] **oprobit**

Aitken, A. C., [R] **reg3**

Aivazian, S. A., [R] **ksmirnov**

Akaike, H., [R] **estimates**, [R] **glm**, [ST] **streg**, [TS] **varsoc**

Albert, P. S., [XT] **xtgee**

Aldenderfer, M. S., [CL] **cluster**

Aldrich, J. H., [R] **logit**, [R] **mlogit**, [R] **probit**

Alexandersson, A., [R] **regress**

Alf, E., [R] **roc**

Alldredge, J. R., [R] **pk**, [R] **pkcross**, [R] **pkequiv**

Allen, M. J., [R] **alpha**

Allison, M. J., [R] **manova**

Allison, P. D., [R] **rologit**, [R] **testnl**, [ST] **discrete**

Altman, D. G., [R] **anova**, [R] **fracpoly**, [R] **kappa**, [R] **kwallis**, [R] **meta**, [R] **mfp**, [R] **nptrend**, [R] **oneway**

Ambler, G., [R] **fracpoly**, [R] **mfp**, [R] **regress**

Amemiya, T., [R] **glogit**, [R] **nlogit**, [R] **tobit**, [TS] **varsoc**, [XT] **intro**, [XT] **xthtaylor**, [XT] **xtivreg**,

Amisano, G., [TS] **var intro**, [TS] **var svar**, [TS] **vargranger**, [TS] **varirf create**, [TS] **varwle**

Anderberg, M. R., [CL] **cluster**

Andersen, E. B., [R] **clogit**

Anderson, E., [P] **matrix eigenvalues**

Anderson, J. A., [R] **ologit**

Anderson, R. E., [R] **rologit**

Anderson, T. W., [R] **manova**, [XT] **xtabond**, [XT] **xtivreg**

Andrews, D. F., [R] **egen**, [R] **manova**, [R] **rreg**

Anscombe, F. J., [R] **glm**

Ansley, C. F., [TS] **arima**

Arbuthnott, J., [R] **signrank**

Arellano, M., [XT] **xtabond**

Arminger, G., [R] **suest**

Armitage, P., [R] **means**, [ST] **ltable**

Armstrong, R. D., [R] **qreg**

Arora, S. S., [XT] **xtivreg**, [XT] **xtreg**

Arthur, B. S., [R] **symmetry**

Atkinson, A. C., [R] **boxcox**, [R] **nl**

Azen, S. P., [R] **anova**

B

Babiker, A., [R] **sampsi**

Bai, Z., [P] **matrix eigenvalues**

Baker, R. J., [R] **glm**

Balakrishnan, N., [R] **functions**

Balanger, A., [R] **sktest**

Balestra, P., [XT] **xtivreg**

Baltagi, B. H., [R] **hausman**, [R] **ivreg**, [XT] **xtabond**, [XT] **xthtaylor**, [XT] **xtivreg**, [XT] **xtreg**, [XT] **xtregar**

Bamber, D., [R] **roc**

Bancroft, T. A., [R] **sw**

Barnard, G. A., [R] **spearman**

Barnow, B., [R] **treatreg**

Barnwell, B. G., [SVY] **svytab**

Barrison, I. G., [R] **binreg**

Bartlett, M. S., [R] **factor**, [R] **oneway**, [TS] **wntestb**

Basmann, R. L., [R] **ivreg**

Bassett, G., Jr., [R] **qreg**

Battese, G. E., [XT] **intro**, [XT] **xtfrontier**

Baum, C. F., [R] **net**, [R] **net search**, [R] **regression diagnostics**, [R] **ssc**, [TS] **arch**, [TS] **arima**, [TS] **dfgls**, [TS] **regression diagnostics**, [TS] **tsset**, [TS] **wntestq**, [XT] **xtgls**, [XT] **xtreg**

Beale, E. M. L., [R] **sw**, [R] **test**, [SVY] **test for svy**

Beaton, A. E., [R] **rreg**

Beck, N., [XT] **xtgls**, [XT] **xtpcse**

Becketti, S., [R] **fracpoly**, [R] **runtest**, [R] **spearman**, [P] **pause**, [TS] **corrgram**

Beerstecher, E., [R] **manova**

Begg, C. B., [R] **meta**

Beggs, S., [R] **rologit**

Belle, G. van, [R] **dstdize**, [ST] **epitab**

Belsley, D. A., [R] **regress**, [R] **regression diagnostics**, [U] **21 Programming Stata**

Bendel, R. B., [R] **sw**

Beniger, J. R., [R] **cumul**

Bera, A. K., [TS] **arch**, [TS] **varnorm**, [XT] **xtreg**

Berk, K. N., [R] **sw**

Berk, R. A., [R] **rreg**

Berkson, J., [R] **logit**, [R] **probit**
Berndt, E. K., [R] **glm**, [TS] **arch**, [TS] **arima**
Berndt, E. R., [R] **treatreg**, [R] **truncreg**
Bernstein, I. H., [R] **alpha**
Berry, G., [R] **means**, [ST] **ltable**
Beyer, W. H., [R] **qc**
Bhargava, A., [XT] **xtregar**
Bibby, J. M., [R] **manova**
Bickel, P. J., [R] **egen**, [R] **rreg**
Bickenböller, H., [R] **symmetry**
Bieler, G. S., [SVY] **svytab**
Binder, D. A., [U] **23 Estimation and post-estimation commands**, [P] **_robust**, [SVY] **svy estimators**
Birdsall, T. G., [R] **logistic**
Bischof, C., [P] **matrix eigenvalues**
Black, F., [TS] **arch**
Black, W. C., [R] **rologit**
Blackford, S., [P] **matrix eigenvalues**
Bland, M., [R] **signrank**
Blashfield, R. K., [CL] **cluster**
Bleda, M. J., [R] **alpha**
Bliss, C. I., [R] **probit**
Bloch, D. A., [R] **brier**
Bloomfield, P., [R] **qreg**
BMDP, [R] **symmetry**
Boice, J. D., [R] **bitest**, [ST] **epitab**
Bollen, K. A., [R] **regression diagnostics**
Bollerslev, T., [TS] **arch**, [TS] **arima**
Bond, S., [XT] **xtabond**
Bortkewitsch, L. von, [R] **poisson**
Bowerman, B., [TS] **tssmooth**, [TS] **tssmooth dexponential**, [TS] **tssmooth exponential**, [TS] **tssmooth hwinters**, [TS] **tssmooth shwinters**
Bowker, A. H., [R] **symmetry**
Box, G. E. P., [R] **anova**, [R] **boxcox**, [R] **lnskew0**, [R] **manova**, [TS] **arch**, [TS] **arima**, [TS] **corrgram**, [TS] **cumsp**, [TS] **dfuller**, [TS] **pergram**, [TS] **pperron**, [TS] **wntestq**, [TS] **xcorr**
Boyd, N. F., [R] **kappa**
Bradburn, M. J., [R] **meta**
Brady, A. R., [R] **logistic**, [R] **spikeplot**
Brady, T., [R] **edit**
Brant, R., [R] **ologit**
Breslow, N. E., [R] **clogit**, [R] **dstdize**, [R] **symmetry**, [ST] **epitab**, [ST] **stcox**, [ST] **sts test**,
Breusch, T., [R] **mvreg**, [R] **regression diagnostics**, [R] **sureg**, [XT] **xtreg**
Brier, G. W., [R] **brier**
Broeck, J. van den, [XT] **xtfrontier**
Brook, R., [R] **brier**
Brown, D. R., [R] **anova**, [R] **loneway**, [R] **oneway**
Brown, J. D., [R] **manova**
Brown, M. B., [R] **sdtest**
Brown, S. E., [R] **symmetry**
Bru, B., [R] **poisson**

Buchner, D. M., [R] **ladder**
Burnam, M. A., [R] **lincom**, [R] **mlogit**, [R] **predictnl**
Burr, I. W., [R] **qc**
Buskens, V., [R] **tabstat**

C

Cain, G., [R] **treatreg**
Caliński, T., [CL] **cluster**, [CL] **cluster stop**
Cameron, A. C., [R] **nbreg**, [R] **poisson**, [R] **regression diagnostics**, [XT] **xtnbreg**, [XT] **xtpoisson**
Campbell, M. J., [R] **estimates**, [R] **kappa**, [R] **logistic**, [R] **poisson**, [R] **tabulate**
Cardell, S., [R] **rologit**
Carlile, T., [R] **kappa**
Carlin, J., [R] **means**, [ST] **epitab**
Carroll, R. J., [R] **boxcox**, [R] **rreg**, [R] **sdtest**
Carter, S. L., [R] **frontier**, [R] **lrtest**, [R] **nbreg**, [ST] **streg**, [XT] **xt**
Caudill, S. B., [R] **frontier**, [XT] **xtfrontier**
Chadwick, J., [R] **poisson**
Chamberlain, G., [R] **clogit**
Chambers, J. M., [R] **diagnostic plots**, [R] **grmeanby**, [R] **lowess**
Chang, Y., [XT] **xtivreg**
Charlett, A., [R] **fracpoly**
Chatfield, C., [TS] **corrgram**, [TS] **pergram**, [TS] **tssmooth**, [TS] **tssmooth dexponential**, [TS] **tssmooth exponential**, [TS] **tssmooth hwinters**, [TS] **tssmooth ma**, [TS] **tssmooth shwinters**
Chatterjee, S., [R] **poisson**, [R] **regress**, [R] **regression diagnostics**, [TS] **prais**
Cheung, Y., [TS] **dfgls**
Chiang, C. L., [ST] **ltable**
Choi, B. C. K., [R] **roc**
Chou, R. Y., [TS] **arch**
Chow, S. C., [R] **pk**, [R] **pkcross**, [R] **pkequiv**, [R] **pkexamine**, [R] **pkshape**
Christakis, N., [R] **rologit**
Christiano, L. J., [TS] **varirf create**
Clark, V. A., [ST] **ltable**
Clarke, M. R. B., [R] **factor**
Clarke, R. D., [R] **poisson**
Clarke-Pearson, D. L., [R] **roc**
Clayton, D., [R] **cloglog**, [R] **cumul**, [R] **egen**, [R] **impute**, [ST] **epitab**, [ST] **stptime**, [ST] **strate**, [ST] **stsplit**, [ST] **sttocc**
Clerget-Darpoux, F., [R] **symmetry**
Cleveland, W. S., [R] **diagnostic plots**, [R] **lowess**
Cleves, M. A., [R] **binreg**, [R] **dstdize**, [R] **logistic**, [R] **logit**, [R] **roc**, [R] **sdtest**, [R] **symmetry**, [ST] **intro**, [ST] **st**, [ST] **stcox**, [ST] **stdes**, [ST] **streg**, [ST] **sts**, [ST] **stset**, [ST] **stsplit**, [ST] **stvary**
Clogg, C. C., [R] **suest**
Cobb, G. W., [R] **anova**

Cochran, W. G., [R] anova, [R] correlate, [R] dstdize,
  [R] means, [R] oneway, [R] poisson
  [R] signrank, [U] 30 Overview of survey
  estimation, [SVY] svy, [SVY] svy estimators,
  [SVY] svymean

Cochrane, D., [TS] prais

Coelli, T. J., [R] frontier, [XT] intro, [XT] xtfrontier

Cohen, J., [R] kappa

Coleman, J. S., [R] poisson

Collett, D., [R] clogit, [R] logistic, [ST] stci,
  [ST] streg, [ST] sts test, [ST] stsplit

Cong, R., [R] tobit, [R] treatreg, [R] truncreg

Conover, W. J., [R] centile, [R] ksmirnov, [R] kwallis,
  [R] nptrend, [R] sdtest, [R] spearman,
  [R] tabulate

Conroy, R. M., [R] adjust

Conway, M. R., [XT] xtlogit, [XT] xtprobit

Cook, R. D., [R] boxcox, [R] regress, [R] regression
  diagnostics, [P] _predict

Cooper, M. C., [CL] cluster, [CL] cluster
  programming subroutines, [CL] cluster stop

Cornfield, J., [ST] epitab

Cornwell, C., [XT] xthtaylor

Cox, D. R., [R] boxcox, [R] lnskew0, [ST] ltable,
  [ST] stcox, [ST] streg, [ST] sts

Cox, N. J., [R] contract, [R] cumul, [R] describe,
  [R] destring, [R] diagnostic plots, [R] drop,
  [R] duplicates, [R] egen, [R] encode,
  [R] kappa, [R] lorenz, [R] net, [R] net search,
  [R] regression diagnostics, [R] rename,
  [R] sample, [R] separate, [R] serrbar,
  [R] smooth, [R] spikeplot, [R] split, [R] ssc,
  [R] tabulate, [U] 1 Read this—it will help,
  [P] matrix define

Cramér, H., [R] tabulate

Cronbach, L. J., [R] alpha

Crowder, M. J., [ST] streg

Crowley, J., [ST] stcox, [ST] stset

Csaki, F., [TS] varsoc

Cui, J., [R] symmetry

Curts-Garcia, J., [R] smooth

Cutler, S. J., [ST] ltable

Cuzick, J., [R] kappa, [R] nptrend

Czekanowski, J., [CL] cluster

D

D'Agostino, R. B., [R] sktest

D'Agostino, R. B., Jr., [R] sktest

Daniel, C., [R] diagnostic plots

Danuso, F., [R] nl

David, H. A., [R] egen

David, J. S., [TS] arima

Davidson, R., [R] boxcox, [R] ivreg, [R] nl,
  [R] regress, [R] truncreg, [TS] arch,
  [TS] arima, [TS] regression diagnostics,
  [TS] varlmar [XT] xtgls, [XT] xtpcse

Davison, A. C., [R] bootstrap

Day, N. E., [R] clogit, [R] dstdize, [R] symmetry,
  [ST] epitab

Day, W. H. E., [CL] cluster, [CL] cluster
  singlelinkage, [CL] cluster programming
  utilities

De Stavola, B. L., [ST] stcox, [ST] stset

Deaton, A., [U] 23 Estimation and post-estimation
  commands

Deeks, J. J., [R] meta

DeLong, D. M., [R] roc

DeLong, E. R., [R] roc

Demmel, J., [P] matrix eigenvalues

Detsky, A. S., [R] meta

Dice, L. R., [CL] cluster

Dickens, R., [TS] prais

Dickey, D. A., [TS] dfgls, [TS] dfuller, [TS] pperron

Diggle, P. J., [TS] arima, [TS] wntestq

DiNardo, J., [R] heckman, [R] ivreg, [R] logit,
  [R] probit, [R] regress, [R] simulate, [R] tobit,
  [TS] arch, [TS] prais, [TS] regression
  diagnostics

Ding, Z., [TS] arch

Dixon, W. J., [R] ttest

Dobson, A., [R] glm

Doll, R., [R] poisson, [ST] epitab

Dongarra, J., [P] matrix eigenvalues

Donner, A., [R] loneway

Dore, C. J., [R] fracpoly

Dorfman, D. D., [R] roc

Draper, N. R., [R] regress, [R] sw

Drukker, D. M., [R] boxcox, [R] frontier, [R] lrtest,
  [R] nbreg, [R] treatreg, [R] tobit, [ST] streg,
  [XT] xt

Du Croz, J., [P] matrix eigenvalues

Duan, N., [R] heckman

Dubes, R. C., [CL] cluster

Duda, R. O., [CL] cluster, [CL] cluster stop

Duncan, A. J., [R] qc

Dunn, G., [R] kappa

Dupont, W. D., [ST] epitab, [ST] stcox, [ST] stir

Durbin, J., [R] regression diagnostics, [TS] prais,
  [TS] regression diagnostics

Duval, R. D., [R] bootstrap, [R] jknife

Dwyer, J., [XT] xtreg

E

Edelsbrunner, H., [CL] cluster, [CL] cluster
  singlelinkage

Ederer, E., [ST] ltable

Edgington, E. S., [R] runtest

Edwards, A. L., [R] anova, [R] correlate

Efron, B., [R] bootstrap, [R] qreg

Efroymson, M. A., [R] sw

Egger, M., [R] meta

Eichenbaum, M., [TS] varirf create

Eisenhart, C., [R] **runtest**

Elliot, G., [TS] **dfgls**

Ellis, C. D., [R] **poisson**

Eltinge, J. L., [R] **test**, [SVY] **svy**, [SVY] **lincom for svy**, [SVY] **svy estimators**, [SVY] **svydes**, [SVY] **svymean**, [SVY] **test for svy**

Emerson, J. D., [R] **lv**, [R] **stem**

Enders, W., [TS] **arch**, [TS] **arima**

Engel, A., [U] **30 Overview of survey estimation**, [SVY] **lincom for svy**, [SVY] **svy estimators**, [SVY] **svydes**, [SVY] **svymean**, [SVY] **svytab**

Engle, R. F., [TS] **arch**, [TS] **arima**, [TS] **regression diagnostics**

Erdreich, L. S., [R] **roc**

Esman, R. M., [R] **egen**

Evans, A. S., [ST] **epitab**

Evans, C. L., [TS] **varirf create**

Evans, M. A., [R] **pk**, [R] **pkcross**, [R] **pkequiv**

Everitt, B. S., [R] **anova**, [R] **glm**, [R] **pca**, [CL] **cluster**, [CL] **cluster stop**

Ewens, W. J., [R] **symmetry**

Ezekiel, M., [R] **regression diagnostics**

F

Feinleib, M., [XT] **xtreg**

Feiveson, A. H., [R] **nlcom**, [R] **signrank**

Feldt, L. S., [R] **anova**

Feller, W., [R] **ci**, [R] **nbreg**, [R] **poisson**, [TS] **wntestb**

Feltbower, R., [ST] **epitab**

Ferri, H. A., [R] **kappa**

Fienberg, S. E., [R] **tabulate**

Filon, L. N. G., [R] **correlate**

Findley, T. W., [R] **ladder**

Finney, D. J., [R] **tabulate**

Fiser, D. H., [R] **logistic**

Fishell, E., [R] **kappa**

Fisher, L. D., [R] **dstdize**, [ST] **epitab**

Fisher, M. R., [XT] **xtcloglog**, [XT] **xtgee**, [XT] **xtintreg**, [XT] **xtlogit**, [XT] **xtprobit**, [XT] **xttobit**

Fisher, R. A., [R] **anova**, [R] **signrank**, [R] **tabulate**, [ST] **streg**

Flannery, B. P., [R] **functions**, [R] **range**, [R] **vwls**, [U] **16 Functions and expressions**, [P] **matrix symeigen**, [TS] **arch**, [TS] **arima**

Fleiss, J. L., [R] **dstdize**, [R] **kappa**, [R] **sampsi**, [ST] **epitab**

Fleming, T. R., [ST] **stcox**, [ST] **sts test**

Ford, J. M., [R] **frontier**, [XT] **xtfrontier**

Forsythe, A. B., [R] **sdtest**

Forthofer, R., [R] **dstdize**

Foster, A., [R] **ivreg**, [R] **regress**

Fourier, J. B. J., [R] **cumul**

Fox, J., [R] **kdensity**, [R] **lv**

Fox, W. C., [R] **logistic**

Francia, R. S., [R] **swilk**

Frankel, M. R., [U] **23 Estimation and post-estimation commands**, [P] **_robust**, [SVY] **svy estimators**

Franzese, R. J., Jr., [XT] **xtpcse**

Franzini, L., [XT] **xtregar**

Frechette, G. R., [XT] **xtprobit**

Freeman, D. H., Jr., [SVY] **svytab**

Freeman, J., [ST] **epitab**

Freeman, J. L., [SVY] **svytab**

Freese, J., [R] **clogit**, [R] **cloglog**, [R] **estimates**, [R] **logistic**, [R] **logit**, [R] **mlogit**, [R] **nbreg**, [R] **ologit**, [R] **oprobit**, [R] **poisson**, [R] **probit**, [R] **regress**, [R] **regression diagnostics**, [R] **zip**, [U] **23 Estimation and post-estimation commands**

Friedman, M., [TS] **arima**

Frison, L., [R] **sampsi**

Frome, E. L., [R] **qreg**

Fu, V. K., [R] **ologit**

Fuller, W. A., [R] **regress**, [U] **23 Estimation and post-estimation commands**, [P] **_robust**, [SVY] **svy estimators**, [SVY] **svytab**, [TS] **dfgls**, [TS] **dfuller**, [TS] **pperron**

Fyler, D. C., [ST] **epitab**

G

Gail, M. H., [U] **23 Estimation and post-estimation commands**, [P] **_robust**

Galbraith, R. F., [R] **meta**

Gall, J. R., Le, [R] **logistic**

Gallant, A. R., [R] **nl**

Gallup, J. L., [R] **estimates**

Galton, F., [R] **correlate**, [R] **cumul**

Gan, F. F., [R] **diagnostic plots**

Gange, S. J., [XT] **xtcloglog**, [XT] **xtgee**, [XT] **xtintreg**, [XT] **xtlogit**, [XT] **xtprobit**, [XT] **xttobit**

Gardiner, J. S., [TS] **tssmooth**, [TS] **tssmooth dexponential**, [TS] **tssmooth exponential**, [TS] **tssmooth hwinters**, [TS] **tssmooth shwinters**

Gardner, E. S., Jr., [TS] **tssmooth dexponential**, [TS] **tssmooth hwinters**

Garrett, J. M., [R] **adjust**, [R] **fracpoly**, [R] **logistic**, [R] **regress**, [ST] **stcox**, [ST] **stphplot**, [SVY] **svy**

Gauvreau, K., [R] **dstdize**, [R] **logistic**, [R] **sampsi**, [R] **sdtest**, [ST] **ltable**, [ST] **sts**

Gehan, E. A., [ST] **sts test**

Geisser, S., [R] **anova**

Gentle, J. E., [R] **anova**, [R] **nl**

Gerkins, V. R., [R] **symmetry**

Giannini, C., [TS] **var intro**, [TS] **var svar**, [TS] **vargranger**, [TS] **varirf create**, [TS] **varwle**

Gibbons, J. D., [R] **ksmirnov**, [R] **spearman**

Glass, R. I., [ST] **epitab**

Gleason, J. R., [R] **anova**, [R] **bootstrap**, [R] **cf**, [R] **ci**, [R] **correlate**, [R] **describe**, [R] **generate**, [R] **infile (fixed format)**, [R] **label**, [R] **loneway**, [R] **notes**, [R] **order**, [R] **summarize**, [R] **ttest**, [U] **16 Functions and expressions**, [ST] **epitab**

Gloeckler, L., [ST] **discrete**

Glosten, L. R., [TS] **arch**

Gnanadesikan, R., [R] **cumul**, [R] **diagnostic plots**, [R] **manova**

Godambe, V. P., [SVY] **svy estimators**

Godfrey, L., [TS] **regression diagnostics**

Goeden, G. B., [R] **kdensity**

Goldberger, A. S., [R] **tobit**, [R] **treatreg**

Goldblatt, A., [ST] **epitab**

Goldstein, R., [R] **brier**, [R] **correlate**, [R] **egen**, [R] **estimates**, [R] **impute**, [R] **lorenz**, [R] **nl**, [R] **ologit**, [R] **oprobit**, [R] **regression diagnostics**, [R] **signrank**

Golub, G. H., [R] **orthog**

González, J. F., Jr., [SVY] **svy estimators**, [SVY] **svymean**

Good, P. I., [R] **permute**, [R] **symmetry**, [R] **tabulate**

Goodall, C., [R] **lowess**, [R] **rreg**

Goodman, L. A., [R] **tabulate**

Gordon, A. D., [CL] **cluster**, [CL] **cluster stop**

Gordon, M. G., [R] **binreg**

Gorman, J. W., [R] **sw**

Gosset, W. S., [R] **ttest**

Gould, W. W., [R] **bootstrap**, [R] **destring**, [R] **egen**, [R] **frontier**, [R] **grmeanby**, [R] **icd9**, [R] **infile (fixed format)**, [R] **jknife**, [R] **kappa**, [R] **logistic**, [R] **maximize**, [R] **mkspline**, [R] **ml**, [R] **net search**, [R] **nlcom**, [R] **ologit**, [R] **oprobit**, [R] **predictnl**, [R] **qreg**, [R] **range**, [R] **reshape**, [R] **rreg**, [R] **simulate**, [R] **sktest**, [R] **smooth**, [R] **swilk**, [R] **testnl**, [U] **21 Programming Stata**, [U] **29 Overview of Stata estimation commands**, [P] **matrix mkmat**, [P] **postfile**, [P] **_robust**, [ST] **intro**, [ST] **stcox**, [ST] **stdes**, [ST] **streg**, [ST] **stset**, [ST] **stsplit**, [ST] **stvary**, [XT] **xtfrontier**

Gourieroux, C., [R] **hausman**, [R] **suest**, [R] **test**, [SVY] **test for svy**, [TS] **arima**

Govindarajulu, Z., [R] **functions**, [U] **16 Functions and expressions**

Gower, J. C., [CL] **cluster**

Grambsch, P. M., [ST] **stcox**

Granger, C. W. J., [TS] **arch**, [TS] **vargranger**

Graubard, B. I., [R] **test**, [SVY] **svy**, [SVY] **svy estimators**, [SVY] **svymean**, [SVY] **svytab**, [SVY] **test for svy**

Graybill, F. A., [R] **centile**

Grazia-Valsecchi, M., [ST] **sts test**

Green, D. M., [R] **logistic**

Green, P. E., [CL] **cluster**

Greenbaum, A., [P] **matrix eigenvalues**

Greene, W. H., [R] **biprobit**, [R] **clogit**, [R] **frontier**, [R] **heckman**, [R] **heckprob**, [R] **hetprob**, [R] **mkspline**, [R] **mlogit**, [R] **nlogit**, [R] **ologit**, [R] **oprobit**, [R] **reg3**, [R] **sureg**, [R] **testnl**, [R] **treatreg**, [R] **truncreg**, [R] **zip**, [P] **matrix accum**, [TS] **arch**, [TS] **arima**, [TS] **regression diagnostics**, [TS] **var**, [XT] **xtgls**, [XT] **xtpcse**, [XT] **xtpoisson**, [XT] **xtrchh**, [XT] **xtreg**

Greenfield, S., [R] **alpha**, [R] **factor**, [R] **lincom**, [R] **mlogit**, [R] **predictnl**

Greenhouse, S., [R] **anova**

Greenland, S., [R] **ci**, [R] **glogit**, [R] **ologit**, [R] **poisson**, [ST] **epitab**

Greenwood, M., [ST] **ltable**, [ST] **sts**

Gregoire, A., [R] **kappa**

Griffith, J. L., [R] **brier**

Griffiths, W. E., [R] **glogit**, [R] **logit**, [R] **lrtest**, [R] **maximize**, [R] **mlogit**, [R] **oneway**, [R] **probit**, [R] **regression diagnostics**, [R] **test**, [XT] **xtgls**, [XT] **xtpcse**, [XT] **xtreg**, [SVY] **test for svy**, [TS] **arch**, [TS] **prais**

Griliches, Z., [XT] **xtgls**, [XT] **xtnbreg**, [XT] **xtpcse**, [XT] **xtpoisson**, [XT] **xtrchh**

Grizzle, J. E., [R] **vwls**

Gronau, R., [R] **heckman**

Gropper, D. M., [R] **frontier**, [XT] **xtfrontier**

Gross, A. J., [ST] **ltable**

Grunfeld, Y., [XT] **xtgls**, [XT] **xtpcse**, [XT] **xtrchh**

Guilford, J. P., [CL] **cluster**

Guilkey, D. K., [XT] **xtprobit**

Gutierrez, R. G., [R] **frontier**, [R] **lrtest**, [R] **nbreg**, [ST] **intro**, [ST] **stcox**, [ST] **stdes**, [ST] **streg**, [ST] **stset**, [ST] **stsplit**, [ST] **stvary**, [XT] **xt**

H

Hadi, A. S., [R] **poisson**, [R] **regress**, [R] **regression diagnostics**, [U] **21 Programming Stata**, [TS] **prais**

Hadorn, D., [R] **brier**

Haenszel, W., [ST] **epitab**, [ST] **sts test**

Hair, J. E., Jr., [R] **rologit**

Hakkio, C. S., [R] **egen**, [TS] **pperron**

Hald, A., [R] **qreg**

Hall, A. D., [R] **frontier**

Hall, B. H., [R] **glm**, [TS] **arch**, [TS] **arima**, [XT] **xtnbreg**, [XT] **xtpoisson**

Hall, P., [R] **bootstrap**

Hall, R. E., [R] **glm**, [TS] **arch**, [TS] **arima**

Hall, W. J., [R] **roc**

Halley, E., [ST] **ltable**

Halvorsen, K., [R] **tabulate**

Hamerle, A., [R] **clogit**

Hamilton, J. D., [P] **matrix eigenvalues**, [TS] **arch**,
　　[TS] **arima**, [TS] **corrgram**, [TS] **dfuller**,
　　[TS] **pergram**, [TS] **pperron**, [TS] **time series**,
　　[TS] **var**, [TS] **var intro**, [TS] **var svar**,
　　[TS] **varfcast compute**, [TS] **vargranger**,
　　[TS] **varirf create**, [TS] **varnorm**, [TS] **varsoc**,
　　[TS] **varstable**, [TS] **varwle**, [TS] **xcorr**

Hamilton, L. C., [R] **bootstrap**, [R] **ci**, [R] **diagnostic
　　plots**, [R] **factor**, [R] **ladder**, [R] **lv**, [R] **mlogit**,
　　[R] **pca**, [R] **regress**, [R] **regression diagnostics**,
　　[R] **rreg**, [R] **simulate**, [R] **summarize**,
　　[R] **xpose**

Hamman, U., [CL] **cluster**

Hammarling, S., [P] **matrix eigenvalues**

Hampel, F. R., [R] **egen**, [R] **rreg**

Hanley, J. A., [R] **roc**

Hansen, L. P., [XT] **xtabond**

Harabasz, J., [CL] **cluster**, [CL] **cluster stop**

Hardin, J. W., [R] **binreg**, [R] **biprobit**, [R] **glm**,
　　[R] **regression diagnostics**, [R] **statsby**,
　　[TS] **newey**, [TS] **prais**, [XT] **xtgee**,
　　[XT] **xtpoisson** [XT] **xtrchh**

Haritou, A., [R] **suest**

Harman, H. H., [R] **factor**

Harrell, F. E., Jr., [R] **ologit**

Harrington, D. P., [ST] **stcox**, [ST] **sts test**

Harris, R. L., [R] **qc**

Harris, T., [R] **qreg**

Harrison, J. A., [R] **dstdize**

Hart, P. E., [CL] **cluster**, [CL] **cluster stop**

Harvey, A. C., [R] **hetprob**, [TS] **arch**, [TS] **arima**,
　　[TS] **prais**, [TS] **var svar**

Hastie, T. J., [R] **grmeanby**

Hauck, W. W., [XT] **xtcloglog**, [XT] **xtlogit**,
　　[XT] **xtprobit**

Hausman, J. A., [R] **glm**, [R] **hausman**, [R] **nlogit**,
　　[R] **rologit**, [R] **suest**, [TS] **arch**, [TS] **arima**,
　　[XT] **intro**, [XT] **xthtaylor**, [XT] **xtnbreg**,
　　[XT] **xtreg**

Haynam, G. E., [R] **functions**, [U] **16 Functions and
　　expressions**

Hays, R. D., [R] **lincom**, [R] **mlogit**, [R] **prcdictnl**

Heckman, J., [R] **biprobit**, [R] **heckman**, [R] **heckprob**

Heinecke, K., [P] **matrix mkmat**

Heinonen, O. P., [ST] **epitab**

Heiss, F., [R] **nlogit**

Henderson, B. E., [R] **symmetry**

Hendrickson, A. E., [R] **factor**

Hendrickx, J., [R] **mlogit**, [R] **xi**

Henry-Amar, M., [ST] **ltable**

Herzberg, A. M., [R] **manova**

Hess, K. R., [ST] **stphplot**

Hickam, D. H., [R] **brier**

Higbee, K. T., [R] **adjust**

Higgins, J. E., [R] **anova**

Higgins, M. L., [TS] **arch**

Hilbe, J., [R] **cloglog**, [R] **glm**, [R] **logistic**,
　　[R] **manova**, [R] **nbreg**, [R] **poisson**, [R] **probit**,
　　[R] **sampsi**, [CL] **cluster**, [XT] **xtgee**,
　　[XT] **xtpoisson**

Hildreth, C., [TS] **prais**, [XT] **xtrchh**

Hill, A. B., [R] **poisson**, [ST] **epitab**

Hill, R. C., [R] **glogit**, [R] **logit**, [R] **lrtest**,
　　[R] **maximize**, [R] **mlogit**, [R] **oneway**,
　　[R] **probit**, [R] **regression diagnostics**, [R] **test**,
　　[SVY] **test for svy**, [TS] **arch**, [TS] **prais**,
　　[XT] **xtgls**, [XT] **xtpcse**, [XT] **xtreg**,

Hills, M., [R] **cloglog**, [R] **cumul**, [R] **egen**,
　　[ST] **epitab**, [ST] **stcox**, [ST] **stptime**,
　　[ST] **strate**, [ST] **stset**, [ST] **stsplit**, [ST] **sttocc**

Hinkley, D. V., [R] **bootstrap**

Hipel, K. W., [TS] **arima**

Hoaglin, D. C., [R] **diagnostic plots**, [R] **lv**,
　　[R] **regression diagnostics**, [R] **smooth**, [R] **stem**

Hochberg, Y., [R] **oneway**

Hocking, R. R., [R] **sw**

Hoel, P. G., [R] **bitest**, [R] **ci**, [R] **sdtest**, [R] **ttest**

Holloway, L., [R] **brier**

Holm, C., [R] **test**, [SVY] **test for svy**

Holmgren, J., [ST] **epitab**

Holt, C. C., [TS] **tssmooth**, [TS] **tssmooth
　　dexponential**, [TS] **tssmooth exponential**,
　　[TS] **tssmooth hwinters**, [TS] **tssmooth
　　shwinters**

Holt, D., [SVY] **svy**, [SVY] **svymean**

Honoré, B., [XT] **xttobit**

Hood, W. C., [R] **ivreg**

Horst, P., [R] **factor**

Hosmer, D. W., Jr., [R] **adjust**, [R] **clogit**, [R] **glm**,
　　[R] **glogit**, [R] **lincom**, [R] **logistic**, [R] **logit**,
　　[R] **lrtest**, [R] **mlogit**, [R] **predictnl**, [R] **sw**,
　　[ST] **stcox**, [ST] **streg**, [XT] **xtgee**

Hossain, K. M. B., [ST] **epitab**

Hotelling, H., [R] **canon**, [R] **hotelling**, [R] **manova**,
　　[R] **pca**, [R] **roc**

Houck, D., [XT] **xtrchh**

Hougaard, P., [ST] **streg**

Hsiao, C., [XT] **xtabond**, [XT] **xtivreg**, [XT] **xtregar**

Hu, M., [ST] **stcox**, [ST] **stset**

Huang, D. S., [R] **sureg**

Huber, P. J., [R] **egen**, [R] **qreg**, [R] **rreg**, [R] **suest**,
　　[U] **23 Estimation and post-estimation
　　commands**, [P] **_robust**

Hughes, J. B., [R] **manova**

Huq, M. I., [ST] **epitab**

Hurd, M., [R] **tobit**

Huynh, H., [R] **anova**

Continued on next page

I

Iglewicz, B., [R] **lv**
Irala-Estévez, J. de, [R] **logistic**
Isaacs, D., [R] **fracpoly**
Ishiguro, M., [R] **estimates**

J

Jaccard, P., [CL] **cluster**
Jackman, R. W., [R] **regression diagnostics**
Jackson, J. E., [R] **pca**
Jacobs, K. B., [R] **symmetry**
Jacobs, M., [R] **duplicates**
Jagannathan, R., [TS] **arch**
Jain, A. K., [CL] **cluster**
Jang, D. S., [SVY] **svymean**
Jarque, C. M., [TS] **varnorm**
Jeffreys, H., [R] **spearman**
Jenkins, G. M., [TS] **arima**, [TS] **corrgram**,
[TS] **cumsp**, [TS] **dfuller**, [TS] **pergram**,
[TS] **pperron**, [TS] **xcorr**
Jenkins, S. P., [R] **lorenz**, [R] **rename**, [ST] **discrete**,
[ST] **stcox**
Jick, H., [ST] **epitab**
Johansen, S., [TS] **varlmar**
Johnson, C. L., [SVY] **svydes**
Johnson, D. E., [R] **anova**, [R] **manova**
Johnson, L. A., [TS] **tssmooth**, [TS] **tssmooth
dexponential**, [TS] **tssmooth exponential**,
[TS] **tssmooth hwinters**, [TS] **tssmooth
shwinters**
Johnson, M. E., [R] **sdtest**
Johnson, M. M., [R] **sdtest**
Johnson, N. L., [R] **functions**, [R] **ksmirnov**,
[R] **means**
Johnson, W., [SVY] **svy estimators**, [SVY] **svymean**
Johnston, J., [R] **heckman**, [R] **ivreg**, [R] **logit**,
[R] **probit**, [R] **regress**, [R] **simulate**, [R] **tobit**,
[TS] **arch**, [TS] **prais**, [TS] **regression
diagnostics**, [XT] **xtrchh**
Jolliffe, D., [R] **ivreg**, [R] **lorenz**, [R] **qreg**, [R] **regress**
Jones, D. R., [R] **meta**
Jones, M. C., [R] **kdensity**
Judge, G. G., [R] **glogit**, [R] **logit**, [R] **lrtest**,
[R] **maximize**, [R] **mlogit**, [R] **oneway**,
[R] **probit**, [R] **regression diagnostics**, [R] **test**,
[SVY] **test for svy**, [TS] **arch**, [TS] **prais**,
[XT] **xtgls**, [XT] **xtpcse**, [XT] **xtreg**
Judson, D. H., [R] **poisson**, [R] **tabulate**

K

Kahn, H. A., [R] **dstdize**, [ST] **ltable**, [ST] **stcox**
Kaiser, H. F., [R] **factor**

Kalbfleisch, J. D., [ST] **ltable**, [ST] **stcox**,
[ST] **stphplot**, [ST] **streg**, [ST] **sts**, [ST] **sts
test**, [ST] **stset**, [XT] **xtcloglog**, [XT] **xtlogit**,
[XT] **xtprobit**
Kalman, R. E., [TS] **arima**
Kaplan, E. L., [ST] **sts**
Katz, J. N., [XT] **xtgls**, [XT] **xtpcse**
Kaufman, L., [CL] **cluster**
Keeler, E. B., [R] **brier**
Keiding, N., [ST] **stsplit**
Kelsey, J. L., [ST] **epitab**
Kempthorne, P. J., [R] **regression diagnostics**
Kendall, M. G., [R] **centile**, [R] **spearman**,
[R] **tabulate**, [R] **tobit**, [CL] **cluster**
Kennedy, W. J., [R] **regress**, [R] **sw**, [SVY] **svytab**
Kennedy, W. J., Jr., [R] **anova**, [R] **nl**, [P] **_robust**
Kent, J. T., [R] **manova**, [U] **23 Estimation and post-
estimation commands**, [P] **_robust**
Keynes, J. M., [R] **means**
Khan, M. R., [ST] **epitab**
Khanti-Akom, S., [XT] **xthtaylor**
Kiernan, M., [R] **kappa**
Kimber, A. C., [ST] **streg**
Kish, L., [R] **loneway**, [U] **23 Estimation and post-
estimation commands**, [P] **_robust**, [SVY] **svy
estimators**, [SVY] **svymean**
Kitagawa, G., [R] **estimates**
Kiviet, J., [XT] **xtabond**
Klar, J., [R] **logistic**
Klein, J. P., [ST] **stci**, [ST] **stcox**, [ST] **streg**, [ST] **sts**,
[ST] **sts test**
Klein, L., [R] **reg3**, [TS] **regression diagnostics**
Klein, M., [R] **binreg**, [R] **clogit**, [R] **logistic**,
[R] **lrtest**, [R] **mlogit**, [R] **ologit**, [XT] **xtgee**
Kleinbaum, D. G., [R] **binreg**, [R] **clogit**, [R] **logistic**,
[R] **lrtest**, [R] **mlogit**, [R] **ologit**, [ST] **epitab**,
[XT] **xtgee**
Kleiner, B., [R] **diagnostic plots**, [R] **lowess**,
[U] **9 Stata's sample datasets**
Kmenta, J., [R] **ivreg**, [R] **reg3**, [R] **regress**, [R] **tobit**,
[TS] **arch**, [TS] **prais**, [XT] **xtpcse**
Koch, G. G., [R] **anova**, [R] **kappa**, [R] **vwls**,
[SVY] **svytab**
Koehler, K. J., [R] **diagnostic plots**
Koenker, R., [R] **qreg**
Kohn, R., [TS] **arima**
Kolmogorov, A. N., [R] **ksmirnov**
Koopmans, T. C., [R] **ivreg**
Korn, E. L., [R] **test**, [SVY] **svy**, [SVY] **svy
estimators**, [SVY] **svymean**, [SVY] **svytab**,
[SVY] **test for svy**
Kott, P. S., [SVY] **svy**
Kotz, S., [R] **functions**, [R] **ksmirnov**, [R] **means**
Krakauer, H., [ST] **ltable**
Krauss, N., [SVY] **svy estimators**, [SVY] **svymean**
Kreidberg, M. B., [ST] **epitab**
Kroner, K. F., [TS] **arch**
Krushelnytskyy, B., [R] **lorenz**, [R] **qreg**

Kruskal, W. H., [R] **kwallis**, [R] **spearman**,
 [R] **tabulate**
Kuder, G. F., [R] **alpha**
Kuehl, R. O., [R] **anova**
Kuh, E., [R] **regress**, [R] **regression diagnostics**,
 [U] **21 Programming Stata**
Kulczynski, S., [CL] **cluster**
Kumbhakar, S. C., [R] **frontier**, [XT] **xtfrontier**
Kung, D. S., [R] **qreg**
Kupper, L. L., [ST] **epitab**
Kutner, M. H., [R] **pkcross**, [R] **pkequiv**, [R] **pkshape**,
 [R] **regression diagnostics**

L

L'Abbé, K. A., [R] **meta**
Lachenbruch, P., [R] **diagnostic plots**
Lafontaine, F., [R] **boxcox**
Lai, K. S., [TS] **dfgls**
Lambert, D., [R] **zip**
Lance, G. N., [CL] **cluster**
Landau, S., [CL] **cluster**, [CL] **cluster stop**
Landis, J. R., [R] **kappa**
Langholz, B., [ST] **sttocc**
Larsen, W. A., [R] **regression diagnostics**
Lauritsen, J. M., [R] **labelbook**, [R] **list**
Lawless, J. F., [ST] **ltable**
Lawley, D. N., [R] **factor**, [R] **manova**
Laynard, R., [XT] **xtabond**
Ledolter, J., [TS] **tssmooth**, [TS] **tssmooth
 dexponential**, [TS] **tssmooth exponential**,
 [TS] **tssmooth hwinters**, [TS] **tssmooth
 shwinters**
Lee, E. S., [R] **dstdize**
Lee, E. T., [R] **roc**, [ST] **streg**
Lee, L., [XT] **xtreg**
Lee, P., [ST] **streg**
Lee, T.-C., [R] **glogit**, [R] **logit**, [R] **lrtest**,
 [R] **maximize**, [R] **mlogit**, [R] **oneway**,
 [R] **probit**, [R] **regression diagnostics**, [R] **test**,
 [SVY] **test for svy**, [TS] **arch**, [TS] **prais**,
 [XT] **xtgls**, [XT] **xtpcse**, [XT] **xtreg**
Lee, W. C., [R] **roc**
Leese, M., [CL] **cluster**, [CL] **cluster stop**
Lemeshow, S., [R] **adjust**, [R] **clogit**, [R] **glm**,
 [R] **glogit**, [R] **lincom**, [R] **logistic**, [R] **logit**,
 [R] **lrtest**, [R] **mlogit**, [R] **predictnl**, [R] **sw**,
 [ST] **stcox**, [ST] **streg**, [SVY] **svy**, [XT] **xtgee**
Leon, A. Ponce de, [R] **roc**
Leone, F. C., [R] **functions**, [U] **16 Functions and
 expressions**
Leroy, A. M., [R] **qreg**, [R] **regression diagnostics**,
 [R] **rreg**
Levene, H., [R] **sdtest**
Levy, P., [SVY] **svy**
Lewis, H., [R] **heckman**
Lewis, I. G., [R] **binreg**

Lewis, J. D., [R] **fracpoly**
Lexis, W., [ST] **stsplit**
Li, G., [R] **rreg**
Li, Q., [XT] **xtivreg**, [XT] **xtreg**, [XT] **xtregar**
Liang, K.-Y., [XT] **xtcloglog**, [XT] **xtgee**, [XT] **xtlogit**,
 [XT] **xtnbreg**, [XT] **xtpoisson**, [XT] **xtprobit**
Likert, R. A., [R] **alpha**
Lilien, D. M., [TS] **arch**
Lilienfeld, D. E., [ST] **epitab**
Lin, D. Y., [U] **23 Estimation and post-estimation
 commands**, [P] **_robust**, [ST] **stcox**
Lindstrom, M. J., [XT] **xtcloglog**, [XT] **xtgee**,
 [XT] **xtintreg**, [XT] **xtlogit**, [XT] **xtprobit**,
 [XT] **xttobit**
Lipset, S. M., [R] **histogram**
Little, R. J. A., [R] **impute**
Liu, J. P., [R] **pk**, [R] **pkcross**, [R] **pkequiv**,
 [R] **pkexamine**, [R] **pkshape**
Ljung, G. M., [TS] **wntestq**
Loan, C. F. van, [R] **orthog**
Long, J. S., [R] **clogit**, [R] **cloglog**, [R] **estimates**,
 [R] **logit**, [R] **mlogit**, [R] **nbreg**, [R] **ologit**,
 [R] **oprobit**, [R] **poisson**, [R] **probit**, [R] **regress**,
 [R] **regression diagnostics**, [R] **testnl**, [R] **tobit**,
 [R] **zip**, [U] **23 Estimation and post-estimation
 commands**
Longley, J. D., [R] **kappa**
López-Vizcaíno, M. E., [ST] **epitab**
Louis, T. A., [R] **tabulate**
Lovell, C. A. K., [R] **frontier**, [XT] **xtfrontier**
Lovie, A. D., [R] **spearman**
Lovie, P., [R] **spearman**
Lu, J. Y., [TS] **prais**
Luce, R. D., [R] **rologit**
Lunt, M., [R] **ologit**, [R] **oprobit**
Lurie, M. B., [R] **manova**
Lütkepohl, H., [R] **glogit**, [R] **logit**, [R] **lrtest**,
 [R] **maximize**, [R] **mlogit**, [R] **oneway**,
 [R] **probit**, [R] **regression diagnostics**, [R] **test**,
 [SVY] **test for svy**, [TS] **arch**, [TS] **dfgls**,
 [TS] **prais**, [TS] **time series**, [TS] **var**,
 [TS] **var intro**, [TS] **var svar**, [TS] **varfcast
 compute**, [TS] **vargranger**, [TS] **varirf create**,
 [TS] **varnorm**, [TS] **varsoc**, [TS] **varstable**,
 [TS] **varwle**, [XT] **xtgls**, [XT] **xtpcse**, [XT] **xtreg**

M

Ma, G., [R] **roc**
Machin, D., [R] **kappa**, [R] **tabulate**
Mack, T. M., [R] **symmetry**
MacKinnon, J. G., [R] **boxcox**, [R] **ivreg**, [R] **nl**,
 [R] **regress**, [R] **truncreg**, [U] **23 Estimation
 and post-estimation commands**, [P] **_robust**,
 [XT] **xtpcse**, [TS] **arch**, [TS] **arima**,
 [TS] **dfuller**, [TS] **pperron**, [TS] **regression
 diagnostics**, [TS] **varlmar**, [XT] **xtgls**
MacMahon, B., [ST] **epitab**

MacRae, K. D., [R] **binreg**
MaCurdy, T., [XT] **intro**, [XT] **xthtaylor**
Madansky, A., [R] **runtest**
Maddala, G. S., [R] **nlogit**, [R] **tobit**, [R] **treatreg**,
    [XT] **xtgls**
Mahalanobis, P. C., [R] **hotelling**
Mallows, C. L., [R] **regression diagnostics**
Mander, A., [R] **impute**, [R] **symmetry**, [ST] **stsplit**
Mann, H. B., [R] **kwallis**, [R] **signrank**
Manning, W. G., [R] **heckman**
Mantel, N., [R] **sw**, [ST] **epitab**, [ST] **sts test**
Marden, J. I., [R] **rologit**
Mardia, K. V., [R] **manova**
Markowski, C. A., [R] **sdtest**
Markowski, E. P., [R] **sdtest**
Marr, J. W., [ST] **stsplit**
Marsaglia, G., [R] **functions**, [U] **16 Functions and**
    expressions
Marschak, J., [R] **ivreg**
Martínez, M. A., [R] **logistic**
Marubini, E., [ST] **sts test**
Massey, F. J., Jr., [R] **ttest**
Massey, J. T., [U] **30 Overview of survey estimation**,
    [SVY] **lincom for svy**, [SVY] **svy estimators**,
    [SVY] **svydes**, [SVY] **svymean**, [SVY] **svytab**
Maurer, K., [U] **30 Overview of survey estimation**,
    [SVY] **lincom for svy**, [SVY] **svy estimators**,
    [SVY] **svydes**, [SVY] **svymean**, [SVY] **svytab**
Maxwell, A. E., [R] **factor**, [R] **symmetry**
Mazumdar, M., [R] **meta**
McCleary, S. J., [R] **regression diagnostics**
McCullagh, P., [R] **glm**, [R] **ologit**, [R] **rologit**,
    [XT] **xtgee**, [XT] **xtpoisson**
McDonald, J., [R] **tobit**
McDowell, A. A., [U] **30 Overview of survey**
    estimation, [SVY] **lincom for svy**, [SVY] **svy**
    estimators, [SVY] **svydes**, [SVY] **svymean**,
    [SVY] **svytab**
McDowell, A. W., [SVY] **svy**, [TS] **arima**
McFadden, D., [R] **clogit**, [R] **hausman**, [R] **nlogit**,
    [R] **suest**
McGilchrist, C. A., [ST] **stcox**, [ST] **streg**
McGinnis, R. E., [R] **symmetry**
McGuire, T. J., [R] **dstdize**
McKelvey, R. D., [R] **ologit**
McKenney, A., [P] **matrix eigenvalues**
McLeod, A. I., [TS] **arima**
McNeil, B. J., [R] **roc**
McNeil, D., [R] **poisson**
McNemar, Q., [ST] **epitab**
Meeusen, W., [XT] **xtfrontier**
Mehta, C. R., [R] **tabulate**
Meier, P., [ST] **sts**
Meiselman, D., [TS] **arima**
Mendenhall, W., [SVY] **svy**
Metz, C. E., [R] **logistic**
Meyer, B. D., [ST] **discrete**

Michels, K. M., [R] **anova**, [R] **loneway**, [R] **oneway**
Michener, C. D., [CL] **cluster**
Mielke, P., [R] **brier**
Miettinen, O. S., [ST] **epitab**
Miller, A. B., [R] **kappa**
Miller, R. G., Jr., [R] **diagnostic plots**, [R] **oneway**
Milligan, G. W., [CL] **cluster**, [CL] **cluster**
    programming subroutines, [CL] **cluster stop**
Milliken, G. A., [R] **anova**, [R] **manova**
Minder, C., [R] **meta**
Moeschberger, M. L., [ST] **stci**, [ST] **stcox**, [ST] **streg**,
    [ST] **sts**, [ST] **sts test**
Moffitt, R., [R] **tobit**
Monfort, A., [R] **hausman**, [R] **suest**, [R] **test**,
    [SVY] **test for svy**, [TS] **arima**
Monson, R. R., [R] **bitest**, [ST] **epitab**
Montgomery, D. C., [TS] **tssmooth**, [TS] **tssmooth**
    dexponential, [TS] **tssmooth exponential**,
    [TS] **tssmooth hwinters**, [TS] **tssmooth**
    shwinters
Mood, A. M., [R] **centile**
Mooney, C. Z., [R] **bootstrap**, [R] **jknife**
Moore, R. J., [R] **functions**
Morgenstern, H., [ST] **epitab**
Morris, J. N., [ST] **stsplit**
Morris, N. F., [R] **binreg**
Morrison, D. F., [R] **manova**
Morrow, A., [ST] **epitab**
Moskowitz, M., [R] **kappa**
Mosteller, F., [R] **jknife**, [R] **regress**, [R] **regression**
    diagnostics, [R] **rreg**
Mullahy, J., [R] **zip**
Mundlak, Y., [XT] **xtregar**
Murphy, A. H., [R] **brier**
Murphy, J. L., [XT] **xtprobit**
Murray-Lyon, I. M., [R] **binreg**

N

Nachtsheim, C. J., [R] **pkcross**, [R] **pkequiv**,
    [R] **pkshape**
Nagler, J., [R] **scobit**
Nardi, G., [ST] **epitab**
Narendranathan, W., [XT] **xtregar**
Narula, S. C., [R] **qreg**
Nash, J. D., [R] **infile (fixed format)**, [R] **merge**
Nee, J. C. M., [R] **kappa**
Neff, R. J., [ST] **epitab**
Nelder, J. A., [R] **glm**, [R] **ologit**, [XT] **xtgee**,
    [XT] **xtpoisson**
Nelson, D. B., [TS] **arch**, [TS] **arima**
Nelson, E. C., [R] **alpha**, [R] **factor**, [R] **lincom**,
    [R] **mlogit**, [R] **predictnl**
Nelson, F. D., [R] **logit**, [R] **mlogit**, [R] **probit**
Nelson, W., [ST] **sts**
Neter, J., [R] **pkcross**, [R] **pkequiv**, [R] **pkshape**,
    [R] **regression diagnostics**

Neuhaus, J. M., [XT] **xtcloglog**, [XT] **xtintreg**,
  [XT] **xtlogit**, [XT] **xtprobit**, [XT] **xttobit**
Newbold, P., [TS] **arima**
Newey, W. K., [R] **glm**, [TS] **newey**
Newman, S. C., [R] **poisson**
Newson, R., [R] **centile**, [R] **glm**, [R] **mkspline**,
  [R] **signrank**, [R] **spearman**, [R] **statsby**,
  [R] **tabulate**
Newton, H. J., [R] **kdensity**, [TS] **arima**,
  [TS] **corrgram**, [TS] **cumsp**, [TS] **dfuller**,
  [TS] **pergram**, [TS] **pperron**, [TS] **wntestb**,
  [TS] **xcorr**, [XT] **xtgee**
Newton, M. A., [XT] **xtcloglog**, [XT] **xtgee**,
  [XT] **xtintreg**, [XT] **xtlogit**, [XT] **xtprobit**,
  [XT] **xttobit**
Ng, S., [TS] **dfgls**
Nickell, S. J., [XT] **xtabond**
Nolan, D., [R] **diagnostic plots**
Nunnally, J. C., [R] **alpha**

O

O'Connell, R., [TS] **tssmooth**, [TS] **tssmooth
  dexponential**, [TS] **tssmooth exponential**,
  [TS] **tssmooth hwinters**, [TS] **tssmooth
  shwinters**
O'Rourke, K., [R] **meta**
Oakes, D., [ST] **ltable**, [ST] **stcox**, [ST] **streg**, [ST] **sts**
Ochiai, A., [CL] **cluster**
Oehlert, G. W., [R] **nlcom**
Olkin, I., [R] **hotelling**, [TS] **wntestb**
Olson, J. M., [R] **symmetry**
Orcutt, G. H., [TS] **prais**
Ord, J. K., [R] **qreg**, [R] **summarize**
Ott, L., [SVY] **svy**
Over, M., [R] **ivreg**, [R] **regress**, [XT] **xtivreg**

P

Pagan, A. R., [R] **frontier**, [R] **mvreg**, [R] **regression
  diagnostics**, [R] **sureg**, [XT] **xtreg**
Pagano, M., [R] **dstdize**, [R] **logistic**, [R] **sampsi**,
  [R] **sdtest**, [R] **tabulate**, [ST] **ltable**, [ST] **sts**
Palta, M., [XT] **xtcloglog**, [XT] **xtgee**, [XT] **xtintreg**,
  [XT] **xtlogit**, [XT] **xtprobit**, [XT] **xttobit**
Panis, C., [R] **mkspline**
Park, H. J., [R] **regress**, [P] **_robust**, [SVY] **svytab**
Park, J. Y., [R] **boxcox**, [R] **nlcom**, [R] **predictnl**,
  [R] **testnl**
Parzen, E., [R] **kdensity**
Patel, N. R., [R] **tabulate**
Paul, C., [R] **logistic**
Pearce, M. S., [R] **logistic**, [ST] **epitab**
Pearson, K., [R] **correlate**, [R] **pca**, [R] **tabulate**
Pendergast, J. F., [XT] **xtcloglog**, [XT] **xtgee**,
  [XT] **xtintreg**, [XT] **xtlogit**, [XT] **xtprobit**,
  [XT] **xttobit**

Peracchi, F., [R] **regress**, [R] **regression diagnostics**
Pérez-Hoyos, S., [R] **lrtest**, [ST] **epitab**
Perkins, A. M., [R] **signrank**
Perrin, E., [R] **alpha**, [R] **factor**, [R] **lincom**,
  [R] **mlogit**, [R] **predictnl**
Perron, P., [TS] **dfgls**, [TS] **pperron**
Peterson, B., [R] **ologit**
Peterson, W. W., [R] **logistic**
Petitclerc, M., [R] **kappa**
Petkova, E., [R] **suest**
Peto, J., [ST] **sts test**
Peto, R., [ST] **stcox**, [ST] **streg**, [ST] **sts test**
Petrov, B., [TS] **varsoc**
Pfeffer, R. I., [R] **symmetry**
Phillips, P. C. B., [R] **boxcox**, [R] **nlcom**, [R] **predictnl**,
  [R] **testnl**, [TS] **pperron**
Piantadosi, S., [U] **23 Estimation and post-estimation
  commands**, [P] **_robust**
Pickles, A., [R] **glm**, [XT] **xtgee**, [XT] **xtreg**
Pierce, D. A., [TS] **wntestq**
Pike, M. C., [R] **symmetry** [ST] **ltable**, [ST] **streg**
Pillai, K. C. S., [R] **manova**
Pindyck, R., [R] **biprobit**, [R] **heckprob**
Pisati, M., [TS] **time series**
Pitblado, J., [SVY] **svy**
Plackett, R. L., [R] **means**, [R] **regress**, [R] **summarize**
Plummer, D., [ST] **epitab**
Pocock, S., [R] **sampsi**
Poirier, D., [R] **biprobit**
Poisson, S. D., [R] **poisson**
Posten, H. O., [R] **functions**
Powers, D. A., [R] **logit**, [R] **probit**
Prais, S. J., [TS] **prais**
Pregibon, D., [R] **glm**, [R] **linktest**, [R] **logistic**,
  [R] **logit**
Prentice, R. L., [ST] **discrete**, [ST] **ltable**, [ST] **stcox**,
  [ST] **stphplot**, [ST] **streg**, [ST] **sts**, [ST] **sts test**,
  [ST] **stset**, [XT] **xtgee**
Press, W. H., [R] **functions**, [R] **range**, [R] **vwls**,
  [U] **16 Functions and expressions**, [P] **matrix
  symeigen**, [TS] **arch**, [TS] **arlma**
Price, B., [R] **poisson**, [R] **regress**, [R] **regression
  diagnostics**, [TS] **prais**
Punj, G. N., [R] **rologit**

Q

Qaqish, B., [XT] **xtgee**

R

Rabe-Hesketh, S., [R] **anova**, [R] **glm**, [R] **pca**,
  [XT] **xtgee**, [XT] **xtreg**
Raftery, A., [R] **estimates**, [R] **glm**
Ramalheira, C., [R] **means**
Ramsey, F., [R] **regress**

Ramsey, J. B., [R] regression diagnostics
Rao, C. R., [R] manova
Rao, D. S. P., [XT] xtfrontier
Rao, J. N. K., [U] 30 Overview of survey estimation, [SVY] svytab
Rao, T. R., [CL] cluster
Ratkowsky, D. A., [R] nl, [R] pk, [R] pkcross, [R] pkequiv
Rawlings, J. O., [R] regress
Redelmeier, D. A., [R] brier
Reichenheim, M. E., [R] kappa, [R] roc
Reilly, M., [R] logistic
Reinsch, C., [P] matrix symeigen
Reinsel, G. C., [TS] arima, [TS] corrgram, [TS] cumsp, [TS] dfuller, [TS] pergram, [TS] pperron, [TS] xcorr
Relles, D. A., [R] rreg
Rencher, A. C., [R] factor, [R] manova, [R] pca, [CL] cluster
Revankar, N., [R] frontier, [XT] xtfrontier
Richardson, M. W., [R] alpha
Riffenburgh, R. H., [R] ksmirnov, [R] kwallis
Riley, A. R., [R] list, [R] net search
Roberson, P. K., [R] logistic
Robins, J. M., [ST] epitab
Robins, R. P., [TS] arch
Robyn, D. L., [R] cumul
Rodríguez, G., [R] nbreg, [R] poisson
Rogers, D. J., [CL] cluster
Rogers, W. H., [R] brier, [R] egen, [R] glm, [R] heckman, [R] lincom, [R] mlogit, [R] nbreg, [R] poisson, [R] predictnl, [R] qreg, [R] regress, [R] rreg, [R] sktest, [R] suest, [U] 23 Estimation and post-estimation commands, [P] _robust, [ST] stcox
Rohlf, F. J., [CL] cluster, [CL] cluster singlelinkage
Ronchetti, E. M., [R] egen
Ronning, G., [R] clogit
Rosner, B., [R] sampsi
Ross, G. J. S., [R] nl
Rothenberg, T., [TS] dfgls, [TS] var svar
Rothman, K. J., [R] ci, [R] dstdize, [R] glogit, [R] poisson, [ST] epitab
Rousseeuw, P. J., [R] egen, [R] qreg, [R] regression diagnostics, [R] rreg, [CL] cluster
Roy, S. N., [R] manova
Royall, R. M., [U] 23 Estimation and post-estimation commands, [P] _robust
Royston, P., [R] centile, [R] cusum, [R] diagnostic plots, [R] dotplot, [R] fracpoly, [R] glm, [R] list, [R] lnskew0, [R] lowess, [R] mfp, [R] nl, [R] range, [R] regress, [R] sampsi, [R] sort, [R] sktest, [R] swilk
Rubin, D. B., [R] impute
Rubinfeld, D., [R] biprobit, [R] heckprob
Runkle, D., [TS] arch
Rupert, P., [XT] xthtaylor
Ruppert, D., [R] boxcox, [R] rreg

Rush, M., [R] egen
Russell, P. F., [CL] cluster
Rutherford, E., [R] poisson
Ruud, B., [R] rologit, [R] suest
Ryan, P., [R] egen, [R] pctile
Ryan, T. P., [R] qc, [R] regression diagnostics

S

Said, S. E., [TS] dfgls
Sakamoto, Y., [R] estimates
Salgado-Ugarte, I. H., [R] kdensity, [R] lowess, [R] smooth
Salim, A., [R] logistic
Sampson, A. R., [R] hotelling
Sanders, F., [R] brier
Sargan, J. D., [TS] prais
Särndal, C.-E., [SVY] svy, [SVY] svy estimators
Sasieni, P., [R] dotplot, [R] list, [R] lowess, [R] memory, [R] nptrend, [R] smooth
Satterthwaite, F. E., [R] ttest, [SVY] svymean
Sauerbrei, W., [R] fracpoly, [R] mfp
Savage, I. R., [ST] sts test
Savin, E., [TS] regression diagnostics
Saw, S. L. C., [R] qc
Schafer, D. W., [R] regress
Schaffer, C. M., [CL] cluster
Scheaffer, R. L., [SVY] svy
Scheffé, H., [R] anova, [R] oneway
Schlesselman, J. J., [R] boxcox, [ST] epitab
Schmidt, C. H., [R] brier
Schmidt, P., [R] frontier, [R] regression diagnostics, [XT] xtfrontier
Schmidt, T. J., [R] egen
Schneider, H., [R] sdtest
Schneider, M., [R] meta
Schnell, D., [R] regress, [P] _robust, [SVY] svytab
Schoenfeld, D., [ST] stcox, [ST] streg
Schwarz, G., [R] estimates
Schwert, G. W., [TS] dfgls
Scott, A. J., [U] 30 Overview of survey estimation, [SVY] svymean, [SVY] svytab
Scott, C., [SVY] svy estimators, [SVY] svymean
Scott, D. W., [R] kdensity
Scotto, M. G., [R] diagnostic plots, [ST] streg
Seber, G. A. F., [R] manova
Seed, P. T., [R] ci, [R] correlate, [R] logistic, [R] roc, [R] sampsi, [R] sdtest, [R] spearman
Selvin, S., [R] poisson, [R] tabulate [ST] ltable, [ST] stcox
Sempos, C. T., [R] dstdize, [ST] ltable, [ST] stcox
Semykina, A., [R] lorenz, [R] qreg
Shah, B. V., [SVY] svytab
Shao, J., [SVY] svy estimators
Shapiro, S., [ST] epitab
Shapiro, S. S., [R] swilk
Sharp, S., [R] meta

Sheldon, T. A., [R] meta
Shewhart, W. A., [R] qc
Shimizu, M., [R] kdensity, [R] lowess
Shumway, R. H., [TS] arima
Sibson, R., [CL] cluster, [CL] cluster singlelinkage
Šidák, Z., [R] oneway
Silverman, B. W., [R] kdensity
Silvey, S. D., [R] ologit, [R] oprobit
Simonoff, J. S., [R] kdensity
Simor, I. S., [R] kappa
Sims, C. A., [TS] var intro, [TS] var svar, [TS] varirf create
Sincich, R., [R] prtest
Skinner, C. J., [SVY] svy, [SVY] svy estimators, [SVY] svymean
Skrondal, A., [R] glm, [XT] xtgee
Slone, D., [ST] epitab
Smeeton, N. C., [R] signrank
Smirnov, N. V., [R] ksmirnov
Smith, B. T., [P] matrix symeigen
Smith, G. D., [R] meta
Smith, H., [R] manova, [R] regress, [R] sw
Smith, J. M., [R] fracpoly
Smith, R. L., [ST] streg
Smith, T. M. F., [SVY] svy
Sneath, P. H. A., [CL] cluster
Snedecor, G. W., [R] anova, [R] correlate, [R] means, [R] oneway, [R] signrank
Snell, E. J., [ST] stcox, [ST] streg
Sokal, R. R., [CL] cluster
Song, F., [R] meta
Soon, T. W., [R] qc
Sorensen, D., [P] matrix eigenvalues
Sørensen, T., [CL] cluster
Sosa-Escudero, W., [XT] xtreg
Späth, H., [CL] cluster
Spearman, C., [R] factor, [R] spearman
Speed, T., [R] diagnostic plots
Sperling, R., [TS] arch, [TS] arima, [TS] dfgls, [TS] wntestq
Spiegelhalter, D. J., [R] brier
Spieldman, R. S., [R] symmetry
Spitzer, J. J., [R] boxcox
Sprent, P., [R] signrank
Sribney, W. M., [R] frontier, [R] maximize, [R] ml, [R] orthog, [R] signrank, [R] sw, [R] test, [U] 30 Overview of survey estimation, [P] matrix mkmat, [P] _robust, [SVY] svy, [SVY] lincom for svy, [SVY] svy estimators, [SVY] svydes, [SVY] svymean, [SVY] svytab, [SVY] test for svy, [XT] xtfrontier
Staelin, R., [R] rologit
Stahel, W. A., [R] egen
Starmer, C. F., [R] vwls
Steel, R. G. D., [R] anova
Stegun, I. A., [R] functions, [R] orthog, [U] 16 Functions and expressions

Steichen, T. J., [R] duplicates, [R] kappa, [R] kdensity, [R] meta
Steiger, W., [R] qreg
Stein, C., [R] bootstrap
Stepniewska, K. A., [R] nptrend
Sterne, J., [R] meta
Stevenson, R. E., [R] frontier
Stewart, J., [ST] ltable
Stewart, M. B., [R] tobit
Stigler, S. M., [R] correlate, [R] qreg
Stine, R., [R] bootstrap
Stock, J. H., [R] ivreg, [TS] dfgls, [TS] time series, [TS] var, [TS] var intro, [TS] var svar, [TS] varirf create, [XT] xthtaylor
Stoll, B. J., [ST] epitab
Stolley, P. D., [ST] epitab
Stoto, M. A., [R] lv
Street, J. O., [R] rreg
Stuart, A., [R] centile, [R] qreg, [R] summarize, [R] symmetry, [R] tobit, [SVY] svy
Sullivan, G., [R] regress, [P] _robust, [SVY] svytab
Sutton, A. J., [R] meta
Svennerholm, A. M., [ST] epitab
Swagel, P., [U] 24 Commands to input data
Swamy, P. A., [XT] xtrchh
Swed, F. S., [R] runtest
Sweeting, T. J., [ST] streg
Swensson, B., [SVY] svy, [SVY] svy estimators
Swets, J. A., [R] logistic
Szroeter, J., [R] regression diagnostics

T

Tamhane, A. C., [R] oneway
Tan, W. Y., [U] 23 Estimation and post-estimation commands, [P] _robust
Tanimoto, T. T., [CL] cluster
Taniuchi, T., [R] kdensity
Tanner, W. P., Jr., [R] logistic
Tapia, R. A., [R] kdensity
Tarlov, A. R., [R] alpha, [R] factor, [R] lincom, [R] mlogit, [R] predictnl
Tarone, R. E., [ST] sts test
Tatham, R. L., [R] rologit
Taub, A. J., [XT] xtreg
Taylor, C., [R] glm, [XT] xtgee, [XT] xtreg
Taylor, W. E., [XT] intro, [XT] xthtaylor
Teukolsky, S. A., [R] functions, [R] range, [R] vwls, [U] 16 Functions and expressions, [P] matrix symeigen, [TS] arch, [TS] arima
Theil, H., [R] ivreg, [R] pcorr, [R] reg3, [TS] prais
Therneau, T. M., [ST] stcox
Thomas, D. C., [ST] sttocc
Thomas, D. R., [SVY] svytab
Thompson, J. C., [R] diagnostic plots
Thompson, J. R., [R] kdensity
Thompson, S. K., [SVY] svy

Thompson, W. D., [ST] **epitab**
Thomson, G. H., [R] **factor**
Thorndike, F., [R] **poisson**
Thurnstone, L. L., [R] **rologit**
Tibshirani, R., [R] **bootstrap**, [R] **qreg**
Tidmarsh, C. E., [R] **fracpoly**
Tilford, J. M., [R] **logistic**
Timm, N. H., [R] **manova**
Tippett, L. H. C., [ST] **streg**
Tobias, A., [R] **alpha**, [R] **estimates**, [R] **logistic**,
    [R] **lrtest**, [R] **meta**, [R] **poisson**, [R] **roc**,
    [R] **sdtest**, [ST] **streg**
Tobin, J., [R] **tobit**
Toman, R. J., [R] **sw**
Toplis, P. J., [R] **binreg**
Torrie, J. H., [R] **anova**
Tosetto, A., [R] **logistic**, [R] **logit**
Trichopoulos, D., [ST] **epitab**
Trivedi, P. K., [R] **nbreg**, [R] **poisson**, [R] **regression
    diagnostics**, [XT] **xtnbreg**, [XT] **xtpoisson**
Tufte, E. R., [R] **stem**
Tukey, J. W., [R] **egen**, [R] **jknife**, [R] **ladder**,
    [R] **linktest**, [R] **lv**, [R] **regress**, [R] **regression
    diagnostics**, [R] **rreg**, [R] **smooth**, [R] **spikeplot**,
    [R] **stem**, [P] **if**
Tukey, P. A., [R] **diagnostic plots**, [R] **lowess**
Tyler, J. H., [R] **estimates**, [R] **regress**

U

Utts, J. M., [R] **ci**

V

Valman, H. B., [R] **fracpoly**
Van de Ven, W. P. M. M., [R] **biprobit**, [R] **heckprob**
Van Kerm, P., [R] **lorenz**
Van Pragg, B. M. S., [R] **biprobit**, [R] **heckprob**
Varadharajan-Krishnakumar, J., [XT] **xtivreg**
Velleman, P. F., [R] **regression diagnostics**, [R] **smooth**
Vetterling, W. T., [R] **functions**, [R] **range**, [R] **vwls**,
    [U] **16 Functions and expressions**, [P] **matrix
    symeigen**, [TS] **arch**, [TS] **arima**
Vidmar, S., [R] **means**, [ST] **epitab**
Von Storch, H., [R] **brier**
Vuong, Q., [R] **zip**

W

Wacholder, S., [R] **binreg**
Wagner, H. M., [R] **qreg**
Walker, A. M., [ST] **epitab**
Wallis, W. A., [R] **kwallis**
Wand, M. P., [R] **kdensity**
Wang, D., [R] **ci**, [R] **dstdize**, [R] **duplicates**,
    [R] **prtest**

Wang, Z., [R] **lrtest**, [R] **sw**, [ST] **epitab**
Ward, J. H., Jr., [CL] **cluster**
Ware, J. E., Jr., [R] **alpha**, [R] **factor**, [R] **lincom**,
    [R] **mlogit**, [R] **predictnl**
Ware, J. H., [ST] **sts test**
Warren, K., [ST] **epitab**
Wasserman, W., [R] **pkcross**, [R] **pkequiv**,
    [R] **pkshape**, [R] **regression diagnostics**
Waterson, E. J., [R] **binreg**
Watson, G. S., [R] **regression diagnostics**, [TS] **prais**
Watson, M. W., [R] **ivreg**, [TS] **dfgls**, [TS] **time
    series**, [TS] **var**, [TS] **var intro**, [TS] **var svar**,
    [TS] **varirf create**
Webster, A. D. B., [R] **fracpoly**
Wedderburn, R. W. M., [R] **glm**, [XT] **xtgee**
Weesie, J., [R] **alpha**, [R] **constraint**, [R] **estimates**,
    [R] **generate**, [R] **hausman**, [R] **joinby**,
    [R] **label**, [R] **labelbook**, [R] **ladder**,
    [R] **list**, [R] **logistic**, [R] **merge**, [R] **order**,
    [R] **pca**, [R] **reg3**, [R] **regress**, [R] **regression
    diagnostics**, [R] **rename**, [R] **reshape**,
    [R] **sample**, [R] **simulate**, [ST] **stsplit**, [R] **suest**,
    [R] **sureg**, [R] **tabstat**, [R] **tabulate**, [R] **test**,
    [U] **23 Estimation and post-estimation
    commands**, [P] **matrix define**, [SVY] **test for
    svy**
Wei, L. J., [U] **23 Estimation and post-estimation
    commands**, [P] **_robust**, [ST] **stcox**
Weibull, W., [ST] **streg**
Weisberg, H. F., [R] **summarize**
Weisberg, S., [R] **boxcox**, [R] **regress**, [R] **regression
    diagnostics**
Welch, B. L., [R] **ttest**
Wellington, J. F., [R] **qreg**
Wells, K. E., [R] **lincom**, [R] **mlogit**, [R] **predictnl**
Welsch, R. E., [R] **regress**, [R] **regression diagnostics**,
    [U] **21 Programming Stata**
Wernow, J. B., [R] **destring**, [R] **encode**
West, K. D., [R] **glm**, [TS] **newey**
West, S., [ST] **epitab**
White, H., [R] **ivreg**, [R] **regress**, [R] **regression
    diagnostics**, [R] **suest**, [U] **23 Estimation
    and post-estimation commands**, [P] **_robust**,
    [TS] **newey**, [TS] **prais**, [XT] **xtivreg**
White, K., [TS] **regression diagnostics**
White, K. J., [R] **boxcox**
White, P. O., [R] **factor**
Whitehouse, E., [R] **lorenz**
Whitney, D. R., [R] **kwallis**, [R] **signrank**
Wichura, M. J., [R] **functions**, [U] **16 Functions and
    expressions**
Wiggins, V. L., [R] **regression diagnostics**, [TS] **arch**,
    [TS] **arima**, [TS] **regression diagnostics**
Wilcoxon, F., [R] **kwallis**, [R] **signrank**, [ST] **sts test**
Wilk, M. B., [R] **cumul**, [R] **diagnostic plots**, [R] **swilk**
Wilkinson, J. H., [P] **matrix symeigen**
Wilks, D. S., [R] **brier**
Wilks, S. S., [R] **canon**, [R] **hotelling**, [R] **manova**

Williams, B., [SVY] **svy**
Williams, W. T., [CL] **cluster**
Wilson, S. R., [R] **bootstrap**
Winer, B. J., [R] **anova**, [R] **loneway**, [R] **oneway**
Winsten, C. B., [TS] **prais**
Winters, P. R., [TS] **tssmooth**, [TS] **tssmooth dexponential**, [TS] **tssmooth exponential**, [TS] **tssmooth hwinters**, [TS] **tssmooth shwinters**
Wolf, I. de, [R] **rologit**
Wolfe, F., [R] **correlate**, [R] **spearman**
Wolfe, R., [R] **ologit**, [R] **oprobit**, [R] **tabulate**
Wolfram, S., [ST] **streg**
Wolfson, C., [R] **kappa**
Wolter, K. M., [SVY] **svy**, [SVY] **svymean**
Wood, F. S., [R] **diagnostic plots**
Woodard, D. E., [R] **manova**
Wooldridge, J. M., [R] **ivreg**, [R] **regress**, [TS] **arch**, [TS] **regression diagnostics**, [XT] **xtreg**
Woolf, B., [ST] **epitab**
Working, H., [R] **roc**
Wretman, J., [SVY] **svy**, [SVY] **svy estimators**
Wright, J. H., [XT] **xthtaylor**
Wright, J. T., [R] **binreg**
Wright, P. G., [R] **ivreg**
Wu, C. F. J., [R] **qreg**
Wu, P. X., [XT] **xtregar**

Zimmerman, F., [R] **regress**
Zubkoff, M., [R] **alpha**, [R] **factor**, [R] **lincom**, [R] **mlogit**, [R] **predictnl**
Zwiers, F. W., [R] **brier**

X

Xie, Y., [R] **logit**, [R] **probit**

Y

Yar, M., [TS] **tssmooth**, [TS] **tssmooth dexponential**, [TS] **tssmooth exponential**, [TS] **tssmooth hwinters**, [TS] **tssmooth shwinters**
Yates, J. F., [R] **brier**
Yellott, J. I., Jr., [R] **rologit**
Yen, S., [ST] **epitab**
Yen, W. M., [R] **alpha**
Yogo, M., [XT] **xthtaylor**
Yule, G. U., [CL] **cluster**

Z

Zakoian, J. M., [TS] **arch**
Zappasodi, P., [R] **manova**
Zavoina, W., [R] **ologit**
Zeger, S. L., [XT] **xtcloglog**, [XT] **xtgee**, [XT] **xtlogit**, [XT] **xtnbreg**, [XT] **xtpoisson**, [XT] **xtprobit**
Zelen, M., [R] **ttest**
Zellner, A., [R] **frontier**, [R] **reg3**, [R] **sureg**, [XT] **xtfrontier**
Zelterman, D., [R] **tabulate**
Zhao, L. P., [XT] **xtgee**

Combined Subject Index

This is the combined subject index for this manual and the *Stata Base Reference Manual*, the *Stata Cluster Analysis Reference Manual*, the *Stata Cross-Sectional Time-Series Reference Manual*, the *Stata Programming Reference Manual*, the *Stata Survey Data Reference Manual*, the *Stata Survival Analysis & Epidemiological Tables Reference Manual*, and the *Stata Time-Series Reference Manual*. Readers interested in graphics topics should see the index in the *Stata Graphics Reference Manual*.

Semicolons set off the most important entries from the rest. Sometimes no entry will be set off with semicolons; this means all entries are equally important.

& (and), *see* logical operators
| (or), *see* logical operators
~ (not), *see* logical operators
! (not), *see* logical operators
!, *see* shell command
== (equality), *see* relational operators
!= (not equal), *see* relational operators
~= (not equal), *see* relational operators
< (less than), *see* relational operators
<= (less than or equal), *see* relational operators
> (greater than), *see* relational operators
>= (greater than or equal), *see* relational operators
* abbreviation character, *see* abbreviation character
* comment indicator, [P] **comments**
~ abbreviation character, *see* abbreviation character
? abbreviation character, *see* abbreviation character
- abbreviation character, *see* abbreviation character
/* */ comment delimiter, [P] **comments**
// comment indicator, [P] **comments**
/// comment indicator, [P] **comments**
., class, [P] **class**
; delimiter, [P] **#delimit**

A

.a, .b, . . . , .z, *see* missing values
Aalen–Nelson cumulative hazard, *see* Nelson–Aalen cumulative hazard
abbrev() string function, [R] **functions**
abbreviations, [U] **14.2 Abbreviation rules**;
　　[U] **14.1.1 varlist**, [U] **14.4 varlists**
　　unabbreviating command names, [P] **unabcmd**
　　unabbreviating variable list, [P] **unab**; [P] **syntax**
aborting command execution, [U] **12 The Break key**,
　　[U] **13 Keyboard use**
about command, [R] **about**
abs() function, [R] **functions**
absolute value dissimilarity measure, [CL] **cluster**
absolute value function, *see* abs() function
absorption in regression, [R] **areg**
ac command, [TS] **corrgram**

accelerated failure-time model, [ST] **streg**
Access, Microsoft, reading data from, [R] **odbc**,
　　[U] **24.4 Transfer programs**
accum, matrix subcommand, [P] **matrix accum**
acos() function, [R] **functions**
acprplot command, [R] **regression diagnostics**
actuarial tables, [ST] **ltable**
add, return subcommand, [P] **return**
added-variable plots, [R] **regression diagnostics**
addition across observations, [R] **egen**
addition across variables, [R] **egen**
addition operator, *see* arithmetic operators
adjust command, [R] **adjust**
adjusted Kaplan–Meier survivor function, [ST] **sts**
adjusted partial residual plot, [R] **regression diagnostics**
.ado filename suffix, [U] **14.6 File-naming conventions**
ado-files, [U] **2.5 The Stata Journal and the Stata Technical Bulletin**, [U] **20 Ado-files**,
　　[U] **21.11 Ado-files**, [P] **sysdir**, [P] **version**
　　adding comments to, [P] **comments**
　　debugging, [P] **trace**
　　downloading, [U] **32 Using the Internet to keep up to date**
　　editing, [R] **doedit**
　　installing, [R] **net**, [R] **sj**, [U] **20.6 How do I install an addition?**
　　location, [U] **20.5 Where does Stata look for ado-files?**
　　long lines, [U] **21.11.2 Comments and long lines in ado-files**, [P] **#delimit**
　　official, [R] **update**, [U] **32 Using the Internet to keep up to date**
　　also see verinst command
adopath command, [U] **20.5 Where does Stata look for ado-files?**, [P] **sysdir**
adosize, [P] **sysdir**; [U] **21.11 Ado-files**, [P] **creturn**,
　　[P] **macro**
　　set adosize, [P] **sysdir**
aggregate functions, [R] **egen**
aggregate statistics, dataset of, [R] **collapse**
agreement, interrater, [R] **kappa**
AIC, [R] **estimates**, [R] **glm**, [ST] **streg**
Akaike Information Criterion (AIC), [R] **estimates**,
　　[R] **glm**, [ST] **streg**
algebraic expressions, functions, and operators,
　　[U] **16 Functions and expressions**,
　　[U] **16.3 Functions**, [P] **matrix define**
_all, [U] **14.1.1 varlist**
alpha coefficient, Cronbach's, [R] **alpha**
alpha command, [R] **alpha**
alphabetizing,
　　observations, [R] **sort**; [R] **gsort**
　　variable names, [R] **order**
alphanumeric variables, *see* string variables
analysis of covariance, *see* ANCOVA
analysis of variance, *see* ANOVA
analysis-of-variance test of normality, [R] **swilk**
analytic weights, [U] **14.1.6 weight**,

[U] **23.16.2 Analytic weights**
ANCOVA, [R] **anova**
and operator, [U] **16.2.4 Logical operators**
Anderberg coefficient similarity measure, [CL] **cluster**
angular similarity measure, [CL] **cluster**
ANOVA, [R] **anova**, [R] **loneway**, [R] **oneway**
  Kruskal–Wallis, [R] **kwallis**
  repeated measures, [R] **anova**
anova command, [R] **anova**; *also see* estimation
    commands
  with string variables, [R] **encode**
any(), egen function, [R] **functions**
aorder command, [R] **order**
append command, [R] **append**; [U] **25 Commands for
    combining data**
_append variable, [R] **append**
appending data, [R] **append**; [U] **25 Commands for
    combining data**
appending files, [R] **copy**
appending rows and columns to matrix, [P] **matrix
    define**
AR, *see* autocorrelation
arc-cosine, arc-sine, and arc-tangent functions,
    [R] **functions**
arch command, [TS] **arch**; *also see* estimation
    commands
ARCH effects,
  testing for, [TS] **regression diagnostics**
  estimation, [TS] **arch**
archlm command, [TS] **regression diagnostics**
areg command, [R] **areg**; *also see* estimation
    commands
Arellano–Bond estimator, [XT] **xtabond**
args command, [P] **syntax**
ARIMA, [TS] **arch**, [TS] **arima**
arima command, [TS] **arima**; *also see* estimation
    commands
arithmetic operators, [U] **16.2.1 Arithmetic operators**,
    [P] **matrix define**
ARMA, [TS] **arch**, [TS] **arima**, *also see* time-series
arrays, class, [P] **class**
.Arrdrop1 built-in class modifier, [P] **class**
.arrindexof built-in class function, [P] **class**
.arrnels built-in class function, [P] **class**
.Arrpop built-in class modifier, [P] **class**
.Arrpush built-in class modifier, [P] **class**
ASCII,
  reading ASCII data, [R] **infile**, [R] **infix (fixed
    format)**, [R] **insheet**, [P] **file**
  saving in ASCII format, [R] **outfile**, [R] **outsheet**
  writing, [P] **file**
asin() function, [R] **functions**
assert command, [R] **assert**
assignment, class, [P] **class**
association, measures of, [R] **tabulate**
association tests, [SVY] **svytab**
asymmetry, [R] **lnskew0**, [R] **lv**, [R] **sktest**,
    [R] **summarize**

atan() function, [R] **functions**
atanh() function, [R] **functions**
attributable proportion, [ST] **epitab**
attribute tables, [R] **table**, [R] **tabsum**, [R] **tabulate**
AUC, [R] **pk**
augmented component-plus-residual plot, [R] **regression
    diagnostics**
augmented partial residual plot, [R] **regression
    diagnostics**
auto.dta, [U] **9 Stata's sample datasets**
autocode() function, [U] **28.1.2 Converting
    continuous to categorical variables**;
    [R] **functions**
autocorrelation, [R] **regression diagnostics**, [TS] **arch**,
    [TS] **arima**, [TS] **corrgram**, [TS] **newey**,
    [TS] **prais**, [TS] **regression diagnostics**,
    [TS] **var**, [TS] **varlmar**, [XT] **xtabond**,
    [XT] **xtgee**, [XT] **xtgls**, [XT] **xtpcse**,
    [XT] **xtregar**, *also see* time-series
autoregression, [TS] **var**, *also see* vector autoregression
autoregressive conditional heteroskedasticity, [TS] **arch**,
    [TS] **regression diagnostics**
autoregressive integrated moving average, [TS] **arch**,
    [TS] **arima**
autoregressive moving average, [TS] **arch**, [TS] **arima**
average linkage clustering, [CL] **cluster**, [CL] **cluster
    averagelinkage**
averagelinkage, cluster subcommand, [CL] **cluster
    averagelinkage**
averages, *see* means
avplot and avplots commands, [R] **regression
    diagnostics**
[aweight=*exp*] modifier, [U] **14.1.6 weight**,
    [U] **23.16.2 Analytic weights**

B

_b[], [U] **16.5 Accessing coefficients and standard
    errors**; [P] **matrix get**
b() function, [R] **functions**, [U] **27.3.2 Specifying
    particular dates (date literals)**
background mode, *see* batch jobs
bar charts, *see Graphics Reference Manual*
Bartlett scoring, [R] **factor**
Bartlett's bands, [TS] **corrgram**
Bartlett's test for equal variances, [R] **oneway**
BASE directory, [U] **20.5 Where does Stata look for
    ado-files?**, [P] **sysdir**
baseline hazard and survivor functions, [ST] **stcox**,
    [ST] **stphplot**
basis, orthonormal, [P] **matrix svd**
batch jobs, *see Getting Started with Stata for Unix* and
    Getting Started with Stata for Windows manuals
Battese-Coelli parameterization, [XT] **xtfrontier**
bcskew0 command, [R] **lnskew0**
bdgodfrey command, [TS] **regression diagnostics**
beta coefficients, [R] **ivreg**, [R] **regress**, [R] **regression
    diagnostics**

beta density function, [R] **functions**

beta density, noncentral, [R] **functions**

beta distribution, cumulative noncentral, [R] **functions**

beta distribution, inverse cumulative noncentral, [R] **functions**

beta function, inverse incomplete, [R] **functions**

betaden() function, [R] **functions**

between estimators, [XT] **xtivreg**, [XT] **xtreg**

between-cell means and variances, [XT] **xtsum**

bias corrected and accelerated, [R] **bootstrap**

BIC, [R] **estimates**

binary files, reading and writing, [P] **file**

binary outcome models, *see* dichotomous outcome models

binomial distribution,
    cdf, [R] **functions**
    confidence intervals, [R] **ci**
    test, [R] **bitest**

binomial family, [R] **binreg**

Binomial() function, [R] **functions**

binorm() function, [R] **functions**

binreg command, [R] **binreg**; *also see* estimation commands

bioequivalence tests, [R] **pk**, [R] **pkequiv**

biprobit command, [R] **biprobit**; *also see* estimation commands

bitest and bitesti commands, [R] **bitest**

bivariate normal function, [R] **functions**

biweight regression estimates, [R] **rreg**

biyear() function, [R] **functions**, [U] **27.3.5 Extracting components of time**

biyearly() function, [R] **functions**, [U] **27.3.6 Creating time variables**

blanks, removing from strings, [R] **functions**

blogit command, [R] **glogit**; *also see* estimation commands

bofd() function, [R] **functions**, [U] **27.3.4 Translating between time units**

Bonferroni adjustment, [R] **test**

Bonferroni multiple comparison test, [R] **oneway**

bootstrap command, [R] **bootstrap**

bootstrap sampling and estimation, [R] **bootstrap**, [R] **qreg**, [R] **simulate**, [P] **postfile**

Box's conservative epsilon, [R] **anova**

box-and-whisker plots, *see Graphics Reference Manual*

Box–Cox power transformation, [R] **lnskew0**

Box–Cox regression, [R] **boxcox**

Box–Jenkins, [TS] **arch**, [TS] **arima**

boxcox command, [R] **boxcox**; *also see* estimation commands

bprobit command, [R] **glogit**; *also see* estimation commands

break command, [P] **break**

Break key, [U] **12 The Break key**, [U] **19.1.4 Error handling in do-files**
    interception, [P] **break**, [P] **capture**

Breusch–Godfrey test, [TS] **regression diagnostics**

Breusch–Pagan Lagrange multiplier test, [XT] **xtreg**

Breusch–Pagan test of independence, [R] **mvreg**, [R] **sureg**

brier command, [R] **brier**

Brier score decomposition, [R] **brier**

browse command, [R] **edit**

bs command, [R] **bootstrap**

bsample command, [R] **bootstrap**

bsqreg command, [R] **qreg**; *also see* estimation commands

bubble plots, *see Graphics Reference Manual*

built-in, class, [P] **class**

built-in variables, [U] **14.3 Naming conventions**, [U] **16.4 System variables (_variables)**

by *varlist*: prefix, [U] **14.5 by varlist: construct**; [R] **by**, [U] **16.7 Explicit subscripting**, [U] **31.2 The by construct**, [P] **byable**

by-groups, [U] **14.5 by varlist: construct**; [R] **by**, [R] **statsby**, [P] **byable**

byable(), [P] **byable**

bysort *varlist*: prefix, [R] **by**

byte, [R] **data types**, [U] **15.2.2 Numeric storage types**

byteorder() function, [R] **functions**

C

c(adopath) c-class value, [P] **creturn**; [P] **sysdir**

c(adosize) c-class value, [P] **creturn**; [P] **sysdir**

c(born_date) c-class value, [P] **creturn**

c(byteorder) c-class value, [P] **creturn**

c(changed) c-class value, [P] **creturn**

c(checksum) c-class value, [P] **creturn**; [R] **checksum**

c(cmdlen) c-class value, [P] **creturn**

c(console) c-class value, [P] **creturn**

c(current_date) c-class value, [P] **creturn**

c(current_time) c-class value, [P] **creturn**

c(dirsep) c-class value, [P] **creturn**

c(dp) c-class value, [P] **creturn**; [R] **format**

c(epsdouble) c-class value, [P] **creturn**

c(epsfloat) c-class value, [P] **creturn**

c(filedate) c-class value, [P] **creturn**

c(filename) c-class value, [P] **creturn**

c(flavor) c-class value, [P] **creturn**

c(graphics) c-class value, [P] **creturn**

c(httpproxy) c-class value, [P] **creturn**; [R] **netio**

c(httpproxyauth) c-class value, [P] **creturn**; [R] **netio**

c(httpproxyhost) c-class value, [P] **creturn**; [R] **netio**

c(httpproxyport) c-class value, [P] **creturn**; [R] **netio**

c(httpproxypw) c-class value, [P] **creturn**; [R] **netio**

c(httpproxyuser) c-class value, [P] **creturn**; [R] **netio**

c(k) c-class value, [P] **creturn**

c(level) c-class value, [P] **creturn**; [R] **level**

c(linegap) c-class value, [P] **creturn**; [R] **set**
c(linesize) c-class value, [P] **creturn**; [R] **log**
c(logtype) c-class value, [P] **creturn**; [R] **log**
c(machine_type) c-class value, [P] **creturn**
c(macrolen) c-class value, [P] **creturn**
c(matsize) c-class value, [P] **creturn**; [R] **matsize**
c(max_cmdlen) c-class value, [P] **creturn**
c(max_k_current) c-class value, [P] **creturn**
c(max_k_theory) c-class value, [P] **creturn**
c(max_macrolen) c-class value, [P] **creturn**
c(max_matsize) c-class value, [P] **creturn**
c(max_N_current) c-class value, [P] **creturn**
c(max_N_theory) c-class value, [P] **creturn**
c(max_width_current) c-class value, [P] **creturn**
c(max_width_theory) c-class value, [P] **creturn**
c(maxbyte) c-class value, [P] **creturn**
c(maxdb) c-class value, [P] **creturn**; [R] **db**
c(maxdouble) c-class value, [P] **creturn**
c(maxfloat) c-class value, [P] **creturn**
c(maxint) c-class value, [P] **creturn**
c(maxlong) c-class value, [P] **creturn**
c(maxstrvarlen) c-class value, [P] **creturn**
c(maxvar) c-class value, [P] **creturn**; [R] **memory**
c(memory) c-class value, [P] **creturn**; [R] **memory**
c(min_matsize) c-class value, [P] **creturn**
c(minbyte) c-class value, [P] **creturn**
c(mindouble) c-class value, [P] **creturn**
c(minfloat) c-class value, [P] **creturn**
c(minint) c-class value, [P] **creturn**
c(minlong) c-class value, [P] **creturn**
c(mode) c-class value, [P] **creturn**
c(more) c-class value, [P] **creturn**; [R] **more**,
    [P] **more**
c(N) c-class value, [P] **creturn**
c(namelen) c-class value, [P] **creturn**
c(os) c-class value, [P] **creturn**
c(osdtl) c-class value, [P] **creturn**
c(pagesize) c-class value, [P] **creturn**; [R] **more**
c(pi) c-class value, [P] **creturn**
c(printcolor) c-class value, [P] **creturn**
c(pwd) c-class value, [P] **creturn**
c(rc) c-class value, [P] **creturn**; [P] **capture**
c(reventries) c-class value, [P] **creturn**; [R] **set**
c(rmsg) c-class value, [P] **creturn**; [P] **rmsg**
c(rmsg_time) c-class value, [P] **creturn**
c(scheme) c-class value, [P] **creturn**
c(scrollbufsize) c-class value, [P] **creturn**; [R] **set**
c(SE) c-class value, [P] **creturn**
c(searchdefault) c-class value, [R] **search**,
    [P] **creturn**
c(seed) c-class value, [P] **creturn**; [R] **generate**
c(stata_version) c-class value, [P] **creturn**
c(sysdir_base) c-class value, [P] **creturn**; [P] **sysdir**
c(sysdir_oldplace) c-class value, [P] **creturn**;
    [P] **sysdir**
c(sysdir_personal) c-class value, [P] **creturn**;
    [P] **sysdir**

c(sysdir_plus) c-class value, [P] **creturn**; [P] **sysdir**
c(sysdir_site) c-class value, [P] **creturn**; [P] **sysdir**
c(sysdir_stata) c-class value, [P] **creturn**;
    [P] **sysdir**
c(sysdir_updates) c-class value, [P] **creturn**;
    [P] **sysdir**
c(timeout1) c-class value, [P] **creturn**; [R] **netio**
c(timeout2) c-class value, [P] **creturn**; [R] **netio**
c(trace) c-class value, [P] **creturn**; [P] **trace**
c(tracedepth) c-class value, [P] **creturn**; [P] **trace**
c(traceexpand) c-class value, [P] **creturn**; [P] **trace**
c(traceindent) c-class value, [P] **creturn**; [P] **trace**
c(tracenumber) c-class value, [P] **creturn**; [P] **trace**
c(tracesep) c-class value, [P] **creturn**; [P] **trace**
c(type) c-class value, [P] **creturn**; [R] **generate**
c(varlabelpos) c-class value, [P] **creturn**; [R] **set**
c(version) c-class value, [P] **creturn**; [P] **version**
c(virtual) c-class value, [P] **creturn**; [R] **memory**
c(width) c-class value, [P] **creturn**
calculator, [R] **display**
Caliński and Harabasz index stopping rules,
    [CL] **cluster**, [CL] **cluster stop**
_caller() function, [R] **functions**
Canberra dissimilarity measure, [CL] **cluster**
canon command, [R] **canon**; *also see* estimation
    commands
canonical correlations, [R] **canon**
capture command, [P] **capture**
case–cohort data, [ST] **sttocc**
case–control data, [R] **clogit**, [R] **logistic**, [R] **mlogit**,
    [R] **rologit**, [ST] **epitab**, [ST] **sttocc**
casewise deletion, [R] **egen**, [P] **mark**
cat command, [R] **type**
categorical data, [R] **anova**, [R] **areg**, [R] **cusum**,
    [R] **egen**, [R] **grmeanby**, [R] **histogram**,
    [R] **manova**, [R] **mlogit**, [R] **ologit**, [R] **oprobit**,
    [R] **recode**, [R] **table**, [R] **tabsum**, [R] **tabulate**,
    [R] **vwls**, [R] **xi**, [U] **28 Commands for dealing
    with categorical variables**, [ST] **epitab**,
    [SVY] **svy estimators**, [SVY] **svytab**
  agreement, measures for, [R] **kappa**
cc and cci commands, [ST] **epitab**
cchart command, [R] **qc**
c-class, [P] **creturn**
cd command, [R] **cd**
cdir, classutil subcommand, [P] **classutil**
ceil() function, [R] **functions**
ceiling function, [R] **functions**
censored-normal regression, [R] **tobit**
centile command, [R] **centile**
centiles, *see* percentiles
central tendency, measures of, [R] **summarize**; [R] **lv**
centroid linkage clustering, [CL] **cluster**, [CL] **cluster
    centroidlinkage**
centroidlinkage, cluster subcommand,
    [CL] **cluster centroidlinkage**
certifying data, [R] **count**, [R] **inspect**, [R] **assert**, *also
    see* reldiff() and mreldiff()

cf command, [R] **cf**

changing data, *see* editing data

changing directories, [R] **cd**

char command, [U] **15.8 Characteristics**, [P] **char**

char() string function, [R] **functions**

character data, *see* string variables

characteristics, [U] **15.8 Characteristics**,
    [U] **21.3.6 Extended macro functions**,
    [U] **21.3.12 Referencing characteristics**,
    [P] **char**

charset, [P] **smcl**

chdir command, [R] **cd**

checksum command, [R] **checksum**

chi-squared,
    cdf, [R] **functions**
    hypothesis test, [R] **sdtest**, [R] **test**
    test of independence, [R] **tabulate**; [ST] **epitab**,
    [SVY] **svytab**

chi2() function, [R] **functions**

chi2tail() function, [R] **functions**

Cholesky decomposition, [P] **matrix define**

cholesky() matrix function, [P] **matrix define**

chop() function, [R] **functions**

Chow test, [R] **anova**

ci and cii commands, [R] **ci**

class command, [P] **class**
    class cdir, [P] **classutil**
    class describe, [P] **classutil**
    class dir, [P] **classutil**
    class drop, [P] **classutil**
    class exit, [P] **class exit**
    class which, [P] **classutil**

class definition, [P] **class**

class instance, [P] **class**

class programming, [P] **class**

class programming utilities, [P] **classutil**

.classmv built-in class function, [P] **class**

.classname built-in class function, [P] **class**

classutil command, [P] **classutil**

classwide, class, [P] **class**

clear command, [R] **drop**

clear, ereturn subcommand, [P] **ereturn**; [P] **return**

clear, _estimates subcommand, [P] **_estimates**

clear option, [U] **14.2 Abbreviation rules**

clear, postutil subcommand, [P] **postfile**

clear, return subcommand, [P] **return**

clear, serset subcommand, [P] **serset**

clear, sreturn subcommand, [P] **return**;
    [P] **program**

clearing estimation results, [R] **estimates**, [P] **ereturn**,
    [P] **_estimates**

clearing memory, [R] **drop**

clip() function, [R] **functions**

clogit command, [R] **clogit**; *also see* estimation
    commands, [R] **mlogit**

cloglog command, [R] **cloglog**; *also see* estimation
    commands

cloglog function, [R] **functions**

close, file subcommand, [P] **file**

cluster analysis, *see Cluster Analysis Reference Manual*

cluster sampling, [R] **areg**, [R] **biprobit**, [R] **bootstrap**,
    [R] **cloglog**, [R] **heckprob**, [R] **hetprob**,
    [R] **ivreg**, [R] **logistic**, [R] **logit**, [R] **probit**,
    [R] **regress**, [R] **rologit**, [R] **scobit**, [R] **tobit**,
    [ST] **stcox**, [ST] **stphplot**, [ST] **streg**, *also*
    see Survey Reference Manual, [XT] **xtabond**,
    [XT] **xtcloglog**, [XT] **xtfrontier**, [XT] **xtgee**,
    [XT] **xtgls**, [XT] **xthtaylor**, [XT] **xtintreg**,
    [XT] **xtivreg** [XT] **xtlogit**, [XT] **xtnbreg**,
    [XT] **xtpcse**, [XT] **xtpoisson**, [XT] **xtprobit**,
    [XT] **xtrchh**, [XT] **xtreg**, [XT] **xtregar**,
    [XT] **xttobit**

cmdlog command, [R] **log**, [U] **18 Printing and
    preserving output**

cnreg command, [R] **tobit**; *also see* estimation
    commands

cnsreg command, [R] **cnsreg**; *also see* estimation
    commands

Cochrane–Orcutt regression, [TS] **prais**

codebook command, [R] **codebook**

_coef[], [U] **16.5 Accessing coefficients and
    standard errors**; [P] **matrix get**

coefficient alpha, [R] **alpha**

coefficients (from estimation),
    accessing, [R] **estimates**, [U] **16.5 Accessing
    coefficients and standard errors**, [P] **ereturn**,
    [P] **matrix get**
    estimated linear combinations, *see* linear
    combinations of estimators

coleq, matrix subcommand, [P] **matrix rowname**

collapse command, [R] **collapse**

collect statistics, [R] **statsby**

collinear variables, removing, [P] **_rmcoll**

collinearity, handling by regress, [R] **regress**

colnames, matrix subcommand, [P] **matrix rowname**

color, *also see Graphics Reference Manual*
    does Stata think you have, [R] **query**
    specifying in programs, [P] **display**

colsof() matrix function, [P] **matrix define**

columns of matrix,
    appending to, [P] **matrix define**
    names, [U] **17.2 Row and column names**;
        [R] **estimates**, [P] **ereturn**, [P] **matrix define**,
        [P] **matrix rowname**
    operators, [P] **matrix define**

comb() function, [R] **functions**

combinatorials, calculating, [R] **functions**

combining datasets, [R] **append**, [R] **cross**, [R] **joinby**,
    [R] **merge**, [U] **25 Commands for combining
    data**

command arguments, [U] **21.4 Program arguments**,
    [P] **gettoken**, [P] **syntax**, [P] **tokenize**

command line, launching dialog box from, [R] **db**

command parsing, [U] **21.4 Program arguments**,
    [P] **gettoken**, [P] **syntax**, [P] **tokenize**

command timings, [U] **11 Error messages and return codes**

commands,
   abbreviating, [U] **14.2 Abbreviation rules**
   aborting, [U] **12 The Break key**, [U] **13 Keyboard use**, [P] **continue**
   adding new, *see* programs and programming
   editing and repeating, [U] **13 Keyboard use**
   immediate, [U] **22 Immediate commands**
   repeating automatically, [R] **by**, [P] **byable**, [P] **continue**, [P] **foreach**, [P] **forvalues**, [P] **while**
   reviewing, [R] **#review**
   unabbreviating names of, [P] **unabcmd**

commas, reading data separated by, [R] **insheet**; [R] **infile (free format)**

comments in programs, do-files, etc.,
   [U] **19.1.2 Comments and blank lines in do-files**, [U] **21.11.2 Comments and long lines in ado-files**
   adding to programs, [P] **comments**

comments with data, [R] **notes**

communalities, [R] **factor**

compare command, [R] **compare**

comparing two files, [R] **cf**

comparing two variables, [R] **compare**

compatibility of Stata programs across releases, [P] **version**

complementary log-log regression, [R] **cloglog**, [R] **glm**, [XT] **xtcloglog**, [XT] **xtgee**

complete linkage clustering, [CL] **cluster**, [CL] **cluster completelinkage**

completelinkage, cluster subcommand, [CL] **cluster completelinkage**

completely determined, [R] **logit**

component analysis, [R] **factor**, [R] **pca**

component-plus-residual plot, [R] **regression diagnostics**

compound double quotes, [P] **macro**

compress command, [R] **compress**

concat(), egen function, [R] **egen**

concatenating strings, [U] **16.2.2 String operators**

cond() function, [R] **functions**

condition statement, [P] **if**

conditional logistic regression, [R] **clogit**, [R] **rologit**, [XT] **xtlogit**

conditional variance, [TS] **arch**

confidence intervals, [R] **level**, [U] **23.6 Specifying the width of confidence intervals**
   for linear combinations of coefficients, [R] **lincom**, [SVY] **lincom for svy**
   for means, proportions, and counts, [R] **ci**, [R] **means**, [R] **ttest**, [SVY] **svymean**
   for medians and percentiles, [R] **centile**
   for nonlinear combinations of coefficients, [R] **nlcom**
   for odds and risk ratios, [R] **lincom**, [R] **nlcom**, [ST] **epitab**, [ST] **stci**

confirm command, [P] **confirm**

conjoint analysis, [R] **rologit**

conren, set subcommand, [R] **set**

console,
   controlling scrolling of output, [R] **more**, [P] **more**
   obtaining input from, [P] **display**

constrained estimation, [R] **constraint**
   linear regression, [R] **cnsreg**
   multinomial logistic regression, [R] **mlogit**
   multivariate regression, [R] **reg3**
   programming, [P] **matrix constraint**
   seemingly unrelated regression, [R] **reg3**
   three-stage least squares, [R] **reg3**

constraint command, [R] **constraint**; [P] **matrix constraint**

contents of data, [R] **describe**; [R] **codebook**, [R] **labelbook**

context, class, [P] **class**

contingency tables, [R] **tabulate**, [ST] **epitab**, [SVY] **svytab**; [R] **table**

continue command, [P] **continue**

contract command, [R] **contract**

control charts, [R] **qc**

convergence criteria, [R] **maximize**

Cook's D, [R] **predict**, [R] **regression diagnostics**

Cook–Weisberg test for heteroskedasticity, [R] **regression diagnostics**

copy and paste, (Macintosh and Windows), *see Getting Started* manual, [R] **edit**, also *see Graphics Reference Manual*

.copy built-in class function, [P] **class**

copy command, [R] **copy**

Cornfield confidence intervals, [ST] **epitab**

corr() matrix function, [P] **matrix define**

corr2data command, [R] **corr2data**

correcting data, *see* editing data

correlate command, [R] **correlate**, [R] **estimates**

correlated errors, *see* robust

correlation, [R] **correlate**
   canonical, [R] **canon**
   data generation, [R] **corr2data**, [R] **drawnorm**
   intracluster, [R] **loneway**
   Kendall's rank correlation, [R] **spearman**
   pairwise, [R] **correlate**
   partial, [R] **pcorr**
   Pearson's product-moment, [R] **correlate**
   similarity measure, [CL] **cluster**
   Spearman's rank correlation, [R] **spearman**

correlation matrices, [R] **correlate**, [R] **vce**, [P] **matrix define**

correlogram, [TS] **corrgram**, [TS] **xcorr**

corrgram command, [TS] **corrgram**

cos() function, [R] **functions**

cosine function, [R] **functions**

cost frontier model, [R] **frontier**, [XT] **xtfrontier**

count command, [R] **count**

count(), egen function, [R] **egen**

count-time data, [ST] **ct**; [R] **nbreg**, [R] **poisson**, [ST] **ctset**, [ST] **cttost**, [ST] **sttoct**, [SVY] **svy estimators**

counts, making dataset of, [R] collapse
courses in Stata, [U] 2.7 NetCourses
covariance matrix of estimators, [R] vce; [R] correlate, [R] estimates, [P] ereturn, [P] matrix get
covariate class, [R] duplicates
covariate patterns, [R] logistic
COVRATIO, [R] regression diagnostics
Cox proportional hazards model, [ST] stcox
    test of assumption, [ST] stcox, [ST] stphplot
Cox–Snell residuals, [ST] stcox, [ST] streg
cprplot command, [R] regression diagnostics
Cramér's V, [R] tabulate
create_cspline, serset subcommand, [P] serset
create, serset subcommand, [P] serset
create_xmedians, serset subcommand, [P] serset
creturn command, [P] creturn
    creturn list, [P] creturn

Cronbach's alpha, [R] alpha
cross command, [R] cross
cross-correlogram, [TS] xcorr
cross-product matrices, [P] matrix accum
cross-sectional time-series data, see xt
cross-tabulations, see tables
crossover designs, [R] pk, [R] pkcross, [R] pkshape
crude estimates, [ST] epitab
cs and csi commands, [ST] epitab
ct, [ST] ct, [ST] ctset, [ST] cttost, [ST] sttoct
ctset command, [ST] ctset
cttost command, [ST] cttost
cubic splines, graphing, see Graphics Reference Manual
cumsp command, [TS] cumsp
cumul command, [R] cumul
cumulative distribution, empirical, [R] cumul, [TS] cumsp
cumulative distribution functions, [R] functions
cumulative hazard function, [ST] sts, [ST] sts generate, [ST] sts graph, [ST] sts list
    graph of, [ST] streg
cumulative incidence data, [R] poisson, [ST] epitab, [SVY] svy estimators
cumulative spectral distribution, empirical, [TS] cumsp
cusum command, [R] cusum
cut(), egen function, [R] egen

D

d() function, [R] functions, [U] 27.2.5 Specifying particular dates (date literals), [U] 27.3.2 Specifying particular dates (date literals)
daily() function, [R] functions, [U] 27.3.6 Creating time variables
data, [U] 15 Data; [R] data types
    appending, see appending data
    autocorrelated, see autocorrelation
    case–cohort, see case–cohort data
    case–control, see case–control data

data, continued
    categorical, see categorical data
    certifying, see certifying data
    characteristics of, see characteristics
    combining, see combining datasets
    contents of, see contents of data
    count-time, see count-time data
    cross-sectional time-series, see xt
    current, [P] creturn
    documenting, see documenting data
    editing, see editing data
    entering, see inputting data interactively
    exporting, see exporting data
    generating, [R] generate; [R] egen
    importing, see importing data
    inputting, see importing data
    labeling, see labeling data
    large, dealing with, see memory
    listing, see listing data
    matched case–control, see matched case–control data
    missing values, see missing values
    panel, see xt
    preserving, [P] preserve
    range of, see range of data
    ranking, [R] rologit
    reading, see reading data from disk
    recoding, [R] recode
    rectangularizing, see rectangularize dataset
    reordering, [R] sort; [R] gsort; [R] order
    reorganizing, see reorganizing data
    sampling, see sampling
    saving, see saving data
    stacking, [R] stack
    strings, see string variables
    summarizing, see summarizing data
    survey, see survey data
    survival-time, see survival analysis
    time-series, see time-series analysis
    transposing, see transposing data
data entry, see inputting data interactively; reading data from disk
data reduction, [R] pca
database, reading data from other software, [R] odbc, [U] 24.4 Transfer programs
dataset, create, [R] corr2data, [R] drawnorm
datasets, sample, [U] 9 Stata's sample datasets
date,
    and time, [P] creturn
    and time stamp, [R] describe
    displaying, [U] 27.2.3 Displaying dates; [U] 15.5.3 Date formats
    elapsed, [U] 27.2.2 Conversion into elapsed dates; [R] functions
    formats, [U] 27.2.3 Displaying dates; [U] 15.5.3 Date formats
    functions, [U] 27.2.2 Conversion into elapsed dates, [U] 27.2.4 Other date functions; [R] functions
    inputting, [U] 27.2.1 Inputting dates
    of year, [P] creturn
    variables, [U] 27 Commands for dealing with dates

date() function, [U] **27.2.2.2 The date() function**;
[R] **functions**

day() function, [U] **27.2.4 Other date functions**;
[R] **functions** [U] **27.3.5 Extracting components
of time**

db command, [R] **db**, [R] **set**

dBASE, reading data from, [U] **24.4 Transfer
programs**

DBETAs, [R] **regression diagnostics**

.dct filename suffix, [R] **infile**, [U] **14.6 File-naming
conventions**

debugging, [P] **trace**; [P] **discard**, [P] **pause**

.Declare built-in class modifier, [P] **class**

declare, class, [P] **class**

decode command, [R] **encode**; [R] **append**

default settings of system parameters, [R] **query**

deff, [R] **loneway**, [SVY] **svy estimators**,
[SVY] **svymean**

define, matrix subcommand, [P] **matrix define**

define, program subcommand, [P] **program**

degree-to-radian conversion, [R] **functions**

delete, cluster subcommand, [CL] **cluster
programming utilities**

deleting,
    casewise, [R] **egen**
    files, [R] **erase**
    variables or observations, [R] **drop**

#delimit command, [P] **#delimit**

delimiter,
    for comments, [P] **comments**
    for lines, [P] **#delimit**

delta beta influence statistic, [R] **logistic**, [R] **regression
diagnostics**

delta chi-squared influence statistic, [R] **logistic**

delta deviance influence statistic, [R] **logistic**

delta method, [R] **nlcom**, [R] **predictnl**, [R] **testnl**

dendrogram, cluster subcommand, [CL] **cluster
dendrogram**

dendrograms, [CL] **cluster**, [CL] **cluster dendrogram**
    applying alternate routine, [CL] **cluster
    programming subroutines**

density estimation, kernel, [R] **kdensity**

derivative of incomplete gamma function, [R] **functions**

derivatives, numerical, [R] **range**, [R] **testnl**

describe, class, [P] **classutil**

describe, classutil subcommand, [P] **classutil**

describe command, [R] **describe**; [U] **15.6 Dataset,
variable, and value labels**

descriptive statistics,
    creating dataset containing, [R] **collapse**
    creating variables containing, [R] **egen**
    displaying, [R] **summarize**, [R] **table**, [R] **tabstat**,
    [R] **tabsum**; [R] **codebook**, [R] **lv**, [R] **pctile**,
    [XT] **xtsum**, [XT] **xttab**

design effects, [R] **loneway**, [SVY] **svy estimators**,
[SVY] **svymean**

destring command, [R] **destring**

destructors, class, [P] **class**

det() matrix function, [P] **matrix define**

determinant of matrix, [P] **matrix define**

deviance residual, [R] **glm**, [R] **logistic**, [ST] **stcox**,
[ST] **streg**

dfbeta command, [R] **regression diagnostics**

DFBETAs, [R] **regression diagnostics**

dfgls command, [TS] **dfgls**

DFITS, [R] **regression diagnostics**

dfuller command, [TS] **dfuller**

dgammapda() function, [R] **functions**

dgammapdada() function, [R] **functions**

dgammapdadx() function, [R] **functions**

dgammapdx() function, [R] **functions**

dgammapdxdx() function, [R] **functions**

diag() matrix function, [P] **matrix define**

diag0cnt() matrix function, [P] **matrix define**

diagnostic codes, [R] **icd9**

diagnostic plots, [R] **diagnostic plots**, [R] **logistic**,
[R] **regression diagnostics**

diagnostics, regression, [R] **regression diagnostics**

diagonals of matrices, [P] **matrix define**

dialog box, [R] **db**, [P] **dialogs**, [P] **window fopen**,
[P] **window stopbox**

dialog programming, [P] **dialogs**; [P] **window menu**

Dice coefficient similarity measure, [CL] **cluster**

dichotomous outcome models, [R] **logistic**; [R] **brier**,
[R] **clogit**, [R] **cloglog**, [R] **glm**, [R] **glogit**,
[R] **hetprob**, [R] **logit**, [R] **probit**, [R] **rologit**,
[R] **scobit**, [XT] **xtcloglog**, [XT] **xtgee**,
[XT] **xtlogit**, [XT] **xtprobit**

dictionaries, [R] **infile**, [R] **infile (fixed format)**,
[R] **infix (fixed format)**, [R] **outfile**

diff(), egen function, [R] **egen**

difference of estimated coefficients, *see* linear
combinations of estimators

difference of subpopulation means, [SVY] **lincom for
svy**, [SVY] **svymean**

digamma() function, [R] **functions**

digits, controlling the number displayed, [R] **format**,
[U] **15.5 Formats: controlling how data are
displayed**

dir command, [R] **dir**

dir, classutil subcommand, [P] **classutil**

dir, cluster subcommand, [CL] **cluster utility**

dir, _estimates subcommand, [P] **_estimates**

dir, macro subcommand, [P] **macro**

dir, matrix subcommand, [P] **matrix utility**

dir, postutil subcommand, [P] **postfile**

dir, program subcommand, [P] **program**

dir, _return subcommand, [P] **_return**

dir, serset subcommand, [P] **serset**

direct standardization, [R] **dstdize**

directories, [U] **14.6 File-naming conventions**,
[U] **21.3.10 Constructing Windows filenames
using macros**
    changing, [R] **cd**
    creating, [R] **mkdir**

directories, *continued*
listing, [R] **dir**
location of ado-files, [U] **20.5 Where does Stata look for ado-files?**
directories and paths, system, [P] **creturn**, [P] **sysdir**
directory, class, [P] **classutil**
discard command, [U] **21.11.3 Debugging ado-files**, [P] **discard**
discrete choice models, [R] **rologit**
discrete survival data, [ST] **discrete**
dispCns, matrix subcommand, [P] **matrix constraint**
dispersion, measures of, [R] **summarize**, [R] **table**; [R] **centile**, [R] **lv**, [R] **pctile**, [XT] **xtsum**
display command, [P] **display**; [P] **macro**
as a calculator, [R] **display**
display, ereturn subcommand, [P] **ereturn**
display formats, [U] **15.5 Formats: controlling how data are displayed**; [R] **describe**, [R] **format**, [U] **27.2.3 Displaying dates**, [P] **macro**, *also see Graphics Reference Manual*
display saved results, [R] **saved results**
display-width and length, [R] **log**
displaying,
contents, [R] **describe**
data, [R] **edit**, [R] **list**
macros, [P] **macro**
matrix, [P] **matrix utility**
output, [P] **display**, [P] **quietly**, [P] **smcl**, [P] **tabdisp**
previously typed lines, [R] **#review**
saved results, [R] **saved results**
scalar expressions, [P] **display**, [P] **scalar**
also see printing
dissimilarity measures, [CL] **cluster**
distances, *see* dissimilarity measures
distributions,
diagnostic plots, [R] **diagnostic plots**
examining, [R] **centile**, [R] **kdensity**, [R] **lv**, [R] **pctile**, [R] **stem**, [R] **summarize**, *also see Graphics Reference Manual*
testing equality, [R] **ksmirnov**, [R] **kwallis**, [R] **signrank**
testing for normality, [R] **sktest**, [R] **swilk**
division operator, *see* arithmetic operators
.do filename suffix, [U] **14.6 File-naming conventions**
do command, [R] **do**, [U] **19 Do-files**
do-files, [R] **do**, [U] **19 Do-files**, [U] **21.2 Relationship between a program and a do-file**, [P] **version**
adding comments to, [P] **comments**
editing, [R] **doedit**
long lines, [U] **21.11.2 Comments and long lines in ado-files**, [P] **#delimit**
documentation, [U] **1 Read this—it will help**
documentation, keyword search on, [R] **search**, [U] **8 Stata's online help and search facilities**
documenting data, [R] **codebook**, [R] **labelbook**, [R] **notes**

doedit command, [R] **doedit**
dofb(), dofd(), dofm(), dofq(), dofw(), and dofy() functions, [R] **functions**, [U] **27.3.4 Translating between time units**
domain sampling, [R] **alpha**
dose–response models, [R] **glm**, [R] **logistic**
dotplot command, [R] **dotplot**
double, [R] **data types**, [U] **15.2.2 Numeric storage types**
double exponential smoothing, [TS] **tssmooth dexponential**
double quotes, [P] **macro**
double-precision floating point number, [U] **15.2.2 Numeric storage types**
dow() date function, [U] **27.2.4 Other date functions**; [R] **functions**
doy() function, [R] **functions**, [U] **27.3.5 Extracting components of time**
dprobit command, [R] **probit**; *also see* estimation commands
drawnorm command, [R] **drawnorm**
drop command, [R] **drop**
drop, class, [P] **classutil**
drop, classutil subcommand, [P] **classutil**
drop, cluster subcommand, [CL] **cluster utility**
drop, _estimates subcommand, [P] **_estimates**
drop, macro subcommand, [P] **macro**
drop, matrix subcommand, [P] **matrix utility**
drop, program subcommand, [P] **program**
drop, _return subcommand, [P] **_return**
drop, serset subcommand, [P] **serset**
dropping programs, [P] **discard**
dropping variables and observations, [R] **drop**
ds command, [R] **describe**
dstdize command, [R] **dstdize**
.dta file suffix, [U] **14.6 File-naming conventions**
technical description, [P] **file formats .dta**
Duda and Hart index stopping rules, [CL] **cluster**, [CL] **cluster stop**
dummy variables, *see* indicator variables
duplicate observations,
dropping, [R] **duplicates**
identifying, [R] **duplicates**
duplicates command, [R] **duplicates**
duplicating observations, [R] **expand**
duration models, *see* survival analysis
Durbin's alternative test, [TS] **regression diagnostics**
Durbin–Watson statistic, [R] **regression diagnostics**, [TS] **prais**, [TS] **regression diagnostics**
durbina command, [TS] **regression diagnostics**
dwstat command, [TS] **regression diagnostics**
dydx command, [R] **range**
dynamic structural simultaneous equations, [TS] **var svar**
.dynamicmv built-in class function, [P] **class**

E

e() scalars, macros, matrices, functions, [R] **saved results**, [R] **functions**, [U] **21.8 Accessing results calculated by other programs**, [U] **21.9 Accessing results calculated by estimation commands**, [U] **21.10.2 Saving results in e()**, [P] **ereturn**, [P] **_estimates**, [P] **return**

e(sample) function, [R] **functions**, [P] **ereturn**, [P] **return**

e-class command, [R] **saved results**, [U] **21.8 Accessing results calculated by other programs**, [P] **program**, [P] **return**

edit command, [R] **edit**

editing,
  ado-files and do-files, [R] **doedit**
  commands, [U] **13 Keyboard use**
  data, [R] **edit**, [R] **generate**, [R] **merge**, [R] **recode**
  files while in Stata, [R] **doedit**, [R] **shell**
  graphs, *see Graphics Reference Manual*
  output, [U] **18 Printing and preserving output**

egen command, [R] **egen**

EGLS (estimated generalized least squares), [XT] **xtgls**, [XT] **xtivreg** [XT] **xtreg**

eigenvalue stability condition, [TS] **varstable**

eigenvalues and eigenvectors, [P] **matrix symeigen**; [R] **pca**, [P] **matrix svd**

eigenvalues, matrix subcommand, [P] **matrix eigenvalues**

eivreg command, [R] **eivreg**; *also see* estimation commands

el() matrix function, [P] **matrix define**

elapsed dates, [U] **27.2.2 Conversion into elapsed dates**; [R] **functions**

elasticity, [R] **mfx**

else command, [P] **if**

empirical cumulative distribution function, [R] **cumul**

encode command, [R] **encode**; [U] **26.2 Categorical string variables**

ending a Stata session, [R] **exit**

endless loop, *see* loop, endless

endogenous treatment, [R] **treatreg**

endogenous variables, [R] **ivreg**, [R] **reg3**, [SVY] **svy estimators**

ends(), egen function, [R] **egen**

Engle's LM test, [TS] **regression diagnostics**

entering data, *see* inputting data interactively; reading data from disk

environment variables (Unix), [P] **macro**, *see Getting Started with Stata for Unix*

Epanechnikov kernel density estimator, [R] **kdensity**

epidemiological tables, [ST] **epitab**; [R] **dstdize**, [R] **tabulate**

EPS files, *see Getting Started manual, also see Graphics Reference Manual*

epsdouble() function, [R] **functions**

epsfloat() function, [R] **functions**

eqany(), egen function, [R] **egen**

equality operator, [U] **16.2.3 Relational operators**

equality tests,
  binomial proportions, [R] **bitest**
  coefficients, [R] **test**, [R] **testnl**
  distributions, [R] **ksmirnov**, [R] **kwallis**, [R] **signrank**
  means, [R] **ttest**; [R] **hotelling**
  medians, [R] **signrank**
  proportions, [R] **bitest**, [R] **prtest**
  survivor functions, [ST] **sts test**
  variances, [R] **sdtest**

equation names of matrix, [P] **matrix rowname**; [U] **17.2 Row and column names**, [P] **ereturn**, [P] **matrix define**

equivalence tests, [R] **pk**, [R] **pkequiv**

erase command, [R] **erase**

erasing files, [R] **erase**

ereturn command, [P] **ereturn**
  ereturn clear, [P] **ereturn**; [P] **return**
  ereturn display, [P] **ereturn**
  ereturn list, [P] **ereturn**; [R] **saved results**
  ereturn local, [P] **ereturn**; [P] **return**
  ereturn matrix, [P] **ereturn**; [P] **return**
  ereturn post, [P] **ereturn**; [P] **matrix constraint**, [P] **return**
  ereturn repost, [P] **ereturn**; [P] **return**
  ereturn scalar, [P] **ereturn**; [P] **return**

error checking, [R] **assert**

error command, [P] **error**

error components model, [XT] **xthtaylor**

error handling, [U] **19.1.4 Error handling in do-files**, [P] **capture**, [P] **confirm**, [P] **error**

error messages and return codes, [P] **error**, [P] **rmsg**; [R] **error messages**, [U] **8.8.5 Return codes**, [U] **11 Error messages and return codes**, *also see* error handling

error-bar charts, [R] **serrbar**

errors-in-variables regression, [R] **eivreg**

estimate linear combinations of coefficients, *see* linear combinations of estimators

estimated generalized least squares, *see* EGLS

_estimates command, [P] **_estimates**
  _estimates clear, [P] **_estimates**
  _estimates dir, [P] **_estimates**
  _estimates drop, [P] **_estimates**
  _estimates hold, [P] **_estimates**
  _estimates uhold, [P] **_estimates**

estimates command, [R] **estimates**, [R] **suest**

estimation commands, [R] **estimates**, [R] **estimation commands**, [U] **21.9 Accessing results calculated by estimation commands**, [U] **23 Estimation and post-estimation commands**, [U] **29 Overview of Stata estimation commands**, [U] **30 Overview of survey estimation**, [P] **ereturn**, [P] **_estimates** *also see* Subject table of contents
  accessing stored information from, [P] **matrix get**
  allowing constraints in, [P] **matrix constraint**
  eliminating stored information from, [P] **discard**

estimation commands, *continued*
  obtaining predictions after, [R] **predictnl**,
    [P] **_predict**
  obtaining robust estimates, [P] **_robust**
  saving results from, [P] **_estimates**
estimation results,
  clearing, [P] **ereturn**, [P] **_estimates**
  listing, [P] **ereturn**, [P] **_estimates**
  restoring, [R] **estimates**
  saving, [P] **ereturn**, [P] **_estimates**
  storing, [R] **estimates**
estimators,
  between, [XT] **xtreg**
  covariance matrix of, [R] **vce** [U] **23.7 Obtaining
    the variance–covariance matrix**; [R] **correlate**,
    [P] **ereturn**, [P] **matrix get**
  linear combinations, [R] **lincom**,
    [U] **23.11 Obtaining linear combinations of
    coefficients**, [SVY] **lincom for svy**
  nonlinear combinations, [R] **nlcom**
  within, [XT] **xtreg**
etiologic fraction, [ST] **epitab**
Euclidean dissimilarity measure, [CL] **cluster**
exact test, Fisher's, [R] **tabulate**; [ST] **epitab**
Excel, Microsoft, reading data from, [R] **odbc**, *also see*
  spreadsheets
excess fraction, [ST] **epitab**
exit class program, [P] **class exit**
exit, class subcommand, [P] **class exit**
exit command, [R] **exit**, [U] **5.3 Exiting Stata**,
    [U] **19.1.4 Error handling in do-files**,
    [P] **capture**, [P] **exit**, *also see Getting Started*
  manuals
exiting Stata, *see* exit command
=*exp*, [U] **14 Language syntax**
exp() function, [R] **functions**
expand command, [R] **expand**; [ST] **epitab**, [ST] **stset**
exponential distribution, [ST] **streg**
exponential function, [R] **functions**
exponential model, stochastic frontier, [R] **frontier**
exponential notation, [U] **15.2 Numbers**
exponential smoothing, [TS] **tssmooth exponential**,
    [TS] **tssmooth**
exponential survival regression, [ST] **streg**
exporting data, [R] **outfile**, [R] **outsheet**;
    [U] **24.4 Transfer programs**
expressions, [U] **16 Functions and expressions**,
    [P] **matrix define**
extended macro funtions, [P] **char**, [P] **display**,
    [P] **macro**, [P] **macro lists**, [P] **serset**
extended memory, [R] **memory**, *also see Getting Started
    with Stata for Unix*
extrapolation, [R] **ipolate**

F

F density, [R] **functions**
F density, noncentral, [R] **functions**

F distribution, reverse cumulative noncentral,
    [R] **functions**
F distribution, inverse reverse cumulative noncentral,
    [R] **functions**
F() distribution function, [R] **functions**
F-keys, [U] **13 Keyboard use**, [U] **13.2 F-keys**
factor analysis, [R] **factor**; [R] **alpha**, [R] **canon**,
    [R] **impute**
factor command, [R] **factor**, [R] **matsize**,
    [R] **maximize**
factorial design, [R] **anova**, [R] **manova**
factorial function, [R] **functions**
failure analysis, *see* survival analysis
failure tables, [ST] **ltable**
failure-time models, [ST] **stcox**, [ST] **stphplot**,
    [ST] **streg**; [R] **nbreg**, [R] **poisson**, [R] **tobit**,
    [R] **zip**, [SVY] **svy estimators**, [XT] **xtpoisson**
FAQs, [U] **2.2 The http://www.stata.com web site**
  search, [R] **search**, [U] **8.8.4 FAQ searches**
fastscroll, [R] **set**
Fden() function, [R] **functions**
feasible generalized least squares, *see* FGLS
fences, [R] **lv**, *also see Graphics Reference Manual*
FGLS (feasible generalized least squares), [R] **reg3**,
    [TS] **dfgls**, [TS] **prais**, [TS] **var**, [XT] **xtgls**,
    [XT] **xtivreg**, [XT] **xtreg**
file command, [P] **file**
  file close, [P] **file**
  file open, [P] **file**
  file query, [P] **file**
  file read, [P] **file**
  file seek, [P] **file**
  file sersetread, [P] **serset**
  file sersetwrite, [P] **serset**
  file set, [P] **file**
  file write, [P] **file**
file, find in path, [P] **findfile**
file format, Stata, [P] **file formats .dta**
files,
  checksum of, [R] **checksum**
  comparison, [R] **cf**
  copying and appending, [R] **copy**
  display contents of, [R] **type**
  downloading, [R] **checksum**, [R] **net**, [R] **sj**,
    [R] **update**, [U] **32 Using the Internet to keep
    up to date**
  erasing, [R] **erase**
  exporting, *see* files, saving, *see* exporting data
  extensions, [U] **14.6 File-naming conventions**
  importing, *see* importing data
  names, [R] **dir**, [U] **14.6 File-naming conventions**,
    [U] **21.3.10 Constructing Windows filenames
    using macros**
  opening, [P] **window fopen**
  reading ASCII text or binary, [P] **file**
  saving, [R] **save**, [P] **window fopen**
  temporary, [P] **macro**, [P] **preserve**, [P] **scalar**
  writing ASCII text or binary, [P] **file**
fill(), egen function, [R] **egen**

fillin command, [R] **fillin**

filters, [TS] **tssmooth**, [TS] **tssmooth dexponential**, [TS] **tssmooth exponential**, [TS] **tssmooth hwinters**, [TS] **tssmooth ma**, [TS] **tssmooth nl**, [TS] **tssmooth shwinters**

findfile command, [P] **findfile**

finding file in path, [P] **findfile**

findit, [R] **search**

first-differenced estimator, [XT] **xtivreg**

Fisher's exact test, [R] **tabulate**; [ST] **epitab**

fixed-effects models, [R] **anova**, [R] **areg**, [R] **clogit**, [XT] **xtivreg**, [XT] **xtlogit**, [XT] **xtnbreg**, [XT] **xtpoisson**, [XT] **xtreg**, [XT] **xtregar**

flexible functional form, [R] **boxcox**, [R] **fracpoly**, [R] **mfp**

float, [R] **data types**, [U] **15.2.2 Numeric storage types**, [U] **16.10 Precision and problems therein**

float() function, [U] **16.10 Precision and problems therein**; [R] **functions**

floor() function, [R] **functions**

%fmts, [R] **format**, [U] **15.5 Formats: controlling how data are displayed**

folders, creating, [R] **mkdir**

fonts, (Macintosh and Windows), *see Getting Started manual, also see Graphics Reference Manual*

foreach command, [P] **foreach**

forecast, [TS] **arch**, [TS] **arima**, [TS] **tsappend**, [TS] **tssmooth**, [TS] **varfcast**
    standard error of, [R] **predict**, [R] **regression diagnostics**

forecast error variance decomposition, [TS] **varbasic**, [TS] **varirf**, [TS] **varirf create**

forecasting, [TS] **arch**, [TS] **arima**, [TS] **tssmooth**, [TS] **tssmooth dexponential**, [TS] **tssmooth exponential**, [TS] **tssmooth hwinters**, [TS] **tssmooth shwinters**, [TS] **varfcast compute**, [TS] **varfcast graph**

format command, [R] **format**

formats, [U] **15.5 Formats: controlling how data are displayed**; [R] **describe**, [R] **format**, [U] **27.2.3 Displaying dates**, *also see Graphics Reference Manual*

formatted data, reading, [R] **infile**, [R] **infile (fixed format)**, [R] **infix (fixed format)**; [U] **24 Commands to input data**

formatting contents of macros, [P] **macro**

formatting statistical output, [R] **format**

forvalues command, [P] **forvalues**

fourfold tables, [ST] **epitab**

FoxPro, reading data from, [U] **24.4 Transfer programs**

fracgen command, [R] **fracpoly**

fracplot command, [R] **fracpoly**

fracpoly command, [R] **fracpoly**; *also see estimation commands*

fraction defective, [R] **qc**

fractional polynomial regression, [R] **fracpoly**
    multivariate, [R] **mfp**

frailty models, [ST] **stcox**, [ST] **streg**; [ST] **discrete**

frequencies,
    creating dataset of, [R] **collapse**, [R] **contract**
    graphical representation, [R] **histogram**, [R] **kdensity**, *also see Graphics Reference Manual*
    table of, [R] **tabulate**; [R] **table**, [R] **tabsum**, [SVY] **svytab**, [XT] **xttab**

frequency domain, [TS] **cumsp**, [TS] **pergram**

frequency weights, [U] **14.1.6 weight**, [U] **23.16.1 Frequency weights**

[frequency=*exp*] modifier, [U] **14.1.6 weight**, [U] **23.16.1 Frequency weights**

frontier command, [R] **frontier**; *also see estimation commands*

frontier models, [R] **frontier**, [XT] **xtfrontier**

Ftail() function, [R] **functions**

functions, [R] **functions**, [U] **16.3 Functions**
    aggregate, [R] **egen**
    combinatorial, [R] **functions**
    creating datasets of, [R] **collapse**, [R] **obs**
    date, [U] **27.2.2 Conversion into elapsed dates**, [U] **27.2.4 Other date functions**; [R] **functions**
    extended macro, [P] **char**, [P] **display**, [P] **macro**, [P] **macro lists**, [P] **serset**
    generate functions, adding, [CL] **cluster programming subroutines**
    graphing, [R] **range**, *also see Graphics Reference Manual*
    link, [R] **glm**
    mathematical, [R] **functions**
    matrix, [U] **17.8 Matrix functions**, [P] **matrix define**
    piecewise linear, [R] **mkspline**
    random number, [R] **generate**
    statistical, [R] **functions**
    string, [R] **functions**
    survivor, *see survivor function*
    time-series, [R] **functions**

[fweight=*exp*] modifier, [U] **14.1.6 weight**, [U] **23.16.1 Frequency weights**

G

g2 inverse of matrix, [P] **matrix define**, [P] **matrix svd**

gamma density, [R] **functions**

gamma density function, incomplete, [R] **functions**

gamma distribution, [R] **functions**

gammaden() function, [R] **functions**

gammap() function, [R] **functions**

GARCH, [TS] **arch**

garch command, [TS] **arch**; *also see estimation commands*

Gauss, reading data from, [U] **24.4 Transfer programs**

Gauss–Hermite quadrature, [XT] **quadchk**

GEE (generalized estimating equations), [XT] **xtgee**; [XT] **xtcloglog**, [XT] **xtintreg**, [XT] **xtlogit**, [XT] **xtnbreg**, [XT] **xtpoisson**, [XT] **xtprobit**, [XT] **xttobit**

generalized autoregressive conditional heteroskedasticity, [TS] **arch**

generalized estimating equations, see GEE
generalized gamma survival regression, [ST] **streg**
generalized inverse of matrix, [P] **matrix define**;
    [P] **matrix svd**
generalized least squares, see FGLS
generalized linear models, see GLM
generalized negative binomial regression, [SVY] **svy**
    estimators
generate, cluster subcommand, [CL] **cluster**
    generate
generate command, [R] **generate**; [R] **egen**
generate functions,
    adding, [CL] **cluster programming subroutines**
    geometric mean, [R] **means**
get() matrix function, [P] **matrix get**; [P] **matrix**
    define
getting started, [U] **1 Read this—it will help**
Getting Started with Stata manuals, [U] **1.1 Getting**
    Started with Stata
    keyword search of, [R] **search**, [U] **8 Stata's online**
    help and search facilities
gettoken command, [P] **gettoken**
gladder command, [R] **ladder**
GLM, [R] **binreg**, [R] **glm**
glm command, [R] **glm**; *also see* estimation commands
Global class prefix operator, [P] **class**
global command, [U] **21.3.2 Global macros**,
    [U] **21.3.9 Advanced global macro**
    manipulation, [P] **macro**
glogit command, [R] **glogit**; *also see* estimation
    commands
glsaccum, matrix subcommand, [P] **matrix accum**
gnbreg command, [R] **nbreg**; *also see* estimation
    commands
Gompertz survival regression, [ST] **streg**
Goodman and Kruskal's gamma, [R] **tabulate**
goodness-of-fit tests, [R] **brier**, [R] **diagnostic plots**,
    [R] **ksmirnov**, [R] **logistic**, [R] **regression**
    diagnostics, [R] **swilk**
Gower coefficient similarity measure, [CL] **cluster**
Gower's method, [CL] **cluster medianlinkage**
.gph files, [U] **14.6 File-naming conventions**; *also see*
    Graphics Reference Manual
gprobit command, [R] **glogit**; *also see* estimation
    commands
Granger causality, [TS] **vargranger**
graph command, see *Graphics Reference Manual*
graphical user interface, [P] **dialogs**
graphs, *also see Graphics Reference Manual*
    added-variable plot, [R] **regression diagnostics**
    adjusted Kaplan–Meier survival curves, [ST] **sts**
    adjusted partial residual plot, [R] **regression**
    diagnostics
    augmented component-plus-residual plot,
    [R] **regression diagnostics**
    augmented partial residual plot, [R] **regression**
    diagnostics
    autocorrelations, [TS] **corrgram**
    baseline hazard and survival, [ST] **stcox**, [ST] **sts**

graphs, *continued*
    binary variable cumulative sum, [R] **cusum**
    component-plus-residual, [R] **regression diagnostics**
    cross-sectional time-series data, [XT] **xtdata**
    cumulative distribution, [R] **cumul**
    cumulative hazard function, [ST] **streg**
    cumulative spectral density, [TS] **cumsp**
    dendrograms, [CL] **cluster**, [CL] **cluster**
    dendrogram
    density, [R] **kdensity**
    derivatives, [R] **range**, [R] **testnl**
    diagnostic, [R] **diagnostic plots**
    dotplot, [R] **dotplot**
    eigenvalues after factor, [R] **factor**
    eigenvalues after pca, [R] **pca**
    error-bar charts, [R] **serrbar**
    fonts, see fonts
    fractional polynomial, [R] **fracpoly**
    functions, [R] **obs**, [R] **range**
    hazard function, [ST] **streg**
    histograms, [R] **histogram**; [R] **kdensity**
    integrals, [R] **range**
    Kaplan–Meier survival curves, [ST] **sts**
    ladder-of-powers histograms, [R] **ladder**
    leverage-versus-(squared)-residual, [R] **regression**
    diagnostics
    log-log curves, [ST] **stphplot**
    logistic diagnostic, [R] **logistic**
    lowess smoothing, [R] **lowess**
    means and medians, [R] **grmeanby**
    normal probability, [R] **diagnostic plots**
    parameterized curves, [R] **range**
    partial residual, [R] **regression diagnostics**
    partial-regression leverage, [R] **regression**
    diagnostics
    periodogram, [TS] **pergram**
    printing, see *Graphics Manual*
    quality control, [R] **qc**
    quantile, [R] **diagnostic plots**
    quantile-normal, [R] **diagnostic plots**
    quantile-quantile, [R] **diagnostic plots**
    regression diagnostic, [R] **regression diagnostics**
    residual versus fitted, [R] **regression diagnostics**
    residual versus predictor, [R] **regression diagnostics**
    ROC curve, [R] **logistic**
    rootograms, [R] **spikeplot**
    smoothing, [R] **kdensity**, [R] **lowess**, [R] **smooth**
    spike plot, [R] **spikeplot**
    stem-and-leaf, [R] **stem**
    survivor function, [ST] **streg**, [ST] **sts graph**
    symmetry, [R] **diagnostic plots**
    time-versus-concentration curve, [R] **pk**,
    [R] **pkexamine**
    trees, [CL] **cluster**, [CL] **cluster dendrogram**
    VAR forecasts, [TS] **varfcast graph**
    VAR impulse-response functions, [TS] **varirf**,
    [TS] **varirf cgraph**, [TS] **varirf graph**,
    [TS] **varirf ograph**
    white noise test, [TS] **wntestb**, [TS] **wntestq**
greater than (or equal) operator, [U] **16.2.3 Relational**
    operators
Greenhouse–Geisser epsilon, [R] **anova**

Greenwood confidence intervals, [ST] **sts**; [ST] **ltable**
greigen command, [R] **factor**, [R] **pca**
grmeanby command, [R] **grmeanby**
group(), egen function, [R] **egen**
group() function, [R] **functions**
grouped data regression, [R] **tobit**
grouping variables, generating, [CL] **cluster generate**
gsort command, [R] **gsort**
GUI,
    examples of, [U] **3 A brief description of Stata**
    programming, [P] **dialogs**, [P] **window menu**

H

hadamard() matrix function, [R] **functions**, [P] **matrix define**
half-normal model, stochastic frontier, [R] **frontier**
halfyear() function, [R] **functions**,
    [U] **27.3.5 Extracting components of time**
Hamman coefficient similarity measure, [CL] **cluster**
harmonic mean, [R] **means**
hat,
    diagonal elements of, [R] **logistic**, [R] **predict**,
      [R] **regression diagnostics**, [R] **rreg**
    obtaining with weighted data, [R] **predict**, [R] **rreg**
hausman command, [R] **hausman**
hausman, running with stored estimates, [R] **estimates**
Hausman specification test, [R] **hausman**, [XT] **xtreg**
Hausman-Taylor estimator, [XT] **xthtaylor**
hazard function, graph of, [ST] **streg**
hazard ratios, models, tables, *see* survival analysis
health ratio, [R] **binreg**
heckman command, [R] **heckman**; *also see* estimation commands
Heckman selection model, [R] **heckman**, [R] **treatreg**, [SVY] **svy estimators**
heckprob command, [R] **heckprob**; *also see* estimation commands
help command, [R] **help**, [U] **8 Stata's online help and search facilities**, [U] **10 –more– conditions**
    writing your own, [U] **21.11.6 Writing online help**
help, set subcommand, [R] **help**, [R] **set**
help—I don't know what to do, [U] **2 Resources for learning and using Stata**
heterogeneity tests, [ST] **epitab**
heteroskedasticity,
    conditional, [TS] **arch**, [TS] **regression diagnostics**
    heteroskedastic probit, [R] **hetprob**
    multiplicative heteroskedastic regression, [TS] **arch**
    panel data, [XT] **xtgls**
    robust variances, *see* robust
    stochastic frontier, [R] **frontier**
    test for, [R] **regression diagnostics**, [TS] **regression diagnostics**
hetprob command, [R] **hetprob**; *also see* estimation commands
hettest command, [R] **regression diagnostics**
hexadecimal report, [P] **hexdump**

hexdump command, [P] **hexdump**
hierarchical cluster analysis, [CL] **cluster**, [CL] **cluster averagelinkage**, [CL] **cluster centroidlinkage**, [CL] **cluster completelinkage**, [CL] **cluster medianlinkage**, [CL] **cluster singlelinkage**, [CL] **cluster wardslinkage**, [CL] **cluster waveragelinkage**
hierarchical regression, [R] **areg**
hierarchical samples, [R] **anova**, [R] **loneway**; [R] **areg**
high-low charts, *see Graphics Reference Manual*
Hildreth–Houck random coefficients model, [XT] **xtrchh**; *also see* estimation commands
Hildreth–Lu regression, [TS] **prais**
histogram command, [R] **histogram**
histograms, [R] **histogram**, *also see Graphics Reference Manual*
    dotplots, [R] **dotplot**
    kernel density estimator, [R] **kdensity**
    ladder-of-powers, [R] **ladder**
    of categorical variables, [R] **histogram**
    stem-and-leaf, [R] **stem**
.hlp files, [U] **8 Stata's online help and search facilities**, [U] **21.11.6 Writing online help**
hold, _estimates subcommand, [P] **_estimates**
hold, _return subcommand, [P] **_return**
Holm adjustment, [R] **test**
homogeneity of variances, [R] **oneway**, [R] **sdtest**
homogeneity tests, [R] **symmetry**, [ST] **epitab**
homoskedasticity tests, [R] **regression diagnostics**
Hosmer–Lemeshow goodness-of-fit test, [R] **logistic**
hotelling command, [R] **hotelling**
Hotelling's generalized T-squared statistic, [R] **hotelling** [R] **manova**
http://www.stata.com, [U] **2.2 The http://www.stata.com web site**
httpproxy, set subcommand, [R] **netio**, [R] **set**
httpproxyauth, set subcommand, [R] **netio**, [R] **set**
httpproxyhost, set subcommand, [R] **netio**, [R] **set**
httpproxyport, set subcommand, [R] **netio**, [R] **set**
httpproxypw, set subcommand, [R] **netio**, [R] **set**
httpproxyuser, set subcommand, [R] **netio**, [R] **set**
Huber weighting, [R] **rreg**
Huber/White/sandwich estimator of variance, *see* robust
Huynh–Feldt epsilon, [R] **anova**
hypertext help, [R] **help**, [U] **8 Stata's online help and search facilities**, [U] **21.11.6 Writing online help**; [P] **smcl**
hypothesis tests, *see* tests

I

I() matrix function, [P] **matrix define**
ibeta() function, [R] **functions**
icd9 command, [R] **icd9**
icd9p command, [R] **icd9**
identifier,
    class, [P] **class**
    unique, [R] **isid**

identity matrix, [P] **matrix define**

idstdize command, [R] **dstdize**

if *exp*, [U] **14 Language syntax**, [P] **syntax**

if programming command, [P] **if**

IIA, [R] **clogit**, [R] **hausman**, [R] **nlogit**

iis command, [XT] **xt**

immediate commands, [U] **22 Immediate commands**;
    [U] **21.4.5 Parsing immediate commands**,
    [P] **display**

implied context, class, [P] **class**

importance weights, [U] **14.1.6 weight**,
    [U] **23.16.4 Importance weights**

importing data, [R] **infile**, [R] **insheet**; [R] **odbc**,
    [U] **24.4 Transfer programs**

impulse-response functions, [TS] **varbasic**, [TS] **varirf**,
    [TS] **varirf add**, [TS] **varirf cgraph**, [TS] **varirf
    create**, [TS] **varirf ctable**, [TS] **varirf describe**,
    [TS] **varirf dir**, [TS] **varirf drop**, [TS] **varirf
    erase**, [TS] **varirf graph**, [TS] **varirf ograph**,
    [TS] **varirf rename**, [TS] **varirf set**, [TS] **varirf
    table**

    cumulative impulse-response functions, [TS] **varirf
    create**

    orthogonalized impulse-response functions,
    [TS] **varirf create**

    cumulative orthogonalized impulse-response
    functions, [TS] **varirf create**

impute command, [R] **impute**

imtest command, [R] **regression diagnostics**

in *range* modifier, [U] **14 Language syntax**, [P] **syntax**

incidence rate and rate ratio, [R] **poisson**, [R] **zip**,
    [ST] **epitab**, [ST] **stci**, [ST] **stir**, [ST] **stptime**,
    [ST] **stsum**, [SVY] **svy estimators**, [XT] **xtgee**,
    [XT] **xtnbreg**, [XT] **xtpoisson**

incidence rate ratio differences, [R] **lincom**, [R] **nlcom**

income tax rate function, [R] **egen**

incomplete beta function, [R] **functions**

incomplete gamma function, [R] **functions**

independence tests, *see* tests

index of probit and logit, [R] **logit**, [R] **predict**

index search, [R] **search**, [U] **8 Stata's online help and
    search facilities**

index, stopping rules, *see* stopping rules

index() string function, [R] **functions**

indicator variables, [R] **anova**, [R] **areg**, [R] **xi**,
    [U] **28.1.3 Converting categorical to indicator
    variables**, [U] **28.2 Using indicator variables in
    estimation**

indirect standardization, [R] **dstdize**

inequality measures, [R] **lorenz**

infile command, [R] **infile (fixed format)**, [R] **infile
    (free format)**; [R] **infile**

infix command, [R] **infix (fixed format)**; [R] **infile**

influence statistics, [R] **logistic**, [R] **predict**,
    [R] **regression diagnostics**, [R] **rreg**

%*infmt*, [R] **infile (fixed format)**

information criteria, [TS] **varsoc**

information matrix, [R] **correlate**, [R] **maximize**,
    [P] **matrix get**

information matrix test, [R] **regression diagnostics**

Informix, reading data from, [U] **24.4 Transfer
    programs**

inheritance, [P] **class**

initialization, class, [P] **class**

inlist() special function, [R] **functions**

inner fence, [R] **lv**

innovation accounting, [TS] **varirf**

input command, [R] **input**

input, matrix subcommand, [P] **matrix define**

input, obtaining from console in programs, *see* console

inputting data from a file, *see* reading data from disk

inputting data interactively, [R] **edit**, [R] **input**; *also see*
    editing data; reading data from disk

inrange() special function, [R] **functions**

insheet command, [R] **insheet**; [R] **infile**

inspect command, [R] **inspect**

installation,
    of official updates, [R] **update**, [U] **32 Using the
    Internet to keep up to date**
    of SJ and STB, [R] **net**, [R] **sj**, [U] **2.6 Updating
    and adding features from the web**,
    [U] **20.6 How do I install an addition?**
    of Stata, *see Getting Started* manual;
    [U] **5.2 Verifying that Stata is correctly
    installed**

instance, class, [P] **class**

instance-specific, class, [P] **class**

.instancemv built-in class function, [P] **class**

instrumental variables, [XT] **xthtaylor**, [XT] **xtivreg**

instrumental variables regression, [R] **ivreg**, [SVY] **svy
    estimators**

int, [R] **data types**, [U] **15.2.2 Numeric storage types**

int() truncation function, [R] **functions**

integ command, [R] **range**

integer truncation function, [R] **functions**

integrals, numerical, [R] **range**

Intercooled Stata, [R] **limits**, [U] **4 Flavors of Stata**

internal consistency, test for, [R] **alpha**

Internet, [U] **2.2 The http://www.stata.com web site**
    commands to control connections to, [R] **netio**
    installation of updates from, [R] **net**, [R] **sj**,
        [R] **update**, [U] **32 Using the Internet to keep
        up to date**
    search, [R] **net search**
    using data over, [R] **webuse**

interpolation, [R] **ipolate**

interquantile regression, [R] **qreg**

interquartile range,
    generating variable containing, [R] **egen**
    making dataset of, [R] **collapse**
    regression, [R] **qreg**
    reporting, [R] **lv**, [R] **table**
    summarizing, [R] **pctile**

interrater agreement, [R] **kappa**

interrupting command execution, [U] **13 Keyboard use**

interval regression, [R] **tobit**
    random-effects, [XT] **xtintreg**

intracluster correlation, [R] **loneway**

intreg command, [R] **tobit**; *also see* estimation commands

inv() matrix function, [P] **matrix define**

invbinomial() function, [R] **functions**

invchi2() function, [R] **functions**

invchi2tail() function, [R] **functions**

invcloglog function, [R] **functions**

inverse beta distribution, [R] **functions**

inverse binomial function, [R] **functions**

inverse F distribution function, [R] **functions**

inverse incomplete gamma function, [R] **functions**

inverse noncentral beta distribution, [R] **functions**

inverse noncentral chi-squared distribution function, [R] **functions**

inverse noncentral F distribution, [R] **functions**

inverse normal distribution function, [R] **functions**

inverse of matrix, [P] **matrix define**, [P] **matrix svd**

inverse t distribution function, [R] **functions**

invF() function, [R] **functions**

invFtail() function, [R] **functions**

invgammap() function, [R] **functions**

invibeta() function, [R] **functions**

invlogit() function, [R] **functions**

invnchi2() function, [R] **functions**

invnFtail() function, [R] **functions**

invnibeta() function, [R] **functions**

invnorm() function, [R] **functions**

invoking Stata, *see Getting Started* manual

invttail() function, [R] **functions**

ipolate command, [R] **ipolate**

IQR, *see* interquartile range

iqr(), egen function, [R] **egen**

iqreg command, [R] **qreg**; *also see* estimation commands

ir and iri commands, [ST] **epitab**

irecode() function, [R] **functions**

IRLS, [R] **glm**, [R] **reg3**

.isa built-in class function, [P] **class**

isid command, [R] **isid**

.isofclass built-in class function, [P] **class**

issym() matrix function, [R] **functions**, [P] **matrix define**

iterate() option, [R] **maximize**

iterated least squares, [R] **reg3**

iterations, controlling the maximum number, [R] **maximize**

ivreg command, [R] **ivreg**

[iweight=*exp*] modifier, [U] **14.1.6 weight**, [U] **23.16.4 Importance weights**

J

J() matrix function, [P] **matrix define**

Jaccard coefficient similarity measure, [CL] **cluster**

jackknife estimation, [R] **jknife**

jackknifed residuals, [R] **predict**, [R] **regression diagnostics**

Jarque-Bera statistic, [TS] **varnorm**

jknife command, [R] **jknife**

joinby command, [R] **joinby**; [U] **25 Commands for combining data**

joining datasets, *see* combining datasets

K

kap command, [R] **kappa**

Kaplan–Meier product-limit estimate, [ST] **sts**, [ST] **sts generate**, [ST] **sts graph**, [ST] **sts list**, [ST] **sts test**

Kaplan–Meier survivor function, [ST] **sts**; [ST] **ltable**, [ST] **stphplot**

kappa command, [R] **kappa**

kapwgt command, [R] **kappa**

kdensity command, [R] **kdensity**

keep command, [R] **drop**

keeping variables or observations, [R] **drop**

Kendall's tau, [R] **spearman**, [R] **tabulate**

kernel density estimator, [R] **kdensity**

keyboard entry, [U] **13 Keyboard use**

keyword search, [R] **search**, [U] **8 Stata's online help and search facilities**

Kish design effects, [R] **loneway**, [SVY] **svy estimators**, [SVY] **svymean**

kmeans, cluster subcommand, [CL] **cluster kmeans**

kmeans clustering, [CL] **cluster**, [CL] **cluster kmeans**

kmedians, cluster subcommand, [CL] **cluster kmedians**

kmedians clustering, [CL] **cluster**, [CL] **cluster kmedians**

Kolmogorov–Smirnov test, [R] **ksmirnov**

KR-20, [R] **alpha**

Kronecker product, [R] **cross**, [P] **matrix define**

Kruskal–Wallis test, [R] **kwallis**

ksmirnov command, [R] **ksmirnov**

ktau command, [R] **spearman**

Kuder–Richardson Formula 20, [R] **alpha**

Kulczynski coefficient similarity measure, [CL] **cluster**

kurt(), egen function, [R] **egen**

kurtosis, [R] **summarize**; [R] **lv**, [R] **sktest**, [TS] **varnorm**

kwallis command, [R] **kwallis**

L

L1-norm models, [R] **qreg**

label command, [R] **label**, [U] **15.6 Dataset, variable, and value labels**

label values, [U] **15.6 Dataset, variable, and value labels**; [U] **16.9 Label values**, [P] **macro**

labelbook command, [R] **labelbook**

labeling data, [R] **describe**, [R] **label**, [R] **notes**, [U] **15.6 Dataset, variable, and value labels**

LAD regression, [R] **qreg**

ladder command, [R] **ladder**

ladder of powers, [R] **ladder**

lag exclusion statistics, [TS] **varwle**

lag-order selection criteria, [TS] **varsoc**

lag-order selection statistics, [TS] **varsoc**

lagged values, [U] **16.7 Explicit subscripting**, [U] **16.7.1 Generating lags and leads**, [U] **16.8.1 Generating lags and leads**

Lagrange multiplier test, [TS] **regression diagnostics**, [TS] **varlmar**

Lance and Williams' formula, [CL] **cluster**

language syntax, [U] **14 Language syntax**, [P] **syntax**

Latin square designs, [R] **anova**, [R] **manova**, [R] **pkshape**

launch dialog box, [R] **db**

launching Stata, *see Getting Started* manual

LAV regression, [R] **qreg**

lead values, *see* lagged values

least absolute deviations, [R] **qreg**

least absolute residuals, [R] **qreg**

least absolute value regression, [R] **qreg**

least squared deviations, [R] **regress**, [R] **regression diagnostics**; [R] **areg**, [R] **cnsreg**, [R] **nl**

least squares, *see* linear regression
generalized, *see* FGLS

length of string function, [R] **functions**

length() string function, [R] **functions**

less than (or equal) operator, [U] **16.2.3 Relational operators**

letter values, [R] **lv**

level command and value, [R] **level**, [R] **query**, [P] **macro**

Levene's robust test statistic, [R] **sdtest**

leverage, [R] **logistic**, [R] **predict**, [R] **regression diagnostics**, [R] **rreg**
obtaining with weighted data, [R] **predict**, [R] **rreg**

leverage-versus-(squared)-residual plot, [R] **regression diagnostics**

lexis command, [ST] **stsplit**

lexis diagram, [ST] **stsplit**

lfit command, [R] **logistic**

life tables, [ST] **ltable**

likelihood, *see* maximum likelihood estimation

likelihood-ratio chi-squared of association, [R] **tabulate**

likelihood-ratio test, [R] **lrtest**

Likert summative scales, [R] **alpha**

limits, [R] **describe**, [R] **limits**, [R] **matsize**, [R] **memory**, [U] **7 Setting the size of memory**
numerical and string, [P] **creturn**
system, [P] **creturn**

lincom command, [R] **lincom**, [R] **test**, [SVY] **lincom for svy**

linear combinations, forming, [P] **matrix score**

linear combinations of estimators, [R] **lincom**, [U] **23.11 Obtaining linear combinations of coefficients**, [SVY] **lincom for svy**

linear interpolation and extrapolation, [R] **ipolate**

linear regression, [R] **regress**, [R] **regression diagnostics**; [R] **anova**, [R] **areg**, [R] **cnsreg**, [R] **eivreg**, [R] **frontier**, [R] **glm**, [R] **heckman**, [R] **ivreg**, [R] **mvreg**, [R] **qreg**, [R] **reg3**, [R] **rreg**, [R] **sureg**, [R] **tobit**, [R] **vwls**, [SVY] **svy estimators**, [TS] **newey**, [TS] **prais**, [TS] **regression diagnostics**, [XT] **xtabond**, [XT] **xtfrontier**, [XT] **xthtaylor**, [XT] **xtivreg**, [XT] **xtpcse**, [XT] **xtreg**, [XT] **xtregar**

linear splines, [R] **mkspline**

linegap, [R] **set**

lines, long, in do-files and ado-files, [U] **21.11.2 Comments and long lines in ado-files**, [P] **#delimit**

linesize, set subcommand, [R] **log**

link function, [R] **glm**, [XT] **xtgee**

linktest command, [R] **linktest**

list, cluster subcommand, [CL] **cluster utility**

list command, [R] **list**; [R] **format**

list, creturn subcommand, [P] **creturn**

list, ereturn subcommand, [P] **ereturn**

list, macro subcommand, [P] **macro**

list manipulation, [P] **macro lists**

list, matrix subcommand, [P] **matrix utility**

list, program subcommand, [P] **program**

list, sysdir subcommand, [P] **sysdir**

listing,
estimation results, [P] **ereturn**, [P] **_estimates**
macro expanded functions, [P] **macro lists**

listing data, [R] **edit**, [R] **list**

listserver, [U] **2.4 The Stata listserver**

ln() function, [R] **functions**

lnfact() function, [R] **functions**

lngamma() function, [R] **functions**

lnskew0 command, [R] **lnskew0**

loading data, *see* inputting data interactively; reading data from disk

local ++ command, [P] **macro**

local -- command, [P] **macro**

Local class prefix operator, [P] **class**

local command, [U] **21.3.1 Local macros**, [U] **21.3.8 Advanced local macro manipulation**, [P] **macro**

local, ereturn subcommand, [P] **ereturn**

local, return subcommand, [P] **return**

local, sreturn subcommand, [P] **return**

locally weighted smoothing, [R] **lowess**

location, measures of, [R] **summarize**, [R] **table**; [R] **lv**

loess, see lowess

log command, [U] **18 Printing and preserving output**; [R] **log**, [R] **view** [U] **18.2 Placing comments in logs**, [U] **19.1.2 Comments and blank lines in do-files**

.log filename suffix, [U] **14.6 File-naming conventions**

log files, *also see* log command
printing, [R] **translate**
reviewing while in Stata, [R] **shell**

log, set subcommand, [R] **log**, [R] **set**

log transformations, [R] **boxcox**, [R] **lnskew0**
log() function, [R] **functions**
log-linear model, [R] **glm**, [R] **poisson**, [R] **zip**,
　　[SVY] **svy estimators**
log-log plots, [ST] **stphplot**
log-logistic survival regression, [ST] **streg**
log-rank test, [ST] **sts test**
log10() function, [R] **functions**
logical operators, [U] **16.2.4 Logical operators**
logistic command, [R] **logistic**; *also see* estimation
　　commands; [R] **brier**
logistic and logit regression, [R] **logistic**, [R] **logit**
　　complementary log-log, [R] **cloglog**
　　conditional, [R] **clogit**, [R] **rologit**
　　fixed-effects, [R] **clogit**, [XT] **xtlogit**
　　generalized estimating equations, [XT] **xtgee**
　　generalized linear model, [R] **glm**
　　multinomial, [R] **mlogit**; [R] **clogit**, [SVY] **svy**
　　　estimators
　　nested, [R] **nlogit**
　　ordered, [R] **ologit**
　　polytomous, [R] **mlogit**, [SVY] **svy estimators**
　　population-averaged, [XT] **xtlogit**; [XT] **xtgee**
　　random-effects, [XT] **xtlogit**
　　rank-ordered, [R] **rologit**
　　skewed logit, [R] **scobit**
　　with grouped data, [R] **glogit**
　　with panel data, [R] **clogit**, [XT] **xtgee**
　　with survey data, [SVY] **svy estimators**
logit command, [R] **logit**; *also see* estimation
　　commands
logit function, [R] **functions**
logit regression, *see* logistic and logit regression
lognormal distribution, [R] **means**
lognormal survival regression, [ST] **streg**
loneway command, [R] **loneway**
long, [R] **data types**, [U] **15.2.2 Numeric storage
　　types**
long lines in ado-files and do-files,
　　[U] **21.11.2 Comments and long lines in ado-
　　files**, [P] **#delimit**
longitudinal data, *see* xt
lookfor command, [R] **describe**
loop, endless, *see* endless loop
looping, [P] **continue**, [P] **foreach**, [P] **forvalues**,
　　[P] **while**
lorenz, [R] **lorenz**
Lotus 1-2-3, reading data from, *see* spreadsheets
lower() string function, [R] **functions**
lowercase-string function, [R] **functions**
lowess command, [R] **lowess**
L-R plot, [R] **regression diagnostics**
LRECLs, [R] **infile (fixed format)**
lroc command, [R] **logistic**
lrtest command, [R] **lrtest**
lrtest, running with stored estimates, [R] **estimates**
ls command, [R] **dir**
lsens command, [R] **logistic**
lstat command, [R] **logistic**

ltable command, [ST] **ltable**
ltolerance() option, [R] **maximize**
ltrim() string function, [R] **functions**
lv command, [R] **lv**
lvalue, class, [P] **class**
lvr2plot command, [R] **regression diagnostics**

M

m() function, [R] **functions**, [U] **27.3.2 Specifying
　　particular dates (date literals)**
MA, [TS] **arch**, [TS] **arima**, *also see* time-series
ma(), egen function, [R] **egen**
Macintosh, *also see* Getting Started with Stata for
　　Macintosh manual
　　copy and paste, [R] **edit**
　　features, [U] **5.4 Features worth learning about**
　　help, [R] **help**
　　identifying in programs, [P] **creturn**
　　keyboard use, [U] **13 Keyboard use**
　　specifying filenames, [U] **14.6 File-naming
　　　conventions**
macro command, [P] **macro**
　　macro dir, [P] **macro**
　　macro drop, [P] **macro**
　　macro list, [P] **macro**
　　macro shift, [P] **macro**
macro substitution, [P] **macro**
　　class, [P] **class**
macros, [U] **21.3 Macros**, [P] **macro**; [P] **creturn**,
　　[P] **scalar**, [P] **syntax**
　　also see S_ macros
macval() macro expansion function, [P] **macro**
MAD regression, [R] **qreg**
mad(), egen function, [R] **egen**
main effects, [R] **anova**, [R] **manova**
makeCns, matrix subcommand, [P] **matrix constraint**
man command, [R] **help**
MANCOVA, [R] **manova**
Mann–Whitney two-sample statistic, [R] **signrank**
MANOVA, [R] **manova**
manova command, [R] **manova**
Mantel–Haenszel test, [ST] **epitab**, [ST] **stir**
mapping strings to numbers, [R] **encode**, [R] **label**
marginal effects, [R] **mfx**, [U] **23.13 Obtaining
　　marginal effects**
marginal homogeneity, test of, [R] **symmetry**
marginal tax rate egen function, [R] **egen**
mark command, [P] **mark**
markin command, [P] **mark**
marking observations, [P] **mark**
markout command, [P] **mark**
marksample command, [P] **mark**
martingale residuals, [ST] **stcox**, [ST] **streg**
match() string function, [R] **functions**
matched case–control data, [R] **clogit**, [R] **symmetry**,
　　[ST] **epitab**

matched-pairs tests, [R] **signrank**, [R] **ttest**;
     [R] **hotelling**
matching coefficient similarity measure, [CL] **cluster**
matcproc command, [P] **matrix constraint**
mathematical functions and expressions, [R] **functions**,
     [U] **16.3 Functions**, [P] **matrix define**
Matlab, reading data from, [U] **24.4 Transfer programs**
matmissing() matrix function, [R] **functions**,
     [P] **matrix define**
matname command, [P] **matrix mkmat**
matrices, [U] **17 Matrix expressions**, [P] **matrix**
     accessing internal, [P] **matrix get**
     accumulating, [P] **matrix accum**
     appending rows and columns, [P] **matrix define**
     Cholesky decomposition, [P] **matrix define**
     coefficient matrices, [P] **ereturn**
     column names, *see* matrices, row and column names
     constrained estimation, [P] **matrix constraint**
     copying, [P] **matrix define**, [P] **matrix get**,
          [P] **matrix mkmat**
     correlation, [P] **matrix define**
     covariance matrix of estimators, [P] **ereturn**,
          [P] **matrix get**
     cross-product, [P] **matrix accum**
     determinant, [P] **matrix define**
     diagonals, [P] **matrix define**
     displaying, [P] **matrix utility**
     dropping, [P] **matrix utility**
     eigenvalues, [P] **matrix eigenvalues**, [P] **matrix
          symeigen**
     eigenvectors, [P] **matrix symeigen**
     elements, [P] **matrix define**
     equation names, *see* matrices, row and column
          names
     estimation results, [P] **ereturn**, [P] **_estimates**
     functions, [R] **functions**, [P] **matrix define**
     identity, [P] **matrix define**
     input, [U] **17.4 Inputting matrices by hand**,
          [P] **matrix define**
     inversion, [P] **matrix define**, [P] **matrix svd**
     Kronecker product, [P] **matrix define**
     labeling rows and columns, *see* matrices, row and
          column names
     linear combinations with data, [P] **matrix score**
     listing, [P] **matrix utility**
     name space and conflicts, [P] **matrix**, [P] **matrix
          define**
     number of rows and columns, [P] **matrix define**
     operators such as addition, etc., [U] **17.7 Matrix
          operators**, [P] **matrix define**
     orthonormal basis, [P] **matrix svd**
     partitioned, [P] **matrix define**
     posting estimation results, [R] **estimates**,
          [P] **ereturn**, [P] **_estimates**
     renaming, [P] **matrix utility**
     row and column names, [R] **estimates**,
          [U] **17.2 Row and column names**, [P] **ereturn**,
          [P] **matrix define**,[P] **matrix mkmat**, [P] **matrix
          rowname**
     rows and columns, [P] **matrix define**
     saving matrix, [P] **matrix mkmat**

matrices, *continued*
     scoring, [P] **matrix score**
     store variables as matrix, [P] **matrix mkmat**
     submatrix extraction, [P] **matrix define**
     submatrix substitution, [P] **matrix define**
     subscripting, [U] **17.9 Subscripting**, [P] **matrix
          define**
     sweep operator, [P] **matrix define**
     temporary names, [P] **matrix**
     trace, [P] **matrix define**
     transposing, [P] **matrix define**
     variables, make into matrix [P] **matrix mkmat**
     zero, [P] **matrix define**
matrix command, [P] **matrix**
     matrix accum, [P] **matrix accum**
     matrix coleq, [P] **matrix rowname**
     matrix colnames, [P] **matrix rowname**
     matrix define, [P] **matrix define**
     matrix dir, [P] **matrix utility**
     matrix dispCns, [P] **matrix constraint**
     matrix drop, [P] **matrix utility**
     matrix eigenvalues, [P] **matrix eigenvalues**
     matrix glsaccum, [P] **matrix accum**
     matrix input, [P] **matrix define**
     matrix list, [P] **matrix utility**
     matrix makeCns, [P] **matrix constraint**
     matrix opaccum, [P] **matrix accum**
     matrix rename, [P] **matrix utility**
     matrix roweq, [P] **matrix rowname**
     matrix rownames, [P] **matrix rowname**
     matrix score, [P] **matrix score**
     matrix svd, [P] **matrix svd**
     matrix symeigen, [P] **matrix symeigen**
     matrix vecaccum, [P] **matrix accum**
matrix, ereturn subcommand, [P] **ereturn**
matrix, return subcommand, [P] **return**
matrix() function, [R] **functions**
matrix() pseudo-function, [P] **matrix define**
matsize [R] **matsize**, [R] **query**, [R] **set**,
     [U] **7.2.2 Advice on setting matsize**
     [U] **17 Matrix expressions**, [P] **creturn**,
     [P] **macro**
matuniform() matrix function, [R] **functions**,
     [P] **matrix define**
max() built-in function, [R] **functions**
max(), egen function, [R] **egen**
maxbyte() function, [R] **functions**
maxdouble() function, [R] **functions**
maxfloat() function, [R] **functions**
maximization technique explained, [R] **maximize**;
     [R] **ml**
maximum function, [R] **egen**, [R] **functions**
maximum likelihood estimation, [R] **maximize**, [R] **ml**;
     [R] **estimation commands**, [SVY] **ml for svy**
maximum number of variables and observations,
     [R] **describe**, [R] **limits**, [R] **memory**,
     [U] **7 Setting the size of memory**
maximum number of variables in a model, [R] **matsize**
maximum size of dataset, [R] **describe**, [R] **memory**,
     [U] **7 Setting the size of memory**,

maximum value dissimilarity measure, [CL] **cluster**

maximums, [R] **limits**

maximums and minimums,
creating dataset of, [R] **collapse**
functions, [R] **egen**, [R] **functions**
reporting, [R] **summarize**, [R] **table**; [R] **lv**

maxint() function, [R] **functions**

maxlong() function, [R] **functions**

maxvar, [R] **set**

mcc and mcci commands, [ST] **epitab**

McFadden's choice model, [R] **clogit**

McNemar's chi-squared test, [R] **clogit**, [ST] **epitab**

md command, [R] **mkdir**

mdev(), egen function, [R] **egen**

mdy() date function, [U] **27.2.2.1 The mdy() function**;
[R] **functions**

mean(), egen function, [R] **egen**

means,
across variables, not observations, [R] **egen**
adjusted, [R] **adjust**
confidence interval and standard error, [R] **ci**
creating dataset of, [R] **collapse**
creating variable containing, [R] **egen**
displaying, [R] **summarize**, [R] **table**, [R] **tabsum**;
[R] **means**, [XT] **xtsum**
geometric and harmonic, [R] **means**
robust, [R] **rreg**
survey data, [SVY] **svymean**
testing equality, [R] **ttest**; [R] **hotelling**
testing equality, sample size or power, [R] **sampsi**
testing equality, weighted, [SVY] **svymean**

means command, [R] **means**

measurement error, [R] **alpha**

measures, cluster subcommand, [CL] **cluster
programming utilities**

measures of association, [R] **tabulate**

measures of inequality, [R] **lorenz**

measures of location, [R] **summarize**; [R] **lv**

median command, [R] **signrank**

median linkage clustering, [CL] **cluster**, [CL] **cluster
medianlinkage**

median regression, [R] **qreg**

median test, [R] **signrank**

median(), egen function, [R] **egen**

medianlinkage, cluster subcommand, [CL] **cluster
medianlinkage**

medians,
creating dataset of, [R] **collapse**
creating variable containing, [R] **egen**
displaying, [R] **summarize**, [R] **table**; [R] **centile**,
[R] **lv**, [R] **pctile**
regression, [R] **qreg**
testing equality, [R] **signrank**

meff, [SVY] **svy estimators**, [SVY] **svymean**

member programs, [P] **class**

member variables, [P] **class**

memory, [U] **7 Setting the size of memory**
clearing, [R] **drop**

memory, *continued*
determining and resetting limits, *see Getting Started*
manuals, [R] **describe**, [R] **memory**; [R] **matsize**,
[R] **query**
loading and saving, [R] **save**
reducing utilization, [R] **compress**, [R] **encode**,
[P] **discard**
setting, [R] **set**, [U] **7.2.3 Advice on setting
memory**
virtual, [U] **7.5 Virtual memory and speed
considerations**

memory command, [R] **memory**; [U] **7 Setting the size
of memory**

memory settings, [P] **creturn**

menus, programming, [P] **window menu**; [P] **dialogs**

merge command, [R] **merge**; [U] **25 Commands for
combining data**

_merge variables, [R] **merge**

merging data, *see* combining datasets

messages and return codes, *see* error messages and
return codes

meta-analysis, [R] **meta**

Metafile, Windows, *see Getting Started with Stata for
Windows* manual

mfp command, [R] **mfp**

mfx command, [R] **mfx**

mhodds command, [ST] **epitab**

mi() function, [R] **functions**

Microsoft Access, reading data from, [R] **odbc**,
[U] **24.4 Transfer programs**

Microsoft Excel, reading data from, [R] **odbc**

Microsoft Windows, *see* Windows

midsummaries, [R] **lv**

mild outliers, [R] **lv**

Mills' ratio, [R] **heckman**, [SVY] **svy estimators**

min() built-in function, [R] **functions**

min(), egen function, [R] **egen**

minbyte() function, [R] **functions**

mindouble() function, [R] **functions**

minfloat() function, [R] **functions**

minimum absolute deviations, [R] **qreg**

minimum squared deviations, [R] **regress**,
[R] **regression diagnostics**; [R] **areg**, [R] **cnsreg**;
[R] **nl**

minimum variance clustering, [CL] **cluster
wardslinkage**

minimums and maximums, *see* maximums and
minimums

minint() function, [R] **functions**

Minkowski dissimilarity measure, [CL] **cluster**

minlong() function, [R] **functions**

missing() function, [R] **functions**

missing values, [R] **missing values**, [U] **15.2.1 Missing
values**, [U] **16 Functions and expressions**
counting, [R] **codebook**, [R] **inspect**
encoding and decoding, [R] **mvencode**
imputing, [R] **impute**
replacing, [R] **merge**

misspecification effects, [SVY] **svy estimators**,
  [SVY] **svymean**
mixed designs, [R] **anova**, [R] **manova**
mkdir command, [R] **mkdir**
mkmat command, [P] **matrix mkmat**
mkspline command, [R] **mkspline**
ml command, [R] **ml**, [SVY] **ml for svy**
mleval command, [R] **ml**
mlmatsum command, [R] **ml**
mlogit command, [R] **mlogit**; *also see* estimation
  commands
mlsum command, [R] **ml**
mlvecsum command, [R] **ml**
mod() function, [R] **functions**
mode(), egen function, [R] **egen**
model sensitivity, [R] **regression diagnostics**, [R] **rreg**
model specification test, [R] **linktest**; [R] **regression
  diagnostics**, [XT] **xtreg**
modifying data, *see* editing data
modulus function, [R] **functions**
modulus transformations, [R] **boxcox**
mofd() function, [R] **functions**, [U] **27.3.4 Translating
  between time units**
Monte Carlo simulations, [R] **permute**, [R] **simulate**,
  [P] **postfile**
month() function, [U] **27.2.4 Other date functions**;
  [R] **functions** [U] **27.3.5 Extracting components
  of time**
monthly() function, [R] **functions**,
  [U] **27.3.6 Creating time variables**
more command and parameter, [R] **more**, [R] **query**,
  [U] **10 –more– conditions**, [P] **macro**, [P] **more**
more condition, [R] **query**, [U] **10 –more– conditions**,
  [U] **19.1.6 Preventing –more– conditions**
move command, [R] **order**
moving average, [R] **egen**, [TS] **arch**, [TS] **arima**,
  [TS] **tssmooth**, [TS] **tssmooth ma**
mreldif() function, [R] **functions**, [P] **matrix define**
mtr(), egen function, [R] **egen**
multinomial logistic regression, [R] **mlogit**; [R] **clogit**,
  [SVY] **svy estimators**
multiple comparison tests, [R] **oneway**
multiple regression, *see* linear regression
multiple testing, [R] **regression diagnostics**, [R] **test**,
  [R] **testnl**
multiplication operator, *see* arithmetic operators
multiplicative heteroskedasticity, [TS] **arch**
multivariable fractional polynomial regression, [R] **mfp**
multivariate analysis, [SVY] **svy estimators**
  canonical correlation, [R] **canon**
  factor analysis, [R] **factor**
  Heckman selection model, [R] **heckman**
  Hotelling's *T*-squared, [R] **hotelling**
  MANOVA, [R] **manova**
  regression, [R] **mvreg**
  Zellner's seemingly unrelated, [R] **sureg**
multivariate analysis of covariance, [R] **manova**
multivariate analysis of variance, [R] **manova**

multivariate time-series, [TS] **var**, [TS] **var svar**,
  [TS] **varbasic**, [TS] **xcorr**
mvdecode and mvencode commands, [R] **mvencode**
mvreg command, [R] **mvreg**; *also see* estimation
  commands
mx_param command, [R] **ml**

N

_n and _N built-in variables, [U] **16.4 System variables
  (_variables)**, [U] **16.7 Explicit subscripting**
n-class command, [R] **saved results**, [P] **program**,
  [P] **return**
name space and conflicts, matrices and scalars,
  [P] **matrix**, [P] **matrix define**
names, [U] **14.3 Naming conventions**
  conflicts, [P] **matrix**, [P] **matrix define**, [P] **scalar**
  matrix row and columns, [R] **estimates**, [P] **ereturn**,
    [P] **matrix**, [P] **matrix define**, [P] **matrix
    rowname**
natural log function, [R] **functions**
nbetaden() function, [R] **functions**
nbreg command, [R] **nbreg**; *also see* estimation
  commands
nchi2() function, [R] **functions**
needle plot, [R] **spikeplot**
negation operator, *see* arithmetic operators
negative binomial regression, [R] **nbreg**; [R] **glm**,
  [SVY] **svy estimators**
  fixed-effects, [XT] **xtnbreg**
  population-averaged, [XT] **xtnbreg**; [XT] **xtgee**
  random-effects, [XT] **xtnbreg**
  zero-inflated, [R] **zip**
Nelson–Aalen cumulative hazard, [ST] **sts**, [ST] **sts
  generate**, [ST] **sts graph**, [ST] **sts list**
neqany(), egen function, [R] **egen**
nested case–control data, [ST] **sttocc**
nested designs, [R] **anova**, [R] **manova**
nested effects, [R] **anova**, [R] **manova**
nested logit, [R] **nlogit**
net command, [R] **net**
net search command, [R] **net search**
net sj, [R] **net**
NetCourses, [U] **2.7 NetCourses**
netio, [R] **set**
.new built-in class function, [P] **class**
newey command, [TS] **newey**; *also see* estimation
  commands, [XT] **xtgls**
Newey–West standard errors, [R] **glm**, [TS] **newey**
newlines, data without, [R] **infile (fixed format)**
news command, [R] **news**
newsletter, [U] **2 Resources for learning and using
  Stata**
Newton–Raphson method, [R] **ml**
nFden() function, [R] **functions**
nFtail() function, [R] **functions**
nibeta() function, [R] **functions**
nl and nlinit commands, [R] **nl**

nlcom command, [R] **nlcom**, [SVY] **nonlinear for svy**

nlogit command, [R] **nlogit**; *also see* estimation commands

nlogitgen command, [R] **nlogit**

nlogittree command, [R] **nlogit**

nobreak command, [P] **break**

noisily prefix, [P] **quietly**

noncentral beta density, [R] **functions**

noncentral beta distribution, [R] **functions**

noncentral chi-squared distribution function, [R] **functions**

noncentral F density, [R] **functions**

noncentral F distribution, [R] **functions**

nonconformities, quality control, [R] **qc**

nonconstant variance, *see* robust

nonlinear combinations of estimators, [R] **nlcom**

nonlinear estimation, [TS] **arch**

nonlinear least squares, [R] **nl**

nonlinear predictions, [R] **predictnl**

nonlinear regression, [R] **boxcox**

nonlinear smoothing, [TS] **tssmooth nl**

nonparametric tests,
    association, [R] **spearman**
    equality of distributions, [R] **ksmirnov**, [R] **kwallis**, [R] **signrank**
    equality of medians, [R] **signrank**
    equality of proportions, [R] **bitest**, [R] **prtest**
    equality of survivor functions, [ST] **sts test**
    percentiles, [R] **centile**
    serial independence, [R] **runtest**
    tables, [R] **tabulate**
    trend, [R] **nptrend**

nonstationary time series, [TS] **dfgls**, [TS] **dfuller**, [TS] **pperron**

norm() function, [R] **functions**

normal distribution and normality,
    bivariate, [R] **functions**
    cdf, [R] **functions**
    density, [R] **functions**
    examining distributions for, [R] **diagnostic plots**, [R] **lv**
    generating multivariate data, [R] **corr2data**, [R] **drawnorm**
    probability and quantile plots, [R] **diagnostic plots**
    test for, [R] **sktest**, [R] **swilk**
    transformations to achieve, [R] **boxcox**, [R] **ladder**, [R] **lnskew0**

normality test after VAR or SVAR, [TS] **varnorm**

normally distributed random numbers, [R] **functions**

normden() function, [R] **functions**

not equal operator, [U] **16.2.3 Relational operators**

not operator, [U] **16.2.4 Logical operators**

note command, [R] **notes**

notes, cluster subcommand, [CL] **cluster notes**

npnchi2() function, [R] **functions**

nptrend command, [R] **nptrend**

nullmat() matrix function, [P] **matrix define**

number to string conversion, *see* string functions

numbers, [U] **15.2 Numbers**
    formatting, [R] **format**
    mapping to strings, [R] **encode**, [R] **label**

numeric list, [U] **14.1.8 numlist**, [P] **numlist**, [P] **syntax**

numeric value labels, [R] **labelbook**

numerical precision, [U] **16.10 Precision and problems therein**

numlabel command, [R] **labelbook**

numlist command, [U] **14.1.8 numlist**, [P] **numlist**

N-way analysis of variance, [R] **anova**

N-way multivariate analysis of variance, [R] **manova**

O

object, [P] **class**

object-oriented programming, [P] **class**

.objkey built-in class function, [P] **class**

.objtype built-in class function, [P] **class**

obs parameter, [R] **obs**; [R] **describe**

observations,
    built-in counter variable, [U] **14.3 Naming conventions**
    creating dataset of, [R] **collapse**
    dropping, [R] **drop**
    dropping duplicate, [R] **duplicates**
    duplicating, [R] **expand**
    identifying duplicate, [R] **duplicates**
    marking for inclusion in programs, [P] **mark**
    maximum number of, [R] **describe**, [R] **memory**, [U] **7 Setting the size of memory**
    ordering, [R] **sort**; [R] **gsort**
    setting number of, [R] **obs**
    transposing with variables, [R] **xpose**

Ochiai coefficient similarity measure, [CL] **cluster**

odbc command, [R] **odbc**

ODBC data source, reading data from, [R] **odbc**, [U] **24.4 Transfer programs**

odds ratio, [R] **binreg**, [R] **clogit**, [R] **cloglog**, [R] **glogit**, [R] **logistic**, [R] **logit**, [R] **mlogit**, [R] **rologit**, [R] **scobit**, [ST] **epitab**, [XT] **xtcloglog**, [XT] **xtgee**, [XT] **xtlogit**, [XT] **xtnbreg**, [XT] **xtpoisson**, [XT] **xtprobit**

odds ratios differences, [R] **lincom**, [R] **nlcom**, [SVY] **lincom for svy**

OLDPLACE directory, [U] **20.5 Where does Stata look for ado-files?**, [P] **sysdir**

ologit command, [R] **ologit**; *also see* estimation commands

OLS regression, *see* linear regression

omitted variables test, [R] **regression diagnostics**

one-way analysis of variance, [R] **oneway**; [R] **kwallis**, [R] **loneway**

one-way scatterplots, *see Graphics Reference Manual*

oneway command, [R] **oneway**

online help, [U] **8 Stata's online help and search facilities**; [R] **help**, [R] **search**, [U] **10 –more– conditions**

opaccum, matrix subcommand, [P] **matrix accum**

open, file subcommand, [P] file
operating system,
    entering temporarily, [R] shell
    identifying in program, [P] creturn
    also see Macintosh, Unix, and Windows
operating system command, [R] cd, [R] copy, [R] dir,
    [R] erase, [R] mkdir, [R] shell, [R] type
operators, [U] 16.2 Operators, [P] matrix define
    order of evaluation, [U] 16.2.5 Order of evaluation
oprobit command, [R] oprobit; *also see* estimation
    commands
options, [U] 14 Language syntax
    in a programming context, [P] syntax, [P] unab,
    [TS] tsrevar
or operator, [U] 16.2.4 Logical operators
ORACLE, reading data from, [R] odbc,
    [U] 24.4 Transfer programs
order command, [R] order
order statistics, [R] egen, [R] lv
ordered logit, [R] ologit
ordered probit, [R] oprobit
ordering observations, [R] sort; [R] gsort
ordering variables, [R] order
ordinal analysis, [R] ologit, [R] oprobit
ordinary least squares, *see* linear regression
orthog and orthpoly commands, [R] orthog
orthogonal polynomials, [R] orthog
orthogonalized impulse-response function, [TS] varirf
orthonormal basis, [P] matrix svd
.out filename suffix, [R] outsheet
outer fence, [R] lv
outer product, [R] cross
outfile command, [R] outfile
outliers, [R] lv, [R] qreg, [R] regression diagnostics,
    [R] rreg
out-of-sample predictions, [R] predict; [R] predictnl,
    [U] 23.8.3 Making out-of-sample predictions
output,
    displaying, [P] display, [P] smcl
    formatting log, *see Getting Started with Stata for
    Unix*
    formatting numbers, [R] format
    printing, [R] translate, [U] 18 Printing and
    preserving output
    suppressing, [P] quietly
output, set subcommand, [P] quietly
output settings, [P] creturn
outsheet command, [R] outsheet
outside values, [R] lv
overloading, class program names, [P] class
ovtest command, [R] regression diagnostics

P

pac command, [TS] corrgram
pagesize, set subcommand, [R] more, [R] set
paging of screen output, controlling, [R] more,
    [P] more

pairwise combinations, [R] cross, [R] joinby
pairwise correlation, [R] correlate
panel data, *see* xt
panel-corrected standard errors, [XT] xtpcse
Paradox, reading data from, [U] 24.4 Transfer
    programs
parameterized curves, [R] range
parameters, system, [P] creturn
parametric survival models, *see* survival analysis
parsedistance, cluster subcommand, [CL] cluster
    programming utilities
parsing, [U] 21.4 Program arguments, [P] syntax;
    [P] gettoken, [P] numlist, [P] tokenize
partial correlation, [R] pcorr
partial regression leverage plot, [R] regression
    diagnostics
partial regression plot, [R] regression diagnostics
partial residual plot, [R] regression diagnostics
partition cluster analysis methods, [CL] cluster kmeans,
    [CL] cluster kmedians
partitioned matrices, [P] matrix define
partitioning memory, [R] memory, [U] 7 Setting the
    size of memory
paths, [U] 14.6 File-naming conventions
paths and directories, [P] creturn
patterns of data, [R] egen
pause command, [P] pause
pausing until key is depressed, [R] more, [P] more
pc(), egen function, [R] egen
pca command, [R] pca
pchart command, [R] qc
pchi command, [R] diagnostic plots
pcorr command, [R] pcorr
PCSE, [XT] xtpcse
pctile and _pctile commands, [R] pctile
pctile(), egen function, [R] egen
PDF files (Macintosh only), [R] translate
Pearson coefficient similarity measure, [CL] cluster
Pearson goodness-of-fit test, [R] logistic, [R] poisson
Pearson product-moment correlation coefficient,
    [R] correlate
Pearson residual, [R] glm, [R] logistic
percentiles,
    create dataset of, [R] collapse
    create variable containing, [R] egen, [R] pctile
    displaying, [R] summarize, [R] table; [R] centile,
    [R] codebook, [R] lv
pergram command, [TS] pergram
periodogram, [TS] pergram
permutation distribution, [R] permute
permutations, [R] permute
permute command, [R] permute
person time, [ST] stptime
PERSONAL directory, [U] 20.5 Where does Stata look
    for ado-files?, [P] sysdir
pharmaceutical statistics, [R] pk, [R] pksumm
Phillips–Perron test, [TS] pperron

_pi built-in variable, [U] **14.3 Naming conventions**,
[U] **16.4 System variables (_variables)**

pi, value of, [U] **14.3 Naming conventions**,
[U] **16.4 System variables (_variables)**

PICT files, see *Getting Started with Stata for Macintosh*
manual

pie charts, see *Graphics Reference Manual*

piecewise linear functions, [R] **mkspline**

Pillai's trace statistic, [R] **manova**

pk (pharmacokinetic data), [R] **pk**, [R] **pkcollapse**,
[R] **pkcross**, [R] **pkequiv**, [R] **pkexamine**,
[R] **pkshape**, [R] **pksumm**

pkcollapse command, [R] **pkcollapse**

pkcross command, [R] **pkcross**

pkequiv command, [R] **pkequiv**

pkexamine command, [R] **pkexamine**

.pkg filename suffix, [R] **net**

pkshape command, [R] **pkshape**

pksumm command, [R] **pksumm**

platforms for which Stata is available,
[U] **4.1 Platforms**

plural() string function, [R] **functions**

PLUS directory, [U] **20.5 Where does Stata look for
ado-files?**, [P] **sysdir**

pnorm command, [R] **diagnostic plots**

poisgof command, [R] **poisson**

poisson command, [R] **poisson**; [R] **nbreg**; *also see*
estimation commands

Poisson distribution,
cdf, [R] **functions**
confidence intervals, [R] **ci**
regression, see Poisson regression

Poisson regression, [R] **poisson**; [R] **glm**, [R] **nbreg**,
[XT] **xtgee**
fixed-effects, [XT] **xtpoisson**
population-averaged, [XT] **xtpoisson**; [XT] **xtgee**
random-effects, [XT] **xtpoisson**
with survey data, [SVY] **svy estimators**
zero-inflated, [R] **zip**

polar coordinates, [R] **range**

polymorphism, [P] **class**

polynomials, orthogonal, [R] **orthog**

polytomous logistic regression, [R] **mlogit**, [SVY] **svy
estimators**

pooled estimates, [ST] **epitab**

population attributable risk, [ST] **epitab**

population-averaged models, [XT] **xtcloglog**,
[XT] **xtgee**, [XT] **xtlogit**, [XT] **xtnbreg**,
[XT] **xtpoisson**, [XT] **xtreg**

populations,
diagnostic plots, [R] **diagnostic plots**
examining, [R] **histogram**, [R] **lv**, [R] **stem**,
[R] **summarize**, [R] **table**
standard, [R] **dstdize**
testing equality, [R] **ksmirnov**, [R] **kwallis**,
[R] **signrank**
testing for normality, [R] **sktest**, [R] **swilk**
totals, [SVY] **svymean**

portmanteau test, [TS] **corrgram**, [TS] **wntestq**

post, ereturn subcommand, [P] **ereturn**, [P] **matrix
constraint**

post hoc tests, [R] **oneway**

post, postclose, and postfile commands,
[P] **postfile**

post-estimation,
after VAR, see VAR, post-estimation
after SVAR, see SVAR, post-estimation

post-estimation command, [R] **adjust**,
[R] **estimates**, [R] **level**, [R] **lincom**, [R] **linktest**,
[R] **lrtest**, [R] **mfx**, [R] **nlcom**, [R] **predict**,
[R] **suest**, [R] **test**, [R] **testnl**, [R] **vce**

postutil command, [P] **postfile**
postutil clear, [P] **postfile**
postutil dir, [P] **postfile**

power of a test, [R] **sampsi**

power, raise to, function, see arithmetic operators

power transformations, [R] **boxcox**; [R] **lnskew0**

pperron command, [TS] **pperron**

prais command, [TS] **prais**; *also see* estimation
commands

Prais–Winsten regression, [TS] **prais**, [XT] **xtpcse**

precision, [U] **16.10 Precision and problems therein**

_predict programmer's command, [P] **_predict**

predict command, [R] **predict**; [R] **regression
diagnostics**, [U] **23.8 Obtaining predicted
values**; [R] **estimation commands**, [P] **ereturn**,
[P] **_estimates**

prediction, standard error of, [R] **glm**, [R] **predict**,
[R] **regression diagnostics**

predictions, [R] **adjust**, [R] **predict**
nonlinear, [R] **predictnl**, [SVY] **nonlinear for svy**
obtaining after estimation, [P] **_predict**

predictnl command, [R] **predictnl**, [SVY] **nonlinear
for svy**

Pregibon delta beta influence statistic, [R] **logistic**

preprocessor commands, [R] **#review**, [P] **#delimit**

preserve command, [P] **preserve**

preserving user's data, [P] **preserve**

prevalence studies, see case–control data

prevented fraction, [ST] **epitab**

principal component analysis, [R] **pca**

principal factors analysis, [R] **factor**

printing,
graphs, see *Graphics Reference Manual*
logs (output), [R] **translate**, [U] **18 Printing and
preserving output**, *also see Getting Started*
manuals
also see displaying

probability weights, [U] **14.1.6 weight**,
[U] **23.16.3 Sampling weights**, *also see* survey
data

probit command, [R] **probit**; *also see* estimation
commands

probit regression, [R] **probit**
generalized estimating equations, [XT] **xtgee**
generalized linear model, [R] **glm**
heteroskedastic, [R] **hetprob**

probit regression, *continued*
  ordered, [R] **oprobit**
  population-averaged, [XT] **xtprobit**; [XT] **xtgee**
  random-effects, [XT] **xtprobit**
  two-equation, [R] **biprobit**
  with grouped data, [R] **glogit**
  with panel data, [XT] **xtprobit**; [XT] **xtgee**
  with sample selection, [R] **heckprob**, [SVY] **svy estimators**
  with survey data, [SVY] **svy estimators**
procedure codes, [R] **icd9**
product-moment correlation, [R] **correlate**
  between ranks, [R] **spearman**
production frontier models, [R] **frontier**
profile.do, *see Getting Started* manuals
program command, [P] **program**
  program define, [P] **program**
  program dir, [P] **program**
  program drop, [P] **program**
  program list, [P] **program**
programming subcommands, [CL] **cluster programming subroutines**
programming utilities, [CL] **cluster programming utilities**
programs,
  adding comments to, [P] **comments**
  debugging, [P] **trace**
  dropping, [P] **discard**
  user written, [R] **sj**
projection matrix, diagonal elements of, [R] **logistic**, [R] **predict**, [R] **regression diagnostics**, [R] **rreg**
  obtaining with weighted data, [R] **predict**, [R] **rreg**
promax rotation, [R] **factor**
proper() string function, [R] **functions**
proportional hazards models, *see* survival analysis
proportional odds model, [R] **ologit**
proportional sampling, [R] **sample**; [R] **bootstrap**
proportions, [R] **ci**, [R] **sampsi**
  adjusted, [R] **adjust**
  survey data, [SVY] **svymean**
  testing equality, [R] **bitest**, [R] **prtest**
prtest command, [R] **prtest**, *also see* bitest command
prtesti command, [R] **prtest**, *also see* bitest command
pseudo *R*-squared, [R] **maximize**
pseudosigmas, [R] **lv**
psi function, [R] **functions**
pwcorr command, [R] **correlate**
pwd command, [R] **cd**
[pweight=*exp*] modifier, [U] **14.1.6 weight**, [U] **23.16.3 Sampling weights**

Q

q() function, [R] **functions**, [U] **27.3.2 Specifying particular dates (date literals)**
Q statistic, [TS] **wntestq**

qchi command, [R] **diagnostic plots**
qnorm command, [R] **diagnostic plots**
qofd() function, [R] **functions**, [U] **27.3.4 Translating between time units**
qqplot command, [R] **diagnostic plots**
_qreg command, [R] **qreg**
qreg command, [R] **qreg**; *also see* estimation commands
quadchk command, [XT] **quadchk**
qualitative dependent variables, [R] **clogit**, [R] **cloglog**, [R] **hetprob**, [R] **logistic**, [R] **logit**, [R] **mlogit**, [R] **ologit**, [R] **oprobit**, [R] **probit**, [R] **rologit**, [R] **scobit**, [SVY] **svy estimators**, [XT] **xtcloglog**, [XT] **xtgee**, [XT] **xtlogit**, [XT] **xtnbreg**, [XT] **xtpoisson**, [XT] **xtprobit**
quality control charts, [R] **qc**; [R] **serrbar**
quantile command, [R] **diagnostic plots**
quantile plots, [R] **diagnostic plots**
quantile regression, [R] **qreg**
quantile–normal plots, [R] **diagnostic plots**
quantile–quantile plot, [R] **diagnostic plots**
quantiles, *see* percentiles
quarter() function, [R] **functions**, [U] **27.3.5 Extracting components of time**
quarterly() function, [R] **functions**, [U] **27.3.6 Creating time variables**
Quattro Pro, reading data from, [U] **24.4 Transfer programs**
query command, [R] **query**
query, cluster subcommand, [CL] **cluster programming utilities**
query, file subcommand, [P] **file**
quick reference, [R] **missing values**
quietly prefix, [P] **quietly**
quitting Stata, *see* exit command
quotes,
  to delimit strings, [U] **21.3.5 Double quotes**
  to expand macros, [U] **21.3.1 Local macros**, [P] **macro**

R

r() function, [R] **functions**, [P] **discard**, [P] **return**
r() saved results, [R] **saved results**, [U] **21.8 Accessing results calculated by other programs**, [U] **21.10.1 Saving results in r()**, [P] **discard**
r-class command, [R] **saved results**, [U] **21.8 Accessing results calculated by other programs**, [P] **program**, [P] **return**
radians, [R] **functions**
raise to a power function, [U] **16.2.1 Arithmetic operators**
Ramsey test, [R] **regression diagnostics**
random number function, [R] **functions**, [R] **generate**

random numbers, normally distributed, *see* random
  number function
random order, test for, [R] **runtest**
random sample, [R] **bootstrap**, [R] **sample**; [U] **24.3 If
  you run out of memory**
random-effects models, [R] **anova**, [XT] **xtabond**,
  [XT] **xtcloglog**, [XT] **xtgee**, [XT] **xthtaylor**,
  [XT] **xtintreg**, [XT] **xtivreg**, [XT] **xtlogit**,
  [XT] **xtnbreg**, [XT] **xtpoisson**, [XT] **xtprobit**,
  [XT] **xtreg**, [XT] **xtregar**, [XT] **xttobit**;
  [R] **loneway**
range chart, [R] **qc**
range command, [R] **range**
range of data, [R] **summarize**, [R] **table**; [R] **codebook**,
  [R] **inspect**, [R] **lv**, [XT] **xtsum**
rank correlation, [R] **spearman**
rank(), egen function, [R] **egen**
rank-order statistics, [R] **signrank**, [R] **spearman**;
  [R] **egen**
rank-ordered logistic regression, [R] **rologit**
ranks of observations, [R] **egen**
ranksum command, [R] **signrank**
ratios, survey data, [SVY] **svymean**
raw data, [U] **15 Data**
.raw filename suffix, [U] **14.6 File-naming conventions**
_rc built-in variable, [U] **16.4 System variables
  (_variables)**, [P] **capture**
rc (return codes), *see* error messages and return codes
rchart command, [R] **qc**
re-expression, [R] **boxcox**, [R] **ladder**, [R] **lnskew0**
read, file subcommand, [P] **file**
reading and writing ASCII text and binary files, [P] **file**
reading console input in programs, *see* console
reading data from disk, [R] **infile**, [U] **24 Commands
  to input data**, [U] **24.4 Transfer programs**;
  also see inputting data interactively; combining
  datasets
real number to string conversion, [R] **functions**
real() string function, [R] **functions**
recase() string function, [R] **functions**
recast command, [R] **recast**
receiver operating characteristic (ROC) analysis,
  [R] **roc**; [R] **logistic**
recode command, [R] **recode**
recode()
  function, [R] **functions**, [U] **28.1.2 Converting
  continuous to categorical variables**
recoding data autocode() function, [R] **functions**
record I/O versus stream I/O, [U] **24 Commands to
  input data**
recording sessions, [U] **18 Printing and preserving
  output**
rectangularize dataset, [R] **fillin**
.ref built-in class function, [P] **class**
.ref_n built-in class function, [P] **class**
references, class, [P] **class**
reg3 command, [R] **reg3**; *also see* estimation
  commands

regress command, [R] **regress**; [U] **28.2 Using
  indicator variables in estimation**; *also see*
  estimation commands
regression (in generic sense),
  accessing coefficients and standard errors,
    [U] **16.5 Accessing coefficients and standard
    errors**, [P] **matrix get**
  basics, *also see* estimation commands
  creating orthogonal polynomials for, [R] **orthog**
  diagnostics, [R] **predict**; [R] **logistic**, [R] **regression
    diagnostics**, [TS] **regression diagnostics**
  dummy variables, with, [R] **anova**, [R] **areg**,
    [R] **xi**, [U] **28.2 Using indicator variables in
    estimation**, [XT] **xtreg**
  fixed-effects, [XT] **xtreg**; [R] **areg**
  fractional polynomial, [R] **fracpoly**
  graphing, [R] **logistic**, [R] **regression diagnostics**
  grouped data, [R] **tobit**
  increasing number of variables allowed, [R] **matsize**
  instrumental variables, [R] **ivreg**, [XT] **xthtaylor**,
    [XT] **xtivreg**
  linear, *see* linear regression
  multivariable fractional polynomial, [R] **mfp**
  random-effects, [XT] **xtgee**, [XT] **xtreg**
  system, [R] **mvreg**, [R] **reg3**, [R] **sureg**
  truncated, [R] **truncreg**
  also see estimation commands
relational operators, [U] **16.2.3 Relational operators**
relative difference function, [R] **functions**
relative risk, [ST] **epitab**; *also see* survival analysis
reldif() function, [R] **functions**
release marker, [P] **version**
reliability, [R] **alpha**, [R] **eivreg**, [R] **factor**,
  [R] **loneway**
reliability theory, *see* survival analysis
remainder function, [R] **functions**
removing files, [R] **erase**
rename command, [R] **rename**
rename, cluster subcommand, [CL] **cluster utility**
rename, matrix subcommand, [P] **matrix utility**
renamevar, cluster subcommand, [CL] **cluster
  utility**
renpfix command, [R] **rename**
reorganizing data, [R] **reshape**, [R] **xpose**
repeated measures ANOVA, [R] **anova**
repeated measures MANOVA, [R] **manova**
repeating and editing commands, [R] **#review**,
  [U] **13 Keyboard use**, [P] **continue**, [P] **foreach**,
  [P] **forvalues**
replace command, [R] **generate**
replace option, [U] **14.2 Abbreviation rules**
replay() function, [R] **estimates**, [R] **functions**,
  [P] **ereturn**, [P] **_estimates**
replicating observations, [R] **expand**
repost, ereturn subcommand, [P] **return**
reserved names, [U] **14.3 Naming conventions**
RESET test, [R] **regression diagnostics**
reset_id, serset subcommand, [P] **serset**
reshape command, [R] **reshape**

residual versus fitted plot, [R] **regression diagnostics**

residual versus predictor plot, [R] **regression diagnostics**

residuals, [R] **logistic**, [R] **predict**, [R] **regression diagnostics**, [R] **rreg**

resistant smoothers, [R] **smooth**

restore command, [P] **preserve**

restore, _return subcommand, [P] **_return**

results,
  clearing, [P] **ereturn**, [P] **_estimates**, [P] **_return**
  listing, [P] **ereturn**, [P] **_estimates**, [P] **_return**
  returning, [P] **_return**, [P] **return**
  saving, [R] **estimates**, [P] **ereturn**, [P] **_estimates**, [P] **postfile**, [P] **_return**, [P] **return**

return codes, *see* error messages and return codes

_return command, [P] **_return**
  _return dir, [P] **_return**
  _return drop, [P] **_return**
  _return hold, [P] **_return**
  _return restore, [P] **_return**

return command, [P] **return**
  return drop, [P] **return**
  return clear, [P] **return**
  return list, [R] **saved results**
  return local, [P] **return**
  return matrix, [P] **return**
  return scalar, [P] **return**

return() function, [R] **functions**

returning results, [P] **return**
  class programs, [P] **class**

reventries, [R] **set**

reverse() string function, [R] **functions**

#review command, [R] **#review**; [U] **13 Keyboard use**, [U] **18 Printing and preserving output**

Review window, saving contents of (Macintosh and Windows), *see Getting Started* manual

rfirst(), egen function, [R] **egen**

risk difference, [R] **binreg**, [ST] **epitab**

risk ratio, [R] **binreg**, [ST] **epitab**, *also see* survival analysis

rlast(), egen function, [R] **egen**

rm command, [R] **erase**

rmax(), egen function, [R] **egen**

_rmcoll command, [P] **_rmcoll**

_rmdcoll command, [P] **_rmcoll**

rmean(), egen function, [R] **egen**

rmin(), egen function, [R] **egen**

rmiss(), egen function, [R] **egen**

rmsg, [P] **rmsg**; [U] **11 Error messages and return codes**, [P] **creturn**, [P] **error**
  set rmsg, [P] **rmsg**

robs(), egen function, [R] **egen**

Continued in next column

robust,
  Huber/White/sandwich estimator of variance,
    [R] **areg**, [R] **biprobit**, [R] **cloglog**,
    [R] **heckman**, [R] **heckprob**, [R] **hetprob**,
    [R] **ivreg**, [R] **logistic**, [R] **logit**, [R] **probit**,
    [R] **regress**, [R] **rologit**, [R] **scobit**,
    [R] **tobit**, [U] **23.14 Obtaining robust variance estimates**, [P] **_robust**, [ST] **stcox**,
    [ST] **streg** [SVY] **svy estimators**, [TS] **newey**,
    [XT] **xtabond**, [XT] **xtcloglog**, [XT] **xtfrontier**,
    [XT] **xtgee**, [XT] **xtgls**, [XT] **xthtaylor**,
    [XT] **xtintreg**, [XT] **xtivreg**, [XT] **xtlogit**,
    [XT] **xtnbreg**, [XT] **xtpcse**, [XT] **xtpoisson**,
    [XT] **xtprobit**,[XT] **xtrchh**, [XT] **xtreg**,
    [XT] **xtregar**, [XT] **xttobit**
  other methods [R] **rreg**; [R] **qreg**, [R] **smooth**

_robust command, [P] **_robust**

robust regression, [R] **regress**, [R] **rreg**

robvar command, [R] **sdtest**

ROC analysis, [R] **roc**; [R] **logistic**

roccomp command, [R] **roc**

rocfit command, [R] **roc**

rocgold command, [R] **roc**

rocplot command, [R] **roc**

roctab command, [R] **roc**

Rogers and Tanimoto coefficient similarity measure, [CL] **cluster**

roh, [R] **loneway**

rologit command, [R] **rologit**

rootogram, [R] **spikeplot**

rotate command, [R] **factor**

round() rounding function, [R] **functions**

round-off error, [U] **16.10 Precision and problems therein**

row operators for data, [R] **egen**

roweq, matrix subcommand, [P] **matrix rowname**

rownames, matrix subcommand, [P] **matrix rowname**

rownumb() matrix function, [P] **matrix define**

rows of matrix,
  appending to, [P] **matrix define**
  names, [R] **estimates**, [P] **ereturn**, [P] **matrix define**, [P] **matrix rowname**
  operators, [P] **matrix define**

rowsof() matrix function, [P] **matrix define**

Roy's largest root test, [R] **manova**

Roy's union-intersection test, [R] **manova**

rreg command, [R] **rreg**; *also see* estimation commands

rsd(), egen function, [R] **egen**

rsum(), egen function, [R] **egen**

rtrim() string function, [R] **functions**

run command, [R] **do**; [U] **19 Do-files**

runtest command, [R] **runtest**

Russell and Rao coefficient similarity measure, [CL] **cluster**

rvalue, class, [P] **class**

rvfplot command, [R] **regression diagnostics**

rvpplot command, [R] **regression diagnostics**

S

s() function, [R] **functions**

s() saved results, [R] **functions**, [R] **saved results**, [U] **21.8 Accessing results calculated by other programs**, [U] **21.10.3 Saving results in s()**, [P] **return**

s-class command, [R] **saved results**, [U] **21.8 Accessing results calculated by other programs**, [P] **program**, [P] **return**

S_ macros, [R] **saved results**, [P] **creturn**, [P] **macro**

sample command, [R] **sample**

sample datasets, [U] **9 Stata's sample datasets**

sample, random, *see* random sample

sample size, [R] **sampsi**

sampling, [R] **bootstrap**, [R] **sample**, *also see* survey sampling

sampling weights, [U] **14.1.6 weight**, [U] **23.16.3 Sampling weights**

sampsi command, [R] **sampsi**

sandwich/Huber/White estimator of variance, *see* robust, Huber/White/sandwich estimator of variance

SAS, reading data from, [U] **24.4 Transfer programs**

save command, [R] **save**

save estimation results, [P] **ereturn**, [P] _**estimates**

saved results, [R] **saved results**, [U] **21.8 Accessing results calculated by other programs**, [U] **21.9 Accessing results calculated by estimation commands**, [U] **21.10 Saving results**, [P] _**return**, [P] **return**

saveold command, [R] **save**

saving data, [R] **save**

saving graphs, *see* Graphics Reference Manual

saving results, [R] **estimates**, [P] **ereturn**, [P] _**estimates**, [P] **postfile**, [P] _**return**, [P] **return**

scalar command and scalar() pseudo-function, [P] **scalar**

scalar, ereturn subcommand, [P] **ereturn**

scalar, return subcommand, [P] **return**

scalar() function, [R] **functions**

scalars, [P] **scalar**

name space and conflicts, [P] **matrix**, [P] **matrix define**

scaled Schoenfeld residuals, [ST] **stcox**, [ST] **streg**

scatterplots, *see* Graphics Reference Manual

Scheffé multiple comparison test, [R] **oneway**

scheme, [R] **set**

Schoenfeld residuals, [ST] **stcox**, [ST] **streg**

Schwarz Information Criterion (BIC), [R] **estimates**

scientific notation, [U] **15.2 Numbers**

scobit command, [R] **scobit**; *also see* estimation commands

scope, class, [P] **class**

score command, [R] **factor**, [R] **pca**

score, matrix subcommand, [P] **matrix score**

scores, obtaining, [U] **23.15 Obtaining scores**

scoring, [R] **factor**, [R] **pca**, [P] **matrix score**

scrollbufsize, [R] **set**

scrolling of output, controlling, [R] **more**, [P] **more**

sd(), egen function, [R] **egen**

sdtest and sdtesti commands, [R] **sdtest**

_se[], [U] **16.5 Accessing coefficients and standard errors**, [P] **matrix get**

search command, [R] **search**; [U] **8 Stata's online help and search facilities**

search Internet, [R] **net search**

searchdefault, [R] **set**

seasonal smoothing, [TS] **tssmooth**, [TS] **tssmooth shwinters**

seed, set subcommand, [R] **generate**, [R] **set**

seek, file subcommand, [P] **file**

seemingly unrelated estimation, [R] **suest**

seemingly unrelated regression, [R] **sureg**; [R] **reg3**

selection models, [R] **heckman**, [R] **heckprob**, [SVY] **svy estimators**

sensitivity, [R] **logistic** model, [R] **regression diagnostics**, [R] **rreg**

separate command, [R] **separate**

separating string variables into parts, [R] **split**

seq(), egen function, [R] **egen**

serial correlation, *see* autocorrelation

serial independence, test for, [R] **runtest**

serrbar command, [R] **serrbar**

serset command, [P] **serset**

serset clear, [P] **serset**

serset create, [P] **serset**

serset create_cspline, [P] **serset**

serset create_xmedians, [P] **serset**

serset dir, [P] **serset**

serset drop, [P] **serset**

serset reset_id, [P] **serset**

serset set, [P] **serset**

serset sort, [P] **serset**

serset summarize, [P] **serset**

serset use, [P] **serset**

sersetread, file subcommand, [P] **serset**

sersetwrite, file subcommand, [P] **serset**

session, recording, [U] **18 Printing and preserving output**

sessions, running simultaneous, (Windows), *see* Getting Started with Stata for Windows manual

set command, [R] **query**, [R] **set**

set adosize, [P] **sysdir**

set checksum, [R] **checksum**

set dp, [R] **format**

set graphics, *see* Graphics Reference Manual

set httpproxy, [R] **netio**

set httpproxyauth, [R] **netio**

set httpproxyhost, [R] **netio**

set httpproxyport, [R] **netio**

set httpproxypw, [R] **netio**

set httpproxyuser, [R] **netio**

set level, [R] **level**

set linesize, [R] **log**

set log, [R] **log**

set logtype, [R] **log**

set matsize, [R] **matsize**

set command, *continued*
  set maxdb, [R] **db**
  set maxvar, [R] **memory**
  set memory, [R] **memory**
  set more, [R] **more**, [U] **10 –more– conditions**
  set obs, [R] **obs**
  set output, [P] **quietly**
  set pagesize, [R] **more**
  set rmsg, [U] **11 Error messages and return
    codes**, [P] **rmsg**
  set scheme, *see Graphics Reference Manual*
  set searchdefault, [R] **search**
  set seed and set seed0, [R] **generate**
  set shell, [R] **shell**
  set timeout1, [R] **netio**
  set timeout2, [R] **netio**
  set trace, [P] **trace**
  set tracedepth, [P] **trace**
  set traceexpand, [P] **trace**
  set traceindent, [P] **trace**
  set tracenumber, [P] **trace**
  set tracesep, [P] **trace**
  set type, [R] **generate**
  set virtual, [R] **memory**
set, cluster subcommand, [CL] **cluster
  programming utilities**
set, file subcommand, [P] **file**
set, serset subcommand, [P] **serset**
set, sysdir subcommand, [P] **sysdir**
settings,
  efficiency, [P] **creturn**
  graphics, [P] **creturn**
  memory, [P] **creturn**
  network, [P] **creturn**
  output, [P] **creturn**
  program debugging, [P] **creturn**
  trace, [P] **creturn**
sfrancia command, [R] **swilk**
Shapiro–Francia test for normality, [R] **swilk**
Shapiro–Wilk test for normality, [R] **swilk**
shared object, [P] **class**
shell command, [R] **shell**
shewhart command, [R] **qc**
shift, macro subcommand, [P] **macro**
shortcuts, (Stata for Windows), *see Getting Started with
  Stata for Windows*
Šidák multiple comparison test, [R] **oneway**
sign() signum function, [R] **functions**
significance levels, [R] **level**, [R] **query**,
  [U] **23.6 Specifying the width of confidence
  intervals**
signrank command, [R] **signrank**
signtest command, [R] **signrank**
signum function, [R] **functions**
similarity measures, [CL] **cluster**
simulate command, [R] **simulate**
simulations, Monte Carlo, [R] **simulate** [R] **permute**,
  [P] **postfile**
simultaneous quantile regression, [R] **qreg**

simultaneous systems, [R] **reg3**
sin() function, [R] **functions**
sine function, [R] **functions**
single linkage clustering, [CL] **cluster**, [CL] **cluster
  singlelinkage**
single-precision floating point number,
  [U] **15.2.2 Numeric storage types**
singlelinkage, cluster subcommand, [CL] **cluster
  singlelinkage**
singular value decomposition, [P] **matrix svd**
SITE directory, [U] **20.5 Where does Stata look for
  ado-files?**, [P] **sysdir**
skew(), egen function, [R] **egen**
skewed logit estimation, [R] **scobit**
skewness, [R] **lv**, [R] **summarize**; [R] **lnskew0**,
  [R] **sktest**, [TS] **varnorm**
sktest command, [R] **sktest**
sleep command, [P] **sleep**
Small Stata, [R] **limits**, [U] **4 Flavors of Stata**
SMCL, [P] **smcl**
smooth command, [R] **smooth**
smoothers, [TS] **tssmooth**, [TS] **tssmooth exponential**,
  [TS] **tssmooth exponential**, [TS] **tssmooth
  hwinters**, [TS] **tssmooth ma**, [TS] **tssmooth nl**,
  [TS] **tssmooth shwinters**
smoothing, [TS] **tssmooth**
smoothing graphs, [R] **kdensity**, [R] **lowess**,
  [R] **smooth**
SMR (standardized mortality ratio), [ST] **epitab**,
  [ST] **stptime**
snapspan command, [ST] **snapspan**
Sneah and Sokal coefficient similarity measure,
  [CL] **cluster**
sort command, [R] **sort**
sort order, [R] **describe**, [P] **byable**, [P] **macro**,
  [P] **sortpreserve**
  for strings, [U] **16.2.3 Relational operators**
sort, serset subcommand, [P] **serset**
sortpreserve option, [P] **sortpreserve**
spearman command, [R] **spearman**
Spearman's rho, [R] **spearman**
Spearman–Brown prophecy formula, [R] **alpha**
specification test, [R] **linktest**; [R] **boxcox**,
  [R] **hausman**, [R] **regression diagnostics**,
  [XT] **xtreg**
specificity, [R] **factor**, [R] **logistic**
spectral distribution, [TS] **cumsp**, [TS] **pergram**
Spiegelhalter's Z statistic, [R] **brier**
spike plot, [R] **spikeplot**
spikeplot command, [R] **spikeplot**
splines, linear, [R] **mkspline**
split command, [R] **split**
split-plot designs, [R] **anova**, [R] **manova**
S-Plus, reading data from, [U] **24.4 Transfer programs**
spread, [R] **lv**
spreadsheet editor, [R] **edit**
  copy and paste, [R] **edit**

spreadsheets,
    transferring from Stata, [R] **outsheet**; [R] **outfile**,
        [U] **24.4 Transfer programs**
    transferring into Stata, [R] **insheet**;
        [U] **24.4 Transfer programs**; [R] **infile**,
        [R] **odbc**, [U] **24 Commands to input data**
SPSS, reading data from, [U] **24.4 Transfer programs**
SQL, [R] **odbc**
sqreg command, [R] **qreg**
sqrt() function, [R] **functions**
square root function, [R] **functions**
sreturn command, [R] **saved results**, [P] **return**
    sreturn clear, [P] **return**
    sreturn local, [P] **return**
ssc, [R] **ssc**
st command, [ST] **st**
st, *see Survival Analysis Reference Manual*
st_ct, [ST] **st_is**
st_is command, [ST] **st_is**
st_is 2 command, [ST] **st_is**
st_show, [ST] **st_is**
stability test after VAR or SVAR, [TS] **varstable**
stack command, [R] **stack**
standard deviations,
    creating dataset of, [R] **collapse**
    creating variable containing, [R] **egen**
    displaying, [R] **summarize**, [R] **table**, [R] **tabsum**;
        [R] **lv**, [XT] **xtsum**
    testing equality, [R] **sdtest**
standard errors,
    accessing, [U] **16.5 Accessing coefficients and
        standard errors**, [P] **matrix get**
    for general predictions, [R] **predictnl**
    forecast, [R] **predict**, [R] **regression diagnostics**
    mean, [R] **ci**
    panel-corrected, [XT] **xtpcse**
    prediction, [R] **glm**, [R] **predict**, [R] **regression
        diagnostics**
    residuals, [R] **predict**, [R] **regression diagnostics**
    robust, *see* robust
standardized,
    mortality ratio, [ST] **epitab**
    rates, [R] **dstdize**, [ST] **epitab**
    residuals, [R] **logistic**, [R] **predict**, [R] **regression
        diagnostics**
    variables, [R] **egen**
starting and stopping Stata, *see Getting Started* manual
Stat/Transfer, [U] **24.4 Transfer programs**
Stata,
    data file format, technical description, [P] **file
        formats .dta**
    description, [U] **3 A brief description of Stata**
    documentation, [U] **1 Read this—it will help**
    exiting, *see* exit command
    graph file format, technical description, *see Graphics
        Reference Manual*
    installation, *see Getting Started* manual;
        [U] **5.2 Verifying that Stata is correctly
        installed**

Stata, *continued*
    Intercooled, *see* Intercooled Stata
    invocation command, details, *see Getting Started
        with Stata for Unix*
    limits, [R] **limits**, [U] **4 Flavors of Stata**
    listserver, [U] **2.4 The Stata listserver**
    Markup and Control Language, [P] **smcl**
    NetCourses, [U] **2.7 NetCourses**
    platforms, [U] **4.1 Platforms**
    sample datasets, [U] **9 Stata's sample datasets**
    Small, *see* Small Stata
    Stata/SE, *see* Stata/SE
    supplementary material, [U] **2 Resources for
        learning and using Stata**
    support, [U] **2 Resources for learning and using
        Stata**
    web site, [U] **2.2 The http://www.stata.com web
        site**
stata command, *see Getting Started with Stata for
        Unix*
STATA directory, [P] **sysdir**
Stata for Macintosh, *see* Macintosh
Stata for Unix, *see* Unix
Stata for Windows, *see* Windows
The Stata Journal and *Stata Technical Bulletin*,
        [U] **2.5 The Stata Journal and the Stata
        Technical Bulletin**
    installation of, [R] **net**, [R] **sj**, [U] **20.6 How do I
        install an addition?**
    keyword search of, [R] **search**, [U] **8 Stata's online
        help and search facilities**
Stata News, [U] **2 Resources for learning and using
        Stata**
Stata/SE, [R] **limits**, [U] **4 Flavors of Stata**
Stata Technical Bulletin Reprints, [U] **2.5 The Stata
        Journal and the Stata Technical Bulletin**
Stata.do Macintosh file, *see Getting Started with Stata
        for Macintosh* manual
stata.key file, [R] **search**
stata.lic Unix file, *see Getting Started with Stata for
        Unix* manual
statalist, [U] **2.4 The Stata listserver**
stationary time series, [TS] **dfgls**, [TS] **dfuller**,
        [TS] **pperron**
statsby command, [R] **statsby**
STB, *see Stata Technical Bulletin*
stbase command, [ST] **stbase**
stci command, [ST] **stci**
stcox command, [ST] **stcox**; *also see* estimation
        commands
stcox, fractional polynomials, [R] **fracpoly**, [R] **mfp**
stcoxkm command, [ST] **stphplot**
stcurve command, [ST] **streg**
std(), egen function, [R] **egen**
stdes command, [ST] **stdes**
stem command, [R] **stem**
stem-and-leaf displays, [R] **stem**
stepwise estimation, [R] **sw**; [R] **estimation commands**
stfill command, [ST] **stfill**

stgen command, [ST] **stgen**

stir command, [ST] **stir**

stjoin command, [ST] **stsplit**

stmc command, [ST] **strate**

stmh command, [ST] **strate**

stochastic frontier models, [R] **frontier**, [XT] **xtfrontier**

stopping command execution, [U] **13 Keyboard use**

stopping rules, [CL] **cluster**, [CL] **cluster stop**
  adding, [CL] **cluster programming subroutines**
  Caliński and Harabasz index, [CL] **cluster**, [CL] **cluster stop**
  Duda and Hart index, [CL] **cluster**, [CL] **cluster stop**

storage types, [U] **15.2.2 Numeric storage types**, [U] **15.4.4 String storage types**; [R] **codebook**, [R] **compress**, [R] **describe**, [R] **encode**, [R] **format**, [R] **generate**, [R] **recast**, [U] **14.4 varlists**

stphplot command, [ST] **stphplot**

stphtest command, [ST] **stcox**

stptime command, [ST] **stptime**

str#, [R] **data types**, [U] **15.4.4 String storage types**

strate command, [ST] **strate**

stratification, [R] **clogit**, [ST] **epitab**, [ST] **stcox**, [ST] **sts**

stratified sampling, *see Survey Reference Manual*

stratified tables, [ST] **epitab**

stratum collapse, [SVY] **svydes**

stream I/O versus record I/O, [U] **24 Commands to input data**

streg command, [ST] **streg**; *also see* estimation commands

string functions, expressions, and operators, [R] **functions**, [U] **15.4 Strings**, [U] **26 Commands for dealing with strings**

string variables, [U] **15.4 Strings**, [U] **26 Commands for dealing with strings**
  converting to numbers, [R] **functions**
  encoding, [R] **encode**
  formatting, [R] **format**
  inputting, [R] **infile**, [U] **24 Commands to input data**
  making from value labels, [R] **encode**
  mapping to numbers, [R] **destring**, [R] **encode**, [R] **label**
  parsing, [P] **gettoken**, [P] **tokenize**
  sort order, [U] **16.2.3 Relational operators**
  splitting into parts, [R] **split**

string() function, [R] **functions**

structural vector autogression, *see* SVAR

sts command, [ST] **sts**, [ST] **sts generate**, [ST] **sts graph**, [ST] **sts list**, [ST] **sts test**, [ST] **stset**, [ST] **stsum**

sts generate command, [ST] **sts**, [ST] **sts generate**

sts graph command, [ST] **sts**, [ST] **sts graph**

sts list command, [ST] **sts**, [ST] **sts list**

sts test command, [ST] **sts**, [ST] **sts test**

stset command, [ST] **stset**

stsplit command, [ST] **stsplit**

stsum command, [ST] **stsum**

sttocc command, [ST] **sttocc**

sttoct command, [ST] **sttoct**

Student's *t* distribution
  cdf, [R] **functions**
  confidence interval for mean, [R] **ci**
  testing equality of means, [R] **ttest**; [R] **hotelling**

studentized residuals, [R] **predict**, [R] **regression diagnostics**

stvary command, [ST] **stvary**

subdirectories, [U] **14.6 File-naming conventions**

subinstr() string function, [R] **functions**

subinword() string function, [R] **functions**

subroutines,
  cluster analysis, [CL] **cluster programming subroutines**
  survival analysis, [ST] **st_is**

subscripting matrices, [P] **matrix define**

subscripts in expressions, [U] **16.7 Explicit subscripting**

substr() function, [R] **functions**

substring function, [R] **functions**

subtraction operator, *see* arithmetic operators

suest, [R] **hausman**
  running with stored estimates, [R] **estimates**

suest command, [R] **suest**

sum(), egen function, [R] **egen**

sum() function, [R] **functions**

summarize command, [R] **summarize**; [R] **format**, [R] **tabsum**

summarize, serset subcommand, [P] **serset**

summarizing data, [R] **summarize**; [R] **codebook**, [R] **inspect**, [R] **lv**, [R] **table**, [R] **tabsum**, [R] **tabulate**, [P] **tabdisp**, [SVY] **svytab**, [XT] **xtsum**

summary statistics, *see* descriptive statistics

summary variables, generating, [CL] **cluster generate**

summative (Likert) scales, [R] **alpha**

sums,
  creating dataset containing, [R] **collapse**
  over observations, [R] **egen**, [R] **functions**, [R] **summarize**
  over variables, [R] **egen**

Super, class prefix operator, [P] **class**

.superclass built-in class function, [P] **class**

support of Stata, [U] **2 Resources for learning and using Stata**

suppressing terminal output, [P] **quietly**

sureg command, [R] **sureg**; *also see* estimation commands

survey data, [U] **30 Overview of survey estimation**, *also see Survey Reference Manual*

survey sampling, [R] **areg**, [R] **biprobit**, [R] **cloglog**, [R] **heckprob**, [R] **hetprob**, [R] **ivreg**, [R] **logistic**, [R] **logit**, [R] **probit**, [R] **regress**, [R] **scobit**, [R] **tobit**, [U] **30 Overview of survey estimation**, [P] **_robust**, [ST] **stcox**, [ST] **streg**, *also see Survey Reference Manual*

survival analysis, *see Survival Analysis Reference Manual*; [R] **nbreg**, [R] **oprobit**, [R] **poisson**, [R] **tobit**, [R] **zip**, [SVY] **svy estimators**, [XT] **xtnbreg**, [XT] **xtpoisson**

survival-time data, *see* survival analysis

survivor function, [ST] **sts**, [ST] **sts generate**, [ST] **sts list**, [ST] **sts test**
    graph of, [ST] **streg**, [ST] **sts graph**

SVAR, [TS] **var svar**; [TS] **var intro**, [TS] **vargranger**, [TS] **varirf create**, [TS] **varlmar**, [TS] **varnorm**, [TS] **varwle**
    post-estimation, [TS] **regression diagnostics**, [TS] **vargranger**, [TS] **varlmar**, [TS] **varnorm**, [TS] **varsoc**, [TS] **varstable**, [TS] **varwle**, [TS] **wntestb**, [TS] **wntestq**

svar command, [TS] **var svar**; *also see* estimation commands

svd, matrix subcommand, [P] **matrix svd**

svmat command, [P] **matrix mkmat**

svydes command, [SVY] **svydes**

svygnbreg command, [SVY] **svy estimators**; estsuf

svyheckman command, [SVY] **svy estimators**; estsuf

svyheckprob command, [SVY] **svy estimators**; estsuf

svyintreg command, [SVY] **svy estimators**; estsuf

svyivreg command, [SVY] **svy estimators**; *also see* estimation commands

svylogit command, [SVY] **svy estimators**; *also see* estimation commands

svymean command, [SVY] **svymean**

svymlogit command, [SVY] **svy estimators**; *also see* estimation commands

svynbreg command, [SVY] **svy estimators**; estsuf

svyologit command, [SVY] **svy estimators**; *also see* estimation commands

svyoprobit command, [SVY] **svy estimators**; *also see* estimation commands

svypoisson command, [SVY] **svy estimators**; *also see* estimation commands

svyprobit command, [SVY] **svy estimators**; *also see* estimation commands

svyprop command, [SVY] **svymean**

svyratio command, [SVY] **svymean**

svyregress command, [SVY] **svy estimators**; *also see* estimation commands

svyset command, [SVY] **svyset**

svytab command, [SVY] **svytab**

svytotal command, [SVY] **svymean**

sweep matrix operator, [R] **functions**, [P] **matrix define**

sweep() matrix function, [P] **matrix define**

swilk command, [R] **swilk**

Sybase, reading data from, [U] **24.4 Transfer programs**

symbolic forms, [R] **anova**

symeigen, matrix subcommand, [P] **matrix symeigen**

syminv() matrix function, [P] **matrix define**

symmetry command, [R] **symmetry**

symmetry plots, [R] **diagnostic plots**

symmetry, test of, [R] **symmetry**

symmi command, [R] **symmetry**

symplot command, [R] **diagnostic plots**

syntax command, [P] **syntax**

syntax diagrams explained, [R] **intro**

syntax of Stata's language, [U] **14 Language syntax**, [P] **syntax**

sysdir command, [U] **20.5 Where does Stata look for ado-files?**, [P] **sysdir**
    sysdir list, [P] **sysdir**
    sysdir set, [P] **sysdir**

Systat, reading data from, [U] **24.4 Transfer programs**

system estimators, [R] **reg3**

system limits, [P] **creturn**

system macros, *see* S_ macros

system parameters, [R] **query**, [R] **netio**, [R] **set**, [P] **creturn**

system values, [P] **creturn**

system variables, [U] **16.4 System variables (_variables)**

sysuse command, [R] **sysuse**

szroeter command, [R] **regression diagnostics**

Szroeter's test for heteroskedasticity, [R] **regression diagnostics**

T

%t formats, [R] **format**, [U] **15.5.4 Time-series formats**, [U] **27.3.3 Time-series formats**

t distribution,
    cdf, [R] **functions**
    confidence interval for mean, [R] **ci**
    testing equality of means, [R] **ttest**; [R] **hotelling**, [SVY] **svymean**

tab characters, show, [R] **type**

tab expansion of variable names, [U] **13.6 Tab expansion of variable names**

tab1 and tab2 commands, [R] **tabulate**

tabdisp command, [P] **tabdisp**

tabi command, [R] **tabulate**

table command, [R] **table**

tables, [TS] **varirf ctable**, [TS] **varirf table**
    contingency, [R] **table**, [R] **tabulate**, [SVY] **svytab**
    epidemiologic, [ST] **epitab**
    formatting numbers in, [R] **format**
    frequency, [R] **tabulate**; [R] **table**, [R] **tabsum**, [SVY] **svytab**, [XT] **xttab**
    life, [ST] **ltable**
    N-way, [R] **table**; [P] **tabdisp**
    of means, [R] **table**, [R] **tabsum**
    of statistics, [R] **table**, [R] **tabstat**, [P] **tabdisp**
    printing, [U] **18 Printing and preserving output**

tabodds command, [ST] **epitab**

tabstat command, [R] **tabstat**

tabulate and tabi commands, [R] **tabulate**

tag(), egen function, [R] **egen**

tan() function, [R] **functions**

tangent function, [R] **functions**

tanh() function, [R] **functions**

tau, [R] **spearman**

taxonomy, [CL] **cluster**

TDT test, [R] **symmetry**

technical support, [U] **2.9 Technical support**

tempfile command, [P] **macro**

tempname, class, [P] **class**

tempname command, [P] **macro**, [P] **matrix**, [P] **scalar**

temporary files, [U] **21.7.3 Temporary files**, [P] **macro**, [P] **preserve**, [P] **scalar**

temporary names, [U] **21.7.2 Temporary scalars and matrices**, [P] **macro**, [P] **matrix**, [P] **scalar**

temporary variables, [U] **21.7.1 Temporary variables**, [P] **macro**

tempvar command, [P] **macro**

termcap(5), [U] **13 Keyboard use**

terminal,
   obtaining input from, [P] **display**
   suppressing output, [P] **quietly**

terminfo(4), [U] **13 Keyboard use**

test command, [R] **test**; [U] **23.10 Performing hypothesis tests on the coefficients**; [R] **anova**, [SVY] **test for svy**

test for normality, [TS] **varnorm**

test-based confidence intervals, [ST] **epitab**

testnl command, [R] **testnl**, [SVY] **nonlinear for svy**

testparm command, [R] **test**, [SVY] **test for svy**

tests,
   association, [R] **tabulate**, [ST] **epitab**, [SVY] **svytab**
   binomial proportion, [R] **bitest**
   Breusch–Pagan, [R] **mvreg**, [R] **sureg**
   Cox proportional hazards assumption, [ST] **stcox**, [ST] **stphplot**
   difference of two means, [SVY] **svymean**
   epidemiological tables, [ST] **epitab**
   equality of coefficients, [R] **test**, [R] **testnl**, [SVY] **test for svy**
   equality of distributions, [R] **ksmirnov**, [R] **kwallis**, [R] **signrank**
   equality of means, [R] **ttest**; [R] **hotelling**
   equality of medians, [R] **signrank**
   equality of proportions, [R] **bitest**, [R] **prtest**
   equality of survivor functions, [ST] **sts test**
   equality of variance, [R] **sdtest**
   equivalence, [R] **pk**, [R] **pkequiv**
   heterogeneity, [ST] **epitab**
   heteroskedasticity, [R] **regression diagnostics**
   homogeneity, [ST] **epitab**
   independence, [R] **tabulate**; [ST] **epitab**, [SVY] **svytab**
   independence of irrelevant alternatives, [R] **clogit**, [R] **hausman**, [R] **nlogit**
   internal consistency, [R] **alpha**
   interrater agreement, [R] **kappa**
   kurtosis, [R] **regression diagnostics**
   likelihood ratio, [R] **lrtest**
   marginal homogeneity, [R] **symmetry**
   model coefficients, [R] **lrtest**, [R] **test**, [R] **testnl**, [SVY] **test for svy**

tests, *continued*
   model specification, [R] **linktest**; [R] **hausman**, [R] **regression diagnostics**, [XT] **xtreg**
   nonlinear, [SVY] **nonlinear for svy**
   normality, [R] **sktest**, [R] **swilk**; [R] **boxcox**, [R] **ladder**
   permutation, [R] **permute**
   serial independence, [R] **runtest**
   skewness, [R] **regression diagnostics**
   symmetry, [R] **symmetry**
   TDT, [R] **symmetry**
   trend, [R] **nptrend**, [R] **symmetry**, [ST] **epitab**, [ST] **sts test**

three-stage least squares, [R] **reg3**

time of day, [P] **creturn**

time stamp, [R] **describe**

timeout1, set subcommand, [R] **netio**, [R] **set**

timeout2, set subcommand, [R] **netio**, [R] **set**

time-series, [U] **27.3 Time-series dates**
   analysis, [R] **egen**, [R] **regression diagnostics**, [P] **matrix accum**, [TS] **corrgram**, [TS] **cumsp**, [TS] **dfuller**, [TS] **pergram**, [TS] **pperron**, [TS] **tsreport**, [TS] **tsset** [TS] **wntestb**, [TS] **wntestq**, [TS] **xcorr**
   estimation, [U] **29.13 Models with time-series data**, [TS] **newey** [TS] **arch**, [TS] **arima**, [TS] **prais**, *also see* xt
   formats, [R] **format**, [U] **15.5.4 Time-series formats**, [U] **27.3.3 Time-series formats**
   functions, [R] **functions**, [U] **27.3 Time-series dates**
   operators, [U] **16.8 Time-series operators**
   operators in variable lists, [TS] **tsrevar**
   unabbreviating varlists, [P] **unab**
   varlists, [U] **14.4.3 Time-series varlists**, [TS] **tsrevar**
   also see Time-Series Reference Manual

time-series operators, [TS] **tsset**

time-varying variance, [TS] **arch**

time-versus-concentration curve, [R] **pk**

tin() function, [R] **functions**, [U] **27.3.8 Selecting periods of time**

tis command, [XT] **xt**

TMPDIR Unix environment variable, [P] **macro**

tobit command, [R] **tobit**; *also see* estimation commands

tobit regression, [R] **tobit**
   random-effects, [XT] **xttobit**

.toc filename suffix, [R] **net**

tokenize command, [P] **tokenize**

tolerance() option, [R] **maximize**

totals, survey data, [SVY] **svymean**

trace(), matrix function, [P] **matrix define**

trace of matrix, [P] **matrix define**

trace, set subcommand, [P] **program**, [P] **trace**

tracedepth, set subcommand, [R] **set**, [P] **trace**

traceexpand, set subcommand, [R] **set**, [P] **trace**

traceindent, set subcommand, [R] **set**, [P] **trace**

tracenumber, set subcommand, [R] **set**, [P] **trace**

tracesep, set subcommand, [R] **set**, [P] **trace**

tracing iterative maximization process, [R] **maximize**

transferring data,
    copying and pasting, [R] **edit**
    from Stata, [R] **outsheet**, [U] **24.4 Transfer programs**; [R] **outfile**
    into Stata, [R] **infile**, [R] **insheet**, [U] **24 Commands to input data**, [U] **24.4 Transfer programs**

transformations,
    log, [R] **lnskew0**
    modulus, [R] **boxcox**
    power, [R] **boxcox**; [R] **lnskew0**
    to achieve normality, [R] **boxcox**; [R] **ladder**
    to achieve zero skewness, [R] **lnskew0**

translate command, [R] **translate**

translator command, [R] **translate**

transmap command, [R] **translate**

transmission-disequilibrium test, [R] **symmetry**

transposing data, [R] **xpose**

transposing matrices, [P] **matrix define**

treatment effects, [R] **treatreg**

treatreg command, [R] **treatreg**; *also see* estimation commands

trees, [CL] **cluster**, [CL] **cluster dendrogram**

trend, test for, [R] **nptrend**, [R] **symmetry**, [ST] **epitab**, [ST] **sts test**

trigamma() function, [R] **functions**

trigonometric functions, [R] **functions**

trim() string function, [R] **functions**

truncated regression, [R] **tobit**, [R] **treatreg**

truncated-normal model, stochastic frontier, [R] **frontier**

truncating,
    real numbers, [R] **functions**
    strings, [R] **functions**

truncreg command, [R] **truncreg**; *also see* estimation commands

tsappend command, [TS] **tsappend**

tsfill command, [TS] **tsset**

tsreport command, [TS] **tsreport**

tsrevar command, [TS] **tsrevar**

tsset command, [TS] **tsset**; [U] **27.3.7 Setting the time variable**

tssmooth dexponential command, [TS] **tssmooth dexponential**

tssmooth exponential command, [TS] **tssmooth exponential**

tssmooth hwinters command, [TS] **tssmooth hwinters**

tssmooth ma command, [TS] **tssmooth ma**

tssmooth nl command, [TS] **tssmooth nl**

tssmooth shwinters command, [TS] **tssmooth shwinters**

tsunab command, [P] **unab**

ttail() function, [R] **functions**

ttest and ttesti commands, [R] **ttest**

tuning constant, [R] **rreg**

tutorials, [U] **9 Stata's sample datasets**

twithin() function, [R] **functions**, [U] **27.3.8 Selecting periods of time**

two-stage least squares, [R] **ivreg**, [R] **regress**, [SVY] **svy estimators**, [XT] **xthtaylor**, [XT] **xtivreg**

two-way analysis of variance, [R] **anova**

two-way multivariate analysis of variance, [R] **manova**

two-way scatterplots, [R] **lowess**, [R] **smooth**, *also see* Graphics Reference Manual

type command, [R] **type**

type parameter, [R] **generate**, [R] **query**, [P] **macro**

U

U statistic, [R] **signrank**

unab command, [P] **unab**

unabbreviate,
    command names, [P] **unabcmd**
    variable list, [P] **unab**; [P] **syntax**

unabcmd command, [P] **unabcmd**

.uname built-in class function, [P] **class**

underlining in syntax diagram, [U] **14 Language syntax**

underscore c() function, [R] **functions**

underscore variables, [U] **16.4 System variables (_variables)**

unhold, _estimates subcommand, [P] **_estimates**

uniform() and uniform0() functions, [R] **functions**, [R] **generate**

uniformly distributed random number function, [R] **functions**, [R] **generate**

unique value labels, [R] **labelbook**

unique values,
    counting, [R] **inspect**, [R] **table**, [R] **tabulate**; [R] **codebook**
    determining, [R] **labelbook**

unit root test, [TS] **dfgls**, [TS] **dfuller**, [TS] **pperron**

univariate distributions, displaying, [R] **cumul**, [R] **diagnostic plots**, [R] **histogram**, [R] **ladder**, [R] **lv**, [R] **stem**

Unix, *also see* Getting Started with Stata for Unix
    environment variables, *see* environment variables
    features, [U] **5.4 Features worth learning about**
    graphics, *see* Getting Started with Stata for Unix
    identifying in programs, [P] **creturn**
    keyboard use, [U] **13 Keyboard use**
    specifying filenames, [U] **14.6 File-naming conventions**
    temporarily entering from Stata, [R] **shell**

Unix equivalent commands, [R] **dir**, [R] **erase**

update command, [R] **update**

UPDATES directory, [U] **20.5 Where does Stata look for ado-files?**, [P] **sysdir**

updates to Stata, [R] **net**, [R] **sj**, [R] **update**, [U] **2.5 The Stata Journal and the Stata Technical Bulletin**, [U] **2.6 Updating and adding features from the web**, [U] **20.6 How do I install an addition?**

upper() string function, [R] **functions**

uppercase-string function, [R] **functions**

use, cluster subcommand, [CL] **cluster utility**

use command, [R] **save**
use, serset subcommand, [P] **serset**
uselabel command, [R] **labelbook**
user interface, [P] **dialogs**
using data, [R] **save**, [R] **sysuse**, [R] **webuse**, [P] **syntax**
utilities, programming, [CL] **cluster utility**

V

value labels, [U] **15.6.3 Value labels**; [R] **codebook**,
    [R] **describe**, [R] **encode**, [R] **inspect**, [R] **label**,
    [R] **labelbook**, [U] **16.9 Label values**, [P] **macro**
    potential problems in, [R] **labelbook**
VAR, [TS] **var**, [TS] **var intro**, [TS] **var svar**,
    [TS] **varbasic**; [TS] **varfcast**, [TS] **varfcast
    clear**, [TS] **varfcast compute**, [TS] **varfcast
    graph**, [TS] **vargranger**, [TS] **varirf create**,
    [TS] **varlmar**, [TS] **varnorm**, [TS] **varsoc**,
    [TS] **varstable**, [TS] **varwle**
    post-estimation, [TS] **regression diagnostics**,
    [TS] **vargranger**, [TS] **varlmar**, [TS] **varnorm**,
    [TS] **varsoc**, [TS] **varstable**, [TS] **varwle**,
    [TS] **wntestb**, [TS] **wntestq**
var command, [TS] **var**; *also see* estimation commands
varbasic command, [TS] **varbasic**
varfcast clear command, [TS] **varfcast clear**
varfcast compute command, [TS] **varfcast compute**
varfcast graph command, [TS] **varfcast graph**
vargranger command, [TS] **vargranger**
variable labels, [U] **15.6.2 Variable labels**;
    [R] **codebook**, [R] **describe**, [R] **label**, [R] **notes**,
    [U] **14.4 varlists**, [P] **macro**
variable lists, *see varlist*
variable types, [R] **describe**, [U] **15.2.2 Numeric
    storage types**, [U] **15.4.4 String storage types**;
    [R] **codebook**, [R] **data types**, [U] **14.4 varlists**,
    [P] **macro**
    class, [P] **class**
_variables, [U] **14.3 Naming conventions**,
    [U] **16.4 System variables (_variables)**
variables,
    categorical, *see* categorical data
    characteristics of, [U] **15.8 Characteristics**,
        [P] **char**, [P] **macro**
    comparing, [R] **compare**
    description, [R] **describe**; [R] **codebook**, [R] **notes**
    displaying contents of, [R] **edit**, [R] **list**
    documenting, [R] **codebook**, [R] **labelbook**,
        [R] **notes**
    dropping, [R] **drop**
    dummy, *see* indicator variables
    generating, summary, or grouping, [CL] **cluster
        generate**
    in dataset, maximum number of, [R] **describe**,
        [R] **memory**, [U] **7 Setting the size of memory**
    in model, maximum number, [R] **matsize**
    listing, [R] **edit** [R] **list**; [R] **codebook**, [R] **describe**,
        [R] **labelbook**

variables, *continued*
    naming, [U] **14.3 Naming conventions**; [R] **rename**,
        [U] **14.2 Abbreviation rules**
    reordering, [R] **order**
    sorting and alphabetizing, [R] **sort**; [R] **gsort**
    standardizing, [R] **egen**
    storage types, *see* storage types
    string *see* string variables
    system, *see* system variables
    tab expansion of, [U] **13.6 Tab expansion of
        variable names**
    temporary, [P] **macro**
    transposing with observations, [R] **xpose**
    unabbreviating, [P] **unab**; [P] **syntax**
    unique values, [R] **inspect**; [R] **codebook**
variance, Huber/White/sandwich estimator, *see* robust
variance, nonconstant, *see* robust
variance analysis, [R] **anova**, [R] **loneway**, [R] **manova**,
    [R] **oneway**
variance inflation factors, [R] **regression diagnostics**
variance stabilizing transformations, [R] **boxcox**
variance–covariance matrix of estimators, [R] **correlate**,
    [R] **vce** [P] **ereturn**, [P] **matrix get**
variance-weighted least squares, [R] **vwls**
variances,
    creating dataset of, [R] **collapse**
    creating variable containing, [R] **egen**
    displaying, [R] **summarize**, [R] **table**, [R] **tabsum**;
        [R] **lv**, [XT] **xtsum**
    testing equality, [R] **sdtest**
varimax rotation, [R] **factor**
varirf add command, [TS] **varirf add**
varirf cgraph command, [TS] **varirf cgraph**
varirf create command, [TS] **varirf create**
varirf ctable command, [TS] **varirf ctable**
varirf describe command, [TS] **varirf describe**
varirf dir command, [TS] **varirf dir**
varirf drop command, [TS] **varirf drop**
varirf erase command, [TS] **varirf erase**
varirf graph command, [TS] **varirf graph**
varirf ograph command, [TS] **varirf ograph**
varirf rename command, [TS] **varirf rename**
varirf set command, [TS] **varirf set**
varirf table command, [TS] **varirf table**
varlabelpos, [R] **set**
varlist, [U] **14.4 varlists**; [U] **14 Language syntax**,
    [P] **syntax**
    existing, [U] **14.4.1 Lists of existing variables**
    new, [U] **14.4.2 Lists of new variables**
    time series, [U] **14.4.3 Time-series varlists**
varlmar command, [TS] **varlmar**
varnorm command, [TS] **varnorm**
varsoc command, [TS] **varsoc**
varstable command, [TS] **varstable**
varwle command, [TS] **varwle**
vce command, [R] **vce**
vec() matrix function, [R] **functions**, [P] **matrix define**
vecaccum, matrix subcommand, [P] **matrix accum**

vecdiag() matrix function, [P] **matrix define**

vector autoregression, *see* VAR

vector autoregression forecast, [TS] **varfcast**

verifying data, *see* certifying data

verinst command, [U] **6.2 verinst problems**, *also see Getting Started* manuals

version command, [U] **19.1.1 Version**, [U] **21.11.1 Version**, [P] **version**
class programming, [P] **class**

version control, *see* version command

version of ado-file, [R] **which**

viewing previously typed lines, [R] **#review**

vif command, [R] **regression diagnostics**

virtual memory, [R] **memory**, [U] **7.5 Virtual memory and speed considerations**

vwls command, [R] **vwls**; *also see* estimation commands

W

w() function, [R] **functions**, [U] **27.3.2 Specifying particular dates (date literals)**

Wald tests, [R] **predictnl**, [R] **test**, [R] **testnl**, [U] **23.10 Performing hypothesis tests on the coefficients**, [U] **23.10.4 Nonlinear Wald tests**, [SVY] **test for svy**

Ward's linkage clustering, [CL] **cluster**, [CL] **cluster wardslinkage**

wardslinkage, cluster subcommand, [CL] **cluster wardslinkage**

waveragelinkage, cluster subcommand, [CL] **cluster waveragelinkage**

web site,
stata.com [U] **2.2 The http://www.stata.com web site**
stata-press.com [U] **2.3 The http://www.stata-press.com web site**

webuse command, [R] **webuse**

week() function, [R] **functions**, [U] **27.3.5 Extracting components of time**

weekly() function, [R] **functions**, [U] **27.3.6 Creating time variables**

Weibull distribution, [ST] **streg**

Weibull survival regression, [ST] **streg**

[weight=exp] modifier, [U] **14.1.6 weight**, [U] **23.16 Weighted estimation**

weighted data, [U] **14.1.6 weight**, [U] **23.16 Weighted estimation**, *also see* survey data

weighted least squares, [R] **ivreg**, [R] **regress**, [R] **regression diagnostics**, [R] **vwls**

weighted moving average, [TS] **tssmooth**, [TS] **tssmooth ma**

weighted-average linkage clustering, [CL] **cluster**, [CL] **cluster waveragelinkage**

weights, probability, *see* Survey Refernce Manual

weights, sampling, *see* weights, probability

Welsch distance, [R] **regression diagnostics**

whelp command, [R] **help**

which, classutil subcommand, [P] **classutil**

which command, [R] **which**; [U] **20.3 How can I tell if a command is built-in or an ado-file**

while command, [P] **while**

white noise test, [TS] **wntestb**, [TS] **wntestq**

White/Huber/sandwich estimator of variance, *see* robust

White's test for heteroskedasticity, [R] **regression diagnostics**

who command, *see* Getting Started with Stata for Unix

Wilcoxon rank-sum test, [R] **signrank**

Wilcoxon signed-ranks test, [R] **signrank**

Wilcoxon test (Wilcoxon–Breslow, Wilcoxon–Gehan, Wilcoxon–Mann–Whitney, [ST] **sts test**

Wilks' lambda, [R] **manova**

Wilks' likelihood-ratio test, [R] **manova**

window fopen command, [P] **window fopen**

window manage command, [P] **window manage**

window menu command, [P] **window menu**

window push command, [P] **window push**

window stopbox command, [P] **window stopbox**

Windows, *also see* Getting Started with Stata for Windows manual
features, [U] **5.4 Features worth learning about**
filenames, [U] **21.3.10 Constructing Windows filenames using macros**
help, [R] **help**
identifying in programs, [P] **creturn**
keyboard use, [U] **13 Keyboard use**
shell, [R] **shell**
specifying filenames, [U] **14.6 File-naming conventions**

winexec command, [R] **shell**

within estimators, [XT] **xthtaylor**, [XT] **xtivreg**, [XT] **xtreg**, [XT] **xtregar**

within-cell means and variances, [XT] **xtsum**

wmf (Windows Metafile), *see* Getting Started with Stata for Windows manual

wntestb command, [TS] **wntestb**

wntestq command, [TS] **wntestq**

wofd() function, [R] **functions**, [U] **27.3.4 Translating between time units**

Woolf confidence intervals, [ST] **epitab**

word() string function, [R] **functions**

wordcount() string function, [R] **functions**

write, file subcommand, [P] **file**

writing data, [R] **outfile**, [R] **outsheet**, [R] **save**

writing and reading ASCII text and binary files, [P] **file**

www.stata.com web site, [U] **2.2 The http://www.stata.com web site**

www.stata-press.com web site, [U] **2.3 The http://www.stata-press.com web site**

X

X Windows, *see* Graphics Reference Manual, *also see* Getting Started with Stata for Unix manual

xchart command, [R] **qc**

xcorr command, [TS] **xcorr**

xi command, [R] xi

xpose command, [R] xpose

xt (panel or cross-sectional data), [R] clogit, *also see* Cross-Sectional Time-Series Reference Manual

xtabond command, [XT] xtabond; [XT] intro, *also see* estimation commands

xtcloglog command, [XT] xtcloglog; [XT] intro, [XT] quadchk, *also see* estimation commands

xtcorr command, [XT] xtgee

xtdata command, [XT] xtdata; [XT] xt

xtdes command, [XT] xtdes; [XT] xt

xtfrontier command, [XT] xtfrontier; [XT] intro, *also see* estimation commands

xtgee command, [XT] xtgee; [XT] intro, *also see* estimation commands

xtgls command, [XT] xtgls; *also see* estimation commands

xthtaylor command, [XT] xthtaylor; [XT] intro, *also see* estimation commands

xtile command, [R] pctile

xtintreg command, [XT] xtintreg; [XT] intro, [XT] quadchk, *also see* estimation commands

xtivreg command, [XT] xtivreg; [XT] intro, *also see* estimation commands

xtlogit command, [XT] xtlogit; [XT] intro, [XT] quadchk, *also see* estimation commands

xtnbreg command, [XT] xtnbreg; *also see* estimation commands

xtpcse command, [XT] xtpcse; [XT] intro, *also see* estimation commands

xtpoisson command, [XT] xtpoisson; [XT] quadchk, *also see* estimation commands

xtprobit command, [XT] xtprobit; [XT] quadchk, *also see* estimation commands

xtrchh command, [XT] xtrchh; *also see* estimation commands

xtreg command, [XT] xtreg; [XT] xt; *also see* estimation commands

xtregar command, [XT] xtregar; [XT] intro, *also see* estimation commands

xtsum command, [XT] xtsum; [XT] xt

xttab command, [XT] xttab; [XT] xt

xttest0 command, [XT] xtreg; [XT] xt

xttobit command, [XT] xttobit; [XT] intro, [XT] quadchk, *also see* estimation commands

xttrans command, [XT] xttab; [XT] xt

Y

y() function, [R] functions, [U] 27.3.2 Specifying particular dates (date literals)

year() function, [R] functions [U] 27.2.4 Other date functions, [U] 27.3.5 Extracting components of time

yearly() function, [R] functions, [U] 27.3.6 Creating time variables

yofd() function, [R] functions, [U] 27.3.4 Translating between time units

Yule coefficient similarity measure, [CL] cluster

Z

Zellner's seemingly unrelated regression, [R] sureg; [R] reg3, [R] suest

zero matrix, [P] matrix define

zero-altered Poisson regression, [R] zip

zero-inflated Poisson regression, [R] zip

zip command, [R] zip; *also see* estimation commands